BASIC MATHEMATICS
A TEXT/WORKBOOK

EIGHTH EDITION

Charles P. McKeague

CUESTA COLLEGE

BROOKS/COLE
CENGAGE Learning

Australia • Brazil • Japan • Korea • Mexico • Singapore • Spain • United Kingdom • United States

Basic Mathematics: A Text/Workbook,
Eighth Edition
Charles P. McKeague

Acquisitions Editor: Marc Bove
Developmental Editor: Shaun Williams
Assistant Editor: Carrie Jones
Editorial Assistant: Zachary Crockett
Media Editor: Bryon Spencer
Marketing Manager: Laura McGinn
Marketing Assistant: Shannon Maier
Marketing Communications Manager:
  Darlene Macanan
Content Project Manager: Jennifer Risden
Design Director: Rob Hugel
Art Director: Vernon Boes
Print Buyer: Becky Cross
Rights Acquisitions Specialist: Don Schlotman
Production Service: XYZ Textbooks
Text Designer: Diane Beasley
Photo Researcher: Bill Smith Group
Copy Editor: Katherine Shields, XYZ Textbooks
Illustrator: Kristina Chung, XYZ Textbooks
Cover Designer: Irene Morris
Cover Image: Pete McArthur
Compositor: Donna Looper, XYZ Textbooks

© 2013, 2010 Brooks/Cole, Cengage Learning

ALL RIGHTS RESERVED. No part of this work covered by the copyright herein may be reproduced, transmitted, stored or used in any form or by any means graphic, electronic, or mechanical, including but not limited to photocopying, recording, scanning, digitizing, taping, Web distribution, information networks, or information storage and retrieval systems, except as permitted under Section 107 or 108 of the 1976 United States Copyright Act, without the prior written permission of the publisher.

Unless otherwise noted, all art is © Cengage Learning.

> For product information and technology assistance, contact us at
> **Cengage Learning Customer & Sales Support, 1-800-354-9706**
> For permission to use material from this text or product,
> submit all requests online at **www.cengage.com/permissions**
> Further permissions questions can be emailed to
> **permissionrequest@cengage.com**

Library of Congress Control Number: 2011930153

ISBN-13: 978-1-133-10362-2
ISBN-10: 1-133-10362-6

Brooks/Cole
20 Davis Drive
Belmont, CA 94002-3098
USA

Cengage Learning is a leading provider of customized learning solutions with office locations around the globe, including Singapore, the United Kingdom, Australia, Mexico, Brazil, and Japan. Locate your local office at: **www.cengage.com/global**

Cengage Learning products are represented in Canada by Nelson Education, Ltd.

To learn more about Brooks/Cole, visit **www.cengage.com/brookscole**
Purchase any of our products at your local college store or at our preferred online store **www.cengagebrain.com**

Printed in the United States of America
1 2 3 4 5 6 7 15 14 13 12 11

Brief Contents

Chapter 1	Whole Numbers	1
Chapter 2	Fractions and Mixed Numbers	103
Chapter 3	Decimals	203
Chapter 4	Ratio and Proportion	279
Chapter 5	Percent	335
Chapter 6	Measurement	407
Chapter 7	Introduction to Algebra	463
Chapter 8	Solving Equations	527
Appendix A	Negative Exponents	597
Appendix B	Scientific Notation	603
Appendix C	Multiplication and Division with Scientific Notation	607
	Solutions to Selected Practice Problems	S-1
	Answers to Odd-Numbered Problems	A-1
	Index	I-1

Contents

1 Whole Numbers

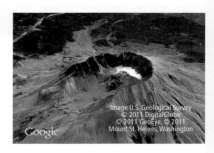

Introduction 1
- 1.1 Place Value and Names for Numbers 3
- 1.2 Addition with Whole Numbers, and Perimeter 13
- 1.3 Rounding Numbers and Estimating Answers 25
- 1.4 Subtraction with Whole Numbers 33
- 1.5 Multiplication with Whole Numbers 41
- 1.6 Division with Whole Numbers 55
- 1.7 Exponents, Order of Operations, and Averages 67
- 1.8 Area and Volume 81

Summary 91

Review 93

Test 94

Projects 95

A Glimpse of Algebra 99

2 Fractions and Mixed Numbers

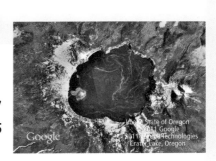

Introduction 103
- 2.1 The Meaning and Properties of Fractions 105
- 2.2 Prime Numbers, Factors, and Reducing to Lowest Terms 117
- 2.3 Multiplication with Fractions, and the Area of a Triangle 125
- 2.4 Division with Fractions 137
- 2.5 Addition and Subtraction with Fractions 145
- 2.6 Mixed-Number Notation 159
- 2.7 Multiplication and Division with Mixed Numbers 165
- 2.8 Addition and Subtraction with Mixed Numbers 171
- 2.9 Combinations of Operations and Complex Fractions 181

Summary 189

Review 193

Cumulative Review 194

Test 195

Projects 197

A Glimpse of Algebra 199

3 Decimals

Introduction 203

3.1 Decimal Notation and Place Value 205

3.2 Addition and Subtraction with Decimals 213

3.3 Multiplication with Decimals; Circumference and Area of a Circle 221

3.4 Division with Decimals 233

3.5 Fractions and Decimals, and the Volume of a Sphere 245

3.6 Square Roots and the Pythagorean Theorem 257

Summary 267

Review 269

Cumulative Review 270

Test 271

Projects 273

A Glimpse of Algebra 275

4 Ratio and Proportion

Introduction 279

4.1 Ratios 281

4.2 Rates and Unit Pricing 289

4.3 Solving Equations by Division 295

4.4 Proportions 301

4.5 Applications of Proportions 307

4.6 Similar Figures 313

Summary 321

Review 323

Cumulative Review 324

Test 325

Projects 327

A Glimpse of Algebra 331

5 Percent

Introduction   335

5.1   Percents, Decimals, and Fractions   337
5.2   Basic Percent Problems   347
5.3   General Applications of Percent   357
5.4   Sales Tax and Commission   363
5.5   Percent Increase or Decrease, and Discount   371
5.6   Interest   379
5.7   Pie Charts   387

Summary   395

Review   397

Cumulative Review   398

Test   399

Projects   401

A Glimpse of Algebra   403

6 Measurement

Introduction   407

6.1   Unit Analysis I: Length   409
6.2   Unit Analysis II: Area and Volume   419
6.3   Unit Analysis III: Weight   429
6.4   Converting Between the Two Systems, and Temperature   435
6.5   Operations with Time and Mixed Units   443

Summary   451

Review   455

Cumulative Review   456

Test   457

Projects   459

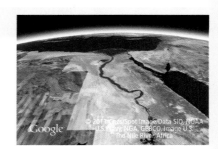

7 Introduction to Algebra

Introduction   463

7.1   Positive and Negative Numbers   465
7.2   Addition with Negative Numbers   475
7.3   Subtraction with Negative Numbers   485
7.4   Multiplication with Negative Numbers   495
7.5   Division with Negative Numbers   505
7.6   Simplifying Algebraic Expressions   513

Summary   519

Review   521

Cumulative Review   522

Test   523

Projects   525

8 Solving Equations

Introduction   527

8.1   The Distributive Property and Algebraic Expressions   529
8.2   The Addition Property of Equality   541
8.3   The Multiplication Property of Equality   549
8.4   Linear Equations in One Variable   557
8.5   Applications   565
8.6   Evaluating Formulas   579

Summary   589

Review   591

Cumulative Review   592

Test   593

Projects   595

Appendix A   Negative Exponents   597
Appendix B   Scientific Notation   603
Appendix C   Multiplication and Division with Scientific Notation   607

Solutions to Selected Practice Problems   S-1

Answers to Odd-Numbered Problems   A-1

Index   I-1

Preface to the Instructor

The Basic Mathematics Course as a Bridge to Further Success

I have a passion for teaching mathematics. That passion carries through to my textbooks. My goal is a textbook that is user-friendly for both students and instructors. For students, this book forms a bridge to beginning algebra with clear, concise writing, continuous review, and interesting applications. For the instructor, I build features into the text that reinforce the habits and study skills we know will bring success to our students.

This course and its syllabus bring the student to the level of ability required of college students, while getting them ready to make a successful start in introductory algebra. After seven successful editions, we have developed several interlocking, proven features that will improve students' chances of success in the course. Here are some of the important success features of the book.

New to This Edition

CHAPTER INTRODUCTIONS
Each chapter opens with a revised introduction in which an exciting real-world application is used to stimulate interest in the chapter. We expand on these opening applications later in the chapter.

KEY WORDS
At the beginning of each chapter, we have provided a concise list of key words and definitions the student will encounter throughout the chapter.

CHAPTER OUTLINES
An outline of objectives arranged by section appears at the beginning of each chapter. This outline helps the student organize expectations for the chapter into short-term goals and prepare for the work ahead.

TICKET TO SUCCESS
Previously included as Getting Ready for Class, these reviewed and revised questions require the student to provide written responses. They reinforce the idea of reading the section before coming to class, as their answers can truly be their ticket to the course. The questions now appear at the beginning of each section to provide the student a chance to become more involved learners as they read through the section, keeping in mind important concepts.

SECTION OPENERS
Similar to the chapter introductions, each section in the book opens with a revised introduction that includes an interesting real-world application. The presence of these applications helps the student engage in and relate to the mathematics they are learning, and feel confident about moving forward. We expand on these opening applications later in the chapter as well.

MOVING TOWARD SUCCESS
Each problem set now starts with this new feature that includes a motivational quote followed by a few questions to focus the student on success. Because good study

habits are essential to the success of developmental math students, this feature is prominently displayed at the start of every problem set.

CHALLENGE PROJECTS

This new feature appears at the end of select chapters in addition to the existing group and research projects. These projects revolve around the use of the Google Earth® online program, and help students apply concepts they have learned in the chapter to real-life locations around the world.

Organization of Problem Sets

The problem sets begin with drill problems that are linked to the section objectives, and are then followed by the categories of problems discussed below.

Applying the Concepts Students are always curious about how the mathematics they are learning can be applied, so we have included inviting applications some with illustrations, in most of the problem sets in the book and have labeled them to show students the array of uses of mathematics.

Getting Ready for the Next Section Many students think of mathematics as a collection of discrete, unrelated topics. As instructors, we know that this is not the case. The Getting Ready for the Next Section problems reinforce the cumulative, connected nature of this course by showing how the concepts and techniques flow one from another. These problems review all of the material that students will need in order to be successful, forming a bridge to the next section and gently preparing students to move forward.

Maintaining Your Skills One of the major themes of our book is continuous review. We strive to continuously hone techniques learned earlier by keeping the important concepts in the forefront of the course. The Maintaining Your Skills problems review material from the previous chapter, or they review problems that form the foundation of the course.

End-of-Chapter Summary, Review, and Assessment

We have learned that students are more comfortable with a chapter that sums up what they have learned thoroughly and accessibly, and reinforces concepts and techniques well. To help students grasp concepts and get more practice, each chapter ends with the following features that together give a comprehensive reexamination of the chapter.

Chapter Summary The chapter summary recaps all main points from the chapter in a visually appealing grid. In the margin, next to each topic, is an example that illustrates the type of problem associated with the topic being reviewed. Our way of summarizing shows students that concepts in mathematics do relate—and that mastering one concept is a bridge to the next. When students prepare for a test, they can use the chapter summary as a guide to the main concepts of the chapter.

Chapter Review Following the chapter summary in each chapter is the chapter review. It contains an extensive set of problems that review all the main topics in the chapter. This feature can be used flexibly, as assigned review, as a recommended self-test for students as they prepare for examinations, or as an in-class quiz or test.

Cumulative Review Starting in Chapter 2, following the chapter review in each chapter is a set of problems that reviews material from all preceding chapters. This keeps students current with past topics and helps them retain the information they study.

Chapter Test A set of problems representative of all the main points of the chapter. These don't contain as many problems as the chapter review, and should be completed in 50 minutes.

Chapter Projects Each chapter closes with a pair of projects. One is a group project, suitable for students to work on in class. Group projects list details about number of participants, equipment, and time, so that instructors can determine how well the project fits into their classroom. The second project is a research project for students to do outside of class and tends to be open ended.

Additional Features of the Book

Blueprint for Problem Solving Found in the main text, this feature is a detailed outline of steps required to successfully work application problems. Intended as a guide to problem solving in general, the blueprint takes the student through the solution process to various kinds of applications.

Facts from Geometry Many of the important facts from geometry are listed under this heading. In most cases, an example or two accompanies each of the facts to give students a chance to see how topics from geometry are related to the algebra they are learning.

A Glimpse of Algebra These sections, found in most chapters, show how some of the material in the chapter looks when it is extended to algebra.

Descriptive Statistics Beginning in Chapter 1 and then continuing through the rest of the book, students are introduced to descriptive statistics. In Chapter 1, we cover tables and bar charts, as well as mean, median, and mode. These topics are carried through the rest of the book. Along the way we add to the list of descriptive statistics by including scatter diagrams and line graphs.

Supplements for the Instructor

Please contact your sales representative.

Annotated Instructor's Edition
ISBN-10: 1-133-11026-6 | ISBN-13: 978-1-133-11026-2
This special instructor's version of the text contains answers next to all exercises and instructor notes at the appropriate location.

Cengage Instructor's Resource Binder for Algebra Activities
ISBN-10: 0-538-73675-5 | ISBN-13: 978-0-538-73675-6
NEW! Each section of the main text is discussed in uniquely designed Teaching Guides containing instruction tips, examples, activities, worksheets, overheads, assessments, and solutions to all worksheets and activities.

Complete Solutions Manual ISBN-10: 1-133-49082-4 | ISBN-13: 978-1-133-49082-1
This manual contains complete solutions for all problems in the text.

Enhanced WebAssign ISBN-10: 0-538-73810-3 | ISBN-13: 978-0-538-73810-1
Exclusively from Cengage Learning, Enhanced WebAssign® combines the exceptional Mathematics content that you know and love with the most powerful online homework solution, WebAssign. Enhanced WebAssign engages students with immediate feedback and rich tutorial content helping students to develop a deeper conceptual understanding of their subject matter. Online assignments can

be built by selecting from thousands of text-specific problems or supplemented with problems from any Cengage Learning textbook.

PowerLecture with ExamView® Algorithmic Equations
ISBN-10: 1-133-49114-6 | ISBN-13: 978-1-133-49114-9
This CD-ROM (or DVD) provides the instructor with dynamic media tools for teaching. Create, deliver, and customize tests (both print and online) in minutes with ExamView® Computerized Testing Featuring Algorithmic Equations. Easily build solution sets for homework or exams using Solution Builder's online solutions manual. Microsoft® PowerPoint® lecture slides and figures from the book are also included on this CD-ROM (or DVD).

Solution Builder This online instructor database offers complete worked solutions to all exercises in the text, allowing you to create customized, secure solutions printouts (in PDF format) matched exactly to the problems you assign in class. Visit http://www.cengage.com/solutionbuilder.

Text-Specific DVDs ISBN-10: 1-133-23171-3 | ISBN-13: 978-1-133-23171-4
This set of text-specific DVDs features segments taught by the author and worked-out solutions to many examples in the book. Available to instructors only.

Supplements for the Student

Enhanced WebAssign ISBN-10: 0-538-73810-3 | ISBN-13: 978-0-538-73810-1
Exclusively from Cengage Learning, Enhanced WebAssign® combines the exceptional Mathematics content that you know and love with the most powerful online homework solution, WebAssign. Enhanced WebAssign engages students with immediate feedback and rich tutorial content helping students to develop a deeper conceptual understanding of their subject matter. Online assignments can be built by selecting from thousands of text-specific problems or supplemented with problems from any Cengage Learning textbook.

Student Solutions Manual ISBN-10: 1-133-11140-8 | ISBN-13: 978-1-133-11140-5
This manual contains complete annotated solutions to all odd problems in the problem sets and all chapter review and chapter test exercises.

Student Workbook ISBN-10: 1-133-52513-X | ISBN-13: 978-1-133-52513-4
Get a head start with this hands-on resource! The Student Workbook is packed with assessments, activities, and worksheets to help you maximize your study efforts.

Acknowledgments

I would like to thank my editor at Cenage Learning, Marc Bove for his help and encouragement and ensuring a good working relationship with the editorial and marketing group at Cengage. Jennifer Risden continues to keep us on track with production and, as always, is the consummate professional. Donna Looper, head of production in my office, along with Staci Truelson did a fantastic job of keeping us organized and efficient. They are both a pleasure to work with. Special thanks to the other members of our team Mary Skutley, Kaela SooHoo, Mike Landrum, Katherine Shields, Kendra Nomoto, and Christina Machado; all of whom played an important roll in the production of this book.

Pat McKeague
September 2011

Preface to the Student

I often find my students asking themselves the question "Why can't I understand this stuff the first time?" The answer is "You're not expected to." Learning a topic in mathematics isn't always accomplished the first time around. There are many instances when you will find yourself reading over new material a number of times before you can begin to work problems. That's just the way things are in mathematics. If you don't understand a topic the first time you see it, that doesn't mean you won't succeed in this course. Understanding mathematics takes time. The process of understanding requires reading the book, studying the examples, working problems, and getting your questions answered.

How to Be Successful in Mathematics

1. **If you are in a lecture class, be sure to attend all class sessions on time.** You cannot know exactly what goes on in class unless you are there. Missing class and then expecting to find out what went on from someone else is not the same as being there yourself.

2. **Read the book.** It is best to read the section that will be covered in class beforehand. Reading in advance, even if you do not understand everything you read, is still better than going to class with no idea of what will be discussed.

3. **Work problems every day and check your answers.** The key to success in mathematics is working problems. The more problems you work, the better you will become at working them. The answers to the odd-numbered problems are given in the back of the book. When you have finished an assignment, be sure to compare your answers with those in the book. If you have made a mistake, find out what it is, and correct it.

4. **Do it on your own.** Don't be misled into thinking someone else's work is your own. Having someone else show you how to work a problem is not the same as working the same problem yourself. It is okay to get help when you are stuck. As a matter of fact, it is a good idea. Just be sure you do the work yourself.

5. **Review every day.** After you have finished the problems your instructor has assigned, take another 15 minutes and review a section you have already completed. The more you review, the longer you will retain the material you have learned.

6. **Don't expect to understand every new topic the first time you see it.** Sometimes you will understand everything you are doing, and sometimes you won't. Expecting to understand each new topic the first time you see it can lead to disappointment and frustration. The process of understanding takes time. You will need to read the book, work problems, and get your questions answered.

7. **Spend as much time as it takes for you to master the material.** No set formula exists for the exact amount of time you need to spend on mathematics to master it. You will find out as you go along what is or isn't enough time for you. If you end up spending 2 or more hours on each section in order to master the material there, then that's how much time it takes; trying to get by with less will not work.

8. **Relax.** Take a deep breath and work each problem one step at a time. It's probably not as difficult as you think.

Whole Numbers

Chapter Outline

1.1 Place Value and Names for Numbers

1.2 Addition with Whole Numbers, and Perimeter

1.3 Rounding Numbers and Estimating Answers

1.4 Subtraction with Whole Numbers

1.5 Multiplication with Whole Numbers

1.6 Division with Whole Numbers

1.7 Exponents, Order of Operations, and Averages

1.8 Area and Volume

The devastating eruption of Mount St. Helens in 1980 was the most catastrophic volcanic eruption in American history. The volcano is located in the state of Washington about 100 miles south of Seattle. Its eruption caused dozens of deaths, and destroyed almost 230 acres of forest and more than 200 homes. The eruption also caused a volcanic debris flow, commonly called a mudflow, to deposit more than 65 million cubic yards of sediment into the Cowlitz and Columbia Rivers. This excess debris and sediment drastically diminished the carrying capacity of the rivers, which then overflowed and flooded the surrounding lands.

	Carrying Capacity of the Cowlitz River (cubic feet per second)	Channel Depth of Columbia River
Prior to 1980	76,000	40 feet
After 1980 Eruption	15,000	14 feet

Source: U.S. Forest Service

In this chapter, we will introduce you to whole numbers and their operations. For instance, looking at the above table, we will learn how to compare the channel depth of the Columbia River from before the 1980 eruption to the channel depth after, and then calculate the difference to be 26 feet.

Preview

Key Words	Definition
Sum	The result of the addition of numbers
Difference	The result of the subtraction of numbers
Product	The result of multiplication
Factors	Numbers that, when multiplied together, give a product
Quotient	The result of division
Mean	The sum of the values in a set divided by the number of values in the set
Median	The number in a set of numbers that separates the higher half of the set from the lower half of the set
Mode	The number that occurs most frequently in a set of numbers

Chapter Outline

1.1 Place Value and Names for Numbers
- A State the place value for numbers in standard notation.
- B Write a whole number in expanded form.
- C Write a number in words.
- D Write a number from words.

1.2 Addition with Whole Numbers, and Perimeter
- A Add whole numbers.
- B Understand the notation and vocabulary of addition.
- C Use the properties of addition.
- D Find a solution to an equation by inspection.
- E Find the perimeter of a figure.

1.3 Rounding Numbers and Estimating Answers
- A Round whole numbers.
- B Estimate the answer to a problem.

1.4 Subtraction with Whole Numbers
- A Understand the notation and vocabulary of subtraction.
- B Subtract whole numbers.
- C Subtract using borrowing.

1.5 Multiplication with Whole Numbers
- A Multiply whole numbers using repeated addition.
- B Understand the notation and vocabulary of multiplication.
- C Identify properties of multiplication.
- D Solve equations with multiplication.
- E Solve applications with multiplication.

1.6 Division with Whole Numbers
- A Understand the notation and vocabulary of division.
- B Divide whole numbers.
- C Solve applications using division.

1.7 Exponents, Order of Operations, and Averages
- A Identify the base and exponent of an expression.
- B Simplify expressions with exponents.
- C Use the rule for order of operations.
- D Find the mean, median, mode, and range of a set of numbers.

1.8 Area and Volume
- A Find the area of a polygon.
- B Find the volume of an object.
- C Find the surface area of an object.

1.1 Place Value and Names for Numbers

OBJECTIVES

A State the place value for numbers in standard notation.

B Write a whole number in expanded form.

C Write a number in words.

D Write a number from words.

TICKET TO SUCCESS

Each section of the book will begin with some problems and questions like the ones below. Think about them while you read through the following section. Before you go to class, answer each problem or question with a written response using complete sentences. Writing about mathematics is a valuable exercise. As with all problems in this course, approach these writing exercises with a positive point of view. You will get better at giving written responses to questions as the course progresses. Even if you never feel comfortable writing about mathematics, just attempting the process will increase your understanding and ability in this course.

Keep these questions in mind as you read through the section. Then respond in your own words and in complete sentences.

1. How would you write the number 742 in expanded form?
2. Place a comma and a hyphen in the appropriate place so that the number 2,345 is written correctly in words below:
   two thousand three hundred forty five
3. Is there a largest whole number?

The table below gives population totals for the listed cities at the time of the 2009 U.S. Census.

City	Population
Harmony, CA	18
San Luis Obispo, CA	44,174
Seattle, WA	616,627
San Francisco, CA	815,358
Los Angeles, CA	3,831,868
New York, NY	8,391,881

Source: U.S. Census Bureau, International Data Base

In this section, we will examine the numbers in our number system and learn about *place values*. For instance, the *digit* 7 in San Luis Obispo's population is in the *tens column*. Keep reading to learn more about what all these terms mean.

NOTE

Next to each example in the text is a practice problem with the same number. After you read through an example, try the practice problem next to it. The answers to the practice problems are at the bottom of the page. Be sure to check your answers as you work these problems. The worked-out solutions to all practice problems with more than one step are given in the back of the book. So if you find a practice problem that you cannot work correctly, you can look up the correct solution to that problem in the back of the book.

PRACTICE PROBLEMS

1. Give the place value of each digit in the number 46,095.

Answer

1. 5 ones, 9 tens, 0 hundreds, 6 thousands, 4 ten thousands

A Place Value

Our number system is based on the number 10 and is, therefore, called a "base 10" number system. We write all numbers in our number system using the *digits* 0, 1, 2, 3, 4, 5, 6, 7, 8, and 9. The positions of the digits in a number determine the values of the digits. For example, the 5 in the number 251 has a different value from the 5 in the number 542.

The *place values* in our number system are as follows: The first digit on the right is in the *ones column*. The next digit to the left of the ones column is in the *tens column*. The next digit to the left is in the *hundreds column*. For a number like 542, the digit 5 is in the hundreds column, the 4 is in the tens column, and the 2 is in the ones column.

If we keep moving to the left, the columns increase in value. Table 1 shows the name and value of each of the first seven columns in our number system. Also, notice how the number 1 is located in the appropriate place value column.

TABLE 1

Millions Column	Hundred Thousands Column	Ten Thousands Column	Thousands Column	Hundreds Column	Tens Column	Ones Column
1,000,000	100,000	10,000	1,000	100	10	1

Let's practice applying place values to a real number.

EXAMPLE 1 Give the place value of each digit in the number 305,964.

SOLUTION Starting with the digit at the right, we have a 4 in the ones column, a 6 in the tens column, a 9 in the hundreds column, a 5 in the thousands column, a 0 in the ten thousands column, and a 3 in the hundred thousands column. ∎

Large Numbers

Place values can be extended to very large numbers as well. The photograph of outer space shown here was taken by the Hubble telescope. The object in the photograph is called the *Cone Nebula*. In astronomy, distances to objects like the Cone Nebula are given in light-years—the distance light travels in a year. If we assume light travels 186,000 miles in one second, then a light-year is 5,865,696,000,000 miles; that is,

5 trillion, 865 billion, 696 million miles

To find the place value of digits in large numbers, such as the one above, we can use Table 2. Note how the Ones, Thousands, Millions, Billions, and Trillions categories are each broken into Ones, Tens, and Hundreds. Therefore, if we write the digits for our light-year in the last row of the table, then it would look like this:

TABLE 2

Trillions			Billions			Millions			Thousands			Ones		
Hundreds	Tens	Ones	Hundreds	Tens	Ones	Hundreds	Tens	Ones	Hundreds	Tens	Ones	Hundreds	Tens	Ones
		5	8	6	5	6	9	6	0	0	0	0	0	0

Using words to read the place values for the digits in Table 2, we say that the first 5 is in the "one trillions" column, the 8 is in the "hundred billions" column, the first 6 is in the "ten billions" column, and so on. Here is another example of a large number to which we can apply place value.

EXAMPLE 2 Give the place value of each digit in the number 73,890,672,540.

SOLUTION The following diagram shows the place value of each digit:

Ten Billions	Billions	Hundred Millions	Ten Millions	Millions	Hundred Thousands	Ten Thousands	Thousands	Hundreds	Tens	Ones
7	3,	8	9	0,	6	7	2,	5	4	0

2. Give the place value of each digit in the number 21,705,328,456.

NOTE
When writing numbers with four or more digits, we use a comma to separate every three digits, starting from the ones column.

B Expanded Form

We can use the idea of place value to write numbers in *expanded form*. For example, the number 542 is written in condensed (standard) form. We can also write it in expanded form as

$$542 = 500 + 40 + 2$$

because the 5 is in the hundreds column, the 4 is in the tens column, and the 2 is in the ones column.

Here are more examples of numbers written in expanded form:

EXAMPLE 3 Write 5,478 in expanded form.

SOLUTION $5{,}478 = 5{,}000 + 400 + 70 + 8$

3. Write 3,972 in expanded form.

We can use money to make the results from Example 3 more intuitive. Suppose you have $5,478 in cash as follows:

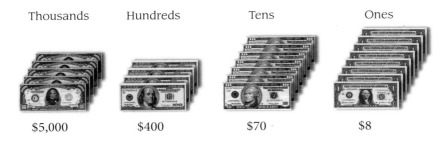

Thousands	Hundreds	Tens	Ones
$5,000	$400	$70	$8

Using this diagram as a guide, we can write

$$\$5{,}478 = \$5{,}000 + \$400 + \$70 + \$8$$

which shows us that our work writing numbers in expanded form is consistent with our intuitive understanding of the different denominations of money. Let's practice more.

Answers
2. 6 ones, 5 tens, 4 hundreds, 8 thousands, 2 ten thousands, 3 hundred thousands, 5 millions, 0 ten millions, 7 hundred millions, 1 billion, 2 ten billions
3. 3,000 + 900 + 70 + 2

4. Write 271,346 in expanded form.

5. Write 71,306 in expanded form.

6. Write 4,003,560 in expanded form.

Chapter 1 Whole Numbers

EXAMPLE 4 Write 354,798 in expanded form.

SOLUTION 354,798 = 300,000 + 50,000 + 4,000 + 700 + 90 + 8

EXAMPLE 5 Write 56,094 in expanded form.

SOLUTION Notice that there is a 0 in the hundreds column. This means we have 0 hundreds. In expanded form, we have

$$56,094 = 50,000 + 6,000 + 90 + 4$$

Note that we don't have to include the 0 hundreds.

EXAMPLE 6 Write 5,070,603 in expanded form.

SOLUTION The columns with 0 in them will not appear in the expanded form.

$$5,070,603 = 5,000,000 + 70,000 + 600 + 3$$

C Writing Numbers in Words

The idea of place value and expanded form can be used to help write the names for numbers. Naming numbers and writing them in words takes some practice. Let's begin by looking at the names of some two-digit numbers. Table 3 lists a few. Notice that the two-digit numbers that do not end in 0 have two parts. These parts are separated by a hyphen.

TABLE 3

Number	In English	Number	In English
25	Twenty-five	30	Thirty
47	Forty-seven	62	Sixty-two
93	Ninety-three	77	Seventy-seven
88	Eighty-eight	50	Fifty

The following examples give the names for some larger numbers. In each case, the names are written according to the place values given in Table 2.

7. Write each number in words.
   a. 724
   b. 595
   c. 307

8. Write each number in words.
   a. 4,758
   b. 62,779
   c. 305,440

EXAMPLE 7 Write each number in words.
   a. 452   b. 397   c. 608

SOLUTION a. Four hundred fifty-two
   b. Three hundred ninety-seven
   c. Six hundred eight

EXAMPLE 8 Write each number in words.
   a. 3,561   b. 53,662   c. 547,801

SOLUTION a. Three thousand, five hundred sixty-one

Notice how the comma separates the thousands from the hundreds.

   b. Fifty-three thousand, six hundred sixty-two
   c. Five hundred forty-seven thousand, eight hundred one

Answers
4. 200,000 + 70,000 + 1,000 + 300 + 40 + 6
5. 70,000 + 1,000 + 300 + 6
6. 4,000,000 + 3,000 + 500 + 60
7. a. Seven hundred twenty-four
   b. Five hundred ninety-five
   c. Three hundred seven
8. a. Four thousand, seven hundred fifty-eight
   b. Sixty-two thousand, seven hundred seventy-nine
   c. Three hundred five thousand, four hundred forty

EXAMPLE 9 Write each number in words.
a. 507,034,005
b. 739,600,075
c. 5,003,007,006

SOLUTION
a. Five hundred seven million, thirty-four thousand, five
b. Seven hundred thirty-nine million, six hundred thousand, seventy-five
c. Five billion, three million, seven thousand, six

The following check is a practical reason for being able to write numbers in word form.

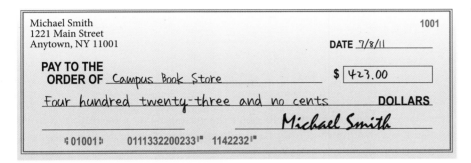

D Writing Numbers from Words

The next examples show how we write a number given in words as a number written with digits.

EXAMPLE 10 Write five thousand, six hundred forty-two, using digits instead of words.

SOLUTION Five thousand, six hundred forty-two
↓ ↓ ↓
5, 6 42 ⇒ 5,642

EXAMPLE 11 Write each number using digits instead of words.
a. Three million, fifty-one thousand, seven hundred
b. Two billion, five
c. Seven million, seven hundred seven

SOLUTION
a. 3,051,700
b. 2,000,000,005
c. 7,000,707

---

9. Write each number in words.
   a. 707,044,002
   b. 452,900,008
   c. 4,008,002,001

10. Write six thousand, two hundred twenty-one using digits instead of words.

11. Write each number with digits instead of words.
    a. Eight million, four thousand, two hundred
    b. Twenty-five million, forty
    c. Nine million, four hundred thirty-one

Answers
9. a. Seven hundred seven million, forty-four thousand, two
   b. Four hundred fifty-two million, nine hundred thousand, eight
   c. Four billion, eight million, two thousand, one
10. 6,221
11. a. 8,004,200   b. 25,000,040
    c. 9,000,431

Sets and the Number Line

In mathematics, a collection of numbers is called a *set*. In this chapter, we will be working with the set of *counting numbers* (also called natural numbers) and the set of *whole numbers*, which are both defined as follows:

Counting numbers = {1, 2, 3, . . .}
Whole numbers = {0, 1, 2, 3, . . .}

The dots mean "and so on," and the braces { } are used to group the numbers in the set together.

Another way to visualize the whole numbers is with a *number line*. To draw a number line, we simply draw a straight line and mark off equally spaced points along the line, as shown in Figure 1. We label the point at the left with 0 and the rest of the points, in order, with the numbers 1, 2, 3, 4, 5, and so on.

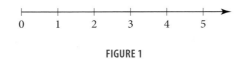

FIGURE 1

The arrow on the right indicates that the number line can continue in that direction forever. When we refer to numbers in this chapter, we will always be referring to the whole numbers.

DESCRIPTIVE STATISTICS

Tables and Charts

Tables and bar charts are two ways to compare numbers. Suppose the data below represents the quantity in milligrams of caffeine in five different beverages. The table and chart give two representations for the same data. The table is a numeric representation of the data, and the bar chart is a visual representation.

TABLE 4

Beverage (6-ounce cup)	Caffeine (in milligrams)
Brewed coffee	100
Instant coffee	70
Tea	50
Cocoa	5
Decaffeinated coffee	4

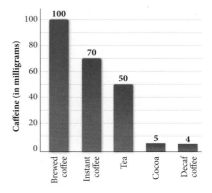

FIGURE 2

The diagram in Figure 2 is called a *bar chart*. The horizontal line below which the drinks are listed is called the *horizontal axis*, while the vertical line that is labeled from 0 to 100 is called the *vertical axis*.

We will use tables and charts quite often in this book to help us visualize the numbers we are working with.

Problem Set 1.1

Moving Toward Success

"You learn to speak by speaking, to study by studying, to run by running, to work by working; in just the same way, you learn to love by loving."

—Anatole France, 1844–1924, French writer and journalist

1. What are study skills? Give examples.
2. Why are study skills necessary for your success in this course?

A Give the place value of each digit in the following numbers. [Examples 1, 2]

1. 78
2. 93
3. 45
4. 79
5. 348
6. 789
7. 608
8. 450
9. 2,378
10. 6,481
11. 273,569
12. 768,253

Give the place value of the 5 in each of the following numbers.

13. 458,992
14. 75,003,782
15. 507,994,787
16. 320,906,050
17. 267,894,335
18. 234,345,678,789
19. 4,569,000
20. 50,000

B Write each of the following numbers in expanded form. [Examples 3–6]

21. 658
22. 479
23. 68
24. 71
25. 4,587
26. 3,762
27. 32,674
28. 54,883
29. 3,462,577
30. 5,673,524
31. 407
32. 508
33. 30,068
34. 50,905
35. 3,004,008
36. 20,088,060

C Write each of the following numbers in words. [Examples 7–9]

37. 29

38. 75

39. 40

40. 90

41. 573

42. 895

43. 707

44. 405

45. 770

46. 450

47. 23,540

48. 56,708

49. 3,004

50. 5,008

51. 3,040

52. 5,080

53. 104,065,780

54. 637,008,500

55. 5,003,040,008

56. 7,050,800,001

57. 2,546,731

58. 6,998,454

D Write each of the following numbers with digits instead of words. [Examples 10, 11]

59. Three hundred twenty-five

60. Forty-eight

61. Five thousand, four hundred thirty-two

62. One hundred twenty-three thousand, sixty-one

63. Eighty-six thousand, seven hundred sixty-two

64. One hundred million, two hundred thousand, three hundred

65. Two million, two hundred

66. Two million, two

67. Two million, two thousand, two hundred

68. Two billion, two hundred thousand, two hundred two

Applying the Concepts

69. The illustration shows the average annual income of workers 18 and older by education level.

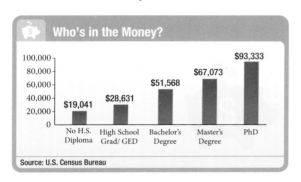

Write the following numbers in words:

a. The average income of someone with no higher than a high school education

b. The average income of someone with a Ph.D.

70. The table from this chapter's introduction shows the effect the 1980 eruption of Mount Helens had on two nearby rivers.

	Carrying Capacity of the Cowlitz River (cubic feet per second)	Channel Depth of Columbia River
Prior to 1980	76,000	40 feet
After 1980 Eruption	15,000	14 feet

Source: U.S. Forest Service

Write the following numbers in words:

a. The carrying capacity of the Cowlitz River prior to 1980

b. The carrying capacity of the Cowlitz River after the 980 eruption

71. MP3s Thanks to new technology, iPods of the future could hold up to 54,000 songs. What is the value of the 5 in the number of songs?

72. YouTube Experts claim that 34,560 hours of video are uploaded to YouTube every day. What is the value of the 4 in the number of hours?

73. Baseball Salaries According to CBSSports.com, Major League Baseball's 2011 average salary is $3,305,393, representing an increase of 0.2% from the previous season's average. Write 3,305,393 in words.

74. Astronomy The distance from the sun to the earth is 92,897,416 miles. Write this number in expanded form.

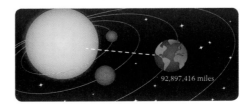

75. School Year Suppose the normal school year at your school is 180 days, or 4,320 hours. Write the number of hours in words.

76. School Year Suppose the normal school year at your school is 180 days, or 259,200 minutes. Write the number of minutes in words.

Writing Checks In each of the checks below, fill in the appropriate space with the dollar amount in either digits or in words.

77.

78.

Populations of Countries The table below gives estimates of the populations of some countries based on the U.S. Census Bureau's International Database for 2010. The first column under *Population* gives the population in digits. The second column gives the population in words. Fill in the blanks.

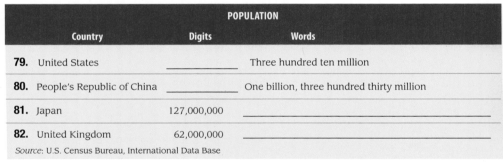

	POPULATION	
Country	Digits	Words
79. United States	_____	Three hundred ten million
80. People's Republic of China	_____	One billion, three hundred thirty million
81. Japan	127,000,000	_____
82. United Kingdom	62,000,000	_____

Source: U.S. Census Bureau, International Data Base

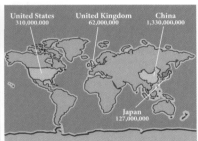

Populations of Cities The table below gives estimates of the populations of some cities based on the 2009 U.S. Census. The first column under *Population* gives the population in digits. The second column gives the population in words. Fill in the blanks.

	POPULATION	
City	Digits	Words
83. Seattle, WA	_____	Six hundred sixteen thousand
84. San Francisco, CA	_____	Eight hundred fifteen thousand
85. Los Angeles, CA	3,831,000	_____
86. New York, NY	8,391,000	_____

Source: U.S. Census Bureau, International Data Base

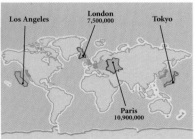

Addition with Whole Numbers, and Perimeter

1.2

OBJECTIVES

- **A** Add whole numbers.
- **B** Understand the notation and vocabulary of addition.
- **C** Use the properties of addition.
- **D** Find a solution to an equation by inspection.
- **E** Find the perimeter of a figure.

TICKET TO SUCCESS

Keep these questions in mind as you read through the section. Then respond in your own words and in complete sentences.

1. What is the definition of the word *sum*?
2. Briefly explain the following two properties:
   a. commutative property of addition
   b. associative property of addition
3. What does it mean to carry when adding numbers?
4. What is the perimeter of a geometric figure?

The table shows baggage fee revenues accrued in 2009 by the following airlines:

Rank	Airline	1st Quarter	2nd Quarter	3rd Quarter	4th Quarter	Full Year
1	Delta	102,838	118,356	129,465	131,060	481,719
2	American	108,117	118,442	119,466	129,159	475,184
3	US Airways	94,227	104,138	111,395	122,520	432,280
4	Northwest	59,786	67,186	78,922	79,931	285,825
5	United	59,102	67,412	77,877	64,586	268,977

Source: Bureau of Transportation Statistics

Recall the number line we introduced in the last section. In this section, we will begin to visualize addition on the number line. This skill will help us further analyze the baggage fee table and gather more information from its data. For instance, you will be able to add the quarterly earnings for American Airlines and find that the total equals the quantity listed in the Full Year column. In other words, you will be able to work the following problem:

$$\begin{array}{r} 108{,}117 \\ 118{,}442 \\ 119{,}466 \\ +\ 129{,}159 \\ \hline 475{,}184 \end{array}$$

Facts of Addition

Using lengths to visualize addition can be very helpful. In mathematics, we generally do so by using the number line. For example, we add 3 and 5 on the number line in Figure 1 like this: Start at 0 and move to 3. From 3, move 5 more units to the right. This brings us to 8. Therefore, 3 + 5 = 8.

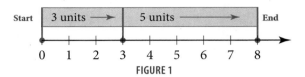

FIGURE 1

If we do this kind of addition on the number line with all combinations of the numbers 0 through 9, we get the results summarized in Table 1.

We call the information in Table 1 our basic addition facts. Your success with the examples and problems in this section depends on knowing the basic addition facts.

TABLE 1

ADDITION TABLE

+	0	1	2	3	4	5	6	7	8	9
0	0	1	2	3	4	5	6	7	8	9
1	1	2	3	4	5	6	7	8	9	10
2	2	3	4	5	6	7	8	9	10	11
3	3	4	5	6	7	8	9	10	11	12
4	4	5	6	7	8	9	10	11	12	13
5	5	6	7	8	9	10	11	12	13	14
6	6	7	8	9	10	11	12	13	14	15
7	7	8	9	10	11	12	13	14	15	16
8	8	9	10	11	12	13	14	15	16	17
9	9	10	11	12	13	14	15	16	17	18

Suppose we want to use Table 1 to find the answer to 3 + 5. We locate the 3 in the left column and the 5 in the top row. We read *across* from the 3 and *down* from the 5. The entry in the table that is across from 3 and below 5 is 8. Therefore, 3 + 5 = 8.

A Adding Whole Numbers

Now let's use the addition facts to help us add whole numbers with more than one digit. To add whole numbers, we add digits within the same place value. First, we add the digits in the ones place, then the tens place, then the hundreds place, and so on.

EXAMPLE 1 Add: 43 + 52.

SOLUTION This type of addition is best done vertically. First, we add the digits in the ones place.

```
  4 3
+ 5 2
-----
    5
```

Then we add the digits in the tens place

```
  4 3
+ 5 2
-----
  9 5
```

> **NOTE**
> Table 1 is a summary of the addition facts that you *must* know in order to make a successful start in your study of basic mathematics. You *must* know how to add any pair of numbers that come from the list. You *must* be fast and accurate. You don't want to have to think about the answer to 7 + 9. You should know it's 16. Memorize these facts now. Don't put it off until later.

PRACTICE PROBLEMS

1. Add: 63 + 25.

> **NOTE**
> To show *why* we add digits with the same place value, we can write each number showing the place value of the digits:
>
>   43 = 4 tens + 3 ones
> + 52 = 5 tens + 2 ones
>   9 tens + 5 ones

Answer
1. 88

EXAMPLE 2 Add: 165 + 801.

SOLUTION Writing the sum vertically, we have

Add ones place.
Add tens place.
Add hundreds place. ∎

Addition with Carrying

In Examples 1 and 2, the sums of the digits with the same place value were always 9 or less. There are many times when the sum of the digits with the same place value will be a number larger than 9. In these cases, we have to do what is called *carrying* in addition. The following examples illustrate this process.

EXAMPLE 3 Add: 197 + 213 + 324.

SOLUTION We write the sum vertically and add digits with the same place value.

$$\begin{array}{r} 1 \\ 197 \\ 213 \\ + 324 \\ \hline 4 \end{array}$$

When we add the ones, we get 7 + 3 + 4 = 14. We write the 4 and carry the 1 to the tens column.

$$\begin{array}{r} 11 \\ 197 \\ 213 \\ + 324 \\ \hline 34 \end{array}$$

We add the tens, including the 1 that was carried over from the last step. We get 13, so we write the 3 and carry the 1 to the hundreds column.

$$\begin{array}{r} 11 \\ 197 \\ 213 \\ + 324 \\ \hline 734 \end{array}$$

We add the hundreds, including the 1 that was carried over from the last step, and write the total. ∎

EXAMPLE 4 Add: 46,789 + 2,490 + 864.

SOLUTION We write the sum vertically—with the digits in the same place value aligned—and then use the shorthand form of addition.

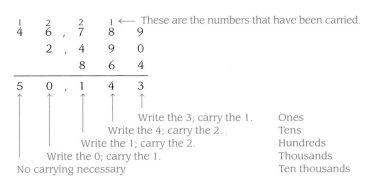

Write the 3; carry the 1.   Ones
Write the 4; carry the 2.   Tens
Write the 1; carry the 2.   Hundreds
Write the 0; carry the 1.   Thousands
No carrying necessary   Ten thousands ∎

---

2. Add: 342 + 605.

3. Add.
  a. 375 + 121 + 473
  b. 495 + 699 + 978

NOTE
Notice that Practice Problem 3 has two parts. Part a is similar to the problem shown in Example 3. Part b is similar also, but a little more challenging in nature. We will do this from time to time throughout the text. If a practice problem contains more parts than the example to which it corresponds, then the additional parts cover the same concept, but are more challenging than Part a.

4. Add.
  a. 57,904 + 7,193 + 655
  b. 68,495 + 7,236 + 878 + 29 + 5

Answers
2. 947
3. a. 969 **b.** 2,172
4. a. 65,752 **b.** 76,643

Adding numbers as we are doing here takes some practice. Most people don't make mistakes in carrying. Most mistakes in addition are made in adding the numbers in the columns. That is why it is so important that you practice and master the basic addition facts given in this chapter.

B Vocabulary

The word we use to indicate addition is *sum*. If we say "the sum of 3 and 5 is 8," what we mean is $3 + 5 = 8$. The word sum always indicates addition. We can state this fact in symbols by using the letters a and b to represent numbers in the following definition.

> **Definition**
> If a and b are any two numbers, then the **sum** of a and b is $a + b$. To find the sum of two numbers, we add them.

Table 2 gives some phrases and sentences in English and their mathematical equivalents written in symbols.

TABLE 2	
In English	In Symbols
The sum of 4 and 1	$4 + 1$
4 added to 1	$1 + 4$
8 more than m	$m + 8$
x increased by 5	$x + 5$
The sum of x and y	$x + y$
The sum of 2 and 4 is 6.	$2 + 4 = 6$

C Properties of Addition

Once we become familiar with addition, we may notice some facts about addition that are true regardless of the numbers involved. The first of these facts involves the number 0 (zero).

Whenever we add 0 to a number, the result is the original number. For example,

$$7 + 0 = 7 \quad \text{and} \quad 0 + 3 = 3$$

Because this fact is true no matter what number we add to 0, we call it a property of 0.

> **Addition Property of 0**
> If we let a represent any number, then it is always true that
> $$a + 0 = a \quad \text{and} \quad 0 + a = a$$
> *In words*: Adding 0 to any number leaves that number unchanged.

A second property we notice by becoming familiar with addition is that the order of two numbers in a sum can be changed without changing the result.

$$3 + 5 = 8 \quad \text{and} \quad 5 + 3 = 8$$
$$4 + 9 = 13 \quad \text{and} \quad 9 + 4 = 13$$

NOTE
When mathematics is used to solve everyday problems, the problems are almost always stated in words. The translation of English to symbols is a very important part of mathematics.

NOTE
When we use letters to represent numbers, as we do when we say, "If a and b are any two numbers," then a and b are called variables, because the values they take on vary. We use the variables a and b in the definitions and properties in this book because we want you to know that the definitions and properties are true for all numbers that you will encounter here.

This fact about addition is true for *all* numbers. The order in which you add two numbers doesn't affect the result. We call this fact the *commutative property of addition,* and we write it in symbols as follows:

Commutative Property of Addition

If *a* and *b* are any two numbers, then it is always true that

$$a + b = b + a$$

In words: Changing the order of two numbers in a sum doesn't change the result.

Let's practice using the commutative property of addition in the next example.

EXAMPLE 5 Use the commutative property of addition to rewrite each sum.

    a. $4 + 6$ **b.** $5 + 9$ **c.** $3 + 0$ **d.** $7 + n$

SOLUTION The commutative property of addition indicates that we can change the order of the numbers in a sum without changing the result. Applying this property we have

    a. $4 + 6 = 6 + 4$
    b. $5 + 9 = 9 + 5$
    c. $3 + 0 = 0 + 3$
    d. $7 + n = n + 7$

Notice that we did not actually add any of the numbers. The instructions were to use the commutative property, and the commutative property involves only the order of the numbers in a sum. ∎

The last property of addition we will consider here has to do with sums of more than two numbers. Suppose we want to find the sum of 2, 3, and 4. We could add 2 and 3 first, and then add 4.

$$(2 + 3) + 4 = 5 + 4 = 9$$

Or, we could add the 3 and 4 together first and then add the 2.

$$2 + (3 + 4) = 2 + 7 = 9$$

The result in both cases is the same. If we try this with any other numbers, the same thing happens. We call this fact about addition the *associative property of addition,* and we write it in symbols as follows:

Associative Property of Addition

If *a*, *b*, and *c* represent any three numbers, then

$$(a + b) + c = a + (b + c)$$

In words: Changing the grouping of three numbers in a sum doesn't change the result.

The next example will give you practice with this property.

5. Use the commutative property of addition to rewrite each sum.
   a. $7 + 9$
   b. $6 + 3$
   c. $4 + 0$
   d. $5 + n$

NOTE
This discussion is here to show why we write the next property the way we do. Sometimes it is helpful to look ahead to the property itself (in this case, the associative property of addition) to see what it is that is being justified.

Answers
5. a. $9 + 7$ **b.** $3 + 6$ **c.** $0 + 4$
    d. $n + 5$

6. Use the associative property of addition to rewrite each sum.
   a. (3 + 2) + 9
   b. (4 + 10) + 1
   c. 5 + (9 + 1)
   d. 3 + (8 + n)

EXAMPLE 6 Use the associative property of addition to rewrite each sum.
   a. (5 + 6) + 7
   b. (3 + 9) + 1
   c. 6 + (8 + 2)
   d. 4 + (9 + n)

SOLUTION The associative property of addition indicates that we are free to regroup the numbers in a sum without changing the result.

   a. (5 + 6) + 7 = 5 + (6 + 7)
   b. (3 + 9) + 1 = 3 + (9 + 1)
   c. 6 + (8 + 2) = (6 + 8) + 2
   d. 4 + (9 + n) = (4 + 9) + n

The commutative and associative properties of addition tell us that when adding whole numbers, we can use any order and grouping. In the next example, you'll see why, when adding several numbers, it is sometimes easier to look for pairs of numbers whose sums are 10, 20, and so on.

EXAMPLE 7 Add: 9 + 3 + 2 + 7 + 1.

SOLUTION We find pairs of numbers that we can add quickly:

$$9 + 3 + 2 + 7 + 1$$
$$= 10 + 10 + 2$$
$$= 22$$

7. Add.
   a. 6 + 2 + 4 + 8 + 3
   b. 24 + 17 + 36 + 13

D Solving Equations

We can use the addition table to help solve some simple equations. If n is used to represent a number, then n is called a *variable*. The equation

$$n + 3 = 5$$

will be true if n is 2. The number 2 is, therefore, called a *solution* to the equation, because, when we replace n with 2, the equation becomes a true statement.

$$2 + 3 = 5$$

Equations like this are really just puzzles or questions. When we say, "Solve the equation $n + 3 = 5$," we are asking the question, "What number do we add to 3 to get 5?"

When we solve equations by reading the equation to ourselves and then stating the solution, as we did with the equation above, we are solving the equation by *inspection*.

NOTE Just as we used *a* and *b* as variables to explain the properties of addition, we are using the letter *n* here as a variable because it also represents a number. In this case, it is the number that is a solution to an equation. We will continue to work with variables throughout this book.

EXAMPLE 8 Find the solution to each equation by inspection.
   a. $n + 5 = 9$
   b. $n + 6 = 12$
   c. $4 + n = 5$
   d. $13 = n + 8$

SOLUTION We find the solution to each equation by using the addition facts given in Table 1.
   a. The solution to $n + 5 = 9$ is 4, because 4 + 5 = 9.
   b. The solution to $n + 6 = 12$ is 6, because 6 + 6 = 12.
   c. The solution to $4 + n = 5$ is 1, because 4 + 1 = 5.
   d. The solution to $13 = n + 8$ is 5, because 13 = 5 + 8.

8. Use inspection to find the solution to each equation.
   a. $n + 9 = 17$
   b. $n + 2 = 10$
   c. $8 + n = 9$
   d. $16 = n + 10$

Answers
6. a. 3 + (2 + 9) b. 4 + (10 + 1)
   c. (5 + 9) + 1 d. (3 + 8) + n
7. a. 23 b. 90
8. a. 8 b. 8 c. 1 d. 6

E Perimeter

We end this section with an introduction to perimeter.

FACTS FROM GEOMETRY Perimeter

Here we will look at several different shapes called *polygons*. A *polygon* is a closed geometric figure, with at least three sides, in which each side is a straight line segment.

The most common polygons are squares, rectangles, and triangles. Examples of these are shown in Figure 2.

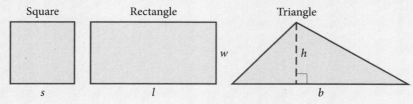

FIGURE 2

In the square, s is the length of the side, and each side has the same length. In the rectangle, l stands for the length, and w stands for the width. The width is usually the lesser of the two. The b and h in the triangle are the base and height, respectively. The height is always perpendicular to the base. That is, the height and base form a 90°, or *right*, angle where they meet.

To find the perimeter of a polygon, we add the lengths of all the sides. For example, the perimeter of the square in Figure 2 would be $s + s + s + s$.

Here is a formal definition to perimeter:

Definition

The **perimeter** of any polygon is the sum of the lengths of the sides, and it is denoted with the letter P.

NOTE
In the triangle, the small square where the broken line meets the base is the notation we use to show that the two line segments meet at right angles. That is, the height h and the base b are perpendicular to each other; the angle between them is 90°.

Let's practice finding the perimeters of some geometric figures.

EXAMPLE 9 Find the perimeter of each geometric figure. Figure a is a square.

a.
15 in.

b.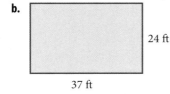
24 ft
37 ft

c.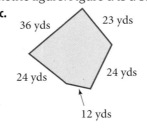
36 yds, 23 yds, 24 yds, 24 yds, 12 yds

SOLUTION In each case, we find the perimeter by adding the lengths of all the sides.

a. The figure is a square. Because the length of each side in the square is the same, the perimeter is
   $P = 15 + 15 + 15 + 15 = 60$ inches

b. In the rectangle, two of the sides are 24 feet long, and the other two are 37 feet long. The perimeter is the sum of the lengths of the sides.
   $P = 24 + 24 + 37 + 37 = 122$ feet

c. For this polygon, we add the lengths of the sides together. The result is the perimeter.
   $P = 36 + 23 + 24 + 12 + 24 = 119$ yards

9. Find the perimeter of each geometric figure. Figure a is a square.

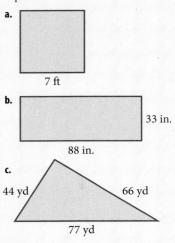

a. 7 ft
b. 33 in., 88 in.
c. 44 yd, 66 yd, 77 yd

Answers
9. a. 28 feet b. 242 inches
   c. 187 yards

Problem Set 1.2

Moving Toward Success

"Get over the idea that only children should spend their time in study. Be a student so long as you still have something to learn, and this will mean all your life."
—Henry L. Doherty, 1870–1931, American businessman

1. What do you do when you hear or read a tip to improve your study skills? Explain.
2. How does applying new study skills help you succeed in this course?

A Find each of the following sums. (Add.) [Examples 1–4]

1. $3 + 5 + 7$
2. $2 + 8 + 6$
3. $1 + 4 + 9$
4. $2 + 8 + 3$
5. $5 + 9 + 4 + 6$
6. $8 + 1 + 6 + 2$
7. $1 + 2 + 3 + 4 + 5$
8. $5 + 6 + 7 + 8 + 9$
9. $9 + 1 + 8 + 2$
10. $7 + 3 + 6 + 4$

A Add each of the following. (There is no carrying involved in these problems.) [Examples 1, 2]

11. 43 + 25
12. 56 + 23
13. 81 + 17
14. 37 + 22
15. 4,281 + 3,016
16. 2,749 + 1,250

17. 3,482 + 3,005
18. 2,496 + 7,503
19. 32 + 21 + 43
20. 521 + 340 + 135
21. 6,245 + 203 + 1,001
22. 27 + 4,510 + 342

A Add each of the following. (All problems involve carrying in at least one column.) [Examples 3, 4]

23. 49 + 16
24. 85 + 29
25. 74 + 28
26. 36 + 46
27. 682 + 193
28. 439 + 270

29. 638 + 191
30. 444 + 595
31. 4,963 + 5,428
32. 8,291 + 7,489
33. 6,205 + 9,999
34. 8,888 + 9,999

35. 56,789 + 98,765
36. 45,678 + 87,654
37. 52,468 + 58,642
38. 13,579 + 97,531
39. 4,296 + 8,720 + 4,375
40. 5,637 + 481 + 7,899

41.	4,994	42.	6,824	43.	12	44.	21	45.	999	46.	646
	449		371		34		43		444		464
	+ 9,449		+ 4,857		56		65		555		525
					+ 78		+ 87		+ 222		+ 252

47.	9,245	48.	45
	672		9,876
	8,341		54
	+ 27		+ 6,789

B Complete the following tables.

49.
First Number a	Second Number b	Their Sum a + b
61	38	
63	36	
65	34	
67	32	

50.
First Number a	Second Number b	Their Sum a + b
10	45	
20	35	
30	25	
40	15	

51.
First Number a	Second Number b	Their Sum a + b
9	16	
36	64	
81	144	
144	256	

52.
First Number a	Second Number b	Their Sum a + b
25	75	
24	76	
23	77	
22	78	

B Write each of the following expressions in words. Use the word *sum* in each case. [Table 2]

53. 4 + 9   **54.** 9 + 4   **55.** 8 + 1

56. 9 + 9   **57.** 2 + 3 = 5   **58.** 8 + 2 = 10

B Write each of the following in symbols. [Table 2]

59. a. The sum of 5 and 2
  b. 3 added to 8

60. a. The sum of a and 4
  b. 6 more than x

61. a. m increased by 1
  b. The sum of m and n

62. a. The sum of 4 and 8 is 12.
  b. The sum of a and b is 6.

C Rewrite each of the following using the commutative property of addition. [Example 5]

63. 5 + 9 **64.** 2 + 1 **65.** 3 + 8 **66.** 9 + 2 **67.** 6 + 4 **68.** 1 + 7

C Rewrite each of the following using the associative property of addition. [Example 6]

69. (1 + 2) + 3 **70.** (4 + 5) + 9 **71.** (2 + 1) + 6 **72.** (2 + 3) + 8

73. 1 + (9 + 1) **74.** 2 + (8 + 2) **75.** (4 + n) + 1 **76.** (n + 8) + 1

D Find a solution for each equation. [Example 8]

77. n + 6 = 10 **78.** n + 4 = 7 **79.** n + 8 = 13 **80.** n + 6 = 15

81. 4 + n = 12 **82.** 5 + n = 7 **83.** 17 = n + 9 **84.** 13 = n + 5

E Find the perimeter of each figure. The first four figures are squares. [Example 9]

85.
3 in.

86.
9 in.

87.
4 ft

88.
2 ft

89.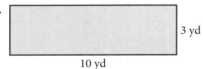
3 yd
10 yd

90.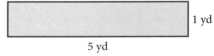
1 yd
5 yd

91.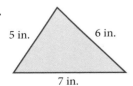
5 in. 6 in.
7 in.

92.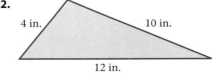
4 in. 10 in.
12 in.

Applying the Concepts

93. Internet usage varies all over the world. The information in the chart shows the number of users by region.

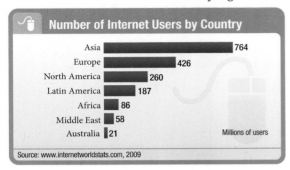

Use the information in the given illustration to write expressions and then find the following:

a. How many more users are there in Latin America as compared to Africa?

b. How many more users are there in Asia as compared to North America?

94. Camera phones have become widely popular. The information in the chart represents the number of picture messages sent by camera phone users over several months.

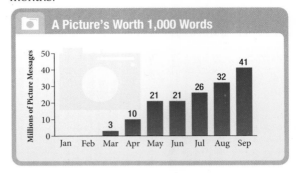

Use the information to find the following:

a. The number of picture messages sent in all nine months

b. The number of picture messages sent in March and April

95. Checkbook Balance On Monday Bob had a balance of $241 in his checkbook. On Tuesday he made a deposit of $108, and on Thursday he wrote a check for $24. What was the balance in his checkbook on Wednesday?

96. Number of Passengers A plane flying from Los Angeles to New York left Los Angeles with 67 passengers on board. The plane stopped in Bakersfield and picked up 28 passengers, and then it stopped again in Dallas where 57 more passengers came on board. How many passengers were on the plane when it landed in New York?

Checked Luggage Recall the chart from the section opening showing the quarterly earnings airlines are receiving for checked luggage. Use it to answer the following questions.

Rank	Airline	1st Quarter	2nd Quarter	3rd Quarter	4th Quarter	Full Year
1	Delta	102,838	118,356	129,465	131,060	481,719
2	American	108,117	118,442	119,466	129,159	475,184
3	US Airways	94,227	104,138	111,395	122,520	432,280
4	Northwest	59,786	67,186	78,922	79,931	285,825
5	United	59,102	67,412	77,877	64,586	268,977

Source: Bureau of Transportation Statistics

97. What were the earnings from checked luggage for US Airways for the first half of the year?

98. What were the combined earnings from checked luggage for Delta, Northwest, and United Airlines for the third quarter?

Rounding Numbers and Estimating Answers

1.3

OBJECTIVES

A Round whole numbers.

B Estimate the answer to a problem.

TICKET TO SUCCESS

Keep these questions in mind as you read through the section. Then respond in your own words and in complete sentences.

1. Briefly explain the strategy for rounding whole numbers.
2. Why do we round the numbers as a last step instead of first?
3. Why is estimating a useful tool?
4. Write an application problem in which estimation could be used.

Many times when we talk about numbers, it is helpful to use numbers that have been *rounded off*, rather than exact numbers. For example, the city where I live has a population of 44,174. But when I tell people how large the city is, I usually say, "The population is about 44,000." The number 44,000 is the original number rounded to the nearest thousand. The number 44,174 is closer to 44,000 than it is to 45,000, so it is rounded to 44,000. We can visualize this situation on the number line.

A Rounding

The steps used in rounding numbers are given below.

Strategy Rounding Whole Numbers

To summarize, we list the following steps:

Step 1 Locate the digit just to the right of the place to which you want to round.

Step 2 If that digit is less than 5, replace it and all digits to its right with zeros.

Step 3 If that digit is 5 or more, replace it and all digits to its right with zeros, and add 1 to the digit to its left.

NOTE
After you have used the steps listed here to work a few problems, you will find that the procedure becomes almost automatic.

PRACTICE PROBLEMS

1. Round 5,742 to the nearest
   a. hundred.
   b. thousand.

2. Round 87 to the nearest
   a. ten.
   b. hundred.

3. Round 980 to the nearest
   a. hundred.
   b. thousand.

4. Round 376,804,909 to the nearest
   a. million.
   b. ten thousand.

Answers
1. a. 5,700 b. 6,000
2. a. 90 b. 100
3. a. 1,000 b. 1,000
4. a. 377,000,000
   b. 376,800,000

Chapter 1 Whole Numbers

You can see from these rules that in order to round a number you must be told what column (or place value) to round to. We will do so in the next examples.

EXAMPLE 1 Round 5,382 to the nearest hundred.

SOLUTION There is a 3 in the hundreds column. We look at the digit just to its right, which is 8. Because 8 is greater than 5, we add 1 to the 3, and we replace the 8 and 2 with zeros.

5,382   is   5,400   to the nearest hundred.

8 is greater than 5.   Add 1 to get 4.   Put zeros here.

EXAMPLE 2 Round 94 to the nearest ten.

SOLUTION There is a 9 in the tens column. To its right is 4. Because 4 is less than 5, we simply replace it with 0.

94   is   90   to the nearest ten.

Less than 5   Replaced with zero

EXAMPLE 3 Round 973 to the nearest hundred.

SOLUTION We have a 9 in the hundreds column. To its right is 7, which is greater than 5. We add 1 to 9 to get 10, and then replace the 7 and 3 with zeros.

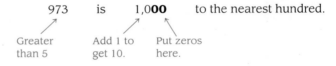

973   is   1,000   to the nearest hundred.

Greater than 5   Add 1 to get 10.   Put zeros here.

EXAMPLE 4 Round 47,256,344 to the nearest million.

SOLUTION We have 7 in the millions column. To its right is 2, which is less than 5. We simply replace all the digits to the right of 7 with zeros to get our answer.

47,256,344   is   47,000,000   to the nearest million.

Less than 5   Leave as is.   Replaced with zeros

Table 1 gives more examples of rounding.

TABLE 1

	Rounded to the Nearest		
Original Number	Ten	Hundred	Thousand
6,914	6,910	6,900	7,000
8,485	8,490	8,500	8,000
5,555	5,560	5,600	6,000
1,234	1,230	1,200	1,000

If we are doing calculations and are asked to round our answer, then it is important to follow this rule.

1.3 Rounding Numbers and Estimating Answers

> **Rule Rounding Numbers**
> Do all arithmetic first and then round the result. That is, the last step is to round the answer; we don't round the numbers first and then do the arithmetic.

EXAMPLE 5   The pie chart shows how a family earning $36,913 a year spends their money.

a. To the nearest hundred dollars, what is the total amount spent on food and entertainment?

b. To the nearest thousand dollars, how much of their income is spent on items other than taxes and savings?

SOLUTION   In each case, we add the numbers in question and round the sum to the indicated place.

a. We add the amounts spent on food and entertainment and then round that result to the nearest hundred dollars.

House Payments $10,200
Taxes $6,137
Miscellaneous $6,142
Entertainment $2,142
Car Expenses $4,847
Savings $2,149
Food $5,296

Food	$5,296
Entertainment	+ 2,142
Total	$7,438 = $7,400 to the nearest hundred dollars

b. We add the numbers for all items except taxes and savings.

House payments	$10,200
Food	5,296
Car expenses	4,847
Entertainment	2,142
Miscellaneous	+ 6,142
Total	$28,627 = $29,000 to the nearest thousand dollars

B Estimating

Now we'll apply what we've learned about rounding numbers to a process called estimation. When we *estimate* the answer to a problem, we simplify the problem so that an approximate answer can be found quickly. There are a number of ways of doing this. One common method is to use rounded numbers to simplify the arithmetic necessary to arrive at an approximate answer. In other words, we will round the numbers prior to calculating our answer. Note this process contrasts with that used in the previous examples.

EXAMPLE 6   Estimate the answer to the following problem by rounding each number to the nearest thousand.

4,872
1,691
777
+ 6,124

---

5. Use the pie chart in Example 5 to answer these questions.
   a. To the nearest ten dollars, what is the total amount spent on food and car expenses?
   b. To the nearest hundred dollars, how much is spent on savings and taxes?
   c. To the nearest thousand dollars, how much is spent on items other than food and entertainment?

6. Estimate the answer by first rounding each number to the nearest thousand.
   a. 5,287
      2,561
      888
      + 4,898

   b.   702
      3,944
      1,001
      + 3,500

Answers
5. a. $10,140   b. $8,300
   c. $29,000
6. a. 14,000   b. 10,000

NOTE

To clarify, in Example 6, we are asked to *estimate* an answer, so it is okay to round the numbers in the problem before adding them. In Example 5, we are asked for a rounded answer, meaning that we are to find the exact answer to the problem and then round to the indicated place. In that case, we must not round the numbers in the problem before adding.

7. Next time you go to the store, the prices have changed. A loaf of bread is now $2.69, milk is $4.26, eggs are $1.99, apples are $1.59, and the cereal is $5.19. Use estimation to determine if your $20 is sill enough to pay for all of the groceries.

SOLUTION We round each of the four numbers in the sum to the nearest thousand. Then we add the rounded numbers.

4,872	rounds to	5,000
1,691	rounds to	2,000
777	rounds to	1,000
+ 6,124	rounds to	+ 6,000
		14,000

We estimate the answer to this problem to be approximately 14,000. The actual answer, found by adding the original unrounded numbers, is 13,464. ∎

Here is a practical application for which the ability to estimate can be a useful tool.

EXAMPLE 7 On the way home from classes you stop at the grocery store to pick up a few things. You know that you have a $20.00 bill in your wallet. You pick up the following items: a loaf of wheat bread for $2.29, a gallon of milk for $3.96, a dozen eggs for $2.18, a pound of apples for $1.19, and a box of your favorite cereal for $4.59. Use estimation to determine if you will have enough to pay for your groceries when you get to the cashier.

SOLUTION We round the items in our grocery cart off to the nearest dollar.

wheat bread for $2.29	rounds to	$2.00
milk for $3.96	rounds to	$4.00
eggs for $2.18	rounds to	$2.00
apples for $1.19	rounds to	$1.00
+ cereal for $4.59	rounds to	+ $5.00
		$14.00

We estimate our total to be $14.00. Thus, $20.00 will be enough to pay for the groceries. (The actual cost of the groceries is $14.21.) ∎

Answer
7. yes

Problem Set 1.3

Moving Toward Success

"The quality of a person's life is in direct proportion to their commitment to excellence, regardless of their chosen field of endeavor."
—Vince Lombardi, 1913–1970, American football coach

1. We recommend you spend two hours on homework for every hour you are in class. What do you think about this schedule?

2. Map out a projected weekly schedule that includes your classes, work shifts, extracurriculars, and any additional obligations. Now fill in the hours you intend to devote to this class and its homework.

A Round each of the numbers to the nearest ten. [Examples 1–5]

1. 42
2. 44
3. 46
4. 48
5. 45
6. 73
7. 77
8. 75
9. 458
10. 455
11. 471
12. 680
13. 56,782
14. 32,807
15. 4,504
16. 3,897

Round each of the numbers to the nearest hundred. [Examples 1–5]

17. 549
18. 954
19. 833
20. 604
21. 899
22. 988
23. 1090
24. 6,778
25. 5,044
26. 56,990
27. 39,603
28. 31,999

Round each of the numbers to the nearest thousand. [Examples 1–5]

29. 4,670   **30.** 9,054   **31.** 9,760   **32.** 4,444   **33.** 978   **34.** 567

35. 657,892   **36.** 688,909   **37.** 509,905   **38.** 608,433   **39.** 3,789,345   **40.** 5,744,500

A Complete the following table by rounding the numbers on the left as indicated by the headings in the table. [Examples 1–5]

Original Number	Rounded to the Nearest		
	Ten	Hundred	Thousand
41. 7,821			
42. 5,945			
43. 5,999			
44. 4,353			
45. 10,985			
46. 11,108			
47. 99,999			
48. 95,505			

B Estimating Estimate the answer to each of the following problems by rounding each number to the indicated place value and then adding. [Example 6]

49. Hundred
750
275
+ 120

50. Thousand
1,891
765
+ 3,223

51. Hundred
472
422
536
+ 511

52. Hundred
399
601
744
+ 298

53. Thousand
25,399
7,601
18,744
+ 6,298

54. Thousand
9,999
8,888
7,777
+ 6,666

55. Hundred
9,999
8,888
7,777
+ 6,666

56. Ten thousand
127,675
72,560
+ 219,065

57. Ten thousand	58. Ten	59. Hundred	60. Hundred
65,000	10,061	20,150	1,950
31,000	10,044	18,250	2,849
15,555	10,035	12,350	3,750
+ 72,000	+ 10,025	+ 30,450	+ 4,649

Applying the Concepts

61. Cars The chart shows which countries manufacture the most cars. Use the chart to answer the following questions.

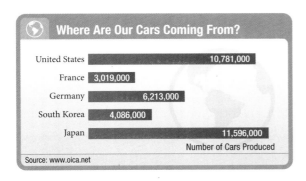

a. What was the total number of cars manufactured?

b. Do you think these numbers are rounded? If so, to what place value?

c. What is the total number of cars manufactured in Europe?

d. What is the total number of cars manufactured in Asia?

62. Business Expenses The pie chart shows one year's worth of expenses for a small business. Use the chart to answer the following questions.

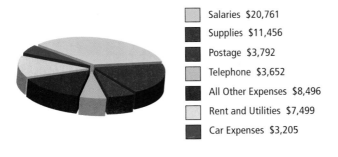

a. To the nearest hundred dollars, how much was spent on postage and supplies?

b. Find the total amount spent, to the nearest hundred dollars, on rent and utilities and car expenses.

c. To the nearest thousand dollars, how much was spent on items other than salaries and rent and utilities?

d. To the nearest thousand dollars, how much was spent on items other than postage, supplies, and car expenses?

Chapter 1 Whole Numbers

63. Mount St. Helens Recall the chart from the chapter opener that shows the changes that occurred to the rivers surrounding Mount St. Helens before and after the 1980 eruption. Use it to answer the following questions.

a. Do you think that the carrying capacity of the Cowlitz river after the 1980 eruption is rounded? If so, to what place value?

	Carrying Capacity of the Cowlitz River (cubic feet per second)	Channel Depth of Columbia River
Prior to 1980	76,000	40 feet
After 1980 Eruption	15,000	14 feet

Source: U.S. Forest Service

b. Do you think that the depth of the Columbia River prior to 1980 is rounded? If so, to what place value?

c. Do you think that the depth of the Columbia River after the 1980 eruption is rounded? If so, to what place value?

64. Fast Food Suppose the number of calories consumed by eating some popular fast foods is listed in the following table. Use the axes in the figure below to construct a bar chart from the information in the table.

CALORIES IN FAST FOOD	
Food	Calories
McDonald's Hamburger	270
Burger King Hamburger	260
Jack in the Box Hamburger	280
McDonald's Big Mac	510
Burger King Whopper	630
Jack in the Box Colossus Burger	940

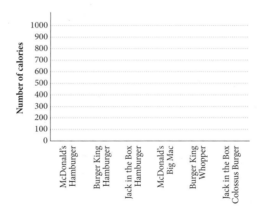

65. Exercise Suppose the following table lists the number of calories burned in 1 hour of exercise by a person who weighs 150 pounds. Use the axes in the figure below to construct a bar chart from the information in the table.

CALORIES BURNED BY A 150-POUND PERSON IN ONE HOUR	
Activity	Calories
Bicycling	374
Bowling	265
Handball	680
Jazzercise	340
Jogging	680
Skiing	544

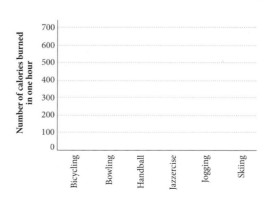

Subtraction with Whole Numbers

1.4

OBJECTIVES

A Understand the notation and vocabulary of subtraction.

B Subtract whole numbers.

C Subtract using borrowing.

TICKET TO SUCCESS

Keep these questions in mind as you read through the section. Then respond in your own words and in complete sentences.

1. Define *difference*.
2. What is an equivalent addition problem for the following subtraction problem: 9 − 3 = 6?
3. When would you use borrowing in a subtraction problem?
4. Describe how you would subtract the number 56 from the number 93.

In business, *subtraction* is used to calculate profit. Profit is found by subtracting costs from revenue. Suppose a children's shoe company tracked its costs and revenue during a 4-week period. The following double bar chart shows the data.

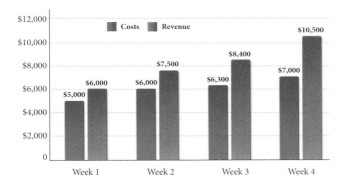

To find the profit for Week 1, we subtract the costs from the revenue.

        Profit = $6,000 − $5,000
        Profit = $1,000

   Subtraction is the opposite operation of addition. If you understand addition and can work simple addition problems quickly and accurately, then subtraction shouldn't be difficult for you.

A Vocabulary

The word *difference* always indicates subtraction. We can state this in symbols by again letting the letters *a* and *b* represent numbers.

> **Definition**
> The **difference** of two numbers a and b is $a - b$.

Table 1 gives some word statements involving subtraction and their mathematical equivalents written in symbols.

TABLE 1	
In English	**In Symbols**
The difference of 9 and 1	$9 - 1$
The difference of 1 and 9	$1 - 9$
The difference of m and 4	$m - 4$
The difference of x and y	$x - y$
3 subtracted from 8	$8 - 3$
2 subtracted from t	$t - 2$
The difference of 7 and 4 is 3.	$7 - 4 = 3$
The difference of 9 and 3 is 6.	$9 - 3 = 6$

B The Meaning of Subtraction

Now, we'll use some vocabulary to further explore the meaning of subtraction. Suppose we want to subtract 3 from 8. We would write

$$8 - 3, \quad 8 \text{ subtract } 3, \quad \text{or} \quad 8 \text{ minus } 3.$$

The answer we are looking for here is the difference between 8 and 3, or the number we add to 3 to get 8. That is,

$$8 - 3 = ? \quad \text{is the same as} \quad ? + 3 = 8$$

In both cases, we are looking for the number we add to 3 to get 8. The number we are looking for is 5. We have two ways to write the same statement:

$$\begin{array}{cc} \text{Subtraction} & \text{Addition} \\ 8 - 3 = 5 \quad \text{or} & 5 + 3 = 8 \end{array}$$

For every subtraction problem, there is an equivalent addition problem. This is a simple, yet important, point to understand. Table 2 lists some examples.

TABLE 2		
Subtraction		**Addition**
$7 - 3 = 4$	because	$4 + 3 = 7$
$9 - 7 = 2$	because	$2 + 7 = 9$
$10 - 4 = 6$	because	$6 + 4 = 10$
$15 - 8 = 7$	because	$7 + 8 = 15$

To subtract numbers with two or more digits, we align the numbers vertically and subtract in columns. Let's practice.

1.4 Subtraction with Whole Numbers

PRACTICE PROBLEMS

EXAMPLE 1 Subtract: 376 − 241.

SOLUTION We write the problem vertically, aligning digits with the same place value. Then we subtract in columns.

$$\begin{array}{r} 376 \\ -241 \\ \hline 135 \end{array}$$ ← Subtract the bottom number in each column from the number above it.

1. Subtract.
  a. 684 − 431
  b. 7,406 − 3,405

EXAMPLE 2 Subtract 503 from 7,835.

SOLUTION In symbols, this statement is equivalent to

$$7,835 - 503$$

To subtract, we write 503 below 7,835 and then subtract in columns.

$$\begin{array}{r} 7,835 \\ -503 \\ \hline 7,332 \end{array}$$

$5 - 3 = 2$ Ones
$3 - 0 = 3$ Tens
$8 - 5 = 3$ Hundreds
$7 - 0 = 7$ Thousands

2. a. Subtract 405 from 6,857.
   b. Subtract 234 from 345.

It is important to note that we always subtract the bottom number from the top number. As you can see, subtraction problems like the ones in Examples 1 and 2 are fairly simple because the digits in the bottom numbers were smaller than the digits in the top numbers with which they aligned. However, this will not always be the case, as we will now discuss.

C Subtraction with Borrowing

Subtraction must involve *borrowing* when the bottom digit in any column is larger than the digit above it. In one sense, borrowing is the reverse of the carrying we did in addition.

EXAMPLE 3 Subtract: 92 − 45.

SOLUTION We write the problem vertically with the place values of the digits showing.

$$\begin{array}{r} 92 = 9 \text{ tens} + 2 \text{ ones} \\ -45 = 4 \text{ tens} + 5 \text{ ones} \\ \hline \end{array}$$

Look at the ones column. We cannot subtract immediately, because 5 is larger than 2. Instead, we borrow 1 ten from the 9 tens in the tens column. We can rewrite the number 92 as

$$9 \text{ tens} + 2 \text{ ones}$$
$$= 8 \text{ tens} + 1 \text{ ten} + 2 \text{ ones}$$
$$= 8 \text{ tens} + 12 \text{ ones}$$

3. Subtract.
  a. 63 − 47
  b. 532 − 403

NOTE
The discussion here shows why borrowing is necessary and how we go about it. To understand borrowing, you should pay close attention to this discussion.

Answers
1. a. 253   **b.** 4,001
2. a. 6,452   **b.** 111
3. a. 16   **b.** 129

Now we are in a position to subtract.

$$92 = 9 \text{ tens} + 2 \text{ ones} = 8 \text{ tens} + 12 \text{ ones}$$
$$-45 = 4 \text{ tens} + 5 \text{ ones} = 4 \text{ tens} + 5 \text{ ones}$$
$$\phantom{-45 = 4 \text{ tens} + 5 \text{ ones} =\ } 4 \text{ tens} + 7 \text{ ones}$$

The result is 4 tens + 7 ones, which can be written in standard form as 47.

Writing the problem out in this way is more trouble than is actually necessary. The shorthand form of the same problem looks like this:

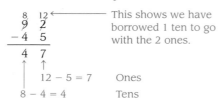

This shows we have borrowed 1 ten to go with the 2 ones.

12 − 5 = 7   Ones
8 − 4 = 4   Tens

This shortcut form shows all the necessary work involved in subtraction with borrowing. We will use it from now on.

The borrowing that changed 9 tens + 2 ones into 8 tens + 12 ones can also be visualized with money.

One $10 bill = Ten $1 bills

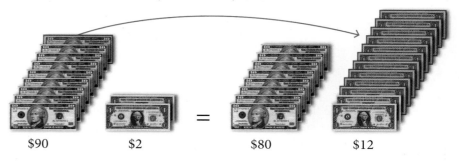

$90 $2   = $80 $12

4. a. Find the difference of 656 and 283.
b. Find the difference of 3,729 and 1,749.

EXAMPLE 4 Find the difference of 549 and 187.

SOLUTION In symbols, the difference of 549 and 187 is written

$$549 - 187$$

Writing the problem vertically so that the digits with the same place value are aligned, we have

$$\begin{array}{r} 549 \\ -187 \end{array}$$

The top number in the tens column is smaller than the number below it. This means that we will have to borrow from the next larger column.

Borrow 1 hundred to go with the 4 tens; note 1 hundred = 10 tens.

9 − 7 = 2   Ones
14 − 8 = 6   Tens
4 − 1 = 3   Hundreds

The actual work we did in borrowing looks like this:

5 hundreds + 4 tens + 9 ones
= 4 hundreds + 1 hundred + 4 tens + 9 ones
= 4 hundreds + 14 tens + 9 ones

Answers
4. a. 373   **b.** 1,980

Problem Set 1.4

Moving Toward Success

"Work while it is called today, for you know not how much you will be hindered tomorrow. One today is worth two tomorrows; never leave that till tomorrow which you can do today."
—Benjamin Franklin, 1706–1790, American statesman and inventor

1. Last section, you created a schedule that included homework time for this class. Was it easy to fit that time into your schedule? Why or why not?
2. What might happen if you don't devote adequate study time for this class?

A Write each of the following expressions in words. Use the word *difference* in each case.

1. $10 - 2$
2. $9 - 5$
3. $a - 6$
4. $7 - x$
5. $8 - 2 = 6$
6. $m - 1 = 4$

7. What number do you subtract from 8 to get 5?
8. What number do you subtract from 6 to get 0?
9. What number do you subtract from 15 to get 7?
10. What number do you subtract from 21 to get 14?
11. What number do you subtract from 35 to get 12?
12. What number do you subtract from 41 to get 11?

A Complete the following tables.

13.

First Number a	Second Number b	The Difference of a and b $a - b$
25	15	
24	16	
23	17	
22	18	

14.

First Number a	Second Number b	The Difference of a and b $a - b$
90	79	
80	69	
70	59	
60	49	

15.

First Number a	Second Number b	The Difference of a and b $a - b$
400	256	
400	144	
225	144	
225	81	

16.

First Number a	Second Number b	The Difference of a and b $a - b$
100	36	
100	64	
25	16	
25	9	

A Write each of the following sentences as mathematical expressions.

17. The difference of 8 and 3

18. The difference of x and 2

19. 9 subtracted from y

20. a subtracted from b

21. The difference of 3 and 2 is 1.

22. The difference of 10 and y is 5.

23. The difference of 37 and $9x$ is 10.

24. The difference of $3x$ and $2y$ is 15.

25. The difference of $2y$ and $15x$ is 24.

26. The difference of $25x$ and $9y$ is 16.

27. The difference of $(x + 2)$ and $(x + 1)$ is 1.

28. The difference of $(x - 2)$ and $(x - 4)$ is 2.

A Perform the indicated operation. [Examples 1, 2]

29. Subtract 24 from 56.

30. Subtract 71 from 89.

31. Subtract 23 from 45.

32. Subtract 97 from 98.

33. Find the difference of 29 and 19.

34. Find the difference of 37 and 27.

35. Find the difference of 126 and 15.

36. Find the difference of 348 and 32.

B Work each of the following subtraction problems. [Examples 1, 2]

37. 975 − 663

38. 480 − 260

39. 904 − 501

40. 657 − 507

41. 9,876 − 8,765

42. 5,008 − 3,002

43. 7,976 − 3,432

44. 6,980 − 470

C Find the difference in each case. (These problems all involve borrowing.) [Example 3]

45. 52 − 37

46. 65 − 48

47. 70 − 37

48. 90 − 21

49. 74 − 69

50. 31 − 28

51. 51 − 18

52. 64 − 58

53. 329 − 234   54. 518 − 492   55. 348 − 196   56. 759 − 661

57.   932   58.   895   59.   647   60.   842
    − 658   − 597   − 159   − 199

61.   905   62.   804   63.   600   64.   800
    − 367   − 238   − 437   − 342

65. 4,583   66. 7,849   67. 79,040   68. 86,492
    − 2,973   − 2,957   − 32,957   − 78,506

Applying the Concepts

Not all of the following application problems involve only subtraction. Some involve addition as well. Be sure to read each problem carefully.

69. Checkbook Balance Diane has $504 in her checking account. If she writes five checks for a total of $249, how much does she have left in her account?

70. Checkbook Balance Larry has $763 in his checking account. If he writes a check for each of the three bills listed below, how much will he have left in his account?

Item	Amount
Rent	$418
Phone	$25
Car repair	$117

71. Home Prices In 1985, Mr. Hicks paid $137,500 for his home. He sold it in 2008 for $310,000. What is the difference between what he sold it for and what he bought it for?

72. Oil Spills In March 1977, an oil tanker hit a reef off Taiwan and spilled 3,134,500 gallons of oil. In March 1989, an oil tanker hit a reef off Alaska and spilled 10,080,000 gallons of oil. How much more oil was spilled in the 1989 disaster?

73. Wind Speeds On April 12, 1934, the wind speed on top of Mount Washington was recorded at 231 miles per hour. When Hurricane Katrina struck on August 28, 2005, the highest recorded wind speed was 140 miles per hour. How much faster was the wind on top of Mount Washington, than the winds from Hurricane Katrina?

74. Concert Attendance Eleven thousand, seven hundred fifty-two people attended a recent concert at the Pepsi Arena in Albany, New York. If the arena holds 17,500 people, how many empty seats were there at the concert?

75. iPod Capacity Suppose your iPod has a capacity of 80 gigabytes and you have used 47 gigabytes for your music. How much space do you have left?

76. State Size Alaska is the largest state in the United States with an area of 663,267 square miles. Rhode Island is the smallest state with an area of 1,545 square miles. How many more square miles does Alaska have compared to Rhode Island?

77. Cars The bar chart below shows the countries producing the most cars.

a. Use the information in the bar chart to fill in the missing entries in the table.

Country	Production
United States	
	3,019,000
Germany	
South Korea	
Japan	

b. How many more cars are produced in Germany than South Korea?

78. Bridge Lengths The bar chart below shows the lengths of some of the longest bridges in the country.

a. Use the information in the bar chart to fill in the missing entries in the table.

Bridge	Length (ft)
Verrazano-Narrows	
Golden Gate	
	3,800
George Washington	

b. How much longer is the Golden Gate Bridge than the George Washington Bridge?

Profit Recall the graph from the section opener that shows the cost and revenue for four weeks for a children's shoe company. Use the graph to answer the following questions.

79. What was the profit in week 4?

80. What was the difference in profit between weeks 2 and 3?

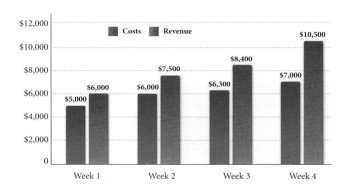

Multiplication with Whole Numbers

1.5

OBJECTIVES

A Multiply whole numbers using repeated addition.

B Understand the notation and vocabulary of multiplication.

C Identify properties of multiplication.

D Solve equations with multiplication.

E Solve applications with multiplication.

TICKET TO SUCCESS

Keep these questions in mind as you read through the section. Then respond in your own words and in complete sentences.

1. How do we think of multiplication as shorthand for repeated addition?
2. Explain the distributive property using words and symbols.
3. Use symbols to explain the following properties:
    a. multiplication property of 0
    b. multiplication property of 1
    c. commutative property of multiplication
    d. associative property of multiplication
4. Define *variable*.

Suppose a supermarket orders 35 cases of a certain soft drink. If each case contains 12 cans of the drink, how many cans were ordered?

To solve this problem and others like it, we must use multiplication. Multiplication is what we will cover in this section.

A Multiplying Whole Numbers

To begin, we can think of multiplication as shorthand for repeated addition. That is, multiplying 3 times 4 can be thought of this way:

$$3 \text{ times } 4 = 4 + 4 + 4 = 12$$

Multiplying 3 times 4 means to add three 4s. We can write 3 times 4 as 3×4, or $3 \cdot 4$.

EXAMPLE 1 Multiply: $3 \cdot 4,000$.

SOLUTION Using the definition of multiplication as repeated addition, we have

$$3 \cdot 4,000 = 4,000 + 4,000 + 4,000$$
$$= 12,000$$

PRACTICE PROBLEMS

1. Multiply.
    a. $4 \cdot 70$
    b. $4 \cdot 700$
    c. $4 \cdot 7,000$

Answers
1. a. 280 b. 2,800 c. 28,000

Here is another way to visualize this process:

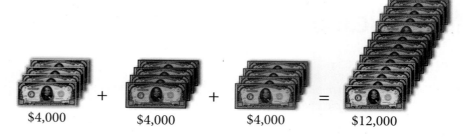

Notice that if we had multiplied 3 and 4 to get 12 and then attached three zeros on the right, the result would have been the same.

Facts of Multiplication

Just as we presented the facts of addition in Section 1.2, here are the basic multiplication facts that will help you in this section.

> **NOTE**
> Memorizing the basic multiplication facts is vital to your success in this course.

×	1	2	3	4	5	6	7	8	9
1	1	2	3	4	5	6	7	8	9
2	2	4	6	8	10	12	14	16	18
3	3	6	9	12	15	18	21	24	27
4	4	8	12	16	20	24	28	32	36
5	5	10	15	20	25	30	35	40	45
6	6	12	18	24	30	36	42	48	54
7	7	14	21	28	35	42	49	56	63
8	8	16	24	32	40	48	56	64	72
9	9	18	27	36	45	54	63	72	81

B Notation and Vocabulary

There are many ways to indicate multiplication. All the following statements are equivalent. They all indicate multiplication with the numbers 3 and 4.

$$3 \cdot 4, \quad 3 \times 4, \quad 3(4), \quad (3)4, \quad (3)(4), \quad \begin{array}{r} 4 \\ \times 3 \end{array}$$

If one or both of the numbers we are multiplying are represented by letters, we may also use the following notation:

> **NOTE**
> The kind of notation we will use to indicate multiplication will depend on the situation. For example, when we are solving equations that involve letters, it is not a good idea to indicate multiplication with the symbol ×, since it could be confused with the variable x. The symbol we will use to indicate multiplication most often in this book is the multiplication dot.

$5n$ means   5 times n

ab means a times b

We use the word *product* to indicate multiplication. If we say, "The product of 3 and 4 is 12," then we mean

$$3 \cdot 4 = 12$$

Both $3 \cdot 4$ and 12 are called the product of 3 and 4. The 3 and 4 are called *factors*.

> **Definition**
> **Factors** are numbers that, when multiplied together, give a product.

Table 1 gives some word statements involving multiplication and their mathematical equivalents written in symbols.

TABLE 1

In English	In Symbols
The product of 2 and 5	$2 \cdot 5$
The product of 5 and 2	$5 \cdot 2$
The product of 4 and n	$4n$
The product of x and y	xy
The product of 9 and 6 is 54.	$9 \cdot 6 = 54$
The product of 2 and 8 is 16.	$2 \cdot 8 = 16$

EXAMPLE 2 Identify the products and factors in the statement

$$9 \cdot 8 = 72$$

SOLUTION The factors are 9 and 8, and the products are $9 \cdot 8$ and 72. ∎

EXAMPLE 3 Identify the products and factors in the statement

$$30 = 2 \cdot 3 \cdot 5$$

SOLUTION The factors are 2, 3, and 5. The products are $2 \cdot 3 \cdot 5$ and 30. ∎

C Properties of Multiplication

To develop an efficient method of multiplication, we need to use what is called the *distributive property*. To begin, consider the following two problems:

Problem 1	Problem 2
$3(4 + 5)$	$3(4) + 3(5)$
$= 3(9)$	$= 12 + 15$
$= 27$	$= 27$

The result in both cases is the same number, 27. This indicates that the original two expressions must have been equal also. That is,

$$3(4 + 5) = 3(4) + 3(5)$$

This is an example of the distributive property. We say that multiplication *distributes* over addition.

$$3(4 + 5) = 3(4) + 3(5)$$

$$3(4+5) \quad = \quad 3 \cdot 4 \quad + \quad 3 \cdot 5$$

2. Identify the products and factors in the statement
$$6 \cdot 7 = 42$$

3. Identify the products and factors in the statement
$$70 = 2 \cdot 5 \cdot 7$$

Answers
2. Factors: 6, 7; products: $6 \cdot 7$ and 42
3. Factors: 2, 5, 7; products: $2 \cdot 5 \cdot 7$ and 70

We can write this property in symbols using the letters *a*, *b*, and *c* to represent any three whole numbers.

> **Distributive Property**
>
> If *a*, *b*, and *c* represent any three whole numbers, then
>
> $$a(b + c) = a(b) + a(c)$$

Suppose we want to find the product 7(65). By writing 65 as 60 + 5 and applying the distributive property, we have

$$
\begin{aligned}
7(65) &= 7(60 + 5) & 65 &= 60 + 5 \\
&= 7(60) + 7(5) & &\text{Distributive property} \\
&= 420 + 35 & &\text{Multiply.} \\
&= 455 & &\text{Add.}
\end{aligned}
$$

We can write the same problem vertically like this:

$$
\begin{array}{r}
60 + 5 \\
\times \quad 7 \\
\hline
35 \quad \leftarrow 7(5) = 35 \\
+ \quad 420 \quad \leftarrow 7(60) = 420 \\
\hline
455
\end{array}
$$

This saves some space in writing. But notice that we can cut down on the amount of writing even more if we write the problem this way:

STEP 2: 7(6) = 42; add the 3 we carried to 42 to get 45.

$$
\begin{array}{r}
\overset{3}{6}5 \\
\times \ 7 \\
\hline
455
\end{array}
$$

STEP 1: 7(5) = 35; write the 5 in the ones column, and then carry the 3 to the tens column.

This shortcut notation takes some practice.

EXAMPLE 4 Multiply: 9(43).

STEP 2: 9(4) = 36; add the 2 we carried to 36 to get 38.

$$
\begin{array}{r}
\overset{2}{4}3 \\
\times \ 9 \\
\hline
387
\end{array}
$$

STEP 1: 9(3) = 27; write the 7 in the ones column, and then carry the 2 to the tens column.

∎

EXAMPLE 5 Multiply: 52(37).

SOLUTION This is the same as 52(30 + 7) or by the distributive property

$$52(30) + 52(7)$$

We can find each of these products by using the shortcut method:

$$
\begin{array}{r}
52 \\
\times \ 30 \\
\hline
1{,}560
\end{array}
\qquad
\begin{array}{r}
\overset{1}{5}2 \\
\times \ 7 \\
\hline
364
\end{array}
$$

The sum of these two numbers is 1,560 + 364 = 1,924. Here is a summary of what we have so far:

$$
\begin{aligned}
52(37) &= 52(30 + 7) & 37 &= 30 + 7 \\
&= 52(30) + 52(7) & &\text{Distributive property} \\
&= 1{,}560 + 364 & &\text{Multiply.} \\
&= 1{,}924 & &\text{Add.}
\end{aligned}
$$

NOTE
When using the distributive property, we multiply *a*(*b*) and *a*(*c*) before adding their products together because of a rule called order of operations. Part of this rule states that once we have worked operations inside parentheses, then we multiply before we add. We will study more about this rule later on in this chapter.

4. Multiply.
  a. 8(57)
  b. 8(570)

5. Multiply.
  a. 45(62)
  b. 45(620)

NOTE
This discussion is to show why we multiply the way we do. You should go over it in detail, so you will understand the reasons behind the process of multiplication.

Answers
4. a. 456 **b.** 4,560
5. a. 2,790 **b.** 27,900

The shortcut form for this problem is

$$\begin{array}{r} 52 \\ \times\ 37 \\ \hline 364 \\ +\ 1{,}560 \\ \hline 1{,}924 \end{array}$$

$364 \longleftarrow 7(52) = 364$
$+\ 1{,}560 \longleftarrow 30(52) = 1{,}560$

In this case, we have not shown any of the numbers we carried, simply because it becomes very messy. ∎

EXAMPLE 6
Multiply: 279(428).

SOLUTION

$$\begin{array}{r} 279 \\ \times\ 428 \\ \hline 2{,}232 \\ 5{,}580 \\ +\ 111{,}600 \\ \hline 119{,}412 \end{array}$$

$2{,}232 \longleftarrow 8(279) = 2{,}232$
$5{,}580 \longleftarrow 20(279) = 5{,}580$
$+\ 111{,}600 \longleftarrow 400(279) = 111{,}600$

∎

USING TECHNOLOGY

Calculators

Here is how we would work the problem shown in Example 6 on a calculator:

Scientific Calculator: 279 [×] 428 [=]
Graphing Calculator: 279 [×] 428 [ENT]

Here are some other important properties of multiplication:

Multiplication Property of 0

If a represents any number, then

$$a \cdot 0 = 0 \quad \text{and} \quad 0 \cdot a = 0$$

In words: Multiplication by 0 always results in 0.

Multiplication Property of 1

If a represents any number, then

$$a \cdot 1 = a \quad \text{and} \quad 1 \cdot a = a$$

In words: Multiplying any number by 1 leaves that number unchanged.

Commutative Property of Multiplication

If a and b are any two numbers, then

$$ab = ba$$

In words: The order of the numbers in a product doesn't affect the result.

---

6. Multiply.
   a. 356(641)
   b. 3,560(641)

NOTE
We can estimate the answer to the problem shown in Example 6 by rounding each number to the nearest hundred and then multiplying the rounded numbers. Doing so would give us this:

$$300(400) = 120{,}000$$

Our estimate of the answer is 120,000, which is close to the actual answer, 119,412. Making estimates is important when we are using calculators; having an estimate of the answer will keep us from making major errors in multiplication.

Answers
6. a. 228,196 **b.** 2,281,960

> **Associative Property of Multiplication**
>
> If a, b, and c represent any three numbers, then
>
> $$(ab)c = a(bc)$$
>
> *In words:* We can change the grouping of the numbers in a product without changing the result.

To visualize the commutative property, we can think of an instructor with 12 students.

4 chairs across, 3 chairs back = 3 chairs across, 4 chairs back

EXAMPLE 7 Use the commutative property of multiplication to rewrite each of the following products:

    a. $7 \cdot 9$ **b.** $4(6)$

SOLUTION Applying the commutative property to each expression, we have

    a. $7 \cdot 9 = 9 \cdot 7$ **b.** $4(6) = 6(4)$ ∎

EXAMPLE 8 Use the associative property of multiplication to rewrite each of the following products:

    a. $(2 \cdot 7) \cdot 9$ **b.** $3 \cdot (8 \cdot 2)$

SOLUTION Applying the associative property of multiplication, we regroup as follows:

    a. $(2 \cdot 7) \cdot 9 = 2 \cdot (7 \cdot 9)$ **b.** $3 \cdot (8 \cdot 2) = (3 \cdot 8) \cdot 2$ ∎

D Solving Equations

Earlier in the chapter, we solved equations including variables by a process called inspection. We will now apply this process to a multiplication problem. If n is used to represent a number, then the equation

$$4 \cdot n = 12$$

is read "4 times n is 12," or "The product of 4 and n is 12." This means that we are looking for the number we multiply by 4 to get 12. The number is 3. Because the equation becomes a true statement if n is 3, we say that 3 is the solution to the equation.

---

7. Use the commutative property of multiplication to rewrite each of the following products.
  a. $5 \cdot 8$
  b. $7(2)$

8. Use the associative property of multiplication to rewrite each of the following products.
  a. $(5 \cdot 7) \cdot 4$
  b. $4 \cdot (6 \cdot 4)$

Answers
7. a. $8 \cdot 5$ **b.** $2(7)$
8. a. $5 \cdot (7 \cdot 4)$ **b.** $(4 \cdot 6) \cdot 4$

EXAMPLE 9
Find the solution to each of the following equations:
a. $6 \cdot n = 24$ b. $4 \cdot n = 36$ c. $15 = 3 \cdot n$ d. $21 = 3 \cdot n$

SOLUTION
a. The solution to $6 \cdot n = 24$ is 4, because $6 \cdot 4 = 24$.
b. The solution to $4 \cdot n = 36$ is 9, because $4 \cdot 9 = 36$.
c. The solution to $15 = 3 \cdot n$ is 5, because $15 = 3 \cdot 5$.
d. The solution to $21 = 3 \cdot n$ is 7, because $21 = 3 \cdot 7$.

E Applications

EXAMPLE 10
A supermarket orders 35 cases of a certain soft drink. If each case contains 12 cans of the drink, how many cans were ordered?

SOLUTION We have 35 cases and each case has 12 cans. The total number of cans is the product of 35 and 12, which is 35(12).

```
     12
   × 35
   ────
     60   ←———— 5(12) = 60
  + 360   ←———— 30(12) = 360
   ────
    420
```

There is a total of 420 cans of the soft drink.

EXAMPLE 11
Shirley earns $12 an hour for the first 40 hours she works each week. If she has $109 deducted from her weekly check for taxes and retirement, how much money will she take home if she works 38 hours this week?

SOLUTION To find the amount of money she earned for the week, we multiply 12 and 38. From that total we subtract 109. The result is her take-home pay. Without showing all the work involved in the calculations, here is the solution:

$38(\$12) = \$456$ Shirley's total weekly earnings
$\$456 - \$109 = \$347$ Her take-home pay

EXAMPLE 12
In 1993, the government standardized the way in which nutrition information is presented on the labels of most packaged food products. Figure 1 shows one of these standardized food labels. It is from a package of corn chips. Approximately how many chips are in the bag, and what is the total number of calories consumed if all the chips in the bag are eaten?

Nutrition Facts

Serving Size 1 oz. (28g/About 32 chips)
Servings Per Container: 3

Amount Per Serving	
Calories 160	Calories from fat 90

	% Daily Value*
Total Fat 10 g	16%
Saturated Fat 1.5g	8%
Cholesterol 0mg	0%
Sodium 160mg	7%
Total Carbohydrate 15g	5%
Dietary Fiber 1g	4%
Sugars 0g	
Protein 2g	

Vitamin A 0%	•	Vitamin C 0%
Calcium 2%	•	Iron 0%

*Percent Daily Values are based on a 2,000 calorie diet

FIGURE 1

---

9. Use multiplication facts to find the solution to each of the following equations.
a. $5 \cdot n = 35$
b. $8 \cdot n = 72$
c. $49 = 7 \cdot n$
d. $27 = 9 \cdot n$

10. If each tablet of vitamin C contains 550 milligrams of vitamin C, what is the total number of milligrams of vitamin C in a bottle that contains 365 tablets?

11. If Shirley works 36 hours the next week and has the same amount deducted from her check for taxes and retirement, how much will she take home?

NOTE
The letter g that is shown after some of the numbers in the nutrition label in Figure 1 stands for grams, a unit used to measure weight. The unit mg stands for milligrams, another smaller unit of weight. We will have more to say about these units later in the book.

12. The amounts given in the middle of the nutrition label in Figure 1 are for one serving of chips. If all the chips in the bag are eaten, how much fat has been consumed? How much sodium?

Answers
9. a. 7 **b.** 9 **c.** 7 **d.** 3
10. 200,750 milligrams
11. $323
12. 30 g of fat, 480 mg of sodium

SOLUTION Reading toward the top of the label, we see that there are about 32 chips in one serving, and 3 servings in the bag. Therefore, the total number of chips in the bag is

$$3(32) = 96 \text{ chips}$$

This is an approximate number, because each serving is approximately 32 chips. Reading further we find that each serving contains 160 calories. Therefore, the total number of calories consumed by eating all the chips in the bag is

$$3(160) = 480 \text{ calories}$$

As we progress through the book, we will study more of the information in nutrition labels. ■

EXAMPLE 13 Suppose the table below lists the number of calories burned in 1 hour of exercise by a person who weighs 150 pounds. Consider a 150-pound person going bowling for 2 hours after having eaten the bag of chips mentioned in Example 12. Will he or she burn all the calories consumed from the chips?

Activity	Calories Burned in 1 Hour by a 150-Pound Person
Bicycling	374
Bowling	265
Handball	680
Jazzercize	340
Jogging	680
Skiing	544

SOLUTION Each hour of bowling burns 265 calories. If the person bowls for 2 hours, a total of

$$2(265) = 530 \text{ calories}$$

will have been burned. Because the bag of chips contained only 480 calories, all of them have been burned with 2 hours of bowling. ■

---

13. If a 150-pound person bowls for 3 hours, will he or she burn all the calories consumed by eating two bags of the chips mentioned in Example 13?

Answer
13. No

Problem Set 1.5

Moving Toward Success

"Your work is to discover your world and then with all your heart give yourself to it."
—Buddha, 563–483 BC, Hindu Prince Gautama Siddharta and founder of Buddhism

1. When examining the rest of your schedule, what priority are you giving to this course? How will that priority affect your success in this course?
2. Why is it important to stick to the schedule you set for this course?

A Multiply each of the following. [Example 1]

1. $3 \cdot 100$
2. $7 \cdot 100$
3. $3 \cdot 200$
4. $4 \cdot 200$
5. $6 \cdot 500$
6. $8 \cdot 400$
7. $5 \cdot 1{,}000$
8. $8 \cdot 1{,}000$
9. $3 \cdot 7{,}000$
10. $6 \cdot 7{,}000$
11. $9 \cdot 9{,}000$
12. $7 \cdot 7{,}000$

B Complete the following tables.

13.

First Number a	Second Number b	Their Product ab
11	11	
11	22	
22	22	
22	44	

14.

First Number a	Second Number b	Their Product ab
25	15	
25	30	
50	15	
50	30	

15.

First Number a	Second Number b	Their Product ab
25	10	
25	100	
25	1,000	
25	10,000	

16.

First Number a	Second Number b	Their Product ab
11	111	
11	222	
22	111	
22	222	

17.

First Number a	Second Number b	Their Product ab
12	20	
36	20	
12	40	
36	40	

18.

First Number a	Second Number b	Their Product ab
10	12	
100	12	
1,000	12	
10,000	12	

B Write each of the following expressions in words, using the word *product*.

19. $6 \cdot 7$

20. $9(4)$

21. $2 \cdot n$

22. $5 \cdot x$

23. $9 \cdot 7 = 63$

24. $(5)(6) = 30$

B Write each of the following in symbols.

25. The product of 7 and n

26. The product of 9 and x

27. The product of 6 and 7 is 42.

28. The product of 8 and 9 is 72.

29. The product of 0 and 6 is 0.

30. The product of 1 and 6 is 6.

B Identify the products in each statement.

31. $9 \cdot 7 = 63$

32. $2(6) = 12$

33. $4(4) = 16$

34. $5 \cdot 5 = 25$

B Identify the factors in each statement.

35. $2 \cdot 3 \cdot 4 = 24$

36. $6 \cdot 1 \cdot 5 = 30$

37. $12 = 2 \cdot 2 \cdot 3$

38. $42 = 2 \cdot 3 \cdot 7$

C Find each of the following products. (Multiply.) In each case use the shortcut method. [Examples 4–6]

39. 25 × 4 **40.** 43 × 9 **41.** 38 × 6 **42.** 45 × 7 **43.** 18 × 2 **44.** 29 × 3

45. 72 × 20 **46.** 68 × 30 **47.** 19 × 50 **48.** 24 × 40 **49.** 69 × 25 **50.** 27 × 36

51. 11 × 11 **52.** 12 × 21 **53.** 97 × 16 **54.** 24 × 39 **55.** 168 × 25 **56.** 452 × 34

57. 728 × 91 **58.** 680 × 76 **59.** 698 × 400 **60.** 879 × 600 **61.** 111 × 111 **62.** 123 × 321

63. 532 × 200 **64.** 277 × 900 **65.** 856 × 232 **66.** 455 × 248 **67.** 976 × 628 **68.** 432 × 555

69. 2,468 × 135 **70.** 2,725 × 324 **71.** 24,563 × 735 **72.** 56,728 × 852 **73.** 44,777 × 5,888 **74.** 33,999 × 2,555

Chapter 1 Whole Numbers

C Rewrite each of the following using the commutative property of multiplication. [Example 7]

75. 5(9) **76.** 4(3) **77.** 6 · 7 **78.** 8 · 3

C Rewrite each of the following using the associative property of multiplication. [Example 8]

79. 2 · (7 · 6) **80.** 4 · (8 · 5) **81.** 3 × (9 × 1) **82.** 5 × (8 × 2)

C Use the distributive property to rewrite each expression, then simplify.

83. 7(2 + 3) **84.** 4(5 + 8) **85.** 9(4 + 7) **86.** 6(9 + 5)

87. 3(x + 1) **88.** 5(x + 8) **89.** 2(x + 5) **90.** 4(x + 3)

D Find a solution for each equation. [Example 9]

91. 4 · n = 12 **92.** 3 · n = 12 **93.** 9 · n = 81 **94.** 6 · n = 36

95. 0 = n · 5 **96.** 6 = 1 · n

Applying the Concepts

Most, but not all, of the application problems that follow require multiplication. Read the problems carefully before trying to solve them.

97. Planning a Trip A family decides to drive their compact car on their vacation. They figure it will require about 130 gallons of gas for the vacation. If each gallon of gas will take them 22 miles, how long is the trip they are planning?

98. Soda A grocery store is restocking its inventory and orders 15 cases of soda. If each case contains 12 cans, how many cans of soda did the store order?

99. Downloading Songs You receive a gift card for the iTunes store for $25.00 and download 18 songs at $1.00 per song. How much is left on your gift card?

100. Cost of Building a Home When you consider building a new home it is helpful to be able to estimate the cost of building that house. A simple way to do this is to multiply the number of square feet under the roof of the house by the average building cost per square foot. Suppose you contact a builder who estimates that, on average, he charges $142.00 per square foot. Determine the cost to build a 2,067 square foot house.

101. World's Busiest Airport Atlanta, Georgia is home to the world's busiest airport, Hartsfield-Jackson Atlanta International Airport. According to the Federal Aviation Administration about 50 jets can land and take off every 15 minutes which is about 200 jets an hour. About how many jets land and take off in the month of July?

102. Flowers It is probably no surprise that Valentine's Day is the busiest day of the year for florists. The Society of American Florists estimates that 214 million roses were produced for Valentine's Day in 2007. If a single rose costs a consumer $3.00, what was the total revenue for the roses produced?

Exercise and Calories The table below is an extension of the table we used in Example 13 of this section. Suppose it gives the amount of energy expended during 1 hour of various activities for people of different weights. The accompanying figure is a nutrition label from a bag of Doritos® tortilla chips. Use the information from the table and the nutrition label to answer Problems 103–108.

CALORIES BURNED THROUGH EXERCISE

Activity	120 Pounds	Calories Per Hour 150 Pounds	180 Pounds
Bicycling	299	374	449
Bowling	212	265	318
Handball	544	680	816
Jazzercise	272	340	408
Jogging	544	680	816
Skiing	435	544	653

Nutrition Facts
Serving Size 1 oz. (28g/About 12 chips)
Servings Per Container About 2

Amount Per Serving
Calories 140   Calories from fat 60

% Daily Value*
Total Fat 7g — 11%
  Saturated Fat 1g — 6%
Cholesterol 0mg — 0%
Sodium 170mg — 7%
Total Carbohydrate 18g — 6%
  Dietary Fiber 1g — 4%
  Sugars less than 1g
Protein 2g

Vitamin A 0% • Vitamin C 0%
Calcium 4% • Iron 2%
*Percent Daily Values are based on a 2,000 calorie diet

103. Suppose you weigh 180 pounds. How many calories would you burn if you play handball for 2 hours and then ride your bicycle for 1 hour?

104. How many calories are burned by a 120-lb person who jogs for 1 hour and then goes bike riding for 2 hours?

105. How many calories would you consume if you ate the entire bag of chips?

106. Approximately how many chips are in the bag?

107. If you weigh 180 pounds, will you burn off the calories consumed by eating 3 servings of tortilla chips if you ride your bike 1 hour?

108. If you weigh 120 pounds, will you burn off the calories consumed by eating 3 servings of tortilla chips if you ride your bike for 1 hour?

Estimating

Mentally estimate the answer to each of the following problems by rounding each number to the indicated place and then multiplying.

109. 750 hundred
      × 12 ten

110. 591 hundred
      × 323 hundred

111. 3,472 thousand
      × 511 hundred

112. 399 hundred
      × 298 hundred

113. 2,399 thousand
      × 698 hundred

114. 9,999 thousand
      × 666 hundred

Extending the Concepts — Number Sequences

A *geometric sequence* is a sequence of numbers in which each number is obtained from the previous number by multiplying by the same number each time. For example, the sequence 3, 6, 12, 24, . . . is a geometric sequence, starting with 3, in which each number comes from multiplying the previous number by 2. Find the next number in each of the following geometric sequences.

115. 5, 10, 20, . . .

116. 10, 50, 250, . . .

117. 2, 6, 18, . . .

118. 12, 24, 48, . . .

Division with Whole Numbers

1.6

OBJECTIVES

A Understand the notation and vocabulary of division.

B Divide whole numbers.

C Solve applications using division.

TICKET TO SUCCESS

Keep these questions in mind as you read through the section. Then respond in your own words and in complete sentences.

1. What are the different types of notation used for division?
2. In the following problem, label the *quotient*, the *dividend*, and the *divisor*: $\frac{15}{3} = 5$.
3. When would we have a remainder in a division problem?
4. Briefly explain the rule for division by 0.

Suppose a sporting goods store purchases 500 basketballs for a total wholesale cost of $2,500. In order to calculate the price of each basketball, we need to use division: dividing 2500 by 500.

As a division problem: As a multiplication problem:
$$2500 \div 500 = 5 \qquad 5 \cdot 500 = 2500$$

As you can see, this problem can be thought of in terms of division, as well as in terms of multiplication. You will see later in this section why this is true.

A Notation

As was the case with multiplication, there are many ways to indicate division. All the following statements are equivalent. They all mean 10 divided by 5.

$$10 \div 5, \quad \frac{10}{5}, \quad 10/5, \quad 5\overline{)10}$$

The kind of notation we use to write division problems will depend on the situation. We will use the notation $5\overline{)10}$ mostly with the long-division problems found in this chapter. The notation $\frac{10}{5}$ will be used in the chapter on fractions and in later chapters. The horizontal line used with the notation $\frac{10}{5}$ is called the *fraction bar*.

Vocabulary

The word *quotient* is used to indicate division. If we say, "The quotient of 10 and 5 is 2," then we mean

$$10 \div 5 = 2 \quad \text{or} \quad \frac{10}{5} = 2$$

The 10 is called the *dividend,* the 5 is called the *divisor,* and the 2 is called the *quotient.* This quotient is a result of dividing 10 by 5 using the expressions $10 \div 5$ or $\frac{10}{5}$.

TABLE 1

In English	In Symbols
The quotient of 15 and 3	$15 \div 3$, or $\frac{15}{3}$, or 15/3
The quotient of 3 and 15	$3 \div 15$, or $\frac{3}{15}$, or 3/15
The quotient of 8 and n	$8 \div n$, or $\frac{8}{n}$, or 8/n
x divided by 2	$x \div 2$, or $\frac{x}{2}$, or x/2
The quotient of 21 and 3 is 7.	$21 \div 3 = 7$, or $\frac{21}{3} = 7$

The Meaning of Division

One way to arrive at an answer to a division problem is by thinking in terms of multiplication. For example, if we want to find the quotient of 32 and 8, we may ask, "What do we multiply by 8 to get 32?"

$$32 \div 8 = ? \quad \text{means} \quad 8 \cdot ? = 32$$

Because we know from our work with multiplication that $8 \cdot 4 = 32$, it must be true that

$$32 \div 8 = 4$$

Table 2 lists some additional examples.

TABLE 2

Division		Multiplication
$18 \div 6 = 3$	because	$6 \cdot 3 = 18$
$32 \div 8 = 4$	because	$8 \cdot 4 = 32$
$10 \div 2 = 5$	because	$2 \cdot 5 = 10$
$72 \div 9 = 8$	because	$9 \cdot 8 = 72$

B Division by One-Digit Numbers

Consider the following division problem:

$$465 \div 5$$

We can think of this problem as asking the question, "How many fives can we subtract from 465?" To answer the question, we begin subtracting multiples of 5.

Here is one way to organize this process:

$$\begin{array}{r} 90 \\ 5\overline{)465} \\ -450 \\ \hline 15 \end{array}$$ ← We first guess that there are at least 90 fives in 465.
← 90(5) = 450
← 15 is left after we subtract 90 fives from 465.

What we have done so far is subtract 90 fives from 465 and found that 15 is still left. Because there are 3 fives in 15, we continue the process.

$$\begin{array}{r} 3 \\ 90 \\ 5\overline{)465} \\ -450 \\ \hline 15 \\ -15 \\ \hline 0 \end{array}$$ ← There are 3 fives in 15.

← 3 · 5 = 15
← The difference is 0.

The total number of fives we have subtracted from 465 is

$90 + 3 = 93$

We now summarize the results of our work.

$465 \div 5 = 93$ We check our answer with multiplication. → $$\begin{array}{r} \overset{1}{9}3 \\ \times\ 5 \\ \hline 465 \end{array}$$

The division problem just shown can be shortened by eliminating the subtraction signs, eliminating the zeros in each estimate, and eliminating some of the numbers that are repeated in the problem.

The shorthand form for this problem →

$$\begin{array}{r} 3 \\ 90 \\ 5\overline{)465} \\ 450 \\ \hline 15 \\ 15 \\ \hline 0 \end{array}$$

looks like this:

$$\begin{array}{r} 93 \\ 5\overline{)465} \\ 45\downarrow \\ \hline 15 \\ 15 \\ \hline 0 \end{array}$$

The arrow indicates that we bring down the 5 after we subtract.

The problem shown above on the right is the shortcut form of what is called *long division*. Here is an example showing this shortcut form of long division from start to finish.

EXAMPLE 1   Divide: 595 ÷ 7.

SOLUTION   Because 7(8) = 56, our first estimate of the number of sevens that can be subtracted from 595 is 80:

$$\begin{array}{r} 8 \\ 7\overline{)595} \\ 56\downarrow \\ \hline 35 \end{array}$$ ← The 8 is placed above the tens column so we know our first estimate is 80.
← 8(7) = 56
← 59 − 56 = 3; then bring down the 5.

Since 7(5) = 35, we have

$$\begin{array}{r} 85 \\ 7\overline{)595} \\ 56\downarrow \\ \hline 35 \\ 35 \\ \hline 0 \end{array}$$ ← There are 5 sevens in 35.

← 5(7) = 35
← 35 − 35 = 0

PRACTICE PROBLEMS

1. Divide.
   a. 296 ÷ 4
   b. 2,960 ÷ 4

Answers
1. a. 74   b. 740

2. Divide.
  a. $6{,}792 \div 24$
  b. $67{,}920 \div 24$

Our result is $595 \div 7 = 85$, which we can check with multiplication.

$$\begin{array}{r} \overset{3}{8}5 \\ \times\ 7 \\ \hline 595 \end{array}$$

Division by Two-Digit Numbers

EXAMPLE 2   Divide: $9{,}380 \div 35$.

SOLUTION   In this case our divisor, 35, is a two-digit number. The process of division is the same. We still want to find the number of thirty-fives we can subtract from 9,380.

$$35\overline{)9380}$$
with quotient 2 ← The 2 is placed above the hundreds column.
70 ← $2(35) = 70$
238 ← $93 - 70 = 23$; then bring down the 8.

We can make a few preliminary calculations to help estimate how many thirty-fives are in 238:

$$5 \times 35 = 175 \qquad 6 \times 35 = 210 \qquad 7 \times 35 = 245$$

Because 210 is the closest to 238 without being larger than 238, we use 6 as our next estimate:

quotient 26 ← 6 in the tens column means this estimate is 60.
$35\overline{)9380}$
70
238
210 ← $6(35) = 210$
280 ← $238 - 210 = 28$; bring down the 0.

Because $35(8) = 280$, we have

quotient 268
$35\overline{)9380}$
70
238
210
280
280 ← $8(35) = 280$
0 ← $280 - 280 = 0$

We can check our result with multiplication.

$$\begin{array}{r} 268 \\ \times\ \ 35 \\ \hline 1{,}340 \\ 8{,}040 \\ \hline 9{,}380 \end{array}$$

Answers
2. a. 283   **b.** 2,830

1.6 Division with Whole Numbers

EXAMPLE 3 Divide: 1,872 by 18.

SOLUTION Here is the first step:

$$\begin{array}{r} 1 \\ 18\overline{)1872} \\ \underline{18} \\ 0 \end{array}$$

⟵ 1 is placed above hundred column.

⟵ Multiply 1(18) to get 18.

⟵ Subtract to get 0.

The next step is to bring down the 7 and divide again.

$$\begin{array}{r} 10 \\ 18\overline{)1872} \\ \underline{18}\downarrow \\ 07 \\ \underline{0} \\ 7 \end{array}$$

⟵ 0 is placed above tens column. 0 is the largest number we can multiply by 18 and not go over 7.

⟵ Multiply 0(18) to get 0.

⟵ Subtract to get 7.

Here is the complete problem:

$$\begin{array}{r} 104 \\ 18\overline{)1872} \\ \underline{18}\downarrow \\ 07 \\ \underline{0}\downarrow \\ 72 \\ \underline{72} \\ 0 \end{array}$$

To show our answer is correct, we multiply.

$$18(104) = 1{,}872$$ ∎

Division with Remainders

Darlene is planning to serve 6-ounce glasses of soda at a party. To see how many glasses she could fill from a 32-ounce bottle, she would divide 32 by 6. If she did so, she would find that she could fill 5 glasses, but after doing so she would have 2 ounces of soda left in the bottle. The 2 ounces is known as the remainder. A diagram of this problem is shown in Figure 1.

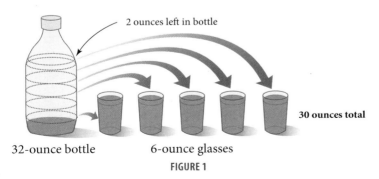

FIGURE 1

Writing the results in the diagram as a division problem looks like this:

$$\begin{array}{r} 5 \\ \text{Divisor} \rightarrow 6\overline{)32} \\ \underline{30} \\ 2 \end{array}$$

⟵ Quotient

⟵ Dividend

⟵ Remainder

3. Divide.
1,872 ÷ 9

Answer
3. 208

4. Divide.
  a. 1,883 ÷ 27
  b. 1,883 ÷ 18

Chapter 1 Whole Numbers

EXAMPLE 4 Divide: 1,690 ÷ 67.

SOLUTION Dividing as we have previously, we get

$$
\begin{array}{r}
25 \\
67{\overline{\smash{\big)}\,1690}} \\
\underline{134{\downarrow}} \\
350 \\
\underline{335} \\
15
\end{array}
$$
← 15 is left over.

We have 15 left, and because 15 is less than 67, no more sixty-sevens can be subtracted. In a situation like this, we call 15 the *remainder* and write

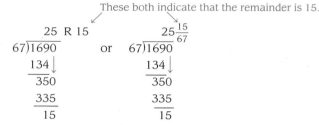

These both indicate that the remainder is 15.

$$25 \text{ R } 15 \quad \text{or} \quad 25\tfrac{15}{67}$$

Both forms of notation shown above indicate that 15 is the remainder. The notation R 15 is the notation we will use in this chapter. The notation $\tfrac{15}{67}$ will be useful in the chapter on fractions.

To check a problem like this, we multiply the divisor and the quotient as usual, and then add the remainder to this result.

$$
\begin{array}{r}
67 \\
\times\, 25 \\
\hline
335 \\
1{,}340 \\
\hline
1{,}675
\end{array}
$$
← Product of divisor and quotient

$$1{,}675 + 15 = 1{,}690$$

Remainder   Dividend

USING TECHNOLOGY

Calculators

Here is how we would work the problem shown in Example 4 on a calculator:

Scientific Calculator: 1690 ÷ 67 =
Graphing Calculator: 1690 ÷ 67 ENT

In both cases the calculator will display 25.223881 (give or take a few digits at the end), which gives the remainder in decimal form. We will discuss decimals later in the book.

Answers
4. a. 69 R 20, or $69\tfrac{20}{27}$
  b. 104 R 11, or $104\tfrac{11}{18}$

C Applications

EXAMPLE 5   A family has an annual income of $35,880. How much is their average monthly income?

SOLUTION   Because there are 12 months in a year and the yearly (annual) income is $35,880, we want to know what $35,880 divided into 12 equal parts is. Therefore, we have

$$\begin{array}{r} 2990 \\ 12\overline{)35880} \\ \underline{24}\downarrow \\ 118 \\ \underline{108}\downarrow \\ 108 \\ \underline{108}\downarrow \\ 00 \end{array}$$

Because $35{,}880 \div 12 = 2{,}990$, the monthly income for this family is $2,990. ∎

EXAMPLE 6   A sporting goods store spends a total of $2,952 on basketballs priced at $6.00 each. How many basketballs did they purchase?

SOLUTION

$$\begin{array}{r} 492 \\ 6\overline{)2952} \\ \underline{24}\downarrow \\ 55 \\ \underline{54}\downarrow \\ 12 \\ \underline{12} \\ 0 \end{array}$$

Because $2{,}952 \div 6 = 492$, they purchased 492 basketballs. ∎

5. A family spends $1,872 on a 12-day vacation. How much did they spend each day on average?

NOTE
To estimate the answer to Example 5 quickly, we can replace 35,880 with 36,000 and mentally calculate
$$36{,}000 \div 12$$
which gives an estimate of 3,000. Our actual answer, 2,990, is close enough to our estimate to convince us that we have not made a major error in our calculation.

6. Suppose the store in Example 6 spent $3,222 on the basketballs. If they are still $6.00 each, how many basketballs did they purchase?

Division by Zero

A final important rule for division is that we cannot divide by 0. That is, we cannot use 0 as a divisor in any division problem. Suppose there was an answer to the problem

$$\frac{8}{0} = ?$$

That would mean that

$$0 \cdot ? = 8$$

But, because of the multiplication property of zero, we know that multiplication by 0 always produces 0. There is no number we can use for the ? to make a true statement out of

$$0 \cdot ? = 8$$

Answers
5. $156
6. 537 basketballs

Because this was equivalent to the original division problem
$$\frac{8}{0} = ?$$
we have no number to associate with the expression $\frac{8}{0}$. It is undefined. Here is the formal rule:

> **Rule Division by Zero**
>
> Division by 0 is undefined. Any expression with a divisor of 0 is undefined. We cannot divide by 0.

Problem Set 1.6

Moving Toward Success

"If I have seen further than others, it is by standing upon the shoulders of giants."
—Isaac Newton, 1642–1727, English mathematician and physicist

1. It is highly encouraged that you seek help in some manner regarding this course. Why is this so?
2. Create a list of names and phone numbers you may need if you require help in this class.

A Write each of the following in symbols.

1. The quotient of 6 and 3
2. The quotient of 3 and 6
3. The quotient of 45 and 9
4. The quotient of 12 and 4
5. The quotient of r and s
6. The quotient of s and r
7. The quotient of 20 and 4 is 5.
8. The quotient of 20 and 5 is 4.

Write a multiplication statement that is equivalent to each of the following division statements.

9. $6 \div 2 = 3$
10. $6 \div 3 = 2$
11. $\dfrac{36}{9} = 4$
12. $\dfrac{36}{4} = 9$
13. $\dfrac{48}{6} = 8$
14. $\dfrac{35}{7} = 5$
15. $28 \div 7 = 4$
16. $81 \div 9 = 9$

B Find each of the following quotients. (Divide.) [Examples 1–3]

17. $25 \div 5$
18. $72 \div 8$
19. $40 \div 5$
20. $12 \div 2$
21. $9 \div 0$
22. $7 \div 1$
23. $360 \div 8$
24. $285 \div 5$

25. $\dfrac{138}{6}$ **26.** $\dfrac{267}{3}$ **27.** $5\overline{)7650}$ **28.** $5\overline{)5670}$

29. $5\overline{)6750}$ **30.** $5\overline{)6570}$ **31.** $3\overline{)54000}$ **32.** $3\overline{)50400}$

33. $3\overline{)50040}$ **34.** $3\overline{)50004}$

Estimating

Work Problems 35 through 38 mentally, without using a calculator.

35. The quotient 845 ÷ 93 is closest to which of the following numbers?
  a. 10   **b.** 100   **c.** 1,000   **d.** 10,000

36. The quotient 762 ÷ 43 is closest to which of the following numbers?
  a. 2   **b.** 20   **c.** 200   **d.** 2,000

37. The quotient 15,208 ÷ 771 is closest to which of the following numbers?
  a. 2   **b.** 20   **c.** 200   **d.** 2,000

38. The quotient 24,471 ÷ 523 is closest to which of the following numbers?
  a. 5   **b.** 50   **c.** 500   **d.** 5,000

Mentally give a one-digit estimate for each of the following quotients. That is, for each quotient, mentally estimate the answer using one of the digits 1, 2, 3, 4, 5, 6, 7, 8, or 9.

39. 316 ÷ 289   **40.** 662 ÷ 289   **41.** 728 ÷ 355   **42.** 728 ÷ 177

43. 921 ÷ 243   **44.** 921 ÷ 442   **45.** 673 ÷ 109   **46.** 673 ÷ 218

B Divide. You shouldn't have any wrong answers because you can always check your results with multiplication. [Examples 1–3]

47. 1,440 ÷ 32   **48.** 1,206 ÷ 67   **49.** $\dfrac{2,401}{49}$ **50.** $\dfrac{4,606}{49}$

51. $28\overline{)12096}$ **52.** $28\overline{)96012}$ **53.** $63\overline{)90594}$ **54.** $45\overline{)17595}$

55. $87\overline{)61335}$ **56.** $79\overline{)48032}$ **57.** $45\overline{)135900}$ **58.** $56\overline{)227920}$

B Complete the following tables.

59.

First Number a	Second Number b	The Quotient of a and b $\dfrac{a}{b}$
100	25	
100	26	
100	27	
100	28	

60.

First Number a	Second Number b	The Quotient of a and b $\dfrac{a}{b}$
100	25	
101	25	
102	25	
103	25	

B The following division problems all have remainders. [Example 4]

61. $6\overline{)370}$ **62.** $8\overline{)390}$ **63.** $3\overline{)271}$ **64.** $3\overline{)172}$

65. $26\overline{)345}$ **66.** $26\overline{)543}$ **67.** $71\overline{)16620}$ **68.** $71\overline{)33240}$

69. $23\overline{)9250}$ **70.** $23\overline{)20800}$ **71.** $169\overline{)5950}$ **72.** $391\overline{)34450}$

Applying the Concepts

The application problems that follow may involve more than merely division. Some may require addition, subtraction, or multiplication, whereas others may use a combination of two or more operations.

73. Monthly Income A family has an annual income of $42,300. How much is their monthly income?

74. Hourly Wages If a man works an 8-hour shift and is paid $96 before deductions, how much does he make for 1 hour?

75. Price per Pound If 6 pounds of sirloin steak cost $48.00, how much does 1 pound cost?

76. Cost of a Dress A dress shop orders 45 dresses for a total of $2,205. If they paid the same amount for each dress, how much was each dress?

77. Basketballs If wholesale basketballs cost $6 each, how many basketballs can you buy with $612?

78. Filling Glasses How many 8-ounce glasses can be filled from three 32-ounce bottles of soda?

three 32-ounce bottles = _____ 8-ounce glasses

79. Filling Glasses How many 5-ounce glasses can be filled from a 32-ounce bottle of milk? How many ounces of milk will be left in the bottle when all the glasses are full?

80. Basketballs How many basketballs can be purchased at $7 each if you have $500 to spend?

81. Payroll The annual payroll for the Philadelphia Phillies for the 2011 season was about $173 million dollars. If there are 40 players on the roster what is the average salary per player for the Phillies?

82. Miles per Gallon A traveling salesman kept track of his mileage for 1 month. He found that he traveled 1,104 miles and used 48 gallons of gas. How many miles did he travel on each gallon of gas?

83. Sports Equipment A softball coach orders a bucket of two dozen softballs for $96. What is the price for each ball?

84. Sports Equipment A soccer coach orders 35 soccer balls for a wholesale price of $140. What is the price for each ball?

Exponents, Order of Operations, and Averages

1.7

OBJECTIVES

A Identify the base and exponent of an expression.

B Simplify expressions with exponents.

C Use the rule for order of operations.

D Find the mean, median, mode, and range of a set of numbers.

TICKET TO SUCCESS

Keep these questions in mind as you read through the section. Then respond in your own words and in complete sentences.

1. What is an exponent?
2. What are the four steps of the order of operations?
3. Briefly define *mean*, *median*, and *mode*.
4. What is the range for a set of numbers?

You've probably encountered exponents in the real world without knowing it. Suppose you are shopping for new hardwood flooring for your home office. The flooring company prices its product by the square foot, which is a measurement that could also be written using the following symbols and an exponent: ft^2. Your office measures 12 feet long by 12 feet wide. To calculate the total square feet of flooring you need, you must multiply 12 by 12. Using the definition for *exponent* we will see later, this multiplication problem could also be written using an exponent: 12^2.

A Exponents

Exponents are a shorthand way of writing repeated multiplication. In the expression 2^3, 2 is called the *base* and 3 is called the *exponent*. The expression 2^3 is read "2 to the third power" or "2 cubed." The exponent 3 tells us to use the base 2 as a multiplication factor three times.

$$2^3 = 2 \cdot 2 \cdot 2 \quad \text{2 is used as a factor three times.}$$

We can simplify the expression by multiplication.

$$2^3 = 2 \cdot 2 \cdot 2$$
$$= 4 \cdot 2$$
$$= 8$$

PRACTICE PROBLEMS

For each expression, name the base and the exponent, and write the expression in words.

1. 5^2

2. 2^3

3. 1^4

Simplify each of the following by using repeated multiplication.

4. 5^2

5. 9^2

6. 2^3

7. 1^4

8. 2^5

Answers
1. See Solutions Section.
2. See Solutions Section.
3. See Solutions Section.
4. 25
5. 81
6. 8
7. 1
8. 32

The expression 2^3 is equal to the number 8. We can summarize this discussion with the following definition.

> **Definition**
> An **exponent** is a whole number that indicates how many times the base is to be used as a factor. Exponents indicate repeated multiplication.

Here are some examples of how to read expressions involving exponents:

EXAMPLE 1 3^2 The base is 3, and the exponent is 2. The expression is read "3 to the second power" or "3 squared." ∎

EXAMPLE 2 3^3 The base is 3, and the exponent is 3. The expression is read "3 to the third power" or "3 cubed." ∎

EXAMPLE 3 2^4 The base is 2, and the exponent is 4. The expression is read "2 to the fourth power." ∎

As you can see from these examples, a base raised to the second power is also said to be *squared,* and a base raised to the third power is also said to be *cubed.* These are the only two exponents (^{2} and ^{3}) that have special names. All other exponents are referred to only as "fourth powers," "fifth powers," "sixth powers," and so on.

B Expressions with Exponents

The next examples show how we can simplify expressions involving exponents by using repeated multiplication.

EXAMPLE 4 $3^2 = 3 \cdot 3 = 9$ ∎

EXAMPLE 5 $4^2 = 4 \cdot 4 = 16$ ∎

EXAMPLE 6 $3^3 = 3 \cdot 3 \cdot 3 = 9 \cdot 3 = 27$ ∎

EXAMPLE 7 $3^4 = 3 \cdot 3 \cdot 3 \cdot 3 = 9 \cdot 9 = 81$ ∎

EXAMPLE 8 $2^4 = 2 \cdot 2 \cdot 2 \cdot 2 = 4 \cdot 4 = 16$ ∎

USING TECHNOLOGY

Calculators

Here is how we use a calculator to evaluate exponents, as we did in Example 8:

Scientific Calculator: 2 $\boxed{x^y}$ 4 $\boxed{=}$ or 2 $\boxed{y^x}$ 4 $\boxed{=}$

Graphing Calculator: 2 $\boxed{\wedge}$ 4 $\boxed{\text{ENT}}$ or 2 $\boxed{y^x}$ 4 $\boxed{\text{ENT}}$ or 2 $\boxed{y^x}$ 4 $\boxed{\text{ENT}}$
(depending on the calculator)

1.7 Exponents, Order of Operations, and Averages

Finally, we should consider what happens when the numbers 0 and 1 are used as exponents. First of all, let's look at the following rule where the number 1 appears as an exponent:

> **Rule Exponent 1**
> Any number raised to the first power is itself. That is, if we let the letter a represent any number, then
> $$a^1 = a$$

To take care of the cases when 0 is used as an exponent, we must use the following rule:

> **Rule Exponent 0**
> Any number other than 0 raised to the 0 power is 1. That is, if a represents any nonzero number, then it is always true that
> $$a^0 = 1$$

Here are some examples where 0 and 1 are exponents:

EXAMPLE 9 $5^1 = 5$ ∎

EXAMPLE 10 $9^1 = 9$ ∎

EXAMPLE 11 $4^0 = 1$ ∎

EXAMPLE 12 $8^0 = 1$ ∎

Simplify each of the following expressions.

9. 7^1
10. 4^1
11. 9^0
12. 1^0

C Order of Operations

The symbols we use to specify operations, $+, -, \cdot, \div$, along with the symbols we use for grouping, ( ) and [ ], serve the same purpose in mathematics as punctuation marks in English. In fact, they may be called the punctuation marks of mathematics.

Consider the following sentence:

> Bob said John is tall.

It can have two different meanings, depending on how we punctuate it:

1. "Bob," said John, "is tall."
2. Bob said, "John is tall."

Without the punctuation marks we don't know which meaning the sentence has.

Now, consider the following mathematical expression:

$$4 + 5 \cdot 2$$

What should we do? Should we add 4 and 5 first, or should we multiply 5 and 2 first? There seem to be two different answers. In mathematics, we want to avoid

Answers
9. 7
10. 4
11. 1
12. 1

NOTE

To help you to remember the order of operations you can use the popular sentence **P**lease **E**xcuse **M**y **D**ear **A**unt **S**ally, or the acronym **PEMDAS**.

Parentheses (or grouping)
Exponents
Multiplication and **D**ivision, from left to right
Addition and **S**ubtraction, from left to right

situations in which two different results are possible. Therefore, we follow the rule for order of operations.

> **Rule** Order of Operations
>
> When evaluating mathematical expressions, we will perform the operations in the following order:
>
> 1. If the expression contains grouping symbols, such as parentheses ( ), brackets [ ], or a fraction bar, then we perform the operations inside the grouping symbols, or above and below the fraction bar, first.
> 2. Then we evaluate, or simplify, any numbers with exponents.
> 3. Then we do all multiplications and divisions in order, starting at the left and moving right.
> 4. Finally, we do all additions and subtractions, from left to right.

According to our rule, the expression $4 + 5 \cdot 2$ would have to be evaluated by multiplying 5 and 2 first, and then adding 4. The correct answer—and the only answer—to this problem is 14.

$$4 + 5 \cdot 2 = 4 + 10 \quad \text{Multiply first, then add.}$$
$$= 14$$

Here are some more examples that illustrate how we apply the rule for order of operations to simplify (or evaluate) expressions.

EXAMPLE 13 Simplify: $4 \cdot 8 - 2 \cdot 6$.

SOLUTION We multiply first and then subtract.

$$4 \cdot 8 - 2 \cdot 6 = 32 - 12 \quad \text{Multiply first, then subtract.}$$
$$= 20$$

EXAMPLE 14 Simplify: $5 + 2(7 - 1)$.

SOLUTION According to the rule for the order of operations, we must do what is inside the parentheses first:

$$5 + 2(7 - 1) = 5 + 2(6) \quad \text{Work inside parentheses first.}$$
$$= 5 + 12 \quad \text{Then multiply.}$$
$$= 17 \quad \text{Then add.}$$

EXAMPLE 15 Simplify: $9 \cdot 2^3 + 36 \div 3^2 - 8$.

SOLUTION
$$9 \cdot 2^3 + 36 \div 3^2 - 8 = 9 \cdot 8 + 36 \div 9 - 8 \quad \text{Work exponents first.}$$
$$= 72 + 4 - 8 \quad \text{Then multiply and divide left to right.}$$
$$= 76 - 8 \quad \Big\} \text{Add and subtract,}$$
$$= 68 \quad \text{left to right.}$$

13. Simplify.
  a. $5 \cdot 7 - 3 \cdot 6$
  b. $5 \cdot 70 - 3 \cdot 60$

14. Simplify: $7 + 3(6 + 4)$.

15. Simplify.
  a. $28 \div 7 - 3$
  b. $6 \cdot 3^2 + 64 \div 2^4 - 2$

Answers
13. a. 17 **b.** 170
14. 37
15. a. 1 **b.** 56

EXAMPLE 16 Simplify: $3 + 2[10 - 3(5 - 2)]$.

SOLUTION The brackets, [ ], are used in the same way as parentheses. In a case like this, we move to the innermost grouping symbols first and begin simplifying.

$$3 + 2[10 - 3(5 - 2)] = 3 + 2[10 - 3(3)]$$
$$= 3 + 2[10 - 9]$$
$$= 3 + 2[1]$$
$$= 3 + 2$$
$$= 5$$

Table 1 lists some English expressions and their corresponding mathematical expressions written in symbols.

TABLE 1

In English	Mathematical Equivalent
5 times the sum of 3 and 8	$5(3 + 8)$
Twice the difference of 4 and 3	$2(4 - 3)$
6 added to 7 times the sum of 5 and 6	$6 + 7(5 + 6)$
The sum of 4 times 5 and 8 times 9	$4 \cdot 5 + 8 \cdot 9$
3 subtracted from the quotient of 10 and 2	$10 \div 2 - 3$

D Average

Next we turn our attention to averages. If we go online, we find the following definition for the word *average* when it is used as a noun:

> **Definition**
>
> **Average** (*noun*) is a single value (as a mean, mode, or median) that summarizes or represents the general significance of a set of unequal values.

In everyday language, the word *average* is a general term that can refer to either the mean, the median, or the mode. We will discuss all three individually now. First, lets define the most common average: *mean*.

Mean

> **Definition**
>
> To find the **mean** for a set of values, we add all the numbers and then divide the sum by the number of values in the set. The mean is sometimes called the **arithmetic mean**.

16. Simplify.
 a. $5 + 3[24 - 5(6 - 2)]$
 b. $50 + 30[240 - 50(6 - 2)]$

Answers
16. a. 17 b. 1,250

17. A woman traveled the following distances on a 5-day business trip: 187 miles, 273 miles, 150 miles, 173 miles, and 227 miles. What was the mean distance the woman traveled each day?

EXAMPLE 17 An instructor at a community college earned the following salaries for the first five years of teaching. Find the mean of these salaries.

$$\$35{,}344 \quad \$38{,}290 \quad \$39{,}199 \quad \$40{,}346 \quad \$42{,}866$$

SOLUTION We add the five salaries and then divide by 5, the number of salaries in the set.

$$\text{Mean} = \frac{35{,}344 + 38{,}290 + 39{,}199 + 40{,}346 + 42{,}866}{5} = \frac{196{,}045}{5} = 39{,}209$$

The instructor's mean salary for the first five years of work is $39,209 per year.

Median

Now let's examine a real-life application of median. The table below can be found on the Bureau of Labor Statistics website. It shows hourly wages for a number of professions.

HOURLY WAGES	
All Jobs	$15.95
Chief Executives	$77.27
Computer Programmers	$34.10
Landscape Architects	$29.12
Librarian	$25.82
Pharmacy Technician	$13.49
Correctional Officer	$18.78
Waiter/Waitress	$8.50

Source: U.S. Bureau of Labor Statistics, median hourly wage estimates for May 2009.

If you look at the type at the bottom of the table, you can see that the numbers are the *median* figures for 2009. The median for a set of numbers is the number such that half of the numbers in the set are above it and half are below it. Here is the exact definition:

> **Definition**
> To find the **median** for a set of values, we write the values in order from smallest to largest. If there is an odd number of values, the median is the middle value. If there is an even number of values, then the median is the mean of the two values in the middle.

18. Find the median for the distances in Practice Problem 17.

EXAMPLE 18 Find the median of the values given in Example 17.

SOLUTION The values in Example 17, written from smallest to largest, are shown below. Because there is an odd number of values in the set, the median is the middle value.

$$35{,}344 \quad 38{,}290 \quad \underset{\text{Median}}{39{,}199} \quad 40{,}346 \quad 42{,}866$$

The instructor's median salary for the first five years of teaching is $39,199.

Answer
17. 202 miles
18. 187 miles

EXAMPLE 19 A teacher at a community college in California will make the following salaries for the first four years she teaches.

$$\$51{,}890 \quad \$53{,}745 \quad \$55{,}601 \quad \$57{,}412$$

Find the mean and the median for the four salaries.

SOLUTION To find the mean, we add the four salaries and then divide by 4:

$$\frac{51{,}890 + 53{,}745 + 55{,}601 + 57{,}412}{4} = \frac{218{,}648}{4} = 54{,}662$$

To find the median, we write the salaries in order from smallest to largest. Then, because there is an even number of salaries, we average the middle two salaries to obtain the median.

$$51{,}890 \quad \underbrace{53{,}745 \quad 55{,}601}_{\text{Median}} \quad 57{,}412$$

$$\frac{53{,}745 + 55{,}601}{2} = 54{,}673$$

The mean is $54,662, and the median is $54,673.

19. A teacher earns the following amounts for the first 4 years he teaches. Find the median.
$40,770 $42,635 $44,475 $46,320

Mode

The mode is best used when we are looking for the frequency of an item, such as the most common eye color in a group of people, the most popular breed of dog in the United States, or the movie that was seen the most often. When we have a set of values in which one value occurs more often than the rest, that value is the *mode*.

> **Definition**
> The **mode** for a set of values is the value that occurs most frequently. If all the values in the set occur the same number of times, there is no mode.

For example, consider this set of iPods:

Given the set of iPods, the most popular color is pink. We call this the mode.

EXAMPLE 20 A math class with 18 students had the grades shown below on their first test. Find the mean, the median, and the mode.

$$77 \quad 87 \quad 100 \quad 65 \quad 79 \quad 87$$
$$79 \quad 85 \quad 87 \quad 95 \quad 56 \quad 87$$
$$56 \quad 75 \quad 79 \quad 93 \quad 97 \quad 92$$

SOLUTION To find the mean, we add all the scores and divide by 18:

$$\text{Mean} = \frac{77+87+100+65+79+87+79+85+87+95+56+87+56+75+79+93+97+92}{18}$$

$$= \frac{1{,}476}{18} = 82$$

20. The students in a small math class have the following scores on their final exam. Find the mode.

56 89 74 68 97
74 68 74 88 45

Answer
19. $43,555
20. 74

To find the median, we must put the test scores in order from smallest to largest; then, because there are an even number of test scores, we must find the mean of the middle two scores.

56 56 65 75 77 79 79 79 85 87 87 87 87 92 93 95 97 100

$$\text{Median} = \frac{85 + 87}{2} = 86$$

The mode is the most frequently occurring score. Because 87 occurs 4 times, and no other scores occur that many times, 87 is the mode.

The mean is 82, the median is 86, and the mode is 87. ∎

More Vocabulary

When we used the word *average* for the first time in this section, we used it as a noun. It can also be used as an adjective and a verb. Below is the definition of the word *average* when it is used as a verb.

> **Definition**
> **Average** (*verb*) is to find the arithmetic mean of a series of unequal quantities.

In everyday language, if you are asked for, or given, the average of a set of numbers, the word average can represent the mean, the median, or the mode. When used in this way, the word average is a noun. However, if you are asked to average a set of numbers, then the word average is a verb, and you are being asked to find the mean of the numbers.

Range

Another way to analyze a set of data is to find the range.

> **Definition**
> The **range** of a set of numbers is the difference between the greatest and least values.

The following chart shows average gas prices around the country.

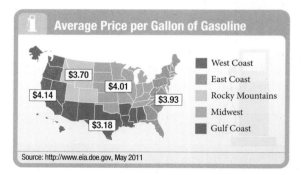

From the information above, we see that the lowest average price was found in the Gulf Coast at $3.18 per gallon. The highest price is on the West Coast at $4.14 per gallon. The range of this set of data is the difference between these two numbers:

$$\$4.14 - \$3.18 = \$0.96$$

We say the country's average gas prices had a range of $0.96.

Problem Set 1.7

Moving Toward Success

"One reason so few of us achieve what we truly want is that we never direct our focus; we never concentrate our power. Most people dabble their way through life, never deciding to master anything in particular."

—Tony Robbins, 1960–present, Author

1. What is the difference between short-term goals and long-term goals?
2. Make a list of your short-term and long-term goals for this class. How do you plan to achieve these goals?

A For each of the following expressions, name the base and the exponent. [Examples 1–3]

1. 4^5
2. 5^4
3. 3^6
4. 6^3
5. 8^2
6. 2^8
7. 9^1
8. 1^9
9. 4^0
10. 0^4

B Use the definition of exponents as indicating repeated multiplication to simplify each of the following expressions. [Examples 4–12]

11. 6^2
12. 7^2
13. 2^3
14. 2^4
15. 1^4
16. 5^1
17. 9^0
18. 27^0
19. 9^2
20. 8^2
21. 10^1
22. 8^1
23. 12^1
24. 16^0
25. 45^0
26. 3^4

C Use the rule for the order of operations to simplify each expression. [Examples 13–16]

27. $16 - 8 + 4$
28. $16 - 4 + 8$
29. $20 \div 2 \cdot 10$
30. $40 \div 4 \cdot 5$
31. $20 - 4 \cdot 4$
32. $30 - 10 \cdot 2$
33. $3 + 5 \cdot 8$
34. $7 + 4 \cdot 9$

35. $3 \cdot 6 - 2$ **36.** $5 \cdot 1 + 6$ **37.** $6 \cdot 2 + 9 \cdot 8$ **38.** $4 \cdot 5 + 9 \cdot 7$

39. $4 \cdot 5 - 3 \cdot 2$ **40.** $5 \cdot 6 - 4 \cdot 3$ **41.** $5^2 + 7^2$ **42.** $4^2 + 9^2$

43. $480 + 12(32)^2$ **44.** $360 + 14(27)^2$ **45.** $3 \cdot 2^3 + 5 \cdot 4^2$ **46.** $4 \cdot 3^2 + 5 \cdot 2^3$

47. $8 \cdot 10^2 - 6 \cdot 4^3$ **48.** $5 \cdot 11^2 - 3 \cdot 2^3$ **49.** $2(3 + 6 \cdot 5)$ **50.** $8(1 + 4 \cdot 2)$

51. $19 + 50 \div 5^2$ **52.** $9 + 8 \div 2^2$ **53.** $9 - 2(4 - 3)$ **54.** $15 - 6(9 - 7)$

55. $4 \cdot 3 + 2(5 - 3)$ **56.** $6 \cdot 8 + 3(4 - 1)$ **57.** $4[2(3) + 3(5)]$ **58.** $3[2(5) + 3(4)]$

59. $(7 - 3)(8 + 2)$ **60.** $(9 - 5)(9 + 5)$ **61.** $3(9 - 2) + 4(7 - 2)$ **62.** $7(4 - 2) - 2(5 - 3)$

63. $18 + 12 \div 4 - 3$ **64.** $20 + 16 \div 2 - 5$ **65.** $4(10^2) + 20 \div 4$ **66.** $3(4^2) + 10 \div 5$

67. $8 \cdot 2^4 + 25 \div 5 - 3^2$ **68.** $5 \cdot 3^4 + 16 \div 8 - 2^2$ **69.** $5 + 2[9 - 2(4 - 1)]$ **70.** $6 + 3[8 - 3(1 + 1)]$

71. $3 + 4[6 + 8(2 - 0)]$ **72.** $2 + 5[9 + 3(4 - 1)]$ **73.** $\dfrac{15 + 5(4)}{17 - 12}$ **74.** $\dfrac{20 + 6(2)}{11 - 7}$

Translate each English expression into an equivalent mathematical expression written in symbols. Then simplify.

75. 8 times the sum of 4 and 2

76. 3 times the difference of 6 and 1

77. Twice the sum of 10 and 3

78. 5 times the difference of 12 and 6

79. 4 added to 3 times the sum of 3 and 4

80. 25 added to 4 times the difference of 7 and 5

81. 9 subtracted from the quotient of 20 and 2

82. 7 added to the quotient of 6 and 2

83. The sum of 8 times 5 and 5 times 4

84. The difference of 10 times 5 and 6 times 2

D Find the mean and the range for each set of numbers. [Examples 17, 19, 20]

85. 1, 2, 3, 4, 5

86. 2, 4, 6, 8, 10

87. 1, 3, 9, 11

88. 5, 7, 9, 12, 12

D Find the median and the range for each set of numbers. [Examples 18–20]

89. 5, 9, 11, 13, 15

90. 42, 48, 50, 64

91. 10, 20, 50, 90, 100

92. 700, 900, 1100

D Find the mode and the range for each set of numbers. [Example 20]

93. 14, 18, 27, 36, 18, 73

94. 11, 27, 18, 11, 72, 11

Applying the Concepts

Suppose the following table lists the number of calories consumed by eating some popular fast foods. Use the table to work Problems 95 and 96.

CALORIES IN FOOD	
Food	Calories
McDonald's hamburger	270
Burger King hamburger	260
Jack in the Box hamburger	280
McDonald's Big Mac	510
Burger King Whopper	630
Jack in the Box Colossus burger	940

95. Suppose a meal of pasta and salad contains 255 calories. Compare the total number of calories in that meal with the number of calories in a McDonald's Big Mac.

96. Suppose a meatball sandwich contains 485 calories. Compare the total number of calories in the sandwich with the number of calories in a Burger King hamburger.

97. Average If a basketball team has scores of 61, 76, 98, 55, 76, and 102 in their first six games, find

  a. the mean score
  b. the median score
  c. the mode of the scores
  d. the range of scores

98. Home Sales Below are listed the prices paid for 10 homes that sold in a city in Texas.

$210,000 $139,000 $122,000 $145,000 $120,000
$540,000 $167,000 $125,000 $125,000 $950,000

a. Find the mean housing price.

b. Find the median housing price.

c. Find the mode of the housing prices.

d. Which measure of "average" best describes the average housing price? Explain your answer.

99. Average Enrollment Suppose the number of students enrolled in a community college during a 5-year period is shown in the table. Find the mean enrollment and the range of enrollments for this 5-year period.

Year	Enrollment
2006	6,789
2007	6,970
2008	7,242
2009	6,981
2010	6,423

100. Car Prices Suppose the prices in the table were listed for Volkswagen Jettas on ebay's car auction site. Use the table to find

a. the mean car price.

b. the median car price.

c. the mode for the car prices.

d. the range of car prices.

CAR PRICES	
Year	Price
1998	$10,000
1999	$14,500
1999	$10,500
1999	$11,700
1999	$15,500
2000	$10,500
2000	$18,200
2001	$19,900

101. Blood Pressure Screening When you have your blood pressure measured, it is written down as two numbers, one over the other. The top number, which is called the systolic pressure, shows the pressure in your arteries when your heart is forcing blood through them. The bottom number, called the diastolic pressure, shows the pressure in your arteries when your heart relaxes. Blood pressure screening is a part of the annual health fair held on your campus. The systolic reading (measured in mmHg) of 10 students were recorded:

140   112   118   120   138   119   130   130   125   128

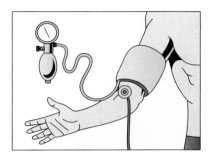

Use this information to find

  a. the mean systolic pressure.

  b. the median systolic pressure.

  c. the mode for the systolic pressure.

  d. the range in the values for systolic pressure.

102. California Counties The table shows those California counties which had a population of more than 1,000,000 in 2006. Use this information to find the following.

  a. The mean population for these counties to the nearest person

  b. The median population

  c. The mode population

  d. The range in the values for the population in these counties

  e. Which measure seems to best describe the average population? Explain your choice.

COUNTY POPULATION	
County	Population
Los Angeles	10,294,280
San Diego	3,120,088
Orange	3,098,183
Riverside	2,070,315
San Bernadino	2,039,467
Santa Clara	1,820,176
Alameda	1,530,620
Sacramento	1,415,117
Contra Costa	1,044,201

Source: U.S. Census Bureau, International Data Base

103. Gasoline Prices The Energy Information Administration (EIA) was created by Congress in 1977 and is a statistical agency of the U.S. Department of Energy. According to the EIA, the average retail prices for regular gasoline in California can be seen in the table. Use this information to find

a. the median price for a gallon of regular gas.

b. the mean price to the nearest tenth of a cent

c. the range in the price of regular gas between March 14th and May 2nd.

AVERAGE PRICE FOR REGULAR GAS IN CALIFORNIA	
Date	Price per Gallon
3/14/2011	$3.954
3/21/2011	$3.966
3/28/2011	$4.028
4/04/2011	$4.057
4/11/2011	$4.161
4/18/2011	$4.205
4/25/2011	$4.217
5/02/2011	$4.257

Source: Energy Information Administration

104. Cell Phones Suppose the table shows the total voice minutes and number of calls sent and received for different age groups. Use this information to find

a. the mean number of minutes used and the mean number of calls sent and received.

b. the range of minutes used and calls sent and received.

c. Based on this information determine the average length of a cell phone call.

CELL PHONE USAGE		
Age	Total Voice Minutes Used	Number of Calls Sent/Received
18-24	1,304	340
25-36	970	246
37-55	726	197
56+	441	119

Extending the Concepts

There is a relationship between the two sequences below. The first sequence is the *sequence of odd numbers*. The second sequence is called the *sequence of squares*.

        1, 3, 5, 7, . . .   The sequence of odd numbers
        1, 4, 9, 16, . . .   The sequence of squares

105. Add the first two numbers in the sequence of odd numbers.

106. Add the first three numbers in the sequence of squares.

107. Add the first four numbers in the sequence of squares.

108. Add the first five numbers in the sequence of odd numbers.

Area and Volume

1.8

OBJECTIVES

A Find the area of a polygon.

B Find the volume of an object.

C Find the surface area of an object.

TICKET TO SUCCESS

Keep these questions in mind as you read through the section. Then respond in your own words and in complete sentences.

1. Define *area*.
2. Define *volume*.
3. How is *surface area* different than area?
4. If the dimensions of a rectangular solid are given in inches, what units will be associated with the surface area?

A Area

At the beginning of the last section, we introduced a measurement called square feet. This measurement is commonly used when calculating *area*. Here is a formal definition:

> **Definition**
> The **area** of a flat object is a measure of the amount of surface the object has.

Suppose you needed to find the area of a tennis court. If you knew that the length of the court for a doubles match was 78 feet and the width was 36 feet, how many square feet would you picture fitting on the court? To solve this problem, we need to multiply the length of the court by the width.

$$\begin{aligned} \text{Area} &= (\text{length} \cdot \text{width}) \\ &= (78 \text{ feet}) \cdot (36 \text{ feet}) \\ &= (78 \cdot 36) \cdot (\text{feet} \cdot \text{feet}) \\ &= 2{,}808 \text{ square feet} \end{aligned}$$

Now let's examine the rectangle in Figure 1 to visualize square feet on a smaller scale.

The area of the rectangle below is 6 square feet, because it takes 6 square feet to cover it.

one square foot	one square foot	one square foot
one square foot	one square foot	one square foot

2 feet

3 feet

FIGURE 1

Again, the area of this rectangle can also be found by multiplying the length and the width.

$$\begin{aligned}\text{Area} &= (\text{length}) \cdot (\text{width}) \\ &= (3 \text{ feet}) \cdot (2 \text{ feet}) \\ &= (3 \cdot 2) \cdot (\text{feet} \cdot \text{feet}) \\ &= 6 \text{ square feet}\end{aligned}$$

From this example, the tennis court example, and others, we conclude that the area of any rectangle is the product of the length and width.

Here are three common geometric figures along with the formula for the area of each one.

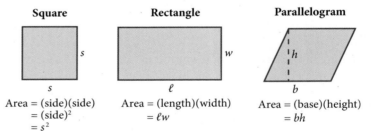

Square
Area = (side)(side)
= (side)^2
= s^2

Rectangle
Area = (length)(width)
= ℓw

Parallelogram
Area = (base)(height)
= bh

The following examples will give you practice finding areas of these common figures.

EXAMPLE 1 The parallelogram below has a base of 5 centimeters and a height of 2 centimeters. Find the area.

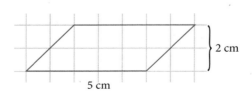

5 cm, 2 cm

SOLUTION If we apply our formula, we have

$$\begin{aligned}\text{Area} &= (\text{base})(\text{height}) \\ A &= bh \\ &= 5 \cdot 2 \\ &= 10 \text{ cm}^2\end{aligned}$$

PRACTICE PROBLEMS

1. Find the area.

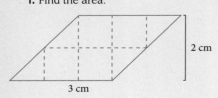

3 cm, 2 cm

Answer
1. 6 cm^2

Or we could simply count the number of square centimeters it takes to cover the object. There are 8 complete squares and 4 half-squares, giving a total of 10 squares for an area of 10 square centimeters. Counting the squares in this manner helps us see why the formula for the area of a parallelogram is the product of the base and the height.

To justify our formula in general, we simply rearrange the parts to form a rectangle.

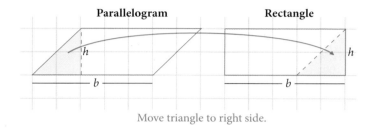

Move triangle to right side.

EXAMPLE 2 Find the area of the following square-shaped stamp.

Each side is 35 millimeters.

SOLUTION Applying our formula for area, we have

$$A = s^2 = (35 \text{ mm})^2 = 1{,}225 \text{ mm}^2$$

EXAMPLE 3 Find the total area of the house and garage shown below.

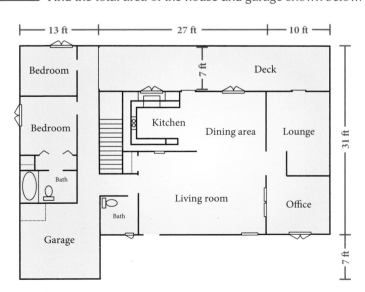

SOLUTION We begin by drawing an additional line, so that the original figure is now composed of two rectangles. Next, we fill in the missing dimensions on the two rectangles.

2. Find the area of a rectangular stamp if it is 35 mm wide and 70 mm long.

3. Find the area of the house without the deck.

Answers
2. 2,450 mm²
3. 1,142 sq ft

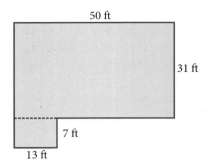

Finally, we calculate the area of the original figure by adding the areas of the individual figures.

$$\begin{aligned}\text{Area} &= \text{Area of the small rectangle} + \text{Area of the large rectangle}\\ &= 13 \cdot 7 + 50 \cdot 31\\ &= 91 + 1{,}550\\ &= 1{,}641 \text{ square feet}\end{aligned}$$

B Volume

Next, we move up one dimension and consider what is called *volume*. Here is a definition for volume:

> **Definition**
> **Volume** is the measure of the space enclosed by a solid.

For instance, if each edge of a cube is 3 feet long, as shown in Figure 2, then we can think of the cube as being made up of a number of smaller cubes, each of which is 1 foot long, 1 foot wide, and 1 foot high. Each of these smaller cubes is called a cubic foot. To count the number of them in the larger cube, think of the large cube as having three layers. You can see that the top layer contains 9 cubic feet. Because there are three layers, the total number of cubic feet in the large cube is $9 \cdot 3 = 27$.

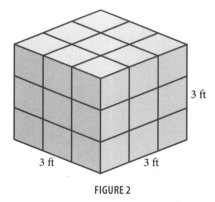

FIGURE 2

On the other hand, if we multiply the length, the width, and the height of the cube, we have the same result:

$$\begin{aligned}\text{Volume} &= (3 \text{ feet})(3 \text{ feet})(3 \text{ feet})\\ &= (3 \cdot 3 \cdot 3)(\text{feet} \cdot \text{feet} \cdot \text{feet})\\ &= 27 \text{ ft}^3 \text{ or } 27 \text{ cubic feet}\end{aligned}$$

For the present, we will confine our discussion of volume to volumes of *rectangular solids*. Rectangular solids are the three-dimensional equivalents of rectangles, where opposite sides are parallel, and any two sides that meet, meet at right angles. A rectangular solid is shown in Figure 3, along with the formula used to calculate its volume.

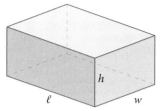

Volume = (length)(width)(height)
$V = \ell wh$

FIGURE 3

EXAMPLE 4 Find the volume of a rectangular solid with length 15 inches, width 3 inches, and height 5 inches.

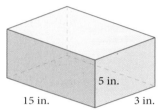

SOLUTION To find the volume, we apply the formula shown in Figure 3.

$$V = l \cdot w \cdot h$$
$$= (15 \text{ in.})(3 \text{ in.})(5 \text{ in.})$$
$$= 225 \text{ in}^3$$

4. A home has a dining room that is 12 feet wide and 15 feet long. If the ceiling is 8 feet high, find the volume of the dining room.

C Surface Area

Now let's shift our discussion to *surface area*.

Definition
Surface area is the total area of all surfaces of an object.

Figure 4 shows a closed box with length l, width w, and height h. The surfaces of the box are labeled as sides, top, bottom, front, and back.

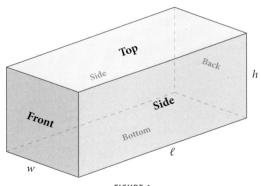

FIGURE 4

Answer
4. 1,440 cubic feet

5. A family is painting a dining room that is 12 feet wide and 15 feet long.
  a. If the ceiling is 8 feet high, find the surface area of the walls and the ceiling, but not the floor.
  b. If a gallon of paint will cover 400 square feet, how many gallons should they buy to paint the walls and the ceiling?

To find the surface area of the box, we add the areas of each of the six surfaces that are labeled in Figure 4.

$$\text{Surface area} = \text{side} + \text{side} + \text{front} + \text{back} + \text{top} + \text{bottom}$$
$$S = (l \cdot h) + (l \cdot h) + (h \cdot w) + (h \cdot w) + (l \cdot w) + (l \cdot w)$$
$$= 2lh + 2hw + 2lw$$

EXAMPLE 5 Find the surface area of the box below.

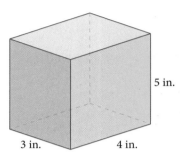

SOLUTION To find the surface area, we find the area of each surface individually, and then we add them together.

$$\text{Surface area} = 2(3 \text{ in.})(4 \text{ in.}) + 2(3 \text{ in.})(5 \text{ in.}) + 2(4 \text{ in.})(5 \text{ in.})$$
$$= 24 \text{ in}^2 + 30 \text{ in}^2 + 40 \text{ in}^2$$
$$= 94 \text{ in}^2$$

The total surface area is 94 square inches. If we calculate the volume enclosed by the box, it is $V = (3 \text{ in.})(4 \text{ in.})(5 \text{ in.}) = 60 \text{ in}^3$. The surface area measures how much material it takes to make the box, whereas the volume measures how much space the box will hold. ∎

Answers
5. a. 612 square feet
   b. 2 gallons will cover everything, with some paint left over.

Problem Set 1.8

Moving Toward Success

"There is nothing too little for so little a creature as man. It is by studying little things that we attain the great art of having as little misery and as much happiness as possible."
—Samuel Johnson, 1709–1784, English poet and critic

1. What was the most important study skill you used while working through Chapter 1?
2. Why should you continue to place an importance on study skills as you work through the rest of the book?

A Find the area enclosed by each figure. [Examples 1–3]

1.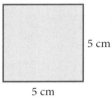

2.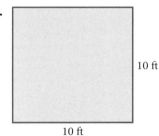

3.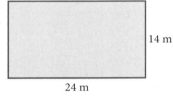

4.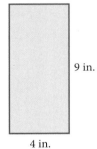

5.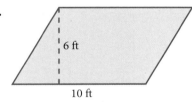

6.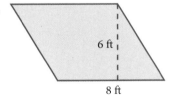

7.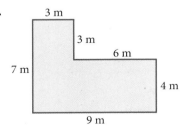

8.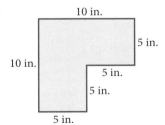

9.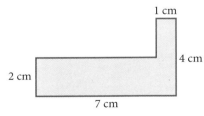

10.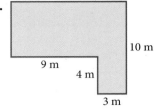

11.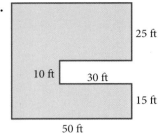

12.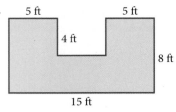

13.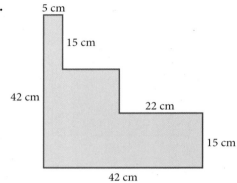

14.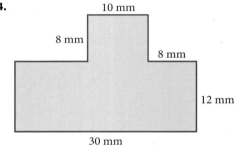

15. Find the area of a square with side 10 inches.

16. Find the area of a square with side 6 centimeters.

B **C** Find the volume and surface area of each figure. [Example 4, 5]

17.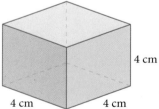

18.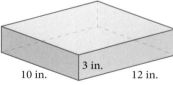

19.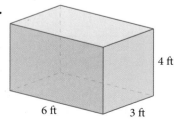

20.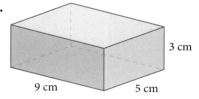

Applying the Concepts

21. Area of an MP3 Player The MP3 player below can hold up to 20,000 songs.

0.24 in. 1.5 in. 3.6 in.

a. Find the area of the face of the MP3 player.

b. Find the area of the side of the MP3 player.

22. Area of an MP3 Player The MP3 player below can hold up to 30,000 songs.

0.41 in. 2.4 in. 4.1 in.

a. Find the area of the face of the MP3 player.

b. Find the area of the side of the MP3 player.

23. Area A swimming pool is 20 feet wide and 40 feet long. If it is surrounded by square tiles, each of which is 1 foot by 1 foot, how many tiles are there surrounding the pool?

24. Area A garden is rectangular with a width of 8 feet and a length of 12 feet. If it is surrounded by a walkway 2 feet wide, how many square feet of area does the walkway cover?

25. Playing Fields The field used in lacrosse is 110 yards long and 60 yards wide. What is area of a lacrosse playing field?

26. Playing Fields The pit for playing horseshoes has a length of 48 feet and a width of 6 feet. What is the area of a horseshoe pit?

27. Comparing Areas The side of a square is 5 feet long. If all four sides are increased by 2 feet, by how much is the area increased?

28. Comparing Areas The length of a side in a square is 20 inches. If all four sides are decreased by 4 inches, by how much is the area decreased?

29. Area of a Euro A 10 euro banknote has a width of 67 millimeters and a length of 127 millimeters. Find the area.

30. Area of a Dollar A $10 bill has a width of 65 millimeters and a length of 156 millimeters. Find the area.

31. Area of a Tennis Court At the beginning of this section we discussed the dimensions of a tennis court for a doubles match. For a singles match the court is smaller measuring 78 feet long by 27 feet wide. Find the area of the singles court.

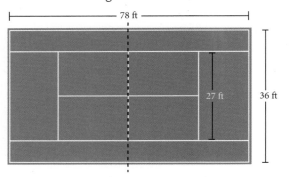

32. Area of a Stamp The stamp shown here shows Italian scientist Enrico Fermi. The image area of the stamp has a width of 21 millimeters and a length of 35 millimeters. Find the area of the image.

33. Hot Air Balloon The woodcut shows the giant hot air balloon known as "Le Geant de Nadar" when it was displayed in the Crystal Palace in England in 1868. The wicker car of the balloon was two stories, consisting of a 6-compartment cottage with a viewing deck on top. If the car was 8 feet high with a square base 13 feet on each side, find the volume.

34. Reading House Plans Find the area of the floor of the house shown here if the garage is not included with the house and find the area if the garage is included with the house.

35. Area of a Square The area of a square is 49 square feet. What is the length of each side?

36. Area of a Square The area of a square is 144 square feet. How long is each side?

37. Area of a Rectangle A rectangle has an area of 36 square feet. If the width is 4 feet, what is the length?

38. Area of a Rectangle A rectangle has an area of 39 square feet. If the length is 13 feet, what is the width?

Chapter 1 Summary

EXAMPLES

The margins of the chapter summaries will be used for examples of the topics being reviewed, whenever convenient.

The numbers in brackets indicate the sections in which the topics were discussed.

Place Values [1.1]

1. The number 42,103,045 written in words is "forty-two million, one hundred three thousand, forty-five."

   The number 5,745 written in expanded form is
   $5{,}000 + 700 + 40 + 5$

The place values for the digits of any base 10 number are as follows:

Trillions			Billions			Millions			Thousands			Ones		
Hundreds	Tens	Ones	Hundreds	Tens	Ones	Hundreds	Tens	Ones	Hundreds	Tens	Ones	Hundreds	Tens	Ones

Vocabulary Associated with Addition, Subtraction, Multiplication, and Division [1.2, 1.4, 1.5, 1.6]

2. The sum of 5 and 2 is $5 + 2$.
   The difference of 5 and 2 is $5 - 2$.
   The product of 5 and 2 is $5 \cdot 2$.
   The quotient of 10 and 2 is $10 \div 2$.

The word *sum* indicates addition.

The word *difference* indicates subtraction.

The word *product* indicates multiplication.

The word *quotient* indicates division.

Properties of Addition and Multiplication [1.2, 1.5]

3. $3 + 2 = 2 + 3$
   $3 \cdot 2 = 2 \cdot 3$
   $(x + 3) + 5 = x + (3 + 5)$
   $(4 \cdot 5) \cdot 6 = 4 \cdot (5 \cdot 6)$
   $3(4 + 7) = 3(4) + 3(7)$

If a, b, and c represent any three numbers, then here are the properties of addition and multiplication used most often:

Commutative property of addition: $a + b = b + a$

Commutative property of multiplication: $a \cdot b = b \cdot a$

Associative property of addition: $(a + b) + c = a + (b + c)$

Associative property of multiplication: $(a \cdot b) \cdot c = a \cdot (b \cdot c)$

Distributive property: $a(b + c) = a(b) + a(c)$

Perimeter of a Polygon [1.2]

4. The perimeter of the rectangle below is
   $P = 37 + 37 + 24 + 24$
   $= 122$ feet

The *perimeter* of any polygon is the sum of the lengths of the sides, and it is denoted with the letter P.

Steps for Rounding Whole Numbers [1.3]

5. 5,482 to the nearest ten is 5,480.
5,482 to the nearest hundred is 5,500.
5,482 to the nearest thousand is 5,000.

1. Locate the digit just to the right of the place to which you want to round.
2. If that digit is less than 5, replace it and all digits to its right with zeros.
3. If that digit is 5 or more, replace it and all digits to its right with zeros, and add 1 to the digit to its left.

Division by 0 (Zero) [1.6]

6. Each expression below is undefined.
$$5 \div 0 \quad \frac{7}{0} \quad 4/0$$

Division by 0 is undefined. We cannot use 0 as a divisor in any division problem.

Order of Operations [1.7]

7. $4 + 6(8 - 2)$
$= 4 + 6(6)$
$= 4 + 36$
$= 40$

To simplify a mathematical expression, perform the operations in the following order:

1. We simplify the expression inside the grouping symbols first. Grouping symbols are parentheses ( ), brackets [ ], or a fraction bar.
2. Then we evaluate any numbers with exponents.
3. We then perform all multiplications and divisions in order, starting at the left and moving right.
4. Finally, we do all the additions and subtractions, from left to right.

Average [1.7]

8. The mean of 4, 7, 9 and 12 is
$(4 + 7 + 9 + 12) \div 4$
$= 32 \div 4$
$= 8$

The *average* for a set of numbers can be the mean, the median, or the mode.

Exponents [1.7]

9. $2^3 = 2 \cdot 2 \cdot 2 = 8$
$5^0 = 1$
$3^1 = 3$

In the expression 2^3, 2 is the *base* and 3 is the *exponent*. An exponent is a shorthand notation for repeated multiplication. The exponent 0 is a special exponent, such that any nonzero number to the 0 power is 1. The exponent 1 is also special because any number raised to the first power is itself.

Formulas for Area, Volume, and Surface Area [1.8]

Square

Area = (side)(side)
= (side)^2
= s^2

Rectangle

Area = (length)(width)
= ℓw

Parallelogram

Area = (base)(height)
= bh

Rectangular Solid

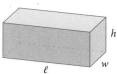

Volume = $V = lwh$
Surface Area = $S = 2lh + 2hw + 2lw$

Chapter 1 Review

The numbers in brackets indicate the sections in which problems of a similar type can be found.

1. One of the largest Pacific blue marlins was caught near Hawaii in 1984. It weighed 1,656 pounds. Write 1,656 in words. [1.1]
2. In 2010, the New York Yankees had the highest home attendance in major league baseball. The attendance that year was 3,765,807. Write 3,765,807 in words. [1.1]

For Problems 3 and 4, write each number with digits instead of words. [1.1]

3. Five million, two hundred forty-five thousand, six hundred fifty-two
4. Twelve million, twelve thousand, twelve
5. In 2010, the Cleveland Indians had the lowest attendance in major league baseball. The attendance that year was 1,394,812. Write 1,394,812 in expanded form. [1.1]
6. According to NumberOf.net, in 2010, there were 201,727 female physicians practicing medicine in the United States. Write 201,727 in expanded form. [1.1]

Identify each of the statements in Problems 7–14 as an example of one of the following properties. [1.2, 1.5]

  a. Addition property of 0
  b. Multiplication property of 0
  c. Multiplication property of 1
  d. Commutative property of addition
  e. Commutative property of multiplication
  f. Associative property of addition
  g. Associative property of multiplication

7. $5 + 7 = 7 + 5$
8. $(4 + 3) + 2 = 4 + (3 + 2)$
9. $6 \cdot 1 = 6$
10. $8 + 0 = 8$
11. $5 \cdot 0 = 0$
12. $4 \cdot 6 = 6 \cdot 4$
13. $5 \cdot (3 \cdot 2) = (5 \cdot 3) \cdot 2$
14. $(6 + 2) + 3 = (2 + 6) + 3$

Find each of the following sums. (Add.) [1.2]

15. 498 + 251
16. 784 + 598
17. 7,384 + 251 + 637
18. 4,901 + 648 + 3,592

Round the number 3,781,092 to the nearest given place value. [1.3]

19. Ten
20. Hundred
21. Hundred thousand
22. Million

Find each of the following differences. (Subtract.) [1.4]

23. 789 − 475
24. 792 − 178
25. 5,908 − 2,759
26. 3,527 − 1,789

Find each of the following products. (Multiply.) [1.5]

27. 8(73)
28. 7(984)
29. 63(59)
30. 49(876)

Find each of the following quotients. (Divide.) [1.6]

31. $692 \div 4$
32. $1,020 \div 15$
33. $36\overline{)15,408}$
34. $286\overline{)21,736}$

Use the rule for the order of operations to simplify each expression as much as possible. [1.7]

35. $4 + 3 \cdot 5^2$
36. $7(9)^2 - 6(4)^3$
37. $3(2 + 8 \cdot 9)$
38. $7 - 2(6 - 4)$
39. $24 \div 6 \cdot 2$
40. $20 \cdot 3 \div 12 \cdot 2$
41. $4(3 - 1)^3$
42. $36 \div 9 \cdot 3^2$

Write an expression using symbols that is equivalent to each of the following expressions; then simplify. [1.7]

43. 3 times the sum of 4 and 6
44. 9 times the difference of 5 and 3
45. Twice the difference of 17 and 5
46. The product of 5 and the sum of 8 and 2
47. The following is a list of rent prices for apartments in San Francisco, CA. Find the mean, median, and range of the prices. Round to the nearest dollar if necessary. [1.7]
$2737 $3746 $5097 $2472 $3630 $2783

Find the perimeter and area of the shapes. [1.2, 1.8]

48.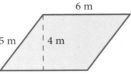

49.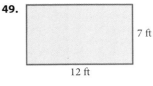

Chapter 1 Test

1. Write the number 20,347 in words. [1.1]

2. Write the number two million, forty-five thousand, six with digits instead of words. [1.1]

3. Write the number 123,407 in expanded form. [1.1]

Identify each of the statements in Problems 4–7 as an example of one of the following properties. [1.2, 1.5]
   a. Addition property of 0
   b. Multiplication property of 0
   c. Multiplication property of 1
   d. Commutative property of addition
   e. Commutative property of multiplication
   f. Associative property of addition
   g. Associative property of multiplication

4. $(5 + 6) + 3 = 5 + (6 + 3)$ 5. $7 \cdot 1 = 7$

6. $9 + 0 = 9$ 7. $5 \cdot 6 = 6 \cdot 5$

Find each of the following sums. (Add.) [1.2]

8. 135
   $+741$

9. $5,401$
   329
   $+10,653$

10. Round the number 516,249 to the nearest ten thousand. [1.3]

Find each of the following differences. (Subtract.) [1.4]

11. 937
    -413

12. $7,052$
    $-3,967$

Find each of the following products. (Multiply.) [1.5]

13. $9(186)$ 14. $62(359)$

Find each of the following quotients. (Divide.) [1.6]

15. $1,105 \div 13$ 16. $583 \overline{)12,243}$

Use the rule for the order of operations to simplify each expression as much as possible. [1.7]

17. $8(5)^2 - 7(3)^3$ 18. $8 - 2(5 - 3)$

19. $7 + 2(53 - 3)$ 20. $3(x - 2)$

21. Twice the sum of 11 and 7 [1.7]

22. The quotient of 20 and 5 increased by 9 [1.7]

23. The following is a list of prices for laptop computers found on buy.com. Find the mean, median, and range of the prices. Round to the nearest cent if necessary.
$599.99 $349.99 $439.95 $657.99 $819.99 $573.90
[1.7]

24. Find the area and perimeter of the parallelogram. [1.2, 1.8]

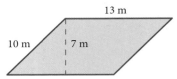

The chart below shows the world's wind electricity generating capacity over several years. Use the information to answer the following questions.

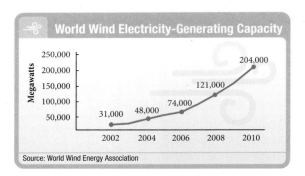

25. How much more wind electricity was generated in 2008 compared to 2004?

26. How much more wind electricity was generated in 2010 compared to 2002?

Chapter 1 Projects

WHOLE NUMBERS

Group Project

Egyptian Numbers

Students and Instructors: The end of each chapter in this book will have two projects. The group projects are intended to be done in class. The research projects are to be completed outside of class. They can be done in groups or individually. Some chapters will also have a third project, called a Challenge Project. These projects revolve around the use of the Google Earth online program.

Number of People 3

Time Needed 10 minutes

Equipment Pencil and paper

Background The Egyptians had a fully-developed number system as early as 3500 BC. Evidence of these very large numbers was recorded in the macehead of Narmer, which is a pear-shaped piece of stone once fixed to a staff and used as a weapon. The Egyptians used a base-ten system, similar to ours. They used a special pictograph to represent each power of ten. The table to the right shows some of these pictographs:

1	10	100	1,000
|	∩	ℯ	⚱
staff	horseshoe	rope	lotus flower

10,000	100,000	1,000,000
𓆑	𓆏	𓀠
bent finger	tadpole or frog	astonished person

Example Usually the direction of writing was from right to left, with the larger units first. Symbols were placed in rows to save lateral space. Here is number 132,146 in Egyptian hieroglyphics:

132,146 = ||| ∩∩∩ ℯ 𓏺𓏺 ||| 𓆏
         ||| ∩

Procedure Again, notice how the tadpole, which represents 100,000, is written on the far right with the smaller place values following to the left. Also, notice how three bent fingers are illustrated to represent a 3 in the ten thousands place value.

Write each of the following Egyptian numbers in our system.

1.

2.

Express each of the given numbers in Egyptian hieroglyphics.

3. 1,842
4. 4,310,175

Research Project

Leopold Kronecker

Leopold Kronecker (1823–1891) was a German mathematician and logician who thought that arithmetic should be based on whole numbers. He is known for the quote, "God made the natural numbers; all else is the work of man." He was openly critical of the efforts of his contemporaries. Kronecker's primary work was in the field of algebraic number theory. Research the life of Leopold Kronecker, or discuss the work of a mathematician who was criticized by Kronecker.

Challenge Project

FOLLOWING BUILDING CODES

Assume you plan to open a restaurant in Santa Monica, CA. To do so, you must follow the city's municipal building codes specific to the type of business you own. For instance, the code for the number of parking spaces for an eat-in restaurant is different than the code for a take-out restaurant. Santa Monica's municipal code for parking spaces requires 1 parking space for every 75 square feet of service and seating area open to customers.

STEP 1 Use Google Earth to find Santa Monica, CA.

STEP 2 Find a square or rectangular shaped building in the city. Let's imagine you will use this building for your restaurant. Use the **Ruler** tool set to **Feet** to measure the length and the width of the building. Round your answers to the nearest foot.

STEP 3 Multiply the length and width measurements to calculate the floor area of your building. Now let's assume that you plan to have $\frac{3}{4}$ of the total floor area open for customers. How many square feet is open to customers?

STEP 4 Based on Santa Monica's municipal code, how many parking spaces will you need for your restaurant?

STEP 5 The Americans with Disabilities Act and Architectural Barriers Act Accessibility Guidelines (ADAAG) for buildings and facilities require accessible parking spaces to be at least 8 feet wide. Imagine you can create a line of parking spaces along one length of your building. Based on the ADAAG, what is the maximum number of spaces you can provide along that length?

Challenge Project

A Glimpse of Algebra

At the end of most chapters of this book, you will find a section like this one. These sections show how some of the material in the chapter looks when it is extended to algebra. If you are planning to take an algebra course after you have finished this one, these sections will give you a head start. If you are not planning to take algebra, these sections will give you an idea of what algebra is like. Who knows? You may decide to take an algebra class after you work through a few of these sections.

In this chapter, we did some work with exponents. We can use the definition of exponents, along with the commutative property of multiplication, to rewrite some expressions that contain variables and exponents.

We can expand the expression $(5x)^2$ using the definition of exponents as

$$(5x)^2 = (5x)(5x)$$

Because the expression on the right is all multiplication, we can rewrite it as

$$(5x)(5x) = 5 \cdot x \cdot 5 \cdot x$$

And because multiplication is a commutative operation, we can rearrange this last expression so that the numbers are grouped together, and the variables are grouped together.

$$5 \cdot x \cdot 5 \cdot x = (5 \cdot 5)(x \cdot x)$$

Now, because

$$5 \cdot 5 = 25 \quad \text{and} \quad x \cdot x = x^2$$

we can rewrite the expression as

$$(5 \cdot 5)(x \cdot x) = 25x^2$$

Here is what the problem looks like when the steps are shown together:

$$(5x)^2 = (5x)(5x) \quad \text{Definition of exponents}$$
$$= (5 \cdot 5)(x \cdot x) \quad \text{Commutative property of multiplication}$$
$$= 25x^2 \quad \text{Multiplication and definition of exponents}$$

We have shown only the important steps in this summary. We rewrite the expression by (1) applying the definition of exponents to expand it, (2) rearranging the numbers and variables by using the commutative property, and then (3) simplifying by multiplication.

Here are some more examples:

EXAMPLE 1 Expand $(7x)^2$ using the definition of exponents, and then simplify the result.

SOLUTION We begin by writing the expression as $(7x)(7x)$, rearranging the numbers and variables, and then simplifying.

$$(7x)^2 = (7x)(7x) \quad \text{Definition of exponents}$$
$$= (7 \cdot 7)(x \cdot x) \quad \text{Commutative property}$$
$$= 49x^2 \quad \text{Multiplication and definition of exponents}$$ ∎

PRACTICE PROBLEMS

1. Expand $(3x)^2$ using the definition of exponents, and then simplify the result.

Answer
1. $9x^2$

2. Expand and simplify: $(2a)^3$.

EXAMPLE 2 Expand and simplify: $(5a)^3$.

SOLUTION We begin by writing the expression as $(5a)(5a)(5a)$.

$$(5a)^3 = (5a)(5a)(5a) \quad \text{Definition of exponents}$$
$$= (5 \cdot 5 \cdot 5)(a \cdot a \cdot a) \quad \text{Commutative property}$$
$$= 125a^3 \quad 5 \cdot 5 \cdot 5 = 125; a \cdot a \cdot a = a^3$$

∎

3. Expand and simplify: $(7xy)^2$.

EXAMPLE 3 Expand and simplify: $(8xy)^2$.

SOLUTION Proceeding as we have above, we have

$$(8xy)^2 = (8xy)(8xy) \quad \text{Definition of exponents}$$
$$= (8 \cdot 8)(x \cdot x)(y \cdot y) \quad \text{Commutative property}$$
$$= 64x^2y^2 \quad 8 \cdot 8 = 64; x \cdot x = x^2; y \cdot y = y^2$$

∎

4. Simplify: $(3x)^2(7xy)^2$.

EXAMPLE 4 Simplify: $(7x)^2(8xy)^2$.

SOLUTION We begin by applying the definition of exponents.

$$(7x)^2(8xy)^2 = (7x)(7x)(8xy)(8xy)$$
$$= (7 \cdot 7 \cdot 8 \cdot 8)(x \cdot x \cdot x \cdot x)(y \cdot y) \quad \text{Commutative property}$$
$$= 3{,}136x^4y^2$$

∎

5. Simplify: $(5x)^3(2x)^2$.

EXAMPLE 5 Simplify: $(2x)^3(4x)^2$.

SOLUTION Proceeding as we have above, we have

$$(2x)^3(4x)^2 = (2x)(2x)(2x)(4x)(4x)$$
$$= (2 \cdot 2 \cdot 2 \cdot 4 \cdot 4)(x \cdot x \cdot x \cdot x \cdot x)$$
$$= 128x^5$$

∎

Answers
2. $8a^3$
3. $49x^2y^2$
4. $441x^4y^2$
5. $500x^5$

A Glimpse of Algebra Problems

Use the definition of exponents to expand each of the following expressions. Apply the commutative property, and simplify the result in each case.

1. $(6x)^2$

2. $(9x)^2$

3. $(4x)^2$

4. $(10x)^2$

5. $(3a)^3$

6. $(6a)^3$

7. $(2ab)^3$

8. $(5ab)^3$

9. $(9xy)^2$

10. $(5xy)^2$

11. $(5xyz)^2$

12. $(7xyz)^2$

13. $(4x)^2(9xy)^2$

14. $(10x)^2(5xy)^2$

15. $(2x)^2(3x)^2(4x)^2$

16. $(5x)^2(2x)^2(10x)^2$

17. $(2x)^3(5x)^2$

18. $(3x)^3(4x)^2$

19. $(2a)^3(3a)^2(10a)^2$

20. $(3a)^3(2a)^2(10a)^2$

21. $(3xy)^3(4xy)^2$

22. $(2xy)^4(3xy)^2$

23. $(5xyz)^2(2xyz)^4$

24. $(6xyz)^2(3xyz)^3$

25. $(xy)^3(xz)^2(yz)^4$

26. $(xy)^4(xz)^2(yz)^3$

27. $(2a^3b^2)^2(3a^2b^3)^4$

28. $(4a^4b^3)^2(5a^2b^4)^2$

29. $(5x^2y^3)(2x^3y^3)^3$

30. $(8x^2y^2)^2(3x^3y^4)^2$

Fractions and Mixed Numbers

2

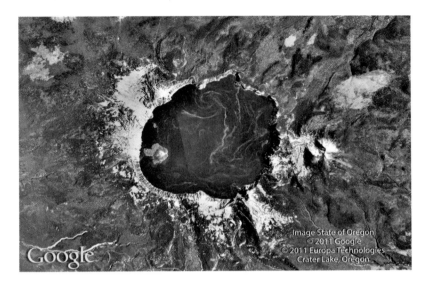

Chapter Outline

2.1 The Meaning and Properties of Fractions

2.2 Prime Numbers, Factors, and Reducing to Lowest Terms

2.3 Multiplication with Fractions, and the Area of a Triangle

2.4 Division with Fractions

2.5 Addition and Subtraction with Fractions

2.6 Mixed-Number Notation

2.7 Multiplication and Division with Mixed Numbers

2.8 Addition and Subtraction with Mixed Numbers

2.9 Combinations of Operations and Complex Fractions

Crater Lake, located in the Cascade Mountain range in Southern Oregon, is 594 meters deep, making it the deepest lake in the United States. Here is a chart showing the depth of Crater Lake and the location of some lakes that are deeper.

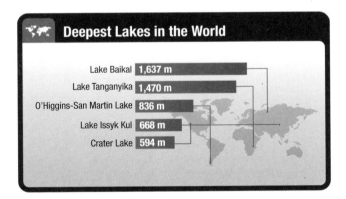

As you can see from the chart, Crater Lake is far from being the deepest lake in the world. We can use fractions to compare the depths of these lakes. For example, Crater Lake is approximately $\frac{2}{5}$ as deep as Lake Tanganyika. In this chapter, we begin our work with fractions to better understand such approximations as this one.

Preview

Key Words	Definition
Numerator	The top number in a fraction
Denominator	The bottom number in a faction
Equivalent Fractions	Fractions representing the same number
Prime Number	Any whole number greater than 1 that has exactly two divisors—1 and itself
Least Common Denominator	The smallest number that is exactly divisible by each denominator in a set of 2 or more fractions

Chapter Outline

2.1 The Meaning and Properties of Fractions
- A Identify the numerator and denominator of a fraction.
- B Identify proper and improper fractions.
- C Write equivalent fractions.
- D Simplify fractions with division.
- E Compare the size of fractions.

2.2 Prime Numbers, Factors, and Reducing to Lowest Terms
- A Identify numbers as prime or composite.
- B Factor a number into the product of prime factors.
- C Write a fraction in lowest terms.
- D Solve applications involving reducing fractions to lowest terms.

2.3 Multiplication with Fractions, and the Area of a Triangle
- A Multiply fractions.
- B Find the area of a triangle.

2.4 Division with Fractions
- A Divide fractions.
- B Simplify order of operations problems involving division of fractions.
- C Solve application problems involving division of fractions.

2.5 Addition and Subtraction with Fractions
- A Add and subtract fractions with a common denominator.
- B Add and subtract fractions with different denominators.

2.6 Mixed-Number Notation
- A Change mixed numbers to improper fractions.
- B Change improper fractions to mixed numbers.

2.7 Multiplication and Division with Mixed Numbers
- A Multiply mixed numbers.
- B Divide mixed numbers.

2.8 Addition and Subtraction with Mixed Numbers
- A Perform addition and subtraction with mixed numbers.
- B Perform subtraction involving borrowing with mixed numbers.

2.9 Combinations of Operations and Complex Fractions
- A Simplify expressions involving fractions and mixed numbers.
- B Simplify complex fractions.

The Meaning and Properties of Fractions

2.1

OBJECTIVES

A Identify the numerator and denominator of a fraction.

B Identify proper and improper fractions.

C Write equivalent fractions.

D Simplify fractions with division.

E Compare the size of fractions.

TICKET TO SUCCESS

Keep these questions in mind as you read through the section. Then respond in your own words and in complete sentences.

1. What is a fraction?
2. What is the difference between an improper fraction and a proper fraction?
3. What makes two fractions equivalent?
4. Briefly explain Property 1 and Property 2 for fractions.

The information in the table below was taken from the website for California Polytechnic State University in San Luis Obispo, CA. The pie chart was created from the table. Both the table and pie chart use fractions to specify how the students at Cal Poly are distributed among the different schools within the university.

CAL POLY ENROLLMENT FOR FALL 2009	
School	**Fraction Of Students**
Agriculture	$\frac{1}{5}$
Architecture and Environmental Design	$\frac{1}{10}$
Business	$\frac{3}{25}$
Engineering	$\frac{7}{25}$
Liberal Arts	$\frac{3}{20}$
Science and Mathematics	$\frac{7}{50}$

Source: California Polytechnic State University

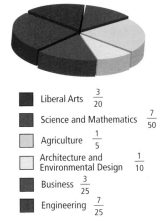

Cal Poly Enrollment for Fall 2009

- Liberal Arts $\frac{3}{20}$
- Science and Mathematics $\frac{7}{50}$
- Agriculture $\frac{1}{5}$
- Architecture and Environmental Design $\frac{1}{10}$
- Business $\frac{3}{25}$
- Engineering $\frac{7}{25}$

Chapter 2 Fractions and Mixed Numbers

From the table, we see that $\frac{1}{5}$ (one-fifth) of the students are enrolled in the School of Agriculture. This means that one out of every five students at Cal Poly is studying agriculture. The fraction $\frac{1}{5}$ tells us we have 1 part of 5 equal parts.

Figure 1 below shows a rectangle that has been divided into equal parts in four different ways. The shaded area for each rectangle is $\frac{1}{2}$ the total area.

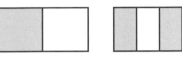

a. $\frac{1}{2}$ is shaded   b. $\frac{2}{4}$ are shaded   c. $\frac{3}{6}$ are shaded   d. $\frac{4}{8}$ are shaded

FIGURE 1

Now that we have an intuitive idea of the meaning of fractions, here are the more formal definitions and vocabulary associated with fractions.

A The Numerator and the Denominator

Definition

A **fraction** is any number that can be put in the form $\frac{a}{b}$ (also sometimes written a/b), where a and b are numbers and b is not 0.

NOTE

Recall from Chapter 1, when we use a letter to represent a number, or a group of numbers, that letter is called a variable. In the definition here, we are restricting the numbers that the variable b can represent to numbers other than 0. As you will see later in the chapter, we do this to avoid writing an expression that would imply division by the number 0.

Some examples of fractions are

$$\frac{1}{2} \qquad \frac{3}{4} \qquad \frac{7}{8} \qquad \frac{9}{5}$$

One-half   Three-fourths   Seven-eighths   Nine-fifths

Definition

For the fraction $\frac{a}{b}$, a and b are called the **terms** of the fraction. More specifically, a is called the **numerator,** and b is called the **denominator.**

$$\text{fraction } \frac{a}{b} \begin{array}{l} \leftarrow \text{numerator} \\ \leftarrow \text{denominator} \end{array}$$

The following example will give you practice identifying the parts of a fraction.

PRACTICE PROBLEMS

1. Name the terms of the fraction $\frac{5}{6}$. Which is the numerator and which is the denominator?

2. Name the numerator and the denominator of the fraction $\frac{x}{3}$.

3. How can the number 9 be considered a fraction?

EXAMPLE 1 The terms of the fraction $\frac{3}{4}$ are 3 and 4. The 3 is called the numerator, and the 4 is called the denominator. ∎

EXAMPLE 2 The numerator of the fraction $\frac{a}{5}$ is a. The denominator is 5. Both a and 5 are called terms. ∎

EXAMPLE 3 The number 7 may also be put in fraction form, because it can be written as $\frac{7}{1}$. In this case, 7 is the numerator and 1 is the denominator. ∎

Answers

1. Terms: 5 and 6; numerator: 5; denominator: 6
2. Numerator: x; denominator: 3
3. Because it can be written $\frac{9}{1}$

B Proper and Improper Fractions

Now let's more closely examine fractions by defining *proper* and *improper* fractions.

> **Definition**
>
> A **proper fraction** is a fraction in which the numerator is less than the denominator. If the numerator is greater than or equal to the denominator, the fraction is called an **improper fraction.**

EXAMPLE 4 The fractions $\frac{3}{4}$, $\frac{1}{8}$, and $\frac{9}{10}$ are all proper fractions, because in each case the numerator is less than the denominator. ∎

EXAMPLE 5 The numbers $\frac{9}{5}$, $\frac{10}{10}$, and 6 are all improper fractions, because in each case the numerator is greater than or equal to the denominator. (Remember that 6 can be written as $\frac{6}{1}$, in which case 6 is the numerator and 1 is the denominator.) ∎

C Equivalent Fractions

Some fractions may look different but they still have the same value. To understand this, we must first explore fractions on the number line. We can give meaning to the fraction $\frac{2}{3}$ by using a number line. If we take that part of the number line from 0 to 1 and divide it into *three equal parts,* we say that we have divided it into *thirds* (see Figure 2). Each of the three segments is $\frac{1}{3}$ (one third) of the whole segment from 0 to 1.

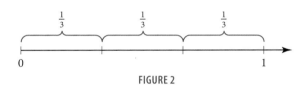

FIGURE 2

Two of these smaller segments together are $\frac{2}{3}$ (two thirds) of the whole segment. And three of them would be $\frac{3}{3}$ (three thirds), or the whole segment, as indicated in Figure 3.

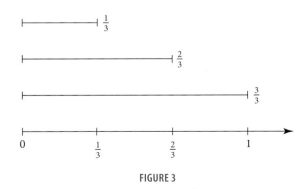

FIGURE 3

4. Which of the following are proper fractions?

$\frac{1}{6}$ $\frac{2}{3}$ $\frac{8}{5}$

5. Which of the following are improper fractions?

$\frac{5}{9}$ $\frac{6}{5}$ $\frac{4}{3}$ 7

> **NOTE**
> Just so you know, there are many ways to give meaning to fractions like $\frac{2}{3}$ other than by using the number line. One popular way is to think of cutting a pie into three equal pieces, as shown below. If you take two of the pieces, you have taken $\frac{2}{3}$ of the pie.
>
>

Answers

4. $\frac{1}{6}, \frac{2}{3}$

5. $\frac{6}{5}, \frac{4}{3}, 7$

Let's do the same thing again with six and twelve equal divisions of the segment from 0 to 1 (see Figure 4).

The same point that we labeled with $\frac{1}{3}$ in Figure 3 is now labeled with $\frac{2}{6}$ and with $\frac{4}{12}$. It must be true then that

$$\frac{4}{12} = \frac{2}{6} = \frac{1}{3}$$

Although these three fractions look different, each names the same point on the number line, as shown in Figure 4. All three fractions have the same value, because they all represent the same number. This also means that these fractions are *equivalent*.

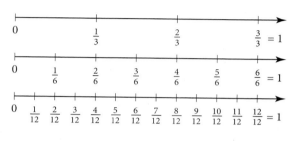

FIGURE 4

Definition

Fractions that represent the same number are said to be **equivalent.**
Equivalent fractions may look different, but they must have the same value.

It is apparent that every fraction has many different representations, each of which is equivalent to the original fraction. The next two properties give us a way of changing the terms of a fraction without changing its value.

Property 1 for Fractions

If a, b, and c are numbers and b and c are not 0, then it is always true that

$$\frac{a}{b} = \frac{a \cdot c}{b \cdot c}$$

In words: If the numerator and the denominator of a fraction are multiplied by the same nonzero number, the resulting fraction is equivalent to the original fraction.

NOTE

Using pie charts again, here is how $\frac{1}{3}$ is equivalent to $\frac{2}{6}$ and to $\frac{4}{12}$. Also how $\frac{2}{3}$ is equivalent to $\frac{4}{6}$ and to $\frac{8}{12}$.

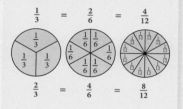

EXAMPLE 6 Write $\frac{3}{4}$ as an equivalent fraction with denominator 20.

SOLUTION The denominator of the original fraction is 4. The fraction we are trying to find must have a denominator of 20. We know that if we multiply 4 by 5, we get 20. Property 1 indicates that we are free to multiply the denominator by 5 so long as we do the same to the numerator.

$$\frac{3}{4} = \frac{3 \cdot \mathbf{5}}{4 \cdot \mathbf{5}} = \frac{15}{20} \quad \text{Multiply numerator and denominator by 5.}$$

The fraction $\frac{15}{20}$ is equivalent to the fraction $\frac{3}{4}$. ∎

6. Write $\frac{2}{3}$ as an equivalent fraction with denominator 12.

Answer

6. $\frac{8}{12}$

2.1 The Meaning and Properties of Fractions

Property 2 for Fractions

If a, b, and c are integers and b and c are not 0, then it is always true that

$$\frac{a}{b} = \frac{a \div c}{b \div c}$$

In words: If the numerator and the denominator of a fraction are divided by the same nonzero number, the resulting fraction is equivalent to the original fraction.

EXAMPLE 7 Write $\frac{10}{12}$ as an equivalent fraction with denominator 6.

SOLUTION If we divide the original denominator 12 by 2, we obtain 6. Property 2 indicates that if we divide both the numerator and the denominator by 2, the resulting fraction will be equivalent to the original fraction.

$$\frac{10}{12} = \frac{10 \div 2}{12 \div 2} = \frac{5}{6}$$

D Fractions and Division

There are two situations involving fractions and the number 1 that occur frequently in mathematics. The first is when the denominator of a fraction is 1. In this case, if we let a represent any number, then

$$\frac{a}{1} = a$$

The second situation occurs when the numerator and the denominator of a fraction are the same nonzero number:

$$\frac{a}{a} = 1$$

EXAMPLE 8 Simplify each expression.

a. $\frac{24}{1}$ b. $\frac{24}{24}$ c. $\frac{48}{24}$ d. $\frac{72}{24}$

SOLUTION In each case, we divide the numerator by the denominator.

a. $\frac{24}{1} = 24$ b. $\frac{24}{24} = 1$ c. $\frac{48}{24} = 2$ d. $\frac{72}{24} = 3$

E Comparing Fractions

When fractions have different denominators, it can be difficult to tell when one fraction is larger or smaller than other. We can compare fractions to see which is larger or smaller when they have the same denominator.

EXAMPLE 9 Write each fraction as an equivalent fraction with denominator 24. Then write them in order from smallest to largest.

$$\frac{5}{8} \quad \frac{5}{6} \quad \frac{3}{4} \quad \frac{2}{3}$$

SOLUTION We begin by writing each fraction as an equivalent fraction with denominator 24.

---

7. Write $\frac{15}{20}$ as an equivalent fraction with denominator 4.

8. Simplify.

a. $\frac{18}{1}$ b. $\frac{18}{18}$

c. $\frac{36}{18}$ d. $\frac{72}{18}$

9. Write each fraction as an equivalent fraction with denominator 12. Then write in order from smallest to largest.

$$\frac{1}{3}, \frac{1}{6}, \frac{1}{4}, \frac{5}{12}$$

Answers

7. $\frac{3}{4}$

8. a. 18 b. 1 c. 2 d. 4

9. $\frac{2}{12}, \frac{3}{12}, \frac{4}{12}, \frac{5}{12}$

$$\frac{5}{8} = \frac{15}{24} \qquad \frac{5}{6} = \frac{20}{24} \qquad \frac{3}{4} = \frac{18}{24} \qquad \frac{2}{3} = \frac{16}{24}$$

Now that they all have the same denominator, the smallest fraction is the one with the smallest numerator and the largest fraction is the one with the largest numerator. Writing them in order from smallest to largest we have

$$\frac{15}{24} < \frac{16}{24} < \frac{18}{24} < \frac{20}{24}$$

or

$$\frac{5}{8} < \frac{2}{3} < \frac{3}{4} < \frac{5}{6}$$ ∎

DESCRIPTIVE STATISTICS

Scatter Diagrams and Line Graphs

The table and bar chart give the daily gain in the price of a certain stock for one week, and stock prices are shown in terms of fractions instead of decimals. Using what we've learned about equivalent fractions, the far right column in the table shows each daily gain with a common denominator of 32. This will help us better understand how we created a bar chart in Figure 5, for instance, how we know $\frac{3}{4}$ is larger than $\frac{9}{16}$.

CHANGE IN STOCK PRICE		
Day	Gain	Common Denominator
Monday	$\frac{3}{4}$	$\frac{24}{32}$
Tuesday	$\frac{9}{16}$	$\frac{18}{32}$
Wednesday	$\frac{3}{32}$	$\frac{3}{32}$
Thursday	$\frac{7}{32}$	$\frac{7}{32}$
Friday	$\frac{1}{16}$	$\frac{2}{32}$

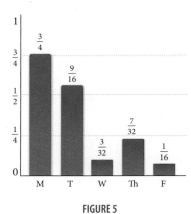

FIGURE 5

Figure 6 below shows another way to visualize the information in the table. It is called a *scatter diagram*. In the scatter diagram, dots are used instead of the bars shown in Figure 5 to represent the gain in stock price for each day of the week. If we connect the dots in Figure 6 with straight lines, we produce the diagram in Figure 7, which is known as a *line graph*.

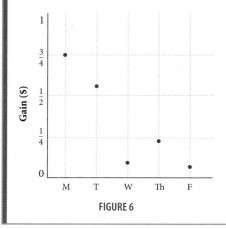

FIGURE 6

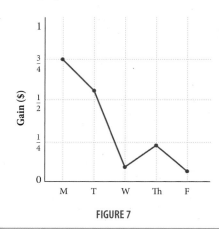

FIGURE 7

Problem Set 2.1

Moving Toward Success

"Don't listen to anyone who tells you that you can't do this or that. That's nonsense. Make up your mind, you'll never use crutches or a stick, then have a go at everything. Go to school, join in all the games you can. Go anywhere you want to. But never, never let them persuade you that things are too difficult or impossible."

—Douglas Bader, 1910–1982, British World War II pilot

1. Rank the following in order of importance to your success in this course:
   a. classroom atmosphere
   b. teaching style
   c. textbook content and format
   d. previous success in math courses
   e. your academic self image
   f. your attitude
   g. any anxiety you may feel about this class
   h. your study habits
2. Which of the above can you control during this course?

A Name the numerator of each fraction. [Examples 1–3]

1. $\dfrac{1}{3}$
2. $\dfrac{1}{4}$
3. $\dfrac{2}{3}$
4. $\dfrac{2}{4}$
5. $\dfrac{x}{8}$
6. $\dfrac{y}{10}$
7. $\dfrac{a}{b}$
8. $\dfrac{x}{y}$

A Name the denominator of each fraction. [Examples 1–3]

9. $\dfrac{2}{5}$
10. $\dfrac{3}{5}$
11. 6
12. 2
13. $\dfrac{a}{12}$
14. $\dfrac{b}{14}$

A Complete the following tables.

15.

Numerator a	Denominator b	Fraction $\dfrac{a}{b}$
3	5	
1		$\dfrac{1}{7}$
	y	$\dfrac{x}{y}$
$x+1$	x	

16.

Numerator a	Denominator b	Fraction $\dfrac{a}{b}$
2	9	
	3	$\dfrac{4}{3}$
1		$\dfrac{1}{x}$
	x	$\dfrac{x}{x+1}$

B [Examples 4,5]

17. For the set of numbers $\left\{\frac{3}{4}, \frac{6}{5}, \frac{12}{3}, \frac{1}{2}, \frac{9}{10}, \frac{20}{10}\right\}$, list all the proper fractions.

18. For the set of numbers $\left\{\frac{1}{8}, \frac{7}{9}, \frac{6}{3}, \frac{18}{6}, \frac{3}{5}, \frac{9}{8}\right\}$, list all the improper fractions.

B Indicate whether each of the following is *True* or *False*.

19. Every whole number greater than 1 can also be expressed as an improper fraction.

20. Some improper fractions are also proper fractions.

C

21. Adding the same number to the numerator and the denominator of a fraction will not change its value.

22. The fractions $\frac{3}{4}$ and $\frac{9}{16}$ are equivalent.

C Divide the numerator and the denominator of each of the following fractions by 2. [Examples 6–7]

23. $\frac{6}{8}$

24. $\frac{10}{12}$

25. $\frac{86}{94}$

26. $\frac{106}{142}$

C Divide the numerator and the denominator of each of the following fractions by 3. [Examples 6–7]

27. $\frac{12}{9}$

28. $\frac{33}{27}$

29. $\frac{39}{51}$

30. $\frac{57}{69}$

C Write each of the following fractions as an equivalent fraction with denominator 6. [Examples 6–7]

31. $\frac{2}{3}$

32. $\frac{1}{2}$

33. $\frac{55}{66}$

34. $\frac{65}{78}$

C Write each of the following fractions as an equivalent fraction with denominator 12. [Examples 6–7]

35. $\frac{2}{3}$

36. $\frac{5}{6}$

37. $\frac{56}{84}$

38. $\frac{143}{156}$

39. One-fourth of the first circle below is shaded. Use the other three circles to show three other ways to shade one-fourth of the circle.

40. The objects below are hexagons, six-sided figures. One-third of the first hexagon is shaded. Shade the other three hexagons to show three other ways to represent one-third.

D Simplify by dividing the numerator by the denominator. [Example 8]

41. $\dfrac{3}{1}$ 42. $\dfrac{3}{3}$ 43. $\dfrac{6}{3}$ 44. $\dfrac{12}{3}$ 45. $\dfrac{37}{1}$ 46. $\dfrac{37}{37}$

47. For each square below, what fraction of the area is given by the shaded region?

a. 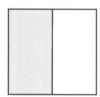   b.   c.   d.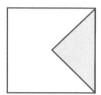

48. For each square below, what fraction of the area is given by the shaded region?

a. 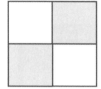   b.   c.   d.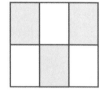

The number line below extends from 0 to 2, with the segment from 0 to 1 and the segment from 1 to 2 each divided into 8 equal parts. Locate each of the following numbers on this number line.

49. $\dfrac{1}{4}$ 50. $\dfrac{1}{8}$ 51. $\dfrac{1}{16}$ 52. $\dfrac{5}{8}$ 53. $\dfrac{3}{4}$

54. $\dfrac{15}{16}$ 55. $\dfrac{3}{2}$ 56. $\dfrac{5}{4}$ 57. $\dfrac{31}{16}$ 58. $\dfrac{15}{8}$

E [Example 9]

59. Write each fraction as an equivalent fraction with denominator 100. Then write the original fractions in order from smallest to largest.

$$\dfrac{3}{10} \quad \dfrac{1}{20} \quad \dfrac{4}{25} \quad \dfrac{2}{5}$$

60. Write each fraction as an equivalent fraction with denominator 30. Then write the original fractions in order from smallest to largest.

$$\dfrac{1}{15} \quad \dfrac{5}{6} \quad \dfrac{7}{10} \quad \dfrac{1}{2}$$

Applying the Concepts

Rainfall Suppose the chart shows the average rainfall for Death Valley in the given months. Use the information to answer the following problems.

61. Write the rainfall for June as an equivalent fraction with a denominator of 100.

62. Write the rainfall for April as an equivalent fraction with a denominator of 200.

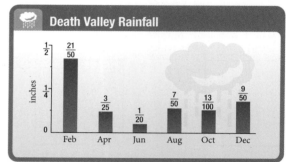

63. **Sending E-mail** The pie chart below shows the fraction of workers who responded to a survey about sending personal e-mail from the office. Use the pie chart to fill in the table.

How Often Workers Send Personal E-mail from the Office	Fraction of Respondents Saying Yes
Never	
1 to 5 times a day	
5 to 10 times a day	
More than 10 times a day	

64. **Surfing the Internet** The pie chart below shows the fraction of workers who responded to a survey about viewing non-work-related sites during working hours. Use the pie chart to fill in the table.

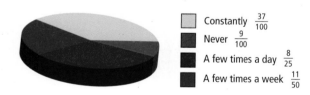

How Often Workers View Non-Work-Related Sites from the Office	Fraction of Respondents Saying Yes
Never	
A few times a week	
A few times a day	
Constantly	

65. Number of Children If there are 3 girls in a family with 5 children, then we say that $\frac{3}{5}$ of the children are girls. If there are 4 girls in a family with 5 children, what fraction of the children are girls?

66. Medical School If 3 out of every 7 people who apply to medical school actually get accepted, what fraction of the people who apply get accepted?

67. Downloaded Songs The iPod Shuffle will hold up to 500 songs. You load 311 of your favorite tunes onto your iPod. Represent the number of songs on your iPod as a fraction of the total number of songs it can hold.

68. Cell Phones In a survey of 1,000 cell phone subscribers, it was determined that 160 subscribers owned more than one cell phone and used different carriers for each phone. Represent the number of cell phone subscribers with more than one carrier as a fraction.

69. College Basketball Recently a basketball team won 19 of the 33 games they played. What fraction represents the number of games won?

70. Score on a Test Your math teacher grades on a point system. You take a test worth 75 points and score a 67 on the test. Represent your score as a fraction.

71. Circles A circle measures 360 degrees, which is commonly written as 360°. The shaded region of each of the circles below is given in degrees. Write a fraction of degrees that represents the area of the shaded region for each of these circles.

a. 90°   b. 45°   c. 180°   d. 270°

72. Carbon Dating All living things contain a small amount of carbon-14, which is radioactive and decays. The half-life of carbon-14 is 5,600 years. During the lifetime of an organism, the carbon-14 is replenished, but after its death the carbon-14 begins to disappear. By measuring the amount left, the amount of time since the death of the organism can be determined with surprising accuracy. The line graph below shows the fraction of carbon-14 remaining after the death of an organism. Use the line graph to complete the table.

CONCENTRATION OF CARBON-14	
Years Since Death of Organism	**Fraction of Carbon-14 Remaining**
0	1
	$\frac{1}{2}$
11,200	
16,800	
	$\frac{1}{16}$

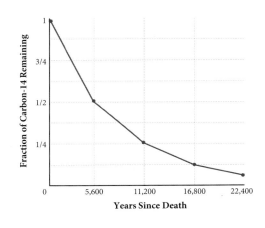

Estimating

73. Which of the following fractions is closest to the number 0?

    a. $\dfrac{1}{2}$ **b.** $\dfrac{1}{3}$ **c.** $\dfrac{1}{4}$ **d.** $\dfrac{1}{5}$

74. Which of the following fractions is closest to the number 0?

    a. $\dfrac{1}{8}$ **b.** $\dfrac{3}{8}$ **c.** $\dfrac{5}{8}$ **d.** $\dfrac{7}{8}$

75. Which of the following fractions is closest to the number 1?

    a. $\dfrac{1}{2}$ **b.** $\dfrac{1}{3}$ **c.** $\dfrac{1}{4}$ **d.** $\dfrac{1}{5}$

76. Which of the following fractions is closest to the number 1?

    a. $\dfrac{1}{8}$ **b.** $\dfrac{3}{8}$ **c.** $\dfrac{5}{8}$ **d.** $\dfrac{7}{8}$

Getting Ready for the Next Section

Multiply.

77. $2 \cdot 2 \cdot 3 \cdot 3 \cdot 3$

78. $2^2 \cdot 3^3$

79. $2^2 \cdot 3 \cdot 5$

80. $2 \cdot 3^2 \cdot 5$

Divide.

81. $12 \div 3$

82. $15 \div 3$

83. $20 \div 4$

84. $24 \div 4$

85. $42 \div 6$

86. $72 \div 8$

87. $102 \div 2$

88. $105 \div 7$

Maintaining Your Skills

The problems below review material covered previously.

Simplify.

89. $3 + 4 \cdot 5$

90. $20 - 8 \cdot 2$

91. $5 \cdot 2^4 - 3 \cdot 4^2$

92. $7 \cdot 8^2 + 2 \cdot 5^2$

93. $4 \cdot 3 + 2(5 - 3)$

94. $6 \cdot 8 + 3(4 - 1)$

95. $18 + 12 \div 4 - 3$

96. $20 + 16 \div 2 - 5$

Prime Numbers, Factors, and Reducing to Lowest Terms

2.2

OBJECTIVES

A Identify numbers as prime or composite.

B Factor a number into the product of prime factors.

C Write a fraction in lowest terms.

D Solve applications involving reducing fractions to lowest terms.

TICKET TO SUCCESS

Keep these questions in mind as you read through the section. Then respond in your own words and in complete sentences.

1. What is a prime number?
2. Why is the number 22 a composite number?
3. Factor 120 into the product of prime factors.
4. What is meant by the phrase "a fraction in lowest possible terms"?

Suppose you and a friend decide to split a medium-sized pizza for lunch. When the pizza is delivered you find that it has been cut into eight equal pieces. If you eat four pieces, you have eaten $\frac{4}{8}$ of the pizza, but you also know that you have eaten $\frac{1}{2}$ of the pizza. The fraction $\frac{4}{8}$ is equivalent to the fraction $\frac{1}{2}$; that is, they both have the same value. The mathematical process we use to rewrite $\frac{4}{8}$ as $\frac{1}{2}$ is called *reducing to lowest terms*. Before we look at that process, we need to define some new terms that will help us understand the components of a fraction in lowest terms. Here is our first one.

A Prime Numbers

> **Definition**
>
> A **prime number** is any whole number greater than 1 that has exactly two divisors—the number 1 and itself. (Remember a number is a divisor of another number if it divides it without a remainder.)

Prime numbers = {2, 3, 5, 7, 11, 13, 17, 19, 23, 29, 31, 37, . . . }

The list goes on indefinitely. Each number in the list has exactly two distinct divisors—the number 1 and itself.

> **Definition**
> Any whole number greater than 1 that is not a prime number is called a **composite number.** A composite number always has at least one divisor other than the number 1 and itself.

EXAMPLE 1 Identify each of the numbers below as either a prime number or a composite number. For those that are composite, give two divisors other than the number 1 or itself.

    a. 43   **b.** 12

SOLUTION **a.** 43 is a prime number, because the only numbers that divide it without a remainder are 43 and 1.

    b. 12 is a composite number, because it can be written as $12 = 4 \cdot 3$, which means that 4 and 3 are divisors of 12. (These are not the only divisors of 12; other divisors are 1, 2, 6, and 12.)

You may have already noticed that the word *divisor* as we are using it here means the same as the word *factor*. A divisor and a factor of a number are the same thing. A number can't be a divisor of another number without also being a factor of it. ∎

B Factoring

Every composite number can be written as the *product of prime factors*. Let's look at the composite number 108. We know we can write 108 as $2 \cdot 54$. The number 2 is a prime number, but 54 is not prime. Because 54 can be written as $2 \cdot 27$, we have

$$108 = 2 \cdot 54$$
$$= 2 \cdot 2 \cdot 27$$

Now the number 27 can be written as $3 \cdot 9$ or $3 \cdot 3 \cdot 3$ (because $9 = 3 \cdot 3$), so

$$108 = 2 \cdot 54$$
$$108 = 2 \cdot 2 \cdot 27$$
$$108 = 2 \cdot 2 \cdot 3 \cdot 9$$
$$108 = 2 \cdot 2 \cdot 3 \cdot 3 \cdot 3$$

This last line is the number 108 written as the product of prime factors. We can use exponents to rewrite the last line:

$$108 = 2^2 \cdot 3^3$$

This process works by writing the original composite number as the product of any two of its factors and then writing any factor that is not prime as the product of any two of its factors. The process is continued until all factors are prime numbers. You do not have to start with the smallest prime factor, as shown above. No matter which factors you start with you will always end up with the same prime factorization of a number.

   There are some "shortcuts" to finding the divisors of a number. For instance, if a number ends in 0 or 5, then it is divisible by 5. If a number ends in an even number (0, 2, 4, 6, or 8), then it is divisible by 2. A number is divisible by 3 if the sum of its digits is divisible by 3. For example, 921 is divisible by 3 because the sum of its digits is $9 + 2 + 1 = 12$, which is divisible by 3.

PRACTICE PROBLEMS

1. Which of the numbers below are prime numbers, and which are composite? For those that are composite, give two divisors other than the number 1 and itself.

37, 39, 51, 59

Answer
1. See Solutions Section.

EXAMPLE 2 Factor 60 into a product of prime factors.

SOLUTION We begin by writing 60 as $6 \cdot 10$ and continue factoring until all factors are prime numbers.

$$60 = 6 \cdot 10$$
$$= 2 \cdot 3 \cdot 2 \cdot 5$$
$$= 2^2 \cdot 3 \cdot 5$$

Notice that if we had started by writing 60 as $3 \cdot 20$, we would have achieved the same result.

$$60 = 3 \cdot 20$$
$$= 3 \cdot 2 \cdot 10$$
$$= 3 \cdot 2 \cdot 2 \cdot 5$$
$$= 2^2 \cdot 3 \cdot 5$$

C Reducing Fractions

We can use the method of factoring numbers into prime factors to help reduce fractions to lowest terms. Here is the definition for *lowest terms*:

Definition
A fraction is said to be in **lowest terms** if the numerator and the denominator have no factors in common other than the number 1.

EXAMPLE 3 The fractions $\frac{1}{2}, \frac{1}{3}, \frac{2}{3}, \frac{1}{4}, \frac{3}{4}, \frac{1}{5}, \frac{2}{5}, \frac{3}{5}$, and $\frac{4}{5}$ are all in lowest terms, because in each case the numerator and the denominator have no factors other than 1 in common. That is, in each fraction, no number other than 1 divides both the numerator and the denominator exactly (without a remainder).

EXAMPLE 4 The fraction $\frac{6}{8}$ is not written in lowest terms, because the numerator and the denominator are both divisible by 2. To write $\frac{6}{8}$ in lowest terms, we apply Property 2 from Section 2.1 and divide both the numerator and the denominator by 2.

$$\frac{6}{8} = \frac{6 \div 2}{8 \div 2} = \frac{3}{4}$$

The fraction $\frac{3}{4}$ is in lowest terms, because 3 and 4 have no factors in common except the number 1.

Reducing a fraction to lowest terms is simply a matter of dividing the numerator and the denominator by all the factors they have in common. We know from property 2 of Section 2.1 that this will produce an equivalent fraction.

EXAMPLE 5 Reduce the fraction $\frac{12}{15}$ to lowest terms by first factoring the numerator and the denominator into prime factors, and then dividing both the numerator and the denominator by the factor they have in common.

SOLUTION The numerator and the denominator factor as follows:

$$12 = 2 \cdot 2 \cdot 3 \quad \text{and} \quad 15 = 3 \cdot 5$$

The factor they have in common is 3. Property 2 tells us that we can divide both terms of a fraction by 3 to produce an equivalent fraction.

2. Factor into a product of prime factors.
   a. 90
   b. 900

3. Which of the following fractions are in lowest terms?
   $\frac{1}{6}, \frac{2}{8}, \frac{15}{25}, \frac{9}{13}$

4. Reduce $\frac{12}{18}$ to lowest terms by dividing the numerator and the denominator by 6.

5. Reduce the fraction $\frac{15}{20}$ to lowest terms by first factoring the numerator and the denominator into prime factors and then dividing out the factors they have in common.

Answers
2. a. $2 \cdot 3^2 \cdot 5$ b. $2^2 \cdot 3^2 \cdot 5^2$
3. $\frac{1}{6}, \frac{9}{13}$
4. $\frac{2}{3}$
5. $\frac{3}{4}$

$$\frac{12}{15} = \frac{2 \cdot 2 \cdot 3}{3 \cdot 5}$$ Factor the numerator and the denominator completely.

$$= \frac{2 \cdot 2 \cdot 3 \div 3}{3 \cdot 5 \div 3}$$ Divide by 3.

$$= \frac{2 \cdot 2}{5} = \frac{4}{5}$$

The fraction $\frac{4}{5}$ is equivalent to $\frac{12}{15}$ and is in lowest terms, because the numerator and the denominator have no factors other than 1 in common.

We can shorten the work involved in reducing fractions to lowest terms by using a slash to indicate division. For example, we can write the above problem as

$$\frac{12}{15} = \frac{2 \cdot 2 \cdot \cancel{3}}{\cancel{3} \cdot 5} = \frac{4}{5}$$

So long as we understand that the slashes through the 3s indicate that we have divided both the numerator and the denominator by 3, we can use this notation.

EXAMPLE 6 Reduce $\frac{6}{42}$ to lowest terms.

SOLUTION We begin by factoring both terms. We then divide through by any factors common to both terms.

$$\frac{6}{42} = \frac{\cancel{2} \cdot \cancel{3}}{\cancel{2} \cdot \cancel{3} \cdot 7} = \frac{1}{7}$$

We must be careful in a problem like this to remember that the slashes indicate division. They are used to indicate that we have divided both the numerator and the denominator by $2 \cdot 3 = 6$. The result of dividing the numerator 6 by $2 \cdot 3$ is 1. It is a very common mistake to call the numerator 0 instead of 1 or to leave the numerator out of the answer.

EXAMPLE 7 Reduce $\frac{4}{40}$ to lowest terms.

SOLUTION $\frac{4}{40} = \frac{\cancel{2} \cdot \cancel{2} \cdot 1}{\cancel{2} \cdot \cancel{2} \cdot 2 \cdot 5} = \frac{1}{10}$

EXAMPLE 8 Reduce $\frac{105}{30}$ to lowest terms.

SOLUTION $\frac{105}{30} = \frac{\cancel{3} \cdot \cancel{5} \cdot 7}{2 \cdot \cancel{3} \cdot \cancel{5}} = \frac{7}{2}$

D Applications

We can now apply what we have learned about reducing fractions to lowest terms to the following real-life problem.

EXAMPLE 9 Laura is having a party. She puts 4 six-packs of soda in a cooler for her guests. At the end of the party, she finds that only 4 sodas have been consumed. What fraction of the sodas are left? Write your answer in lowest terms.

SOLUTION She had 4 six-packs of soda, which is $4(6) = 24$ sodas. Only 4 were consumed at the party, so 20 are left. The fraction of sodas left is

$$\frac{20}{24}$$

Factoring 20 and 24 completely and then dividing out both the factors they have in common gives us

$$\frac{20}{24} = \frac{\cancel{2} \cdot \cancel{2} \cdot 5}{\cancel{2} \cdot \cancel{2} \cdot 2 \cdot 3} = \frac{5}{6}$$

---

6. Reduce to lowest terms.
  a. $\frac{8}{72}$ **b.** $\frac{16}{144}$

NOTE
The slashes in Example 9 indicate that we have divided both the numerator and the denominator by $2 \cdot 2$, which is equal to 4. With some fractions it is apparent at the start what number divides the numerator and the denominator. For instance, you may have recognized that both 20 and 24 in Example 9 are divisible by 4. We can divide both terms by 4 without factoring first, just as we did in Section 2.1. Property 2 guarantees that dividing both terms of a fraction by 4 will produce an equivalent fraction:

$$\frac{20}{24} = \frac{20 \div 4}{24 \div 4} = \frac{5}{6}$$

Reduce each fraction to lowest terms.

7. $\frac{5}{50}$

8. $\frac{120}{25}$

9. Reduce to lowest terms.
  a. $\frac{30}{35}$ **b.** $\frac{300}{350}$

Answers
6. Both are $\frac{1}{9}$.
7. $\frac{1}{10}$
8. $\frac{24}{5}$
9. Both are $\frac{6}{7}$.

Problem Set 2.2

Moving Toward Success

"Inaction breeds doubt and fear. Action breeds confidence and courage. If you want to conquer fear, do not sit home and think about it. Go out and get busy."
—Dale Carnegie, 1888–1955, American lecturer and author

1. What does the phrase "academic self image" mean?
2. Does your level of success achieved in previous math classes have an affect on your success in this one? Explain.

A Identify each of the numbers below as either a prime number or a composite number. For those that are composite, give at least one divisor (factor) other than the number 1 or the number itself. [Example 1]

1. 11
2. 23
3. 105
4. 41
5. 81
6. 50
7. 13
8. 219

B Factor each of the following into the product of prime factors. [Example 2]

9. 12
10. 8
11. 81
12. 210
13. 215
14. 75
15. 15
16. 42

C Reduce each fraction to lowest terms. [Examples 4–8]

17. $\dfrac{5}{10}$
18. $\dfrac{3}{6}$
19. $\dfrac{4}{6}$
20. $\dfrac{4}{10}$

21. $\dfrac{8}{10}$
22. $\dfrac{6}{10}$
23. $\dfrac{36}{20}$
24. $\dfrac{32}{12}$

25. $\dfrac{42}{66}$
26. $\dfrac{36}{60}$
27. $\dfrac{24}{40}$
28. $\dfrac{50}{75}$

29. $\dfrac{14}{98}$
30. $\dfrac{12}{84}$
31. $\dfrac{70}{90}$
32. $\dfrac{80}{90}$

33. $\dfrac{42}{30}$
34. $\dfrac{60}{36}$
35. $\dfrac{18}{90}$
36. $\dfrac{150}{210}$

37. $\dfrac{110}{70}$ **38.** $\dfrac{45}{75}$ **39.** $\dfrac{180}{108}$ **40.** $\dfrac{105}{30}$

41. $\dfrac{96}{108}$ **42.** $\dfrac{66}{84}$ **43.** $\dfrac{126}{165}$ **44.** $\dfrac{210}{462}$

45. $\dfrac{102}{114}$ **46.** $\dfrac{255}{285}$ **47.** $\dfrac{294}{693}$ **48.** $\dfrac{273}{385}$

49. Reduce each fraction to lowest terms.
  a. $\dfrac{6}{51}$ **b.** $\dfrac{6}{52}$ **c.** $\dfrac{6}{54}$ **d.** $\dfrac{6}{56}$ **e.** $\dfrac{6}{57}$

50. Reduce each fraction to lowest terms.
  a. $\dfrac{6}{42}$ **b.** $\dfrac{6}{44}$ **c.** $\dfrac{6}{45}$ **d.** $\dfrac{6}{46}$ **e.** $\dfrac{6}{48}$

51. Reduce each fraction to lowest terms.
  a. $\dfrac{2}{90}$ **b.** $\dfrac{3}{90}$ **c.** $\dfrac{5}{90}$ **d.** $\dfrac{6}{90}$ **e.** $\dfrac{9}{90}$

52. Reduce each fraction to lowest terms.
  a. $\dfrac{3}{105}$ **b.** $\dfrac{5}{105}$ **c.** $\dfrac{7}{105}$ **d.** $\dfrac{15}{105}$ **e.** $\dfrac{21}{105}$

53. The answer to each problem below is wrong. Give the correct answer in lowest terms.
  a. $\dfrac{5}{15} = \dfrac{\cancel{5}}{3 \cdot \cancel{5}} = \dfrac{0}{3}$
  b. $\dfrac{5}{6} = \dfrac{3 + \cancel{2}}{4 + \cancel{2}} = \dfrac{3}{4}$
  c. $\dfrac{6}{30} = \dfrac{\cancel{2} \cdot \cancel{3}}{\cancel{2} \cdot \cancel{3} \cdot 5} = 5$

54. The answer to each problem below is wrong. Give the correct answer in lowest terms.
  a. $\dfrac{10}{20} = \dfrac{7 + \cancel{3}}{17 + \cancel{3}} = \dfrac{7}{17}$
  b. $\dfrac{9}{36} = \dfrac{\cancel{3} \cdot \cancel{3}}{2 \cdot 2 \cdot \cancel{3} \cdot \cancel{3}} = \dfrac{0}{4}$
  c. $\dfrac{4}{12} = \dfrac{\cancel{2} \cdot \cancel{2}}{\cancel{2} \cdot \cancel{2} \cdot 3} = 3$

55. Which of the fractions $\dfrac{6}{8}, \dfrac{15}{20}, \dfrac{9}{16}, $ and $\dfrac{21}{28}$ does not reduce to $\dfrac{3}{4}$?

56. Which of the fractions $\dfrac{4}{9}, \dfrac{10}{15}, \dfrac{8}{12}, $ and $\dfrac{6}{12}$ does not reduce to $\dfrac{2}{3}$?

The number line below extends from 0 to 2, with the segment from 0 to 1 and the segment from 1 to 2 each divided into 8 equal parts. Locate each of the following numbers on this number line.

57. $\dfrac{1}{2}, \dfrac{2}{4}, \dfrac{4}{8},$ and $\dfrac{8}{16}$
58. $\dfrac{3}{2}, \dfrac{6}{4}, \dfrac{12}{8},$ and $\dfrac{24}{16}$
59. $\dfrac{5}{4}, \dfrac{10}{8},$ and $\dfrac{20}{16}$
60. $\dfrac{1}{4}, \dfrac{2}{8},$ and $\dfrac{4}{16}$

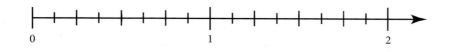

Applying the Concepts

61. Tower Heights The Eiffel Tower is 1,060 feet tall and the Stratosphere Tower in Las Vegas is 1,150 feet tall. Write the height of the Eiffel tower over the height of the Stratosphere Tower and then reduce to lowest terms.

62. Relief Pitchers The chart below shows the number of saves by major league pitchers as of the 2010 season. Write the saves for Billy Wagner over the number of saves for Trevor Hoffman, and then reduce to lowest terms.

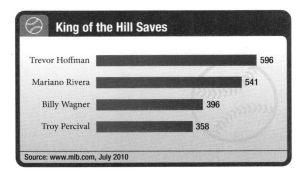

63. Hours and Minutes There are 60 minutes in 1 hour. What fraction of an hour is 20 minutes? Write your answer in lowest terms.

64. Final Exam Suppose 33 people took the final exam in a math class. If 11 people got an A on the final exam, what fraction of the students did not get an A on the exam? Write your answer in lowest terms.

65. Driving Distractions Many of us focus our attention on things other than driving when we are behind the wheel of our car. In a survey of 150 drivers, it was noted that 48 drivers spend time reading or writing while they are driving. Represent the number of drivers who spend time reading or writing while driving as a fraction in lowest terms.

66. Watching Television According to the U.S. Census Bureau, it is estimated that the average person watches 4 hours of TV each day. Represent the number of hours of TV watched each day as a fraction in lowest terms.

67. Pizza You and a friend are sharing a pizza cut into 8 equal slices. Suppose you and your friend each eat 3 slices. What fraction represents the total number of slices eaten? Write your answer in lowest terms.

68. Gasoline Tax Suppose a gallon of regular gas costs $3.99, and 54 cents of this goes to pay state gas taxes. What fractional part of the cost of a gallon of gas goes to state taxes? Write your answer in lowest terms.

69. On-Time Record A random check of Delta airline flights for one month showed that of the 350 flights scheduled, 185 left on time. Represent the number of on time flights as a fraction in lowest terms.

70. Internet Users Based on the most recent data available, there are approximately 1,800,000,000 internet users in the world. Africa makes up about 90,000,000 of this total. Represent the number of internet users in Africa as a fraction of the total in lowest terms.

Nutrition The nutrition labels below are from two different granola bars.

71. What fraction of the calories in Bar 1 comes from fat?

72. What fraction of the calories in Bar 2 comes from fat?

73. What fraction of the total fat in Bar 1 is from saturated fat?

74. What fraction of the total carbohydrates in Bar 1 is from sugar?

GRANOLA BAR 1

Nutrition Facts
Serving Size 2 bars (47g)
Servings Per Container: 6

Amount Per Serving	
Calories 210	Calories from fat 70

	% Daily Value*
Total Fat 8g	12%
Saturated Fat 1g	5%
Cholesterol 0mg	0%
Sodium 150mg	6%
Total Carbohydrate 32g	11%
Fiber 2g	10%
Sugars 12g	
Protein 4g	

*Percent Daily Values are based on a 2,000 calorie diet. Your daily values may be higher or lower depending on your calorie needs.

GRANOLA BAR 2

Nutrition Facts
Serving Size 1 bar (21g)
Servings Per Container: 8

Amount Per Serving	
Calories 80	Calories from fat 15

	% Daily Value*
Total Fat 1.5g	2%
Saturated Fat 0g	0%
Cholesterol 0mg	0%
Sodium 60mg	3%
Total Carbohydrate 16g	5%
Fiber 1g	4%
Sugars 5g	
Protein 2g	

*Percent Daily Values are based on a 2,000 calorie diet. Your daily values may be higher or lower depending on your calorie needs.

Getting Ready for the Next Section

Multiply.

75. $1 \cdot 3 \cdot 1$

76. $2 \cdot 4 \cdot 5$

77. $3 \cdot 5 \cdot 3$

78. $1 \cdot 4 \cdot 1$

79. $5 \cdot 5 \cdot 1$

80. $6 \cdot 6 \cdot 2$

Factor into prime factors.

81. 60

82. 72

83. $15 \cdot 4$

84. $8 \cdot 9$

Expand and multiply.

85. 3^2

86. 4^2

87. 5^2

88. 6^2

Maintaining Your Skills

Simplify.

89. $16 - 8 + 4$

90. $16 - 4 + 8$

91. $24 - 14 + 8$

92. $24 - 16 + 6$

93. $36 - 6 + 12$

94. $36 - 9 + 20$

95. $48 - 12 + 17$

96. $48 - 13 + 15$

2.3 Multiplication with Fractions, and the Area of a Triangle

OBJECTIVES

A Multiply fractions.

B Find the area of a triangle.

TICKET TO SUCCESS

Keep these questions in mind as you read through the section. Then respond in your own words and in complete sentences.

1. Briefly explain in words and symbols the rule for multiplying fractions.
2. When we multiply the fractions $\frac{3}{5}$ and $\frac{2}{7}$, the numerator in the answer will be what number?
3. True or false? Reducing to lowest terms before you multiply two fractions will give the same answer as if you were to reduce after you multiply.
4. How do you use fractions to find the area of a triangle with base x and height y?

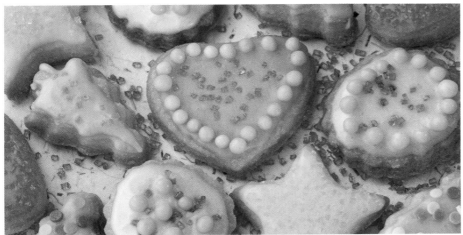

A cookie recipe calls for $\frac{3}{4}$ cup of flour. If you are making only $\frac{1}{2}$ the recipe, how much flour do you use? This question can be answered by multiplying $\frac{1}{2}$ and $\frac{3}{4}$. Here is the problem written with numbers:

$$\frac{1}{2} \cdot \frac{3}{4} = \frac{3}{8}$$

As you can see from this example, to multiply two fractions, we multiply the numerators and then multiply the denominators. We begin this section with the rule for multiplication of fractions.

NOTE You may wonder why we did not divide the amount needed by 2. In fact, we did. Dividing by 2 is the same as multiplying by $\frac{1}{2}$.

A Multiplying Fractions

> **Rule** Product of Two Fractions
>
> If a, b, c, and d represent any numbers and b and d are not zero, then
>
> $$\frac{a}{b} \cdot \frac{c}{d} = \frac{a \cdot c}{b \cdot d}$$
>
> *In words:* The product of two fractions is the fraction whose numerator is the product of the two numerators and whose denominator is the product of the two denominators.

Let's practice this rule now.

Chapter 2 Fractions and Mixed Numbers

PRACTICE PROBLEMS

1. Multiply: $\dfrac{2}{3} \cdot \dfrac{5}{9}$

EXAMPLE 1 Multiply: $\dfrac{3}{5} \cdot \dfrac{2}{7}$.

SOLUTION Using our rule for multiplication, we multiply the numerators and multiply the denominators.

$$\frac{3}{5} \cdot \frac{2}{7} = \frac{3 \cdot 2}{5 \cdot 7} = \frac{6}{35}$$

The product of $\dfrac{3}{5}$ and $\dfrac{2}{7}$ is the fraction $\dfrac{6}{35}$. The numerator 6 is the product of 3 and 2, and the denominator 35 is the product of 5 and 7. ∎

EXAMPLE 2 Multiply: $\dfrac{3}{8} \cdot 5$.

SOLUTION The number 5 can be written as $\dfrac{5}{1}$. That is, 5 can be considered a fraction with numerator 5 and denominator 1. Writing 5 this way enables us to apply the rule for multiplying fractions.

2. Multiply: $\dfrac{2}{5} \cdot 7$

$$\frac{3}{8} \cdot 5 = \frac{3}{8} \cdot \frac{5}{1}$$
$$= \frac{3 \cdot 5}{8 \cdot 1}$$
$$= \frac{15}{8}$$
∎

EXAMPLE 3 Multiply: $\dfrac{1}{2}\left(\dfrac{3}{4} \cdot \dfrac{1}{5}\right)$.

SOLUTION We find the product inside the parentheses first and then multiply the result by $\dfrac{1}{2}$.

3. Multiply: $\dfrac{1}{3}\left(\dfrac{4}{5} \cdot \dfrac{1}{3}\right)$

$$\frac{1}{2}\left(\frac{3}{4} \cdot \frac{1}{5}\right) = \frac{1}{2}\left(\frac{3 \cdot 1}{4 \cdot 5}\right)$$
$$= \frac{1}{2}\left(\frac{3}{20}\right)$$
$$= \frac{1 \cdot 3}{2 \cdot 20}$$
$$= \frac{3}{40}$$
∎

The properties of multiplication that we developed in Chapter 1 for whole numbers apply to fractions as well. That is, if a, b, and c are fractions, then

$a \cdot b = b \cdot a$ Multiplication with fractions is commutative.

$a \cdot (b \cdot c) = (a \cdot b) \cdot c$ Multiplication with fractions is associative.

To demonstrate the associative property for fractions, let's do Example 3 again, but this time we will apply the associative property first.

$$\frac{1}{2}\left(\frac{3}{4} \cdot \frac{1}{5}\right) = \left(\frac{1}{2} \cdot \frac{3}{4}\right) \cdot \frac{1}{5} \quad \text{Associative property}$$
$$= \left(\frac{1 \cdot 3}{2 \cdot 4}\right) \cdot \frac{1}{5}$$
$$= \left(\frac{3}{8}\right) \cdot \frac{1}{5}$$
$$= \frac{3 \cdot 1}{8 \cdot 5}$$
$$= \frac{3}{40}$$

Answers
1. $\dfrac{10}{27}$
2. $\dfrac{14}{5}$
3. $\dfrac{4}{45}$

The result is identical to that of Example 3.

2.3 Multiplication with Fractions, and the Area of a Triangle

Here is another example that involves the associative property. Problems like this will be useful when we solve equations.

EXAMPLE 4 Multiply: $\frac{1}{3}\left(\frac{3}{5} \cdot \frac{1}{2}\right)$.

SOLUTION

$$\frac{1}{3}\left(\frac{3}{5} \cdot \frac{1}{2}\right) = \left(\frac{1}{3} \cdot \frac{3}{5}\right) \cdot \frac{1}{2}$$
$$= \left(\frac{1 \cdot 3}{3 \cdot 5}\right) \cdot \frac{1}{2}$$
$$= \frac{3}{15} \cdot \frac{1}{2}$$
$$= \frac{3 \cdot 1}{15 \cdot 2}$$
$$= \frac{3}{30}$$
$$= \frac{1}{10}$$

The answers to all the examples so far in this section have been in lowest terms. Let's see what happens when we multiply two fractions to get a product that is not in lowest terms.

EXAMPLE 5 Multiply: $\frac{15}{8} \cdot \frac{4}{9}$.

SOLUTION Multiplying the numerators and multiplying the denominators, we have

$$\frac{15}{8} \cdot \frac{4}{9} = \frac{15 \cdot 4}{8 \cdot 9}$$
$$= \frac{60}{72}$$

The product is $\frac{60}{72}$, which can be reduced to lowest terms by factoring 60 and 72 and then dividing out any factors they have in common.

$$\frac{60}{72} = \frac{\not{2} \cdot \not{2} \cdot \not{3} \cdot 5}{\not{2} \cdot \not{2} \cdot 2 \cdot \not{3} \cdot 3}$$
$$= \frac{5}{6}$$

We can actually save ourselves some time by factoring before we multiply. Here's how it is done:

$$\frac{15}{8} \cdot \frac{4}{9} = \frac{15 \cdot 4}{8 \cdot 9}$$
$$= \frac{(3 \cdot 5) \cdot (2 \cdot 2)}{(2 \cdot 2 \cdot 2) \cdot (3 \cdot 3)}$$
$$= \frac{\not{3} \cdot 5 \cdot \not{2} \cdot \not{2}}{\not{2} \cdot \not{2} \cdot 2 \cdot \not{3} \cdot 3}$$
$$= \frac{5}{6}$$

The result is the same in both cases. Reducing to lowest terms before we multiply actually takes less time. ∎

Here are some additional examples:

4. Multiply: $\frac{1}{4}\left(\frac{2}{3} \cdot \frac{1}{2}\right)$

5. Multiply.
  a. $\frac{12}{25} \cdot \frac{5}{6}$
  b. $\frac{12}{25} \cdot \frac{50}{60}$

Answers
4. $\frac{1}{12}$
5. Both are $\frac{2}{5}$.

6. Multiply.

a. $\dfrac{8}{3} \cdot \dfrac{9}{24}$

b. $\dfrac{8}{30} \cdot \dfrac{90}{24}$

NOTE
Although $\dfrac{2}{1}$ is in lowest terms, it is still simpler to write the answer as just 2. We will always do this when the denominator is the number 1.

7. Multiply: $\dfrac{3}{4} \cdot \dfrac{8}{3} \cdot \dfrac{1}{6}$.

Apply the definition of exponents, and then multiply.

8. $\left(\dfrac{2}{3}\right)^2$

9. a. $\left(\dfrac{3}{4}\right)^2 \cdot \dfrac{1}{2}$

b. $\left(\dfrac{2}{3}\right)^3 \cdot \dfrac{9}{8}$

Answers
6. Both are 1.
7. $\dfrac{1}{3}$
8. $\dfrac{4}{9}$
9. a. $\dfrac{9}{32}$ **b.** $\dfrac{1}{3}$

Chapter 2 Fractions and Mixed Numbers

EXAMPLE 6 Multiply: $\dfrac{9}{2}\left(\dfrac{8}{18}\right)$.

SOLUTION
$$\dfrac{9}{2}\left(\dfrac{8}{18}\right) = \dfrac{9 \cdot 8}{2 \cdot 18}$$
$$= \dfrac{(3 \cdot 3) \cdot (2 \cdot 2 \cdot 2)}{2 \cdot (2 \cdot 3 \cdot 3)}$$
$$= \dfrac{\cancel{3} \cdot \cancel{3} \cdot \cancel{2} \cdot \cancel{2} \cdot 2}{\cancel{2} \cdot \cancel{2} \cdot \cancel{3} \cdot \cancel{3}}$$
$$= \dfrac{2}{1}$$
$$= 2$$

EXAMPLE 7 Multiply: $\dfrac{2}{3} \cdot \dfrac{6}{5} \cdot \dfrac{5}{8}$.

SOLUTION
$$\dfrac{2}{3} \cdot \dfrac{6}{5} \cdot \dfrac{5}{8} = \dfrac{2 \cdot 6 \cdot 5}{3 \cdot 5 \cdot 8}$$
$$= \dfrac{2 \cdot (2 \cdot 3) \cdot 5}{3 \cdot 5 \cdot (2 \cdot 2 \cdot 2)}$$
$$= \dfrac{\cancel{2} \cdot \cancel{2} \cdot \cancel{3} \cdot \cancel{5}}{\cancel{3} \cdot \cancel{5} \cdot \cancel{2} \cdot \cancel{2} \cdot 2}$$
$$= \dfrac{1}{2}$$

In Chapter 1, we did some work with exponents. We can extend our work with exponents to include fractions, as the following examples indicate.

EXAMPLE 8 Expand and multiply: $\left(\dfrac{3}{4}\right)^2$.

SOLUTION
$$\left(\dfrac{3}{4}\right)^2 = \left(\dfrac{3}{4}\right)\left(\dfrac{3}{4}\right)$$
$$= \dfrac{3 \cdot 3}{4 \cdot 4}$$
$$= \dfrac{9}{16}$$

EXAMPLE 9 Expand and multiply: $\left(\dfrac{5}{6}\right)^2 \cdot \dfrac{1}{2}$.

SOLUTION
$$\left(\dfrac{5}{6}\right)^2 \cdot \dfrac{1}{2} = \dfrac{5}{6} \cdot \dfrac{5}{6} \cdot \dfrac{1}{2}$$
$$= \dfrac{5 \cdot 5 \cdot 1}{6 \cdot 6 \cdot 2}$$
$$= \dfrac{25}{72}$$

The word *of* used in connection with fractions indicates multiplication. If we want to find $\dfrac{1}{2}$ of $\dfrac{2}{3}$, then what we do is multiply $\dfrac{1}{2}$ and $\dfrac{2}{3}$.

2.3 Multiplication with Fractions, and the Area of a Triangle

EXAMPLE 10 Find $\frac{1}{2}$ of $\frac{2}{3}$.

SOLUTION Knowing the word *of*, as used here, indicates multiplication, we have

$$\frac{1}{2} \text{ of } \frac{2}{3} = \frac{1}{2} \cdot \frac{2}{3}$$
$$= \frac{1 \cdot \cancel{2}}{\cancel{2} \cdot 3}$$
$$= \frac{1}{3}$$

This seems to make sense. Logically, $\frac{1}{2}$ of $\frac{2}{3}$ should be $\frac{1}{3}$, as Figure 1 shows.

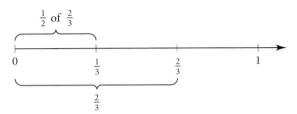

FIGURE 1

10. a. Find $\frac{2}{3}$ of $\frac{1}{2}$.
b. Find $\frac{3}{5}$ of 15.

> **NOTE**
> As you become familiar with multiplying fractions, you may notice shortcuts that reduce the number of steps in the problems. It's okay to use these shortcuts if you understand why they work and are consistently getting correct answers. If you are using shortcuts and not consistently getting correct answers, then go back to showing all the work until you completely understand the process.

EXAMPLE 11 What is $\frac{3}{4}$ of 12?

SOLUTION Again, *of* means multiply.

$$\frac{3}{4} \text{ of } 12 = \frac{3}{4}(12)$$
$$= \frac{3}{4}\left(\frac{12}{1}\right)$$
$$= \frac{3 \cdot 12}{4 \cdot 1}$$
$$= \frac{3 \cdot \cancel{2} \cdot \cancel{2} \cdot 3}{\cancel{2} \cdot \cancel{2} \cdot 1}$$
$$= \frac{9}{1}$$
$$= 9$$

11. a. What is $\frac{2}{3}$ of 12?
b. What is $\frac{2}{3}$ of 120?

B The Area of a Triangle

FACTS FROM GEOMETRY **The Area of a Triangle**
The formula for the area of a triangle is one application of multiplication with fractions. Figure 2 shows a triangle with base *b* and height *h*. Below the triangle is the formula for its area. As you can see, it is a product containing the fraction $\frac{1}{2}$.

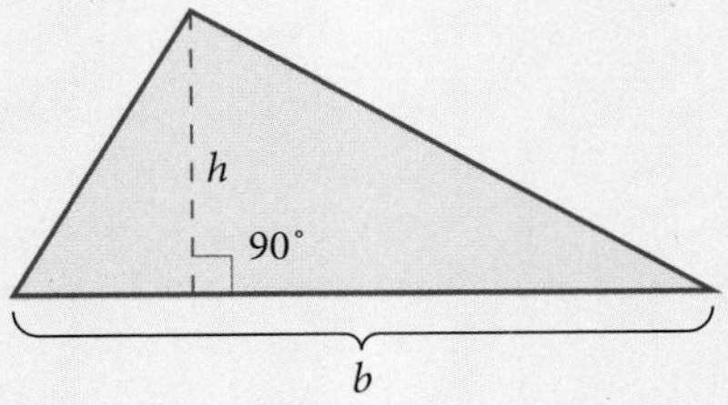

Area = $\frac{1}{2}$ (base)(height)

$A = \frac{1}{2} bh$

FIGURE 2

> **NOTE**
> The height *h* of a triangle is made by drawing a straight line from the apex of the triangle to its base. This line creates a 90° angle with the base, otherwise known as a right angle.

Answers
10. a. $\frac{1}{3}$ **b.** 9
11. a. 8 **b.** 80

12. Find the area of the triangle below.

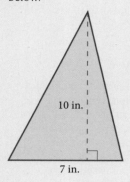

NOTE
How did we get in.² as the final units in Example 12? In this problem,
$A = \frac{1}{2}bh$
$= \frac{1}{2} \cdot 10 \text{ inches} \cdot 7 \text{ inches}$
$= 5 \text{ in.} \cdot 7 \text{ in.} = 35 \text{ in.}^2$

13. Find the total area enclosed by the figure.

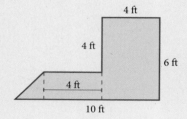

NOTE
This is just a reminder about unit notation. In Example 12, we wrote our final units as **in.²** but could have just as easily written them as **sq in**. In Example 13, we wrote our final units as **sq ft** but could have just as easily written them as **ft²**.

EXAMPLE 12 Find the area of the triangle.

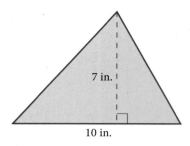

SOLUTION Applying the formula for the area of a triangle, we have
$$A = \frac{1}{2}bh = \frac{1}{2} \cdot 10 \cdot 7 = 5 \cdot 7 = 35 \text{ in.}^2$$

EXAMPLE 13 Find the area of the figure.

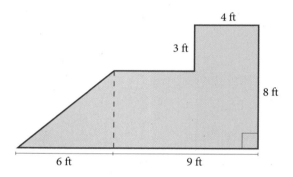

SOLUTION We divide the figure into three parts and then find the area of each part (see Figure 3). The area of the whole figure is the sum of the areas of its parts.

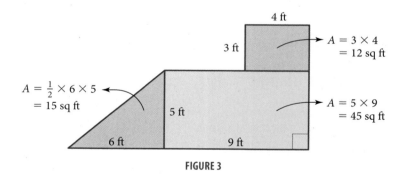

FIGURE 3

Total area = 12 + 45 + 15
= 72 sq ft

Answers
12. 35 in²
13. 34 ft²

Problem Set 2.3

Moving Toward Success

"You can teach a student a lesson for a day; but if you can teach him to learn by creating curiosity, he will continue the learning process as long as he lives."
— Clay P. Bedford, 1903–1991, American industrialist and businessman

1. What is your attitude toward this course?
2. Is this attitude helping or hindering your success in this course?

A Find each of the following products. (Multiply.) [Examples 1–7]

1. $\dfrac{2}{3} \cdot \dfrac{4}{5}$
2. $\dfrac{5}{6} \cdot \dfrac{7}{4}$
3. $\dfrac{1}{2} \cdot \dfrac{7}{4}$
4. $\dfrac{3}{5} \cdot \dfrac{4}{7}$
5. $\dfrac{5}{3} \cdot \dfrac{3}{5}$
6. $\dfrac{4}{7} \cdot \dfrac{7}{4}$

7. $\dfrac{3}{4} \cdot 9$
8. $\dfrac{2}{3} \cdot 5$
9. $\dfrac{6}{7}\left(\dfrac{7}{6}\right)$
10. $\dfrac{2}{9}\left(\dfrac{9}{2}\right)$
11. $\dfrac{1}{2} \cdot \dfrac{1}{3} \cdot \dfrac{1}{4}$
12. $\dfrac{2}{3} \cdot \dfrac{4}{5} \cdot \dfrac{1}{3}$

13. $\dfrac{2}{5} \cdot \dfrac{3}{5} \cdot \dfrac{4}{5}$
14. $\dfrac{1}{4} \cdot \dfrac{3}{4} \cdot \dfrac{3}{4}$
15. $\dfrac{3}{2} \cdot \dfrac{5}{2} \cdot \dfrac{7}{2}$
16. $\dfrac{4}{3} \cdot \dfrac{5}{3} \cdot \dfrac{7}{3}$

A Complete the following tables.

17.

First Number x	Second Number y	Their Product xy
$\dfrac{1}{2}$	$\dfrac{2}{3}$	
$\dfrac{2}{3}$	$\dfrac{3}{4}$	
$\dfrac{3}{4}$	$\dfrac{4}{5}$	
$\dfrac{5}{a}$	$\dfrac{a}{6}$	

18.

First Number x	Second Number y	Their Product xy
12	$\dfrac{1}{2}$	
12	$\dfrac{1}{3}$	
12	$\dfrac{1}{4}$	
12	$\dfrac{1}{6}$	

19.

First Number x	Second Number y	Their Product xy
$\frac{1}{2}$	30	
$\frac{1}{5}$	30	
$\frac{1}{6}$	30	
$\frac{1}{15}$	30	

20.

First Number x	Second Number y	Their Product xy
$\frac{1}{3}$	$\frac{3}{5}$	
$\frac{3}{5}$	$\frac{5}{7}$	
$\frac{5}{7}$	$\frac{7}{9}$	
$\frac{7}{b}$	$\frac{b}{11}$	

A Multiply each of the following. Be sure all answers are written in lowest terms. [Examples 1–7]

21. $\frac{9}{20} \cdot \frac{4}{3}$ **22.** $\frac{135}{16} \cdot \frac{2}{45}$ **23.** $\frac{3}{4} \cdot 12$ **24.** $\frac{3}{4} \cdot 20$

25. $\frac{1}{3}(3)$ **26.** $\frac{1}{5}(5)$ **27.** $\frac{2}{5} \cdot 20$ **28.** $\frac{3}{5} \cdot 15$

29. $\frac{72}{35} \cdot \frac{55}{108} \cdot \frac{7}{110}$ **30.** $\frac{32}{27} \cdot \frac{72}{49} \cdot \frac{1}{40}$

A Expand and simplify each of the following. [Examples 8, 9]

31. $\left(\frac{2}{3}\right)^2$ **32.** $\left(\frac{3}{5}\right)^2$ **33.** $\left(\frac{3}{4}\right)^2$ **34.** $\left(\frac{2}{7}\right)^2$

35. $\left(\frac{1}{2}\right)^2$ **36.** $\left(\frac{1}{3}\right)^2$ **37.** $\left(\frac{2}{3}\right)^3$ **38.** $\left(\frac{3}{5}\right)^3$

39. $\left(\frac{3}{4}\right)^2 \cdot \frac{8}{9}$ **40.** $\left(\frac{5}{6}\right)^2 \cdot \frac{12}{15}$ **41.** $\left(\frac{1}{2}\right)^2 \left(\frac{3}{5}\right)^2$ **42.** $\left(\frac{3}{8}\right)^2 \left(\frac{4}{3}\right)^2$

43. $\left(\frac{1}{2}\right)^2 \cdot 8 + \left(\frac{1}{3}\right)^2 \cdot 9$ **44.** $\left(\frac{2}{3}\right)^2 \cdot 9 + \left(\frac{1}{2}\right)^2 \cdot 4$

A [Examples 10,11]

45. Find $\frac{3}{8}$ of 64.

46. Find $\frac{2}{3}$ of 18.

47. What is $\frac{1}{3}$ of the sum of 8 and 4?

48. What is $\frac{3}{5}$ of the sum of 8 and 7?

49. Find $\frac{1}{2}$ of $\frac{3}{4}$ of 24.

50. Find $\frac{3}{5}$ of $\frac{1}{3}$ of 15.

Find the mistakes in Problems 51 and 52. Correct the right-hand side of each one.

51. $\frac{1}{2} \cdot \frac{3}{5} = \frac{4}{10}$

52. $\frac{2}{7} \cdot \frac{3}{5} = \frac{5}{35}$

53. a. Complete the following table.

Number x	Square x^2
1	
2	
3	
4	
5	
6	
7	
8	

b. Using the results of part a, fill in the blank in the following statement:
For numbers larger than 1, the square of the number is _____ than the number.

54. a. Complete the following table.

Number x	Square x^2
$\frac{1}{2}$	
$\frac{1}{3}$	
$\frac{1}{4}$	
$\frac{1}{5}$	
$\frac{1}{6}$	
$\frac{1}{7}$	
$\frac{1}{8}$	

b. Using the results of part a, fill in the blank in the following statement:
For numbers between 0 and 1, the square of the number is _____ than the number.

B [Examples 12, 13]

55. Find the area of the triangle with base 19 inches and height 14 inches.

56. Find the area of the triangle with base 13 inches and height 8 inches.

57. The base of a triangle is $\frac{4}{3}$ feet and the height is $\frac{2}{3}$ feet. Find the area.

58. The base of a triangle is $\frac{8}{7}$ feet and the height is $\frac{14}{5}$ feet. Find the area.

Find the area of each figure.

59.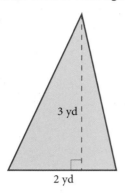

60.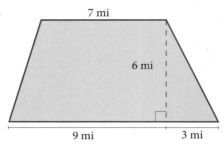

61.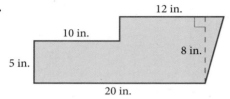

62.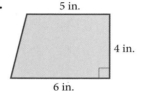

Applying the Concepts

63. Rainfall The chart shows the average rainfall for Death Valley in the given months. Use this chart to answer the questions below.

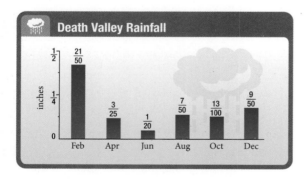

a. How many inches of rain is 5 times the average for December?

b. How many inches of rain is 7 times the average for August?

c. How many inches of rain is 20 times the average for June?

64. Lake Depths The chart from this chapter's introduction shows the depths of some of the deepest lakes in the world. Use the chart to answer the questions below.

a. How many meters is $\frac{1}{2}$ the depth of Lake Tanganyika?

b. How many meters is $\frac{3}{4}$ the depth of Lake Issyk Kul?

c. How many meters is $\frac{2}{3}$ the depth of Crater Lake?

65. Hot Air Balloon Aerostar International makes a hot air balloon called the Rally 105 that has a volume of 105,400 cubic feet. Another balloon, the Rally 126, was designed with a volume that is approximately $\frac{6}{5}$ the volume of the Rally 105. Find the volume of the Rally 126 to the nearest hundred cubic feet.

66. Bicycle Safety The National Safe Kids Campaign and Bell Sports sponsored a study that surveyed 8,159 children ages 5 to 14 who were riding bicycles. Approximately $\frac{2}{5}$ of the children were wearing helmets, and of those, only $\frac{13}{20}$ were wearing the helmets correctly. About how many of the children were wearing helmets correctly?

67. Health Care According to a study reported on MSNBC, almost one-third of people diagnosed with diabetes don't seek proper medical care. If there are 12 million Americans with diabetes, about how many of them are seeking proper medical care?

68. Working Students Studies indicate that approximately $\frac{3}{4}$ of all undergraduate college students work while attending school. A local community college has a student enrollment of 8,500 students. How many of these students work while attending college?

69. Cigarette Tax In a recent survey of 1,410 adults, it was determined that $\frac{3}{5}$ of those surveyed favored raising the tax on cigarettes as a way to discourage young people from smoking. What number of adults believe that this would reduce the number of young people who smoke?

70. Shared Rent You and three friends decide to rent an apartment for the academic year rather than to live in the dorms. The monthly rent is $1250. If you and your friends split the rent equally, what is your share of the monthly rent?

71. Importing Oil According to the U.S. Department of Energy, we imported approximately 8,340,000 barrels of oil in November 2007, which represents a typical month. We import a little over $\frac{1}{5}$ of our oil from Canada, approximately $\frac{3}{20}$ of our oil from Venezuela, and less than $\frac{1}{10}$ of our oil from Iraq. Determine the amount of oil we imported from each of these countries.

72. Cooking If a dozen peanut butter truffles require $\frac{3}{4}$ cup of powdered sugar, how much powdered sugar would be needed to make 48 truffles?

Geometric Sequences Recall that a geometric sequence is a sequence in which each term comes from the previous term by multiplying by the same number each time. For example, the sequence $1, \frac{1}{2}, \frac{1}{4}, \frac{1}{8}, \ldots$ is a geometric sequence in which each term is found by multiplying the previous term by $\frac{1}{2}$. By observing this fact, we know that the next term in the sequence will be $\frac{1}{8} \cdot \frac{1}{2} = \frac{1}{16}$.

Find the next number in each of the geometric sequences below.

73. $1, \frac{1}{3}, \frac{1}{9}, \ldots$

74. $1, \frac{1}{4}, \frac{1}{16}, \ldots$

75. $\frac{3}{2}, 1, \frac{2}{3}, \frac{4}{9}, \ldots$

76. $\frac{2}{3}, 1, \frac{3}{2}, \frac{9}{4}, \ldots$

Estimating For each problem below, mentally estimate which of the numbers 0, 1, 2, or 3 is closest to the answer. Make your estimate without using pencil and paper or a calculator.

77. $\dfrac{11}{5} \cdot \dfrac{19}{20}$

78. $\dfrac{3}{5} \cdot \dfrac{1}{20}$

79. $\dfrac{16}{5} \cdot \dfrac{23}{24}$

80. $\dfrac{9}{8} \cdot \dfrac{31}{32}$

Getting Ready for the Next Section

In the next section we will do division with fractions. As you already know, division and multiplication are closely related. These review problems are intended to let you see more of the relationship between multiplication and division. Perform the indicated operations.

81. $8 \div 4$

82. $8 \cdot \dfrac{1}{4}$

83. $15 \div 3$

84. $15 \cdot \dfrac{1}{3}$

85. $18 \div 6$

86. $18 \cdot \dfrac{1}{6}$

For each number below, find a number to multiply it by to obtain 1.

87. $\dfrac{3}{4}$

88. $\dfrac{9}{5}$

89. $\dfrac{1}{3}$

90. $\dfrac{1}{4}$

91. 7

92. 2

Maintaining Your Skills

Simplify.

93. $20 \div 2 \cdot 10$

94. $40 \div 4 \cdot 5$

95. $24 \div 8 \cdot 3$

96. $24 \div 4 \cdot 6$

97. $36 \div 6 \cdot 3$

98. $36 \div 9 \cdot 2$

99. $48 \div 12 \cdot 2$

100. $48 \div 8 \cdot 3$

2.4 Division with Fractions

OBJECTIVES

A Divide fractions.

B Simplify order of operations problems involving division of fractions.

C Solve application problems involving division of fractions.

TICKET TO SUCCESS

Keep these questions in mind as you read through the section. Then respond in your own words and in complete sentences.

1. Define *reciprocal*.
2. True or false? The quotient of $\frac{3}{5}$ and $\frac{3}{8}$ is the same as the product of $\frac{3}{5}$ and $\frac{8}{3}$.
3. Use symbols to show how division is defined as multiplication by the reciprocal.
4. Dividing by $\frac{19}{9}$ is the same as multiplying by what number?

A few years ago a 4-H club was making blankets to keep their lambs clean at the county fair. Each blanket required $\frac{3}{4}$ yard of material. They had 9 yards of material left over from the year before. To see how many blankets they could make, they divided 9 by $\frac{3}{4}$. The result was 12, meaning that they could make 12 lamb blankets out of the 9 remaining yards.

Before we define division with fractions, we must first introduce the idea of *reciprocals*. Look at the following multiplication problems:

$$\frac{3}{4} \cdot \frac{4}{3} = \frac{12}{12} = 1 \qquad \frac{7}{8} \cdot \frac{8}{7} = \frac{56}{56} = 1$$

In each case the product is 1. Whenever the product of two numbers is 1, we say the two numbers are reciprocals.

> **Definition**
>
> Two numbers whose product is 1 are said to be **reciprocals.** In symbols, the reciprocal of $\frac{a}{b}$ is $\frac{b}{a}$, because
>
> $$\frac{a}{b} \cdot \frac{b}{a} = \frac{a \cdot b}{b \cdot a} = \frac{a \cdot b}{a \cdot b} = 1 \qquad (a \neq 0, b \neq 0)$$

Every number has a reciprocal except 0. The reason that 0 does not have a reciprocal is because the product of *any* number with 0 is 0. It can never be 1. Reciprocals of whole numbers are fractions with 1 as the numerator. For example, the reciprocal of 5 is $\frac{1}{5}$, because

$$5 \cdot \frac{1}{5} = \frac{5}{1} \cdot \frac{1}{5} = \frac{5}{5} = 1$$

Table 1 lists some numbers and their reciprocals.

TABLE 1

Number	Reciprocal	Reason
$\frac{3}{4}$	$\frac{4}{3}$	Because $\frac{3}{4} \cdot \frac{4}{3} = \frac{12}{12} = 1$
$\frac{9}{5}$	$\frac{5}{9}$	Because $\frac{9}{5} \cdot \frac{5}{9} = \frac{45}{45} = 1$
$\frac{1}{3}$	3	Because $\frac{1}{3} \cdot 3 = \frac{1}{3} \cdot \frac{3}{1} = \frac{3}{3} = 1$
7	$\frac{1}{7}$	Because $7 \cdot \frac{1}{7} = \frac{7}{1} \cdot \frac{1}{7} = \frac{7}{7} = 1$

A Dividing Fractions

Division with fractions is accomplished by using reciprocals. More specifically, we can define division by a fraction to be the same as multiplication by its reciprocal. Here is the precise rule:

Rule Dividing by a Fraction

When dividing by a fraction, if a, b, c, and d are numbers and b, c, and d are not equal to 0, then

$$\frac{a}{b} \div \frac{c}{d} = \frac{a}{b} \cdot \frac{d}{c}$$

This rule states that dividing by the fraction $\frac{c}{d}$ is exactly the same as multiplying by its reciprocal $\frac{d}{c}$. Because we developed the rule for multiplying fractions in Section 2.3, we do not need a new rule for division. We simply replace the divisor by its reciprocal and multiply. Here are some examples to illustrate the procedure.

EXAMPLE 1 Divide: $\frac{1}{2} \div \frac{1}{4}$.

SOLUTION The divisor is $\frac{1}{4}$, and its reciprocal is $\frac{4}{1}$. Applying the definition of division for fractions, we have

$$\frac{1}{2} \div \frac{1}{4} = \frac{1}{2} \cdot \frac{4}{1}$$
$$= \frac{1 \cdot 4}{2 \cdot 1}$$
$$= \frac{1 \cdot \cancel{2} \cdot 2}{\cancel{2} \cdot 1}$$
$$= \frac{2}{1}$$
$$= 2$$

The quotient of $\frac{1}{2}$ and $\frac{1}{4}$ is 2. Or, $\frac{1}{4}$ "goes into" $\frac{1}{2}$ two times. Logically, our definition for division of fractions seems to be giving us answers that are consistent with what we know about fractions from previous experience. Because 2 times $\frac{1}{4}$ is $\frac{2}{4}$ or $\frac{1}{2}$, it seems logical that $\frac{1}{2}$ divided by $\frac{1}{4}$ should be 2.

NOTE

Defining division to be the same as multiplication by the reciprocal does make sense. If we divide 6 by 2, we get 3. On the other hand, if we multiply 6 by $\frac{1}{2}$ (the reciprocal of 2), we also get 3. Whether we divide by 2 or multiply by $\frac{1}{2}$, we get the same result.

PRACTICE PROBLEMS

1. Divide.

    a. $\frac{1}{3} \div \frac{1}{6}$

    b. $\frac{1}{30} \div \frac{1}{60}$

Answers

1. a. 2 b. 2

EXAMPLE 2 Divide: $\frac{3}{8} \div \frac{9}{4}$.

SOLUTION Dividing by $\frac{9}{4}$ is the same as multiplying by its reciprocal, which is $\frac{4}{9}$.

$$\frac{3}{8} \div \frac{9}{4} = \frac{3}{8} \cdot \frac{4}{9}$$

$$= \frac{\cancel{3} \cdot \cancel{2} \cdot \cancel{2}}{\cancel{2} \cdot \cancel{2} \cdot 2 \cdot \cancel{3} \cdot 3}$$

$$= \frac{1}{6}$$

The quotient of $\frac{3}{8}$ and $\frac{9}{4}$ is $\frac{1}{6}$. ∎

EXAMPLE 3 Divide: $\frac{2}{3} \div 2$.

SOLUTION The reciprocal of 2 is $\frac{1}{2}$. Applying the definition for division of fractions, we have

$$\frac{2}{3} \div 2 = \frac{2}{3} \cdot \frac{1}{2}$$

$$= \frac{\cancel{2} \cdot 1}{3 \cdot \cancel{2}}$$

$$= \frac{1}{3}$$ ∎

EXAMPLE 4 Divide: $2 \div \left(\frac{1}{3}\right)$.

SOLUTION We replace $\frac{1}{3}$ by its reciprocal, which is 3, and multiply.

$$2 \div \left(\frac{1}{3}\right) = 2(3)$$

$$= 6$$ ∎

Here are some further examples of division with fractions. Notice in each case that the first step is the only new part of the process.

EXAMPLE 5 Divide: $\frac{4}{27} \div \frac{16}{9}$.

SOLUTION $\frac{4}{27} \div \frac{16}{9} = \frac{4}{27} \cdot \frac{9}{16}$

$$= \frac{\cancel{4} \cdot \cancel{9}}{3 \cdot \cancel{9} \cdot \cancel{4} \cdot 4}$$

$$= \frac{1}{12}$$ ∎

In Example 5, we did not factor the numerator and the denominator completely in order to reduce to lowest terms because, as you have probably already noticed, it is not necessary to do so. We need to factor only enough to show what numbers are common to the numerator and the denominator. If we factored completely in the second step, it would look like this:

$$= \frac{\cancel{2} \cdot \cancel{2} \cdot \cancel{3} \cdot \cancel{3}}{\cancel{3} \cdot \cancel{3} \cdot 3 \cdot \cancel{2} \cdot \cancel{2} \cdot 2 \cdot 2}$$

$$= \frac{1}{12}$$

The result is the same in both cases. From now on, we will factor numerators and denominators only enough to show the factors we are dividing out.

2. Divide: $\frac{5}{9} \div \frac{10}{3}$.

3. Divide.
   a. $\frac{3}{4} \div 3$
   b. $\frac{3}{5} \div 3$
   c. $\frac{3}{7} \div 3$

4. Divide: $4 \div \frac{1}{5}$.

5. Find each quotient.
   a. $\frac{5}{32} \div \frac{10}{42}$
   b. $\frac{15}{32} \div \frac{30}{42}$

Answers
2. $\frac{1}{6}$
3. a. $\frac{1}{4}$ b. $\frac{1}{5}$ c. $\frac{1}{7}$
4. 20
5. Both are $\frac{21}{32}$.

Chapter 2 Fractions and Mixed Numbers

6. Divide.

   a. $\dfrac{12}{25} \div 6$

   b. $\dfrac{24}{25} \div 6$

EXAMPLE 6 Divide: $\dfrac{16}{35} \div 8$.

SOLUTION
$$\dfrac{16}{35} \div 8 = \dfrac{16}{35} \cdot \dfrac{1}{8}$$
$$= \dfrac{2 \cdot \cancel{8} \cdot 1}{35 \cdot \cancel{8}}$$
$$= \dfrac{2}{35}$$

7. Divide.

   a. $12 \div \dfrac{4}{3}$

   b. $12 \div \dfrac{4}{5}$

   c. $12 \div \dfrac{4}{7}$

EXAMPLE 7 Divide: $27 \div \left(\dfrac{3}{2}\right)$.

SOLUTION
$$27 \div \left(\dfrac{3}{2}\right) = 27 \cdot \left(\dfrac{2}{3}\right)$$
$$= \dfrac{\cancel{3} \cdot 9 \cdot 2}{\cancel{3}}$$
$$= 18$$

B Fractions and the Order of Operations

The next two examples combine what we have learned about division of fractions with the rule for order of operations.

8. The quotient of $\dfrac{5}{4}$ and $\dfrac{1}{8}$ is increased by 8. What number results?

EXAMPLE 8 The quotient of $\dfrac{8}{3}$ and $\dfrac{1}{6}$ is increased by 5. What number results?

SOLUTION Translating to symbols, we have
$$\dfrac{8}{3} \div \dfrac{1}{6} + 5 = \dfrac{8}{3} \cdot \dfrac{6}{1} + 5$$
$$= 16 + 5$$
$$= 21$$

9. Simplify:

$18 \div \left(\dfrac{3}{5}\right)^2 + 48 \div \left(\dfrac{2}{5}\right)^2$

EXAMPLE 9 Simplify: $32 \div \left(\dfrac{4}{3}\right)^2 + 75 \div \left(\dfrac{5}{2}\right)^2$.

SOLUTION According to the rule for order of operations, we must first evaluate the numbers with exponents, then divide, and finally, add.

$$32 \div \left(\dfrac{4}{3}\right)^2 + 75 \div \left(\dfrac{5}{2}\right)^2 = 32 \div \dfrac{16}{9} + 75 \div \dfrac{25}{4}$$
$$= 32 \cdot \dfrac{9}{16} + 75 \cdot \dfrac{4}{25}$$
$$= 18 + 12$$
$$= 30$$

C Applications

10. How many blankets can the 4-H club make with 12 yards of material, if each blanket requires $\dfrac{3}{4}$ yard of material?

EXAMPLE 10 A 4-H club is making blankets to keep their lambs clean at the county fair. If each blanket requires $\dfrac{2}{3}$ yard of material, how many blankets can they make from 12 yards of material?

SOLUTION To answer this question we must divide 12 by $\dfrac{2}{3}$.

$$12 \div \dfrac{2}{3} = 12 \cdot \dfrac{3}{2}$$
$$= 6 \cdot 3$$
$$= 18$$

They can make 18 blankets from the 12 yards of material.

Answers

6. a. $\dfrac{2}{25}$ b. $\dfrac{4}{25}$
7. a. 9 b. 15 c. 21
8. 18
9. 350
10. 16 blankets

Problem Set 2.4

Moving Toward Success

"I am a member of a team, and I rely on the team, I defer to it and sacrifice for it, because the team, not the individual, is the ultimate champion."

—Mia Hamm, 1972–present, American female soccer player

1. How is your instructor a vital resource for your success in this course?
2. Utilize your instructor's office hours. Why is this important?

A Find the quotient in each case by replacing the divisor by its reciprocal and multiplying. Reduce to lowest terms. [Examples 1–7]

1. $\dfrac{3}{4} \div \dfrac{1}{5}$
2. $\dfrac{1}{3} \div \dfrac{1}{2}$
3. $\dfrac{2}{3} \div \dfrac{1}{2}$
4. $\dfrac{5}{8} \div \dfrac{1}{4}$

5. $6 \div \left(\dfrac{2}{3}\right)$
6. $8 \div \left(\dfrac{3}{4}\right)$
7. $20 \div \dfrac{1}{10}$
8. $16 \div \dfrac{1}{8}$

9. $\dfrac{3}{4} \div 2$
10. $\dfrac{3}{5} \div 2$
11. $\dfrac{7}{8} \div \dfrac{7}{8}$
12. $\dfrac{4}{3} \div \dfrac{4}{3}$

13. $\dfrac{7}{8} \div \dfrac{8}{7}$
14. $\dfrac{4}{3} \div \dfrac{3}{4}$
15. $\dfrac{9}{16} \div \dfrac{3}{4}$
16. $\dfrac{25}{36} \div \dfrac{5}{6}$

17. $\dfrac{25}{46} \div \dfrac{40}{69}$
18. $\dfrac{25}{24} \div \dfrac{15}{36}$
19. $\dfrac{13}{28} \div \dfrac{39}{14}$
20. $\dfrac{28}{125} \div \dfrac{5}{2}$

21. $\dfrac{27}{196} \div \dfrac{9}{392}$
22. $\dfrac{16}{135} \div \dfrac{2}{45}$
23. $\dfrac{25}{18} \div 5$
24. $\dfrac{30}{27} \div 6$

25. $6 \div \dfrac{4}{3}$
26. $12 \div \dfrac{4}{3}$
27. $\dfrac{4}{3} \div 6$
28. $\dfrac{4}{3} \div 12$

29. $\dfrac{3}{4} \div \dfrac{1}{2} \cdot 6$
30. $12 \div \dfrac{6}{7} \cdot 7$
31. $\dfrac{2}{3} \cdot \dfrac{3}{4} \div \dfrac{5}{8}$
32. $4 \cdot \dfrac{7}{6} \div 7$

33. $\dfrac{35}{110} \cdot \dfrac{80}{63} \div \dfrac{16}{27}$

34. $\dfrac{20}{72} \cdot \dfrac{42}{18} \div \dfrac{20}{16}$

B Simplify each expression as much as possible. [Examples 8, 9]

35. $10 \div \left(\dfrac{1}{2}\right)^2$

36. $12 \div \left(\dfrac{1}{4}\right)^2$

37. $\dfrac{18}{35} \div \left(\dfrac{6}{7}\right)^2$

38. $\dfrac{48}{55} \div \left(\dfrac{8}{11}\right)^2$

39. $\dfrac{4}{5} \div \dfrac{1}{10} + 5$

40. $\dfrac{3}{8} \div \dfrac{1}{16} + 4$

41. $10 + \dfrac{11}{12} \div \dfrac{11}{24}$

42. $15 + \dfrac{13}{14} \div \dfrac{13}{42}$

43. $24 \div \left(\dfrac{2}{5}\right)^2 + 25 \div \left(\dfrac{5}{6}\right)^2$

44. $18 \div \left(\dfrac{3}{4}\right)^2 + 49 \div \left(\dfrac{7}{9}\right)^2$

45. $100 \div \left(\dfrac{5}{7}\right)^2 + 200 \div \left(\dfrac{2}{3}\right)^2$

46. $64 \div \left(\dfrac{8}{11}\right)^2 + 81 \div \left(\dfrac{9}{11}\right)^2$

47. What is the quotient of $\dfrac{3}{8}$ and $\dfrac{5}{8}$?

48. Find the quotient of $\dfrac{4}{5}$ and $\dfrac{16}{25}$.

49. If the quotient of 18 and $\dfrac{3}{5}$ is increased by 10, what number results?

50. If the quotient of 50 and $\dfrac{5}{3}$ is increased by 8, what number results?

51. Show that multiplying 3 by 5 is the same as dividing 3 by $\dfrac{1}{5}$.

52. Show that multiplying 8 by $\dfrac{1}{2}$ is the same as dividing 8 by 2.

Applying the Concepts

53. Pyramids The Luxor Hotel in Las Vegas is $\frac{5}{7}$ the original height of the Great Pyramid of Giza. If the hotel is 350 feet tall, what was the original height of the Great Pyramid of Giza?

54. Skyscrapers The Emerates Tower Two in Dubai is $\frac{2}{3}$ the height of the Taipei 101 building shown in the diagram below. How tall is the Emerates Tower Two?

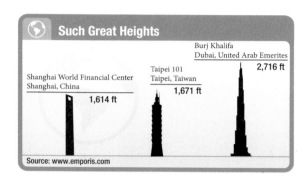

55. Sewing If $\frac{6}{7}$ yard of material is needed to make a blanket, how many blankets can be made from 12 yards of material?

56. Manufacturing A clothing manufacturer is making scarves that require $\frac{3}{8}$ yard of material each. How many can be made from 27 yards of material?

57. Grilling A man is making a marinade for some meat he plans to grill. The marinade recipe calls for $\frac{3}{4}$ teaspoon of salt. If the only measuring spoon he can find is a $\frac{1}{8}$ teaspoon, how many of these will he have to fill with salt in order to have a total of $\frac{3}{4}$ teaspoon of salt?

58. Cooking A cake recipe calls for $\frac{1}{2}$ cup of sugar. If the only measuring cup available is a $\frac{1}{8}$ cup, how many of these will have to be filled to make a total of $\frac{1}{2}$ cup of sugar?

59. Cartons of Milk If a small carton of milk holds exactly $\frac{1}{2}$ pint, how many of the $\frac{1}{2}$-pint cartons can be filled from a 14-pint container?

60. Pieces of Pipe How many pieces of $\frac{2}{3}$-foot pipe must be laid together to make a pipe 16 feet long?

61. Lot Size A land developer wants to subdivide 5 acres of property into lots suitable for building a home. If each lot is to be $\frac{1}{4}$ of an acre in size how many lots can be made?

62. House Plans If $\frac{1}{8}$ inch represents 1 ft on a drawing of a new home, determine the dimensions of a bedroom that measures 2 inches by 2 inches on the drawing.

Getting Ready for the Next Section

Write each fraction as an equivalent fraction with denominator 6.

63. $\dfrac{1}{2}$ **64.** $\dfrac{1}{3}$ **65.** $\dfrac{3}{2}$ **66.** $\dfrac{2}{3}$

Write each fraction as an equivalent fraction with denominator 12.

67. $\dfrac{1}{3}$ **68.** $\dfrac{1}{2}$ **69.** $\dfrac{2}{3}$ **70.** $\dfrac{3}{4}$

Write each fraction as an equivalent fraction with denominator 30.

71. $\dfrac{7}{15}$ **72.** $\dfrac{3}{10}$ **73.** $\dfrac{3}{5}$ **74.** $\dfrac{1}{6}$

Write each fraction as an equivalent fraction with denominator 24.

75. $\dfrac{1}{2}$ **76.** $\dfrac{1}{4}$ **77.** $\dfrac{1}{6}$ **78.** $\dfrac{1}{8}$

Write each fraction as an equivalent fraction with denominator 36.

79. $\dfrac{5}{12}$ **80.** $\dfrac{7}{18}$ **81.** $\dfrac{1}{4}$ **82.** $\dfrac{1}{6}$

Maintaining Your Skills

83. Fill in the table by rounding the numbers.

	Rounded to the Nearest Number		
	Ten	Hundred	Thousand
74			
747			
474			

84. Fill in the table by rounding the numbers.

	Rounded to the Nearest Number		
	Ten	Hundred	Thousand
63			
636			
363			

85. Estimating The quotient $253 \div 24$ is closer to which of the following?

a. 5 **b.** 10 **c.** 15 **d.** 20

86. Estimating The quotient $1{,}000 \div 47$ is closer to which of the following?

a. 5 **b.** 10 **c.** 15 **d.** 20

Addition and Subtraction with Fractions

2.5

OBJECTIVES

A Add and subtract fractions with a common denominator.

B Add and subtract fractions with different denominators.

TICKET TO SUCCESS

Keep these questions in mind as you read through the section. Then respond in your own words and in complete sentences.

1. Using symbols, how do you add two fractions with same denominator?
2. What does LCD stand for and what does it mean?
3. What is the first step when adding or subtracting any two fractions?
4. When adding fractions, what is the last step?

The Atacama Desert on the Pacific side of the Andes Mountains in Chile is arguably the driest place in the world. Some sources say that it rains here approximately $\frac{3}{100}$ inches per year. Others report that a measurable amount of rainfall has never been recorded. The land is similar to that found on the planet Mars, and NASA uses the desert to test instruments for future space missions.

The table below compares annual rainfall quantities for some of the driest places on Earth, including the Atacama Desert.

Location	Annual Rainfall (inches)
Atacama Desert, Chile	$\frac{3}{100}$
Wadi Halfa, Sudan	$\frac{1}{10}$
South Pole, Antarctica	$\frac{4}{5}$
Death Valley, CA	$\frac{8}{5}$
Aden, Yemen	$\frac{9}{5}$

Source: www.worldclimate.com

In this section, we will learn how to add and subtract fractions similar to those found in the table. If we needed to determine how much more rain Death Valley, CA received than the South Pole, we need to understand how to work with fractions that have a common denominator. Let's begin.

A Combining Fractions with a Common Denominator

Adding and subtracting fractions is actually just another application of the distributive property. The distributive property looks like this:

$$a(b + c) = a(b) + a(c)$$

where a, b, and c may be whole numbers or fractions. We will want to apply this property to expressions like

$$\frac{2}{7} + \frac{3}{7}$$

But before we do, we must make one additional observation about fractions. The fraction $\frac{2}{7}$ can be written as $2 \cdot \frac{1}{7}$, because

$$2 \cdot \frac{1}{7} = \frac{2}{1} \cdot \frac{1}{7} = \frac{2}{7}$$

Likewise, the fraction $\frac{3}{7}$ can be written as $3 \cdot \frac{1}{7}$, because

$$3 \cdot \frac{1}{7} = \frac{3}{1} \cdot \frac{1}{7} = \frac{3}{7}$$

In general, we can say that the fraction $\frac{a}{b}$ can always be written as $a \cdot \frac{1}{b}$, because

$$a \cdot \frac{1}{b} = \frac{a}{1} \cdot \frac{1}{b} = \frac{a}{b}$$

To add the fractions $\frac{2}{7}$ and $\frac{3}{7}$, we simply rewrite each of them as we have done above and apply the distributive property. Here is how it works:

$$\frac{2}{7} + \frac{3}{7} = 2 \cdot \frac{1}{7} + 3 \cdot \frac{1}{7} \quad \text{Rewrite each fraction.}$$

$$= (2 + 3) \cdot \frac{1}{7} \quad \text{Apply the distributive property.}$$

$$= 5 \cdot \frac{1}{7} \quad \text{Add 2 and 3 to get 5.}$$

$$= \frac{5}{7} \quad \text{Rewrite } 5 \cdot \frac{1}{7} \text{ as } \frac{5}{7}.$$

We can visualize the process shown above by using circles that are divided into 7 equal parts.

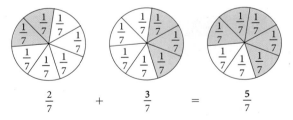

$$\frac{2}{7} \quad + \quad \frac{3}{7} \quad = \quad \frac{5}{7}$$

The fraction $\frac{5}{7}$ is the sum of $\frac{2}{7}$ and $\frac{3}{7}$. The steps and diagrams above show why we add numerators *but do not add denominators*. Using this example as justification, we can write a rule for adding two fractions that have the same denominator.

> **Rule** Addition with Common Denominator
>
> To add two fractions that have the same denominator, we add their numerators to get the numerator of the answer. The denominator in the answer is the same denominator as in the original fractions.

2.5 Addition and Subtraction with Fractions

What we have here is the sum of the numerators placed over the *common denominator*. In symbols, we have the following:

> **Property** Addition and Subtraction of Fractions
>
> If a, b, and c are numbers, and c is not equal to 0, then
> $$\frac{a}{c} + \frac{b}{c} = \frac{a+b}{c}$$
> This rule holds for subtraction as well. That is,
> $$\frac{a}{c} - \frac{b}{c} = \frac{a-b}{c}$$

EXAMPLE 1 Add: $\frac{3}{8} + \frac{1}{8}$.

SOLUTION
$\frac{3}{8} + \frac{1}{8} = \frac{3+1}{8}$ Add numerators; keep the same denominator.

$= \frac{4}{8}$ The sum of 3 and 1 is 4.

$= \frac{1}{2}$ Reduce to lowest terms.

EXAMPLE 2 Subtract: $\frac{a+5}{8} - \frac{3}{8}$.

SOLUTION
$\frac{a+5}{8} - \frac{3}{8} = \frac{a+5-3}{8}$ Combine numerators; keep the same denominator.

$= \frac{a+2}{8}$ The difference of 5 and 3 is 2.

EXAMPLE 3 Subtract: $\frac{9}{5} - \frac{3}{5}$.

SOLUTION
$\frac{9}{5} - \frac{3}{5} = \frac{9-3}{5}$ Subtract numerators; keep the same denominator.

$= \frac{6}{5}$ The difference of 9 and 3 is 6.

EXAMPLE 4 Add: $\frac{3}{7} + \frac{2}{7} + \frac{9}{7}$.

SOLUTION
$\frac{3}{7} + \frac{2}{7} + \frac{9}{7} = \frac{3+2+9}{7}$

$= \frac{14}{7}$

$= 2$

As Examples 1–4 indicate, addition and subtraction are simple, straightforward processes when all the fractions have the same denominator.

PRACTICE PROBLEMS

Find the sum or difference. Reduce all answers to lowest terms.

1. $\frac{3}{10} + \frac{1}{10}$

2. $\frac{a+5}{12} + \frac{3}{12}$

3. $\frac{8}{7} - \frac{5}{7}$

4. $\frac{5}{9} + \frac{8}{9} + \frac{5}{9}$

Answers
1. $\frac{2}{5}$
2. $\frac{a+8}{12}$
3. $\frac{3}{7}$
4. 2

5. a. Find the LCD for the fractions:
$$\frac{5}{18} \text{ and } \frac{3}{14}$$

b. Find the LCD for the fractions:
$$\frac{5}{36} \text{ and } \frac{3}{28}$$

NOTE
The ability to find least common denominators is very important in mathematics. The discussion here is a detailed explanation of how to find an LCD.

6. Add.
a. $\frac{5}{18} + \frac{3}{14}$
b. $\frac{5}{36} + \frac{3}{28}$

Answers
5. a. 126 **b.** 252
6. a. $\frac{31}{63}$ **b.** $\frac{31}{126}$

B The Least Common Denominator or LCD

We will now turn our attention to the process of adding fractions that have different denominators. Look back at the table that began this section. How would we determine the difference in rainfall between Death Valley, California, and Wodi Halfa, Sudan? To begin answering this question, we need the following definition:

> **Definition**
> The **least common denominator** (LCD) for a set of denominators is the smallest number that is exactly divisible by each denominator. (Note that, in some books, the least common denominator is also called the *least common multiple*.)

In other words, all the denominators of the fractions involved in a problem must divide into the least common denominator exactly. That is, they divide it without leaving a remainder.

EXAMPLE 5 Find the LCD for the fractions $\frac{5}{12}$ and $\frac{7}{18}$.

SOLUTION The least common denominator for the denominators 12 and 18 must be the smallest number divisible by both 12 and 18. We can factor 12 and 18 completely and then build the LCD from these factors. Factoring 12 and 18 completely gives us

$$12 = 2 \cdot 2 \cdot 3 \qquad 18 = 2 \cdot 3 \cdot 3$$

Now, if 12 is going to divide the LCD exactly, then the LCD must have factors of $2 \cdot 2 \cdot 3$. If 18 is to divide it exactly, it must have factors of $2 \cdot 3 \cdot 3$. We don't need to repeat the factors that 12 and 18 have in common.

$$\left.\begin{array}{l}12 = 2 \cdot 2 \cdot 3 \\ 18 = 2 \cdot 3 \cdot 3\end{array}\right\} \qquad \text{LCD} = 2 \cdot 2 \cdot 3 \cdot 3 = 36$$

12 divides the LCD.
18 divides the LCD.

The LCD for 12 and 18 is 36. It is the smallest number that is divisible by both 12 and 18; 12 divides it exactly three times, and 18 divides it exactly two times. ■

We can visualize the results in Example 5 with the diagram below. It shows that 36 is the smallest number that both 12 and 18 divide evenly. As you can see, 12 divides 36 exactly 3 times, and 18 divides 36 exactly 2 times.

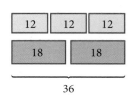

EXAMPLE 6 Add: $\frac{5}{12} + \frac{7}{18}$.

SOLUTION We can add fractions only when they have the same denominators. In Example 5, we found the LCD for $\frac{5}{12}$ and $\frac{7}{18}$ to be 36. We change $\frac{5}{12}$ and $\frac{7}{18}$ to equivalent fractions that have 36 for a denominator by applying Property 1 from Section 2.1 for fractions.

$$\frac{5}{12} = \frac{5 \cdot 3}{12 \cdot 3} = \frac{15}{36}$$

$$\frac{7}{18} = \frac{7 \cdot 2}{18 \cdot 2} = \frac{14}{36}$$

The fraction $\frac{15}{36}$ is equivalent to $\frac{5}{12}$, because it was obtained by multiplying both the numerator and the denominator by 3. Likewise, $\frac{14}{36}$ is equivalent to $\frac{7}{18}$, because it was obtained by multiplying the numerator and the denominator by 2. All we have left to do is to add numerators.

$$\frac{15}{36} + \frac{14}{36} = \frac{29}{36}$$

The sum of $\frac{5}{12}$ and $\frac{7}{18}$ is the fraction $\frac{29}{36}$. Let's write the complete problem again step by step.

$$\frac{5}{12} + \frac{7}{18} = \frac{5 \cdot 3}{12 \cdot 3} + \frac{7 \cdot 2}{18 \cdot 2}$$ Rewrite each fraction as an equivalent fraction with denominator 36.

$$= \frac{15}{36} + \frac{14}{36}$$

$$= \frac{29}{36}$$ Add numerators; keep the common denominator. ∎

EXAMPLE 7 Find the LCD for $\frac{3}{4}$ and $\frac{1}{6}$.

SOLUTION We factor 4 and 6 into products of prime factors and build the LCD from these factors.

$$\left.\begin{array}{l} 4 = 2 \cdot 2 \\ 6 = 2 \cdot 3 \end{array}\right\} \text{LCD} = 2 \cdot 2 \cdot 3 = 12$$

The LCD is 12. Both denominators divide it exactly; 4 divides 12 exactly 3 times, and 6 divides 12 exactly 2 times. ∎

EXAMPLE 8 Add: $\frac{3}{4} + \frac{1}{6}$.

SOLUTION In Example 7, we found that the LCD for these two fractions is 12. We begin by changing $\frac{3}{4}$ and $\frac{1}{6}$ to equivalent fractions with denominator 12.

$$\frac{3}{4} = \frac{3 \cdot 3}{4 \cdot 3} = \frac{9}{12}$$

$$\frac{1}{6} = \frac{1 \cdot 2}{6 \cdot 2} = \frac{2}{12}$$

The fraction $\frac{9}{12}$ is equal to the fraction $\frac{3}{4}$, because it was obtained by multiplying the numerator and the denominator of $\frac{3}{4}$ by 3. Likewise, $\frac{2}{12}$ is equivalent to $\frac{1}{6}$, because it was obtained by multiplying the numerator and the denominator of $\frac{1}{6}$ by 2. To complete the problem we add numerators.

$$\frac{9}{12} + \frac{2}{12} = \frac{11}{12}$$

The sum of $\frac{3}{4}$ and $\frac{1}{6}$ is $\frac{11}{12}$. Here is how the complete problem looks:

$$\frac{3}{4} + \frac{1}{6} = \frac{3 \cdot 3}{4 \cdot 3} + \frac{1 \cdot 2}{6 \cdot 2}$$ Rewrite each fraction as an equivalent fraction with denominator 12.

$$= \frac{9}{12} + \frac{2}{12}$$

$$= \frac{11}{12}$$ Add numerators; keep the same denominator. ∎

7. **a.** Find the LCD for $\frac{2}{9}$ and $\frac{4}{15}$.
   b. Find the LCD for $\frac{2}{27}$ and $\frac{4}{45}$.

8. Add.
   a. $\frac{2}{9} + \frac{4}{15}$
   b. $\frac{2}{27} + \frac{4}{45}$

NOTE
We can visualize the work in Example 8 using circles and shading:

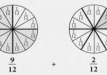

Answers
7. **a.** 45 **b.** 135
8. **a.** $\frac{22}{45}$ **b.** $\frac{22}{135}$

9. Subtract: $\dfrac{8}{25} - \dfrac{3}{20}$.

Chapter 2 Fractions and Mixed Numbers

EXAMPLE 9 Subtract: $\dfrac{7}{15} - \dfrac{3}{10}$.

SOLUTION Let's factor 15 and 10 completely and use these factors to build the LCD.

$$\left.\begin{array}{l} 15 = 3 \cdot 5 \\ 10 = 2 \cdot 5 \end{array}\right\} \text{LCD} = 2 \cdot 3 \cdot 5 = 30$$

15 divides the LCD.
10 divides the LCD.

Changing to equivalent fractions and subtracting, we have

$$\begin{aligned} \frac{7}{15} - \frac{3}{10} &= \frac{7 \cdot \mathbf{2}}{15 \cdot \mathbf{2}} - \frac{3 \cdot \mathbf{3}}{10 \cdot \mathbf{3}} && \text{Rewrite as equivalent fractions with the LCD for the denominator.} \\ &= \frac{14}{30} - \frac{9}{30} \\ &= \frac{5}{30} && \text{Subtract numerators; keep the LCD.} \\ &= \frac{1}{6} && \text{Reduce to lowest terms.} \end{aligned}$$

As a summary of what we have done so far, and as a guide to working other problems, we now list the steps involved in adding and subtracting fractions with different denominators.

> **Strategy Adding or Subtracting Any Two Fractions**
>
> **Step 1** Factor each denominator completely, and use the factors to build the LCD. (Remember, the LCD is the smallest number divisible by each of the denominators in the problem.)
>
> **Step 2** Rewrite each fraction as an equivalent fraction that has the LCD for its denominator. This is done by multiplying both the numerator and the denominator of the fraction in question by the appropriate whole number.
>
> **Step 3** Add or subtract the numerators of the fractions produced in Step 2. This is the numerator of the sum or difference. The denominator of the sum or difference is the LCD.
>
> **Step 4** Reduce the fraction produced in Step 3 to lowest terms if it is not already in lowest terms.

The idea behind adding or subtracting fractions is really very simple. We can only add or subtract fractions that have the same denominators. If the fractions we are trying to add or subtract do not have the same denominators, we rewrite each of them as an equivalent fraction with the LCD for a denominator.

Here are some additional examples of sums and differences of fractions.

10. Subtract: $\dfrac{3}{4} - \dfrac{1}{5}$.

EXAMPLE 10 Subtract: $\dfrac{3}{5} - \dfrac{1}{6}$.

SOLUTION The LCD for 5 and 6 is their product, 30. We begin by rewriting each fraction with this common denominator.

$$\begin{aligned} \frac{3}{5} - \frac{1}{6} &= \frac{3 \cdot \mathbf{6}}{5 \cdot \mathbf{6}} - \frac{1 \cdot \mathbf{5}}{6 \cdot \mathbf{5}} \\ &= \frac{18}{30} - \frac{5}{30} \\ &= \frac{13}{30} \end{aligned}$$

Answers

9. $\dfrac{17}{100}$

10. $\dfrac{11}{20}$

2.5 Addition and Subtraction with Fractions

EXAMPLE 11 Add: $\frac{1}{6} + \frac{1}{8} + \frac{1}{4}$.

SOLUTION We begin by factoring the denominators completely and building the LCD from the factors that result.

$$6 = 2 \cdot 3$$
$$8 = 2 \cdot 2 \cdot 2$$
$$4 = 2 \cdot 2$$

8 divides the LCD
LCD $= 2 \cdot 2 \cdot 2 \cdot 3 = 24$
4 divides the LCD   6 divides the LCD

We then change to equivalent fractions and add as usual.

$$\frac{1}{6} + \frac{1}{8} + \frac{1}{4} = \frac{1 \cdot \mathbf{4}}{6 \cdot \mathbf{4}} + \frac{1 \cdot \mathbf{3}}{8 \cdot \mathbf{3}} + \frac{1 \cdot \mathbf{6}}{4 \cdot \mathbf{6}}$$

$$= \frac{4}{24} + \frac{3}{24} + \frac{6}{24}$$

$$= \frac{13}{24}$$ ∎

EXAMPLE 12 Subtract: $3 - \frac{5}{6}$.

SOLUTION The denominators are 1 (because $3 = \frac{3}{1}$) and 6. The smallest number divisible by both 1 and 6 is 6.

$$3 - \frac{5}{6} = \frac{3}{1} - \frac{5}{6} = \frac{3 \cdot \mathbf{6}}{1 \cdot \mathbf{6}} - \frac{5}{6}$$

$$= \frac{18}{6} - \frac{5}{6}$$

$$= \frac{13}{6}$$ ∎

EXAMPLE 13 Find the next number in each sequence.

a. $\frac{1}{2}, 0, -\frac{1}{2}, \ldots$ b. $\frac{1}{2}, 1, \frac{3}{2}, \ldots$ c. $\frac{1}{2}, \frac{1}{4}, \frac{1}{8}, \ldots$

SOLUTION

a. $\frac{1}{2}, 0, -\frac{1}{2}, \ldots$: Adding $-\frac{1}{2}$ to each term produces the next term. The fourth term will be $-\frac{1}{2} + \left(-\frac{1}{2}\right) = -1$. This is an arithmetic sequence.

b. $\frac{1}{2}, 1, \frac{3}{2}, \ldots$: Each term comes from the term before it by adding $\frac{1}{2}$. The fourth term will be $\frac{3}{2} + \frac{1}{2} = 2$. This sequence is also an arithmetic sequence.

c. $\frac{1}{2}, \frac{1}{4}, \frac{1}{8}, \ldots$: This is a geometric sequence in which each term comes from the term before it by multiplying by $\frac{1}{2}$ each time. The next term will be $\frac{1}{8} \cdot \frac{1}{2} = \frac{1}{16}$. ∎

---

11. Add.
  a. $\frac{1}{9} + \frac{1}{4} + \frac{1}{6}$
  b. $\frac{1}{90} + \frac{1}{40} + \frac{1}{60}$

12. Subtract: $2 - \frac{3}{4}$.

13. Find the next number in each sequence.
  a. $\frac{1}{3}, 0, -\frac{1}{3}, \ldots$
  b. $\frac{1}{3}, \frac{2}{3}, 1, \ldots$
  c. $1, \frac{1}{3}, \frac{1}{9}, \ldots$

Answers
11. a. $\frac{19}{36}$ b. $\frac{19}{360}$
12. $\frac{5}{4}$
13. a. $-\frac{2}{3}$ b. $\frac{4}{3}$ c. $\frac{1}{27}$

Problem Set 2.5

Moving Toward Success

"Failure is not falling down but refusing to get up."
—Chinese proverb

1. When you find a solution to a math problem, why should you compare your answer to those in the back of the book?
2. What should you do if you have made a mistake?

A Find the following sums and differences, and reduce to lowest terms. (Add or subtract as indicated.) [Examples 1–4]

1. $\dfrac{3}{6} + \dfrac{1}{6}$
2. $\dfrac{2}{5} + \dfrac{3}{5}$
3. $\dfrac{5}{8} - \dfrac{3}{8}$
4. $\dfrac{6}{7} - \dfrac{1}{7}$

5. $\dfrac{3}{4} - \dfrac{1}{4}$
6. $\dfrac{7}{9} - \dfrac{4}{9}$
7. $\dfrac{2}{3} - \dfrac{1}{3}$
8. $\dfrac{9}{8} - \dfrac{1}{8}$

9. $\dfrac{1}{4} + \dfrac{2}{4} + \dfrac{3}{4}$
10. $\dfrac{2}{5} + \dfrac{3}{5} + \dfrac{4}{5}$
11. $\dfrac{x+7}{2} - \dfrac{1}{2}$
12. $\dfrac{x+5}{4} - \dfrac{3}{4}$

13. $\dfrac{1}{10} + \dfrac{3}{10} + \dfrac{4}{10}$
14. $\dfrac{3}{20} + \dfrac{1}{20} + \dfrac{4}{20}$
15. $\dfrac{1}{3} + \dfrac{4}{3} + \dfrac{5}{3}$
16. $\dfrac{5}{4} + \dfrac{4}{4} + \dfrac{3}{4}$

B Complete the following tables. [Examples 5–12]

17.

First Number a	Second Number b	The Sum of a and b $a+b$
$\dfrac{1}{2}$	$\dfrac{1}{3}$	
$\dfrac{1}{3}$	$\dfrac{1}{4}$	
$\dfrac{1}{4}$	$\dfrac{1}{5}$	
$\dfrac{1}{5}$	$\dfrac{1}{6}$	

18.

First Number a	Second Number b	The Sum of a and b $a+b$
1	$\dfrac{1}{2}$	
1	$\dfrac{1}{3}$	
1	$\dfrac{1}{4}$	
1	$\dfrac{1}{5}$	

19.

First Number a	Second Number b	The Sum of a and b $a+b$
$\frac{1}{12}$	$\frac{1}{2}$	
$\frac{1}{12}$	$\frac{1}{3}$	
$\frac{1}{12}$	$\frac{1}{4}$	
$\frac{1}{12}$	$\frac{1}{6}$	

20.

First Number a	Second Number b	The Sum of a and b $a+b$
$\frac{1}{8}$	$\frac{1}{2}$	
$\frac{1}{8}$	$\frac{1}{4}$	
$\frac{1}{8}$	$\frac{1}{16}$	
$\frac{1}{8}$	$\frac{1}{24}$	

B Find the LCD for each of the following; then use the methods developed in this section to add or subtract as indicated. [Examples 5–12]

21. $\frac{4}{9} + \frac{1}{3}$

22. $\frac{1}{2} + \frac{1}{4}$

23. $2 + \frac{1}{3}$

24. $3 + \frac{1}{2}$

25. $\frac{3}{4} + 1$

26. $\frac{3}{4} + 2$

27. $\frac{1}{2} + \frac{2}{3}$

28. $\frac{1}{8} + \frac{3}{4}$

29. $\frac{1}{4} + \frac{1}{5}$

30. $\frac{1}{3} + \frac{1}{5}$

31. $\frac{1}{2} + \frac{1}{5}$

32. $\frac{1}{2} - \frac{1}{5}$

33. $\frac{5}{12} + \frac{3}{8}$

34. $\frac{9}{16} + \frac{7}{12}$

35. $\frac{8}{30} - \frac{1}{20}$

36. $\frac{9}{40} - \frac{1}{30}$

37. $\frac{3}{10} + \frac{1}{100}$

38. $\frac{9}{100} + \frac{7}{10}$

39. $\frac{10}{36} + \frac{9}{48}$

40. $\frac{12}{28} + \frac{9}{20}$

41. $\dfrac{17}{30} + \dfrac{11}{42}$

42. $\dfrac{19}{42} + \dfrac{13}{70}$

43. $\dfrac{25}{84} + \dfrac{41}{90}$

44. $\dfrac{23}{70} + \dfrac{29}{84}$

45. $\dfrac{13}{126} - \dfrac{13}{180}$

46. $\dfrac{17}{84} - \dfrac{17}{90}$

47. $\dfrac{3}{4} + \dfrac{1}{8} + \dfrac{5}{6}$

48. $\dfrac{3}{8} + \dfrac{2}{5} + \dfrac{1}{4}$

49. $\dfrac{3}{10} + \dfrac{5}{12} + \dfrac{1}{6}$

50. $\dfrac{5}{21} + \dfrac{1}{7} + \dfrac{3}{14}$

51. $\dfrac{1}{2} + \dfrac{1}{3} + \dfrac{1}{4} + \dfrac{1}{6}$

52. $\dfrac{1}{8} + \dfrac{1}{4} + \dfrac{1}{5} + \dfrac{1}{10}$

53. $10 - \dfrac{2}{9}$

54. $9 - \dfrac{3}{5}$

55. $\dfrac{1}{10} + \dfrac{4}{5} - \dfrac{3}{20}$

56. $\dfrac{1}{2} + \dfrac{3}{4} - \dfrac{5}{8}$

57. $\dfrac{1}{4} - \dfrac{1}{8} + \dfrac{1}{2} - \dfrac{3}{8}$

58. $\dfrac{7}{8} - \dfrac{3}{4} + \dfrac{5}{8} - \dfrac{1}{2}$

There are two ways to work the problems below. You can combine the fractions inside the parentheses first and then multiply; or you can apply the distributive property first, then add.

59. $15\left(\dfrac{2}{3} + \dfrac{3}{5}\right)$

60. $15\left(\dfrac{4}{5} - \dfrac{1}{3}\right)$

61. $4\left(\dfrac{1}{2} + \dfrac{1}{4}\right)$

62. $6\left(\dfrac{1}{3} + \dfrac{1}{2}\right)$

63. Find the sum of $\dfrac{3}{7}$, 2, and $\dfrac{1}{9}$.

64. Find the sum of 6, $\dfrac{6}{11}$, and 11.

65. Find the difference of $\dfrac{7}{8}$ and $\dfrac{1}{4}$.

66. Find the difference of $\dfrac{9}{10}$ and $\dfrac{1}{100}$.

Applying the Concepts

Some of the application problems below involve multiplication or division, while others involve addition or subtraction.

67. Rainfall Recall the table from earlier in this chapter. How much total rainfall did Death Valley get during the months of April and October?

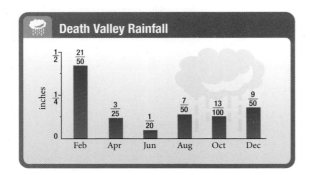

68. Rainfall Recall the table from the beginning of this section. How much more rainfall did Death Valley get than the South Pole?

Location	Annual Rainfall (inches)
Atacama Desert, Chile	$\frac{3}{100}$
Wadi Halfa, Sudan	$\frac{1}{10}$
South Pole, Antarctica	$\frac{4}{5}$
Death Valley, CA	$\frac{8}{5}$
Aden, Yemen	$\frac{9}{5}$

Source: www.worldclimate.com

69. Capacity One carton of milk contains $\frac{1}{2}$ pint while another contains 4 pints. How much milk is contained in both cartons?

70. Baking A recipe calls for $\frac{2}{3}$ cup of flour and $\frac{3}{4}$ cup of sugar. What is the total amount of flour and sugar called for in the recipe?

71. Budgeting A student earns $2,500 a month while working in college. She sets aside $\frac{1}{20}$ of this money for gas to travel to and from campus, $\frac{1}{16}$ for food, and $\frac{1}{25}$ for savings. What fraction of her income does she plan to spend on these three items?

72. Popular Majors Enrollment figures show that the most popular programs at a local college are liberal arts studies and business programs. The liberal arts studies program accounts for $\frac{1}{5}$ of the student enrollment while business programs account for $\frac{1}{10}$ of the enrollment. What fraction of students choose one of these two areas of study?

73. Exercising According to national studies, obesity in America is on the rise. Doctors recommend a minimum of 30 minutes of exercise three times a week to help keep us fit. Suppose during a given week you walk for $\frac{1}{4}$ hour one day, $\frac{2}{3}$ of an hour a second day and $\frac{3}{4}$ of an hour on a third day. Find the total number of hours you have walked as a fraction.

74. Cooking You are making pancakes for breakfast and need $\frac{3}{4}$ of a cup of milk for your batter. You discover that you only have $\frac{1}{2}$ cup of milk in the refrigerator. How much more milk do you need?

75. Conference Attendees At a recent mathematics conference $\frac{1}{3}$ of the attendees were teachers, $\frac{1}{4}$ were software salespersons, and $\frac{1}{12}$ were representatives from various book publishing companies. The remainder of the people in the conference center were employees of the center. What fraction represents the employees of the conference center?

76. Painting Recently you purchased a $\frac{1}{2}$ gallon of paint to paint your dorm room. Once the job was finished you realized that you only used $\frac{1}{3}$ of a gallon. What fractional amount of paint is left in your can?

77. Subdivision A 6-acre piece of land is subdivided into $\frac{3}{5}$-acre lots. How many lots are there?

78. Cutting Wood A 12-foot piece of wood is cut into shelves. If each is $\frac{3}{4}$ foot in length, how many shelves are there?

Find the perimeter of each figure.

79.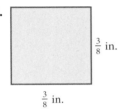

80.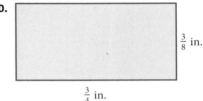

81.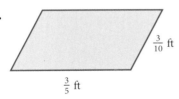

82.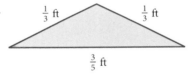

Extending the Concepts

Arithmetic Sequences An arithmetic sequence is a sequence in which each term comes from the previous term by adding the same number each time. For example, the sequence $1, \frac{3}{2}, 2, \frac{5}{2}, \ldots$ is an arithmetic sequence that starts with the number 1. Then each term after that is found by adding $\frac{1}{2}$ to the previous term. By observing this fact, we know that the next term in the sequence will be $\frac{5}{2} + \frac{1}{2} = \frac{6}{2} = 3$.
Find the next number in each arithmetic sequence below.

83. $1, \frac{4}{3}, \frac{5}{3}, 2, \ldots$

84. $1, \frac{5}{4}, \frac{3}{2}, \frac{7}{4}, \ldots$

85. $\frac{3}{2}, 2, \frac{5}{2}, \ldots$

86. $\frac{2}{3}, 1, \frac{4}{3}, \ldots$

Getting Ready for the Next Section

Simplify.

87. $9 \cdot 6 + 5$

88. $4 \cdot 6 + 3$

89. Write 2 as a fraction with denominator 8.

90. Write 2 as a fraction with denominator 4.

91. Write 1 as a fraction with denominator 8.

92. Write 5 as a fraction with denominator 4.

Add.

93. $\dfrac{8}{4} + \dfrac{3}{4}$

94. $\dfrac{16}{8} + \dfrac{1}{8}$

95. $2 + \dfrac{1}{8}$

96. $2 + \dfrac{3}{4}$

97. $1 + \dfrac{1}{8}$

98. $5 + \dfrac{3}{4}$

Divide.

99. $11 \div 4$

100. $10 \div 3$

101. $208 \div 24$

102. $207 \div 26$

Maintaining Your Skills

Multiply or divide as indicated.

103. $\dfrac{3}{4} \div \dfrac{5}{6}$

104. $12 \div \dfrac{1}{2}$

105. $12 \cdot \dfrac{2}{3}$

106. $12 \cdot \dfrac{3}{4}$

107. $4 \cdot \dfrac{3}{4}$

108. $4 \cdot \dfrac{1}{2}$

109. $\dfrac{7}{6} \div \dfrac{7}{12}$

110. $\dfrac{9}{10} \div \dfrac{7}{10}$

111. $\dfrac{2}{3} \cdot \dfrac{3}{4} \cdot \dfrac{4}{5} \cdot \dfrac{5}{6} \cdot \dfrac{6}{7}$

112. $\dfrac{11}{12} \cdot \dfrac{10}{11} \cdot \dfrac{9}{10} \cdot \dfrac{8}{9} \cdot \dfrac{7}{8}$

113. $\dfrac{35}{110} \cdot \dfrac{80}{63} \div \dfrac{16}{27}$

114. $\dfrac{20}{72} \cdot \dfrac{42}{18} \div \dfrac{20}{16}$

Mixed-Number Notation

2.6

OBJECTIVES

A Change mixed numbers to improper fractions.

B Change improper fractions to mixed numbers.

TICKET TO SUCCESS

Keep these questions in mind as you read through the section. Then respond in your own words and in complete sentences.

1. What is a mixed number?
2. How do you change a mixed number to an improper fraction?
3. The expression $5\frac{3}{4}$ is equivalent to what addition problem?
4. Why is $\frac{13}{5}$ an improper fraction, but $\frac{3}{5}$ is not an improper fraction?

If you are interested in the stock market, you know that prior to the year 2000, stock prices were given in eighths. For example, at one point in 1990 one share of Intel Corporation was selling at $\$73\frac{5}{8}$, or seventy-three and five-eighths dollars. The number $73\frac{5}{8}$ is called a *mixed number*. It is the sum of a whole number, 73, and a proper fraction, $\frac{5}{8}$, written without a + sign. With mixed-number notation, we leave out the addition sign.

Notation

Here are some further examples of mixed number notation:

$$2\frac{1}{8} = 2 + \frac{1}{8}, \quad 6\frac{5}{9} = 6 + \frac{5}{9}, \quad 11\frac{2}{3} = 11 + \frac{2}{3}$$

The notation used in writing mixed numbers (writing the whole number and the proper fraction next to each other) must always be interpreted as addition. It is a mistake to read $5\frac{3}{4}$ as meaning 5 times $\frac{3}{4}$. If we want to indicate multiplication, we must use parentheses or a multiplication symbol. That is,

$5\frac{3}{4}$ is **not** the same as $5\left(\frac{3}{4}\right)$.

This implies addition.

This implies multiplication.

$5\frac{3}{4}$ is **not** the same as $5 \cdot \frac{3}{4}$.

2.6 Mixed-Number Notation

159

A Changing Mixed Numbers to Improper Fractions

To change a mixed number to an improper fraction, we write the mixed number with the + sign showing and then add the two numbers, as we did earlier.

EXAMPLE 1 Change $2\frac{3}{4}$ to an improper fraction.

SOLUTION

$2\frac{3}{4} = 2 + \frac{3}{4}$ Write the mixed number as a sum.

$= \frac{2}{1} + \frac{3}{4}$ Show that the denominator of 2 is 1.

$= \frac{4 \cdot 2}{4 \cdot 1} + \frac{3}{4}$ Multiply the numerator and the denominator of $\frac{2}{1}$ by 4 so both fractions will have the same denominator.

$= \frac{8}{4} + \frac{3}{4}$

$= \frac{11}{4}$ Add the numerators; keep the common denominator.

The mixed number $2\frac{3}{4}$ is equal to the improper fraction $\frac{11}{4}$. The diagram that follows further illustrates the equivalence of $2\frac{3}{4}$ and $\frac{11}{4}$.

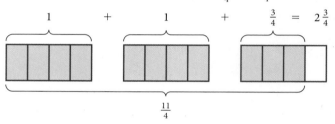

EXAMPLE 2 Change $2\frac{1}{8}$ to an improper fraction.

SOLUTION

$2\frac{1}{8} = 2 + \frac{1}{8}$ Write as addition.

$= \frac{2}{1} + \frac{1}{8}$ Write the whole number 2 as a fraction.

$= \frac{8 \cdot 2}{8 \cdot 1} + \frac{1}{8}$ Change $\frac{2}{1}$ to a fraction with denominator 8.

$= \frac{16}{8} + \frac{1}{8}$

$= \frac{17}{8}$ Add the numerators.

If we look closely at Examples 1 and 2, we can see the following shortcut that will let us change a mixed number to an improper fraction without so many steps.

> **Strategy Changing a Mixed Number to an Improper Fraction (Shortcut)**
>
> **Step 1** Multiply the whole number part of the mixed number by the denominator.
>
> **Step 2** Add your answer to the numerator of the fraction.
>
> **Step 3** Put your new number over the original denominator.

PRACTICE PROBLEMS

1. Change $5\frac{2}{3}$ to an improper fraction.

2. Change $3\frac{1}{6}$ to an improper fraction.

Answers
1. $\frac{17}{3}$
2. $\frac{19}{6}$

EXAMPLE 3 Use the shortcut to change $5\frac{3}{4}$ to an improper fraction.

SOLUTION
1. First, we multiply 4×5 to get 20.
2. Next, we add 20 to 3 to get 23.
3. The improper fraction equal to $5\frac{3}{4}$ is $\frac{23}{4}$.

Here is a diagram showing what we have done:

Step 1 Multiply $4 \times 5 = 20$.

Step 2 Add $20 + 3 = 23$.

Mathematically, our shortcut is written like this:

$$5\frac{3}{4} = \frac{(4 \cdot 5) + 3}{4} = \frac{20 + 3}{4} = \frac{23}{4}$$

The result will always have the same denominator as the original mixed number.

The shortcut shown in Example 3 works because the whole-number part of a mixed number can always be written with a denominator of 1. Therefore, the LCD for a whole number and fraction will always be the denominator of the fraction. That is why we multiply the whole number by the denominator of the fraction.

$$5\frac{3}{4} = 5 + \frac{3}{4} = \frac{5}{1} + \frac{3}{4} = \frac{\mathbf{4 \cdot 5}}{\mathbf{4 \cdot 1}} + \frac{3}{4} = \frac{4 \cdot 5 + 3}{4} = \frac{23}{4}$$

EXAMPLE 4 Change $6\frac{5}{9}$ to an improper fraction.

SOLUTION Using the first method, we have

$$6\frac{5}{9} = 6 + \frac{5}{9} = \frac{6}{1} + \frac{5}{9} = \frac{\mathbf{9 \cdot 6}}{\mathbf{9 \cdot 1}} + \frac{5}{9} = \frac{54}{9} + \frac{5}{9} = \frac{59}{9}$$

Using the shortcut method, we have

$$6\frac{5}{9} = \frac{(9 \cdot 6) + 5}{9} = \frac{54 + 5}{9} = \frac{59}{9}$$

B Changing Improper Fractions to Mixed Numbers

To change an improper fraction to a mixed number, we divide the numerator by the denominator. The result is used to write the mixed number.

EXAMPLE 5 Change $\frac{11}{4}$ to a mixed number.

SOLUTION Dividing 11 by 4 gives us

$$\begin{array}{r} 2 \\ 4\overline{)11} \\ \underline{8} \\ 3 \end{array}$$

We see that 4 goes into 11 two times with 3 for a remainder. We write this as

$$\frac{11}{4} = 2 + \frac{\mathbf{3}}{4} = 2\frac{3}{4}$$

The improper fraction $\frac{11}{4}$ is equivalent to the mixed number $2\frac{3}{4}$.

---

3. Use the shortcut to change $5\frac{2}{3}$ to an improper fraction.

4. Change $6\frac{4}{9}$ to an improper fraction.

CALCULATOR NOTE
The sequence of keys to press on a calculator to obtain the numerator in Example 4 looks like this:

$9 \boxed{\times} 6 \boxed{+} 5 \boxed{=}$

5. Change $\frac{11}{3}$ to a mixed number.

NOTE
This division process shows us how many ones are in $\frac{11}{4}$ and, when the ones are taken out, how many fourths are left.

Answers
3. $\frac{17}{3}$
4. $\frac{58}{9}$
5. $3\frac{2}{3}$

An easy way to visualize the results in Example 5 is to imagine having 11 quarters. Your 11 quarters are equivalent to $\frac{11}{4}$ dollars. In dollars, your quarters are worth 2 dollars plus 3 quarters, or $2\frac{3}{4}$ dollars.

Change each improper fraction to a mixed number.

6. $\frac{14}{5}$

EXAMPLE 6 Change $\frac{10}{3}$ to a mixed number.

SOLUTION Dividing 10 by 3 gives us $3\overline{)10}$ with quotient 3 remainder 1

so $\frac{10}{3} = 3 + \frac{1}{3} = 3\frac{1}{3}$

7. $\frac{207}{26}$

EXAMPLE 7 Change $\frac{208}{24}$ to a mixed number.

SOLUTION $24\overline{)208}$ quotient 8 remainder 16, so $\frac{208}{24} = 8 + \frac{16}{24} = 8 + \frac{2}{3} = 8\frac{2}{3}$

Reduce to lowest terms.

Long Division, Remainders, and Mixed Numbers

Mixed numbers give us another way of writing the answers to long division problems that contain remainders. Here is how we divided 1,690 by 67 in Chapter 1:

$$67\overline{)1690} \quad 25 \text{ R } 15$$
$$\underline{134}$$
$$350$$
$$\underline{335}$$
$$15$$

The answer is 25 with a remainder of 15. Using mixed numbers, we can now write the answer as $25\frac{15}{67}$. That is,

$$\frac{1690}{67} = 25\frac{15}{67}$$

The quotient of 1,690 and 67 is $25\frac{15}{67}$.

Answers

6. $2\frac{4}{5}$
7. $7\frac{25}{26}$

Problem Set 2.6

Moving Toward Success

"There are no secrets to success. It is the result of preparation, hard work, and learning from failure."

—Colin Powell, 1937–present, Chairman of the U.S. Joint Chiefs of Staff and U.S. Secretary of State

1. Why are mistakes important in mathematics?
2. What will you do to make sure you will not make the same mistake again?

A Change each mixed number to an improper fraction. [Examples 1–4]

1. $4\frac{2}{3}$
2. $3\frac{5}{8}$
3. $5\frac{1}{4}$
4. $7\frac{1}{2}$
5. $1\frac{5}{8}$
6. $1\frac{6}{7}$

7. $15\frac{2}{3}$
8. $17\frac{3}{4}$
9. $4\frac{20}{21}$
10. $5\frac{18}{19}$
11. $12\frac{31}{33}$
12. $14\frac{29}{31}$

B Change each improper fraction to a mixed number. [Examples 5–7]

13. $\frac{9}{8}$
14. $\frac{10}{9}$
15. $\frac{19}{4}$
16. $\frac{23}{5}$
17. $\frac{29}{6}$
18. $\frac{7}{2}$

19. $\frac{13}{4}$
20. $\frac{41}{15}$
21. $\frac{109}{27}$
22. $\frac{319}{23}$
23. $\frac{428}{15}$
24. $\frac{769}{27}$

Getting Ready for the Next Section

Change to improper fractions.

25. $2\frac{3}{4}$ **26.** $3\frac{1}{5}$ **27.** $4\frac{5}{8}$ **28.** $1\frac{3}{5}$ **29.** $2\frac{4}{5}$ **30.** $5\frac{9}{10}$

Find the following products. (Multiply.)

31. $\frac{3}{8} \cdot \frac{3}{5}$ **32.** $\frac{11}{4} \cdot \frac{16}{5}$ **33.** $\frac{2}{3}\left(\frac{9}{16}\right)$ **34.** $\frac{7}{10}\left(\frac{5}{21}\right)$

Find the quotients. (Divide.)

35. $\frac{4}{5} \div \frac{7}{8}$ **36.** $\frac{3}{4} \div \frac{1}{2}$ **37.** $\frac{8}{5} \div \frac{14}{5}$ **38.** $\frac{59}{10} \div 2$

Maintaining Your Skills

Perform the indicated operations.

39. $\frac{2}{3} \cdot \frac{3}{4} \div \frac{5}{8}$

40. $4 \cdot \frac{7}{6} \div 7$

41. $\frac{3}{4} \div \frac{1}{2} \cdot 6$

42. $\frac{2}{3} \div \frac{1}{6} \cdot 12$

43. $12 \div \frac{6}{7} \cdot 7$

44. $15 \div \frac{5}{8} \cdot 16$

Multiplication and Division with Mixed Numbers

2.7

OBJECTIVES

A Multiply mixed numbers.

B Divide mixed numbers.

TICKET TO SUCCESS

Keep these questions in mind as you read through the section. Then respond in your own words and in complete sentences.

1. What is the first step when multiplying or dividing mixed numbers?
2. What is the reciprocal of $2\frac{4}{5}$?
3. When dividing mixed numbers, why should we change all mixed numbers to improper fractions before doing the division?
4. Dividing $5\frac{9}{10}$ by 2 is equivalent to multiplying $5\frac{9}{10}$ by what number?

The figure here shows one of the nutrition labels we worked with in Chapter 1. It is from a can of Italian tomatoes. Notice toward the top of the label, the number of servings in the can is $3\frac{1}{2}$. The number $3\frac{1}{2}$ is a *mixed number*. If we want to know how many calories are in the whole can of tomatoes, we must be able to multiply $3\frac{1}{2}$ by 25 (the number of calories per serving). Multiplication with mixed numbers is one of the topics we will cover in this section.

The procedures for multiplying and dividing mixed numbers are the same as those we used in Sections 2.3 and 2.4 to multiply and divide fractions. The only additional work involved is changing the mixed numbers to improper fractions before we actually multiply or divide.

CANNED ITALIAN TOMATOES

PRACTICE PROBLEMS

1. Multiply: $2\frac{3}{4} \cdot 4\frac{1}{3}$.

A Multiplying Mixed Numbers

EXAMPLE 1 Multiply: $2\frac{3}{4} \cdot 3\frac{1}{5}$.

SOLUTION We begin by changing each mixed number to an improper fraction.

$$2\frac{3}{4} = \frac{11}{4} \quad \text{and} \quad 3\frac{1}{5} = \frac{16}{5}$$

Using the resulting improper fractions, we multiply as usual. (That is, we multiply numerators and multiply denominators.)

$$\frac{11}{4} \cdot \frac{16}{5} = \frac{11 \cdot 16}{4 \cdot 5}$$
$$= \frac{11 \cdot \cancel{4} \cdot 4}{\cancel{4} \cdot 5}$$
$$= \frac{44}{5} \quad \text{or} \quad 8\frac{4}{5} \quad \blacksquare$$

2. Multiply: $2 \cdot 3\frac{5}{8}$.

EXAMPLE 2 Multiply: $3 \cdot 4\frac{5}{8}$.

SOLUTION Writing each number as an improper fraction, we have

$$3 = \frac{3}{1} \quad \text{and} \quad 4\frac{5}{8} = \frac{37}{8}$$

The complete problem looks like this:

$$3 \cdot 4\frac{5}{8} = \frac{3}{1} \cdot \frac{37}{8} \quad \text{Change to improper fractions.}$$
$$= \frac{111}{8} \quad \text{Multiply numerators and multiply denominators.}$$
$$= 13\frac{7}{8} \quad \text{Write the answer as a mixed number.} \quad \blacksquare$$

NOTE
As you can see, once you have changed each mixed number to an improper fraction, you multiply the resulting fractions the same way you did in Section 2.3.

B Dividing Mixed Numbers

Dividing mixed numbers also requires that we change all mixed numbers to improper fractions before we actually do the division.

3. Divide: $1\frac{3}{5} \div 3\frac{2}{5}$.

EXAMPLE 3 Divide: $1\frac{3}{5} \div 2\frac{4}{5}$.

SOLUTION We begin by rewriting each mixed number as an improper fraction.

$$1\frac{3}{5} = \frac{8}{5} \quad \text{and} \quad 2\frac{4}{5} = \frac{14}{5}$$

We then divide using the same method we used in Section 2.4: multiply by the reciprocal of the divisor. Here is the complete problem:

$$1\frac{3}{5} \div 2\frac{4}{5} = \frac{8}{5} \div \frac{14}{5} \quad \text{Change to improper fractions.}$$
$$= \frac{8}{5} \cdot \frac{5}{14} \quad \text{To divide by } \frac{14}{5}, \text{ multiply by } \frac{5}{14}.$$
$$= \frac{4 \cdot \cancel{2} \cdot \cancel{5}}{\cancel{5} \cdot \cancel{2} \cdot 7} \quad \text{Divide out factors common to the numerator and denominator.}$$
$$= \frac{4}{7} \quad \text{Answer in lowest terms.} \quad \blacksquare$$

Answers
1. $11\frac{11}{12}$
2. $7\frac{1}{4}$
3. $\frac{8}{17}$

Problem Set 2.7

Moving Toward Success

"The purpose of learning is growth, and our minds, unlike our bodies, can continue growing as we continue to live."
—Mortimer Adler, 1902–2001, American philosopher, educator, and author

1. Why should you keep a list of difficult problems?

2. How will you use your list of difficult problems to further your success in this course?
   a. Rework the problems on your list as you progress through the course.
   b. Add a problem or two to the list but never access it again.
   c. Use the list to help you study for upcoming exams.
   d. Answers a and c

A Write your answers as proper fractions or mixed numbers, not as improper fractions. Find the following products. (Multiply.) [Examples 1, 2]

1. $3\frac{2}{5} \cdot 1\frac{1}{2}$
2. $2\frac{1}{3} \cdot 6\frac{3}{4}$
3. $5\frac{1}{8} \cdot 2\frac{2}{3}$
4. $1\frac{5}{6} \cdot 1\frac{4}{5}$

5. $2\frac{1}{10} \cdot 3\frac{3}{10}$
6. $4\frac{7}{10} \cdot 3\frac{1}{10}$
7. $1\frac{1}{4} \cdot 4\frac{2}{3}$
8. $3\frac{1}{2} \cdot 2\frac{1}{6}$

9. $2 \cdot 4\frac{7}{8}$
10. $10 \cdot 1\frac{1}{4}$
11. $\frac{3}{5} \cdot 5\frac{1}{3}$
12. $\frac{2}{3} \cdot 4\frac{9}{10}$

13. $2\frac{1}{2} \cdot 3\frac{1}{3} \cdot 1\frac{1}{2}$
14. $3\frac{1}{5} \cdot 5\frac{1}{6} \cdot 1\frac{1}{8}$
15. $\frac{3}{4} \cdot 7 \cdot 1\frac{4}{5}$
16. $\frac{7}{8} \cdot 6 \cdot 1\frac{5}{6}$

B Find the following quotients. (Divide.) [Example 3]

17. $3\frac{1}{5} \div 4\frac{1}{2}$
18. $1\frac{4}{5} \div 2\frac{5}{6}$
19. $6\frac{1}{4} \div 3\frac{3}{4}$
20. $8\frac{2}{3} \div 4\frac{1}{3}$

21. $10 \div 2\frac{1}{2}$
22. $12 \div 3\frac{1}{6}$
23. $8\frac{3}{5} \div 2$
24. $12\frac{6}{7} \div 3$

25. $\left(\frac{3}{4} \div 2\frac{1}{2}\right) \div 3$
26. $\frac{7}{8} \div \left(1\frac{1}{4} \div 4\right)$
27. $\left(8 \div 1\frac{1}{4}\right) \div 2$
28. $8 \div \left(1\frac{1}{4} \div 2\right)$

29. $2\frac{1}{2} \cdot \left(3\frac{2}{5} \div 4\right)$
30. $4\frac{3}{5} \cdot \left(2\frac{1}{4} \div 5\right)$

31. Find the product of $2\frac{1}{2}$ and 3.

32. Find the product of $\frac{1}{5}$ and $3\frac{2}{3}$.

33. What is the quotient of $2\frac{3}{4}$ and $3\frac{1}{4}$?

34. What is the quotient of $1\frac{1}{5}$ and $2\frac{2}{5}$?

Applying the Concepts

35. Cooking A certain recipe calls for $2\frac{3}{4}$ cups of sugar. If the recipe is to be doubled, how much sugar should be used?

36. Cooking A recipe calls for $3\frac{1}{4}$ cups of flour. If Diane is using only half the recipe, how much flour should she use?

37. Number Problem Find $\frac{3}{4}$ of $1\frac{7}{9}$. (Remember that *of* means multiply.)

38. Number Problem Find $\frac{5}{6}$ of $2\frac{4}{15}$.

39. Cost of Gasoline If a gallon of gas costs $335\frac{9}{10}$¢, how much do 8 gallons cost?

40. Cost of Gasoline If a gallon of gas costs $353\frac{9}{10}$¢, how much does $\frac{1}{2}$ gallon cost?

41. Distance Traveled If a car can travel $32\frac{3}{4}$ miles on a gallon of gas, how far will it travel on 5 gallons of gas?

42. Sewing If it takes $1\frac{1}{2}$ yards of material to make a pillow cover, how much material will it take to make 3 pillow covers?

43. Buying Stocks Assume that you have $1000 to invest in the stock market. Because you own an iPod™ and an iPhone™, you decide to buy Apple stock. It is currently selling at a cost of $\$150\frac{7}{8}$ per share. At this price how many shares can you buy?

44. Subdividing Land A local developer owns $145\frac{3}{4}$ acres of land that he hopes to subdivide into $2\frac{1}{2}$ acre home site lots to sell. How many home sites can be developed from this tract of land?

45. Selling Stocks You inherit 100 shares of Cisco stock that has a current value of $\$25\frac{1}{6}$ per share. How much will you receive when you sell the stock?

46. Gas Mileage You won a new car and are anxious to see what kind of gas mileage you get. You travel $427\frac{1}{5}$ miles before needing to fill your tank. You purchase $13\frac{3}{4}$ gallons of gas. How many miles were you able to travel on a single gallon of gas?

47. Area Find the area of a bedroom that measures $11\frac{1}{2}$ ft by $15\frac{7}{8}$ ft.

48. Building Shelves You are building a small bookcase. You need three shelves, each with a length of $4\frac{7}{8}$ ft. You bought a piece of wood that is 15 ft long. Will this board be long enough?

49. The Google Earth image shows some fields in the midwestern part of the United States. The rectangle outlines a corn field and gives the dimensions in miles. Find the area of the corn field written as a mixed number.

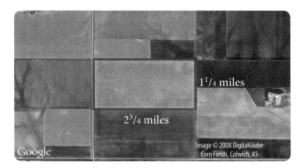

50. Recall our discussion of Crater Lake from this chapter's introduction. The Google Earth map below shows Crater Lake National Park in Oregon. If Crater Lake is roughly the shape of a circle with a radius of $2\frac{1}{2}$ miles, how long is the shoreline? Use the formula $2\pi r$ for the circumference where r is the radius and $\frac{22}{7}$ is π.

Nutrition The figure below shows nutrition labels for two different cans of Italian tomatoes.

CANNED TOMATOES 1

Nutrition Facts
Serving Size 1/2 cup (121g)
Servings Per Container: about 3 1/2

Amount Per Serving	
Calories 45	Calories from fat 0

	% Daily Value*
Total Fat 0g	0%
Saturated Fat 0g	0%
Cholesterol 0mg	0%
Sodium 560mg	23%
Total Carbohydrate 11g	4%
Dietary Fiber 2g	8%
Sugars 9g	
Protein 1g	

| Vitamin A 10% | • | Vitamin C 25% |
| Calcium 2% | • | Iron 2% |

*Percent Daily Values are based on a 2,000 calorie diet.

CANNED TOMATOES 2

Nutrition Facts
Serving Size 1/2 cup (121g)
Servings Per Container: about 3 1/2

Amount Per Serving	
Calories 25	Calories from fat 0

	% Daily Value*
Total Fat 0g	0%
Saturated Fat 0g	0%
Cholesterol 0mg	0%
Sodium 300mg	12%
Potassium 145mg	4%
Total Carbohydrate 4g	2%
Dietary Fiber 1g	4%
Sugars 4g	
Protein 1g	

| Vitamin A 20% | • | Vitamin C 15% |
| Calcium 4% | • | Iron 15% |

*Percent Daily Values are based on a 2,000 calorie diet. Your daily values may be higher or lower depending on your calorie needs.

51. Compare the total number of calories in the two cans of tomatoes.

52. Compare the total amount of sugar in the two cans of tomatoes.

53. Compare the total amount of sodium in the two cans of tomatoes.

54. Compare the total amount of protein in the two cans of tomatoes.

Getting Ready for the Next Section

55. Write as equivalent fractions with denominator 15.
  a. $\dfrac{2}{3}$ b. $\dfrac{1}{5}$ c. $\dfrac{3}{5}$ d. $\dfrac{1}{3}$

56. Write as equivalent fractions with denominator 12.
  a. $\dfrac{3}{4}$ b. $\dfrac{1}{3}$ c. $\dfrac{5}{6}$ d. $\dfrac{1}{4}$

57. Write as equivalent fractions with denominator 20.
  a. $\dfrac{1}{4}$ b. $\dfrac{3}{5}$ c. $\dfrac{9}{10}$ d. $\dfrac{1}{10}$

58. Write as equivalent fractions with denominator 24.
  a. $\dfrac{3}{4}$ b. $\dfrac{7}{8}$ c. $\dfrac{5}{8}$ d. $\dfrac{3}{8}$

Maintaining Your Skills

Add or subtract the following fractions, as indicated.

59. $\dfrac{2}{3} + \dfrac{1}{5}$

60. $\dfrac{3}{4} + \dfrac{5}{6}$

61. $\dfrac{2}{3} + \dfrac{8}{9}$

62. $\dfrac{1}{4} + \dfrac{3}{5} + \dfrac{9}{10}$

63. $\dfrac{9}{10} - \dfrac{3}{10}$

64. $\dfrac{7}{10} - \dfrac{3}{5}$

65. $\dfrac{1}{14} + \dfrac{3}{21}$

66. $\dfrac{5}{12} - \dfrac{1}{3}$

Extending the Concepts

To find the square of a mixed number, we first change the mixed number to an improper fraction, and then we square the result. For example,

$$\left(2\dfrac{1}{2}\right)^2 = \left(\dfrac{5}{2}\right)^2 = \dfrac{25}{4}$$

If we are asked to write our answer as a mixed number, we write it as $6\dfrac{1}{4}$.

Find each of the following squares, and write your answers as mixed numbers.

67. $\left(1\dfrac{1}{2}\right)^2$

68. $\left(3\dfrac{1}{2}\right)^2$

69. $\left(1\dfrac{3}{4}\right)^2$

70. $\left(2\dfrac{3}{4}\right)^2$

Addition and Subtraction with Mixed Numbers

2.8

OBJECTIVES

A Perform addition and subtraction with mixed numbers.

B Perform subtraction involving borrowing with mixed numbers.

TICKET TO SUCCESS

Keep these questions in mind as you read through the section. Then respond in your own words and in complete sentences.

1. Looking at Example 1, why would you use Method 1 instead of Method 2 to add the mixed numbers?
2. How is subtraction of mixed numbers similar to addition of mixed numbers?
3. Is it necessary to "borrow" when subtracting $1\frac{3}{10}$ from $3\frac{9}{10}$?
4. To subtract $1\frac{2}{7}$ from 10, it is necessary to rewrite 10 as what mixed number?

Suppose you are shopping for a new car that gets better gas mileage than your current car. You can drive your current car $285\frac{1}{5}$ miles before having to refill your gas tank. You are considering two new cars: one will travel $330\frac{3}{4}$ miles before needing more gas, and the second will travel $319\frac{3}{8}$ miles. How many more miles will each of the new cars travel than your current car? To answer this question, we must be able to subtract mixed numbers.

The notation we use for mixed numbers is especially useful for addition and subtraction. When adding and subtracting mixed numbers, we will assume you recall how to go about finding a least common denominator (LCD). (If you don't remember, then review Section 2.5.)

A Combining Mixed Numbers

EXAMPLE 1 Add: $3\frac{2}{3} + 4\frac{1}{5}$.

SOLUTION **Method 1** We begin by writing each mixed number showing the + sign. We then apply the commutative and associative properties to rearrange the order and grouping.

$$3\frac{2}{3} + 4\frac{1}{5} = 3 + \frac{2}{3} + 4 + \frac{1}{5}$$ Expand each number to show the + sign.

PRACTICE PROBLEMS

1. Add: $3\frac{2}{3} + 2\frac{1}{4}$.

Answer

1. $5\frac{11}{12}$

Chapter 2 Fractions and Mixed Numbers

$$= 3 + 4 + \frac{2}{3} + \frac{1}{5} \quad \text{Commutative property}$$

$$= (3 + 4) + \left(\frac{2}{3} + \frac{1}{5}\right) \quad \text{Associative property}$$

$$= 7 + \left(\frac{\mathbf{5} \cdot 2}{\mathbf{5} \cdot 3} + \frac{\mathbf{3} \cdot 1}{\mathbf{3} \cdot 5}\right) \quad \text{Add } 3 + 4 = 7; \text{ then multiply to get the LCD.}$$

$$= 7 + \left(\frac{10}{15} + \frac{3}{15}\right) \quad \text{Write each fraction with the LCD.}$$

$$= 7 + \frac{13}{15} \quad \text{Add the numerators.}$$

$$= 7\frac{13}{15} \quad \text{Write the answer in mixed-number notation.}$$

Method 2 As you can see, we obtain our result by adding the whole-number parts $(3 + 4 = 7)$ and the fraction parts $\left(\frac{2}{3} + \frac{1}{5} = \frac{13}{15}\right)$ of each mixed number. Knowing this, we can save ourselves some writing by doing the same problem in columns.

$$3\frac{2}{3} = 3\frac{2 \cdot \mathbf{5}}{3 \cdot \mathbf{5}} = 3\frac{10}{15} \quad \text{Add whole numbers.}$$

$$+ 4\frac{1}{5} = 4\frac{1 \cdot \mathbf{3}}{5 \cdot \mathbf{3}} = 4\frac{3}{15} \quad \text{Then add fractions.}$$

$$7\frac{13}{15}$$

Write each fraction with LCD 15. ■

The second method shown above requires less writing and lends itself to mixed-number notation. We will use this method for the rest of this section.

> **NOTE**
> You should try both methods given in Example 1 on Practice Problem 1.

2. Add: $5\frac{3}{4} + 6\frac{4}{5}$.

EXAMPLE 2   Add: $5\frac{3}{4} + 9\frac{5}{6}$.

SOLUTION   The LCD for 4 and 6 is 12. Writing the mixed numbers in a column and then adding looks like this:

$$5\frac{3}{4} = 5\frac{3 \cdot 3}{4 \cdot 3} = 5\frac{9}{12}$$

$$+ 9\frac{5}{6} = 9\frac{5 \cdot 2}{6 \cdot 2} = 9\frac{10}{12}$$

$$14\frac{19}{12}$$

The fraction part of the answer is an improper fraction. We rewrite it as a whole number and a proper fraction.

$$14\frac{19}{12} = 14 + \frac{19}{12} \quad \text{Write the mixed number with a + sign.}$$

$$= 14 + 1\frac{7}{12} \quad \text{Write } \tfrac{19}{12} \text{ as a mixed number.}$$

$$= 15\frac{7}{12} \quad \text{Add 14 and 1.} \quad ■$$

> **NOTE**
> Once you see how to change from a whole number and an improper fraction to a whole number and a proper fraction, you will be able to do this step without showing any work.

Answer
2. $12\frac{11}{20}$

2.8 Addition and Subtraction with Mixed Numbers

EXAMPLE 3 Add: $5\frac{2}{3} + 6\frac{8}{9}$.

SOLUTION

$$5\frac{2}{3} = 5\frac{2 \cdot 3}{3 \cdot 3} = 5\frac{6}{9}$$

$$+ 6\frac{8}{9} = 6\frac{8}{9} = 6\frac{8}{9}$$

$$11\frac{14}{9} = 12\frac{5}{9}$$

The last step involves writing $\frac{14}{9}$ as $1\frac{5}{9}$ and then adding 11 and 1 to get 12. ■

3. Add: $6\frac{3}{4} + 2\frac{7}{8}$.

EXAMPLE 4 Add: $3\frac{1}{4} + 2\frac{3}{5} + 1\frac{9}{10}$.

SOLUTION The LCD is 20. We rewrite each fraction as an equivalent fraction with denominator 20 and add.

$$3\frac{1}{4} = 3\frac{1 \cdot 5}{4 \cdot 5} = 3\frac{5}{20}$$

$$2\frac{3}{5} = 2\frac{3 \cdot 4}{5 \cdot 4} = 2\frac{12}{20}$$

$$+ 1\frac{9}{10} = 1\frac{9 \cdot 2}{10 \cdot 2} = 1\frac{18}{20}$$

$$6\frac{35}{20} = 7\frac{15}{20} = 7\frac{3}{4} \quad \text{Reduce to lowest terms.}$$

$$\frac{35}{20} = 1\frac{15}{20}$$

Change to a mixed number. ■

4. Add: $2\frac{1}{3} + 1\frac{1}{4} + 3\frac{11}{12}$.

We should note here that we could have worked each of the first four examples in this section by first changing each mixed number to an improper fraction and then adding as we did in Section 2.5. To illustrate, if we were to work Example 4 this way, it would look like this:

$$3\frac{1}{4} + 2\frac{3}{5} + 1\frac{9}{10} = \frac{13}{4} + \frac{13}{5} + \frac{19}{10} \quad \text{Change to improper fractions.}$$

$$= \frac{13 \cdot 5}{4 \cdot 5} + \frac{13 \cdot 4}{5 \cdot 4} + \frac{19 \cdot 2}{10 \cdot 2} \quad \text{LCD is 20.}$$

$$= \frac{65}{20} + \frac{52}{20} + \frac{38}{20} \quad \text{Equivalent fractions}$$

$$= \frac{155}{20} \quad \text{Add numerators.}$$

$$= 7\frac{15}{20} = 7\frac{3}{4} \quad \text{Change to a mixed number, and reduce.}$$

As you can see, the result is the same as the result we obtained in Example 4.

There are advantages to both methods. The method just shown works well when the whole-number parts of the mixed numbers are small. The vertical method shown in Examples 1–4 works well when the whole-number parts of the mixed numbers are large.

Answers

3. $9\frac{5}{8}$

4. $7\frac{1}{2}$

5. Subtract: $4\frac{7}{8} - 1\frac{5}{8}$.

Subtraction with mixed numbers is very similar to addition with mixed numbers.

EXAMPLE 5 Subtract: $3\frac{9}{10} - 1\frac{3}{10}$.

SOLUTION Because the denominators are the same, we simply subtract the whole numbers and subtract the fractions.

$$\begin{array}{r} 3\frac{9}{10} \\ -1\frac{3}{10} \\ \hline 2\frac{6}{10} = 2\frac{3}{5} \end{array}$$

Reduce to lowest terms.

An easy way to visualize the results in Example 5 is to imagine 3 dollar bills and 9 dimes in your pocket. If you spend 1 dollar and 3 dimes, you will have 2 dollars and 6 dimes left.

6. Subtract: $12\frac{7}{10} - 7\frac{2}{5}$.

EXAMPLE 6 Subtract: $12\frac{7}{10} - 8\frac{3}{5}$.

SOLUTION The common denominator is 10. We must rewrite $\frac{3}{5}$ as an equivalent fraction with denominator 10.

$$\begin{array}{rcccc} 12\frac{7}{10} & = & 12\frac{7}{10} & = & 12\frac{7}{10} \\ -8\frac{3}{5} & = & -8\frac{3 \cdot 2}{5 \cdot 2} & = & -8\frac{6}{10} \\ \hline & & & & 4\frac{1}{10} \end{array}$$

B Borrowing with Mixed Numbers

Sometimes we'll encounter subtraction problem with mixed numbers where it is necessary to borrow a 1 from one of the whole numbers. Let's work three examples of borrowing with mixed numbers.

7. Subtract: $10 - 5\frac{4}{7}$.

EXAMPLE 7 Subtract: $10 - 5\frac{2}{7}$.

SOLUTION In order to have a fraction from which to subtract $\frac{2}{7}$, we borrow 1 from 10 and rewrite the 1 we borrow as $\frac{7}{7}$. The process looks like this:

$$\begin{array}{rl} 10 = & 9\frac{7}{7} \quad \text{We rewrite 10 as } 9 + 1, \text{ which is } 9 + \frac{7}{7} = 9\frac{7}{7}. \\ -5\frac{2}{7} = & -5\frac{2}{7} \quad \text{Then we can subtract as usual.} \\ \hline & 4\frac{5}{7} \end{array}$$

NOTE
Convince yourself that 10 is the same as $9\frac{7}{7}$. The reason we choose to write the 1 we borrowed as $\frac{7}{7}$ is that the fraction we eventually subtracted from $\frac{7}{7}$ was $\frac{2}{7}$. Both fractions must have the same denominator, 7, so that we can subtract.

Answers

5. $3\frac{1}{4}$

6. $5\frac{3}{10}$

7. $4\frac{3}{7}$

2.8 Addition and Subtraction with Mixed Numbers

EXAMPLE 8 Subtract: $8\frac{1}{4} - 3\frac{3}{4}$.

SOLUTION Because $\frac{3}{4}$ is larger than $\frac{1}{4}$, we again need to borrow 1 from the whole number. The 1 that we borrow from the 8 is rewritten as $\frac{4}{4}$, because 4 is the denominator of both fractions.

$$\begin{aligned} 8\frac{1}{4} &= 7\frac{5}{4} \\ -3\frac{3}{4} &= -3\frac{3}{4} \\ \hline &\ 4\frac{2}{4} = 4\frac{1}{2} \end{aligned}$$

Borrow 1 in the form $\frac{4}{4}$; then $\frac{4}{4} + \frac{1}{4} = \frac{5}{4}$.

Reduce to lowest terms.

8. Subtract: $6\frac{1}{3} - 2\frac{2}{3}$.

EXAMPLE 9 Subtract: $4\frac{3}{4} - 1\frac{5}{6}$.

SOLUTION This is about as complicated as it gets with subtraction of mixed numbers. We begin by rewriting each fraction with the common denominator 12.

$$\begin{aligned} 4\frac{3}{4} &= 4\frac{3 \cdot 3}{4 \cdot 3} = 4\frac{9}{12} \\ -1\frac{5}{6} &= -1\frac{5 \cdot 2}{6 \cdot 2} = -1\frac{10}{12} \end{aligned}$$

Because $\frac{10}{12}$ is larger than $\frac{9}{12}$, we must borrow 1 from 4 in the form $\frac{12}{12}$ before we subtract.

$$\begin{aligned} 4\frac{9}{12} &= 3\frac{21}{12} \\ -1\frac{10}{12} &= -1\frac{10}{12} \\ \hline &\ 2\frac{11}{12} \end{aligned}$$

$4 = 3 + 1 = 3 + \frac{12}{12}$, so $4\frac{9}{12} = \left(3 + \frac{12}{12}\right) + \frac{9}{12}$
$= 3 + \left(\frac{12}{12} + \frac{9}{12}\right)$
$= 3 + \frac{21}{12}$
$= 3\frac{21}{12}$

9. Subtract: $6\frac{3}{4} - 2\frac{5}{6}$.

Answers
8. $3\frac{2}{3}$
9. $3\frac{11}{12}$

Problem Set 2.8

Moving Toward Success

"You learn something every day if you pay attention."

—Ray LaBlond, Author

1. When you do your homework, you usually work a number of similar problems at a time. On a test, the problems will be varied. Why is it important to pay attention to the instructions on your homework *and* on a test?
2. Give an example where not reading instructions could lead to a wrong answer.

A Add and subtract the following mixed numbers as indicated. [Examples 1–6]

1. $2\frac{1}{5} + 3\frac{3}{5}$
2. $8\frac{2}{9} + 1\frac{5}{9}$
3. $4\frac{3}{10} + 8\frac{1}{10}$
4. $5\frac{2}{7} + 3\frac{3}{7}$

5. $6\frac{8}{9} - 3\frac{4}{9}$
6. $12\frac{5}{12} - 7\frac{1}{12}$
7. $9\frac{1}{6} + 2\frac{5}{6}$
8. $9\frac{1}{4} + 5\frac{3}{4}$

9. $3\frac{5}{8} - 2\frac{1}{4}$
10. $7\frac{9}{10} - 6\frac{3}{5}$
11. $11\frac{1}{3} + 2\frac{5}{6}$
12. $1\frac{5}{8} + 2\frac{1}{2}$

13. $7\frac{5}{12} - 3\frac{1}{3}$
14. $7\frac{3}{4} - 3\frac{5}{12}$
15. $6\frac{1}{3} - 4\frac{1}{4}$
16. $5\frac{4}{5} - 3\frac{1}{3}$

17. $10\frac{5}{6} + 15\frac{3}{4}$
18. $11\frac{7}{8} + 9\frac{1}{6}$
19. $5\frac{2}{3} + 6\frac{1}{3}$
20. $8\frac{5}{6} + 9\frac{5}{6}$

21. $10\dfrac{13}{16}$
   $-\ 8\dfrac{5}{16}$

22. $17\dfrac{7}{12}$
   $-\ 9\dfrac{5}{12}$

23. $6\dfrac{1}{2}$
   $+\ 2\dfrac{5}{14}$

24. $9\dfrac{11}{12}$
   $+\ 4\dfrac{1}{6}$

25. $1\dfrac{5}{8}$
   $+\ 1\dfrac{3}{4}$

26. $7\dfrac{6}{7}$
   $+\ 2\dfrac{3}{14}$

27. $4\dfrac{2}{3}$
   $+\ 5\dfrac{3}{5}$

28. $9\dfrac{4}{9}$
   $+\ 1\dfrac{1}{6}$

29. $5\dfrac{4}{10}$
   $-\ 3\dfrac{1}{3}$

30. $12\dfrac{7}{8}$
   $-\ 3\dfrac{5}{6}$

31. $10\dfrac{1}{20}$
   $+\ 11\dfrac{4}{5}$

32. $18\dfrac{7}{8}$
   $+\ 19\dfrac{1}{12}$

A Find the following sums. (Add.) [Examples 1–4]

33. $1\dfrac{1}{4} + 2\dfrac{3}{4} + 5$

34. $6 + 5\dfrac{3}{5} + 8\dfrac{2}{5}$

35. $7\dfrac{1}{10} + 8\dfrac{3}{10} + 2\dfrac{7}{10}$

36. $5\dfrac{2}{7} + 8\dfrac{1}{7} + 3\dfrac{5}{7}$

37. $\dfrac{3}{4} + 8\dfrac{1}{4} + 5$

38. $\dfrac{5}{8} + 1\dfrac{1}{8} + 7$

39. $3\dfrac{1}{2} + 8\dfrac{1}{3} + 5\dfrac{1}{6}$

40. $4\dfrac{1}{5} + 7\dfrac{1}{3} + 8\dfrac{1}{15}$

41. $8\dfrac{2}{3}$
   $9\dfrac{1}{8}$
   $+\ 6\dfrac{1}{4}$

42. $7\dfrac{3}{5}$
   $8\dfrac{2}{3}$
   $+\ 1\dfrac{1}{5}$

43. $6\dfrac{1}{7}$
   $9\dfrac{3}{14}$
   $+\ 12\dfrac{1}{2}$

44. $1\dfrac{5}{6}$
   $2\dfrac{3}{4}$
   $+\ 5\dfrac{1}{2}$

B The following problems all involve the concept of borrowing. Subtract in each case. [Examples 7–9]

45. $8 - 1\dfrac{3}{4}$

46. $5 - 3\dfrac{1}{3}$

47. $15 - 5\dfrac{3}{10}$

48. $24 - 10\dfrac{5}{12}$

49. $8\dfrac{1}{4} - 2\dfrac{3}{4}$ **50.** $12\dfrac{3}{10} - 5\dfrac{7}{10}$ **51.** $9\dfrac{1}{3} - 8\dfrac{2}{3}$ **52.** $7\dfrac{1}{6} - 6\dfrac{5}{6}$

53. $4\dfrac{1}{4} - 2\dfrac{1}{3}$ **54.** $6\dfrac{1}{5} - 1\dfrac{2}{3}$ **55.** $9\dfrac{2}{3} - 5\dfrac{3}{4}$ **56.** $12\dfrac{5}{6} - 8\dfrac{7}{8}$

57. $16\dfrac{3}{4} - 10\dfrac{4}{5}$ **58.** $18\dfrac{5}{12} - 9\dfrac{3}{4}$ **59.** $10\dfrac{3}{10} - 4\dfrac{4}{5}$ **60.** $9\dfrac{4}{7} - 7\dfrac{2}{3}$

61. $13\dfrac{1}{6} - 12\dfrac{5}{8}$ **62.** $21\dfrac{2}{5} - 20\dfrac{5}{6}$ **63.** $15\dfrac{3}{10} - 11\dfrac{4}{5}$ **64.** $19\dfrac{3}{15} - 10\dfrac{2}{3}$

Applying the Concepts

Stock Prices In March 1995, rumors that after a two-season absence Michael Jordan would return to basketball sent stock prices for the companies whose products he endorses higher. Suppose the table at the right gives some of the details of those increases. Use the table to work Problems 65–70.

STOCK PRICES FOR COMPANIES WITH MICHAEL JORDAN ENDORSEMENTS			
Company 3/8/95–3/13/95	Product Endorsed	Stock Price (Dollars) Before Rumors	After Rumors
Nike	Air Jordans	$74\dfrac{7}{8}$	$77\dfrac{3}{8}$
Quaker Oats	Gatorade	$32\dfrac{1}{4}$	$32\dfrac{5}{8}$
General Mills	Wheaties	$60\dfrac{1}{2}$	$63\dfrac{3}{8}$
McDonald's	All	$32\dfrac{7}{8}$	$34\dfrac{3}{8}$

65. Find the difference in the price of Nike stock between March 13 and March 8.

66. Find the difference in price of General Mills stock between March 13 and March 8.

67. If you owned 100 shares of Nike stock, how much more are the 100 shares worth on March 13 than on March 8?

68. If you owned 1,000 shares of General Mills stock on March 8, how much more would they be worth on March 13?

69. If you owned 200 shares of McDonald's stock on March 8, how much more would they be worth on March 13?

70. If you owned 100 shares of McDonald's stock on March 8, how much more would they be worth on March 13?

71. Area and Perimeter The diagrams below show the dimensions of playing fields for the National Football League (NFL), the Canadian Football League, and Arena Football.

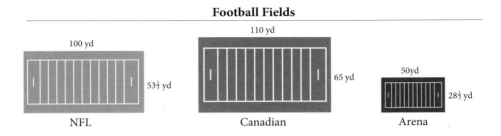

a. Find the perimeter of each football field.

b. Find the area of each football field.

72. Triple Crown The three races that constitute the Triple Crown in horse racing are shown in the table. The information comes from the ESPN website.

a. Write the distances in order from smallest to largest.

b. How much longer is the Belmont Stakes race than the Preakness Stakes?

Race	Distance (miles)
Kentucky Derby	$1\frac{1}{4}$
Preakness Stakes	$1\frac{3}{16}$
Belmont Stakes	$1\frac{1}{2}$

Source: ESPN

73. Length of Jeans A pair of jeans is $32\frac{1}{2}$ inches long. How long are the jeans after they have been washed if they shrink $1\frac{1}{3}$ inches?

74. Manufacturing A sports jersey manufacturer has two rolls of cloth. One roll is $35\frac{1}{2}$ yards, and the other is $62\frac{5}{8}$ yards. What is the total number of yards in the two rolls?

75. Surfboard A surfboard manufacturer decides that the surfboard would be more challenging to ride if its length were decreased by $2\frac{3}{5}$ inches. If the original length was 7 feet $4\frac{1}{2}$ inches, what will be the new length of the board in inches?

76. Wakeboard A wakeboard manufacturer decides that a board will be more efficient in the water if the length is increased by $5\frac{2}{3}$ centimeters. If the original length of the board was $133\frac{2}{5}$ centimeters, what will be the length of the new board in centimeters?

Getting Ready for the Next Section

Multiply or divide as indicated.

77. $\dfrac{11}{8} \cdot \dfrac{29}{8}$ **78.** $\dfrac{3}{4} \div \dfrac{5}{6}$ **79.** $\dfrac{7}{6} \cdot \dfrac{12}{7}$ **80.** $10\dfrac{1}{3} \div 8\dfrac{2}{3}$

Combine.

81. $\dfrac{3}{4} + \dfrac{5}{8}$ **82.** $\dfrac{1}{2} + \dfrac{2}{3}$ **83.** $2\dfrac{3}{8} + 1\dfrac{1}{4}$ **84.** $3\dfrac{2}{3} + 4\dfrac{1}{3}$

Maintaining Your Skills

Use the rule for order of operations to combine the following.

85. $3 + 2 \cdot 7$ **86.** $8 \cdot 3 - 2$ **87.** $4 \cdot 5 - 3 \cdot 2$ **88.** $9 \cdot 7 + 6 \cdot 5$

89. $3 \cdot 2^3 + 5 \cdot 4^2$ **90.** $6 \cdot 5^2 + 2 \cdot 3^3$ **91.** $3[2 + 5(6)]$ **92.** $4[2(3) + 3(5)]$

93. $(7 - 3)(8 + 2)$ **94.** $(9 - 5)(9 + 5)$

Extending the Concepts

95. Find the difference between $6\dfrac{1}{5}$ and $2\dfrac{7}{10}$.

96. Give the difference between $5\dfrac{1}{3}$ and $1\dfrac{5}{6}$.

97. Find the sum of $3\dfrac{1}{8}$ and $2\dfrac{3}{5}$.

98. Find the sum of $1\dfrac{5}{6}$ and $3\dfrac{4}{9}$.

99. Improving Your Quantitative Literacy A column on horse racing in the Daily News in Los Angeles reported that the horse Action This Day ran 3 furlongs in $35\dfrac{1}{5}$ seconds, and another horse, Halfbridled, went two-fifths of a second faster. How many seconds did it take Halfbridled to run 3 furlongs?

Combinations of Operations and Complex Fractions

2.9

OBJECTIVES

A Simplify expressions involving fractions and mixed numbers.

B Simplify complex fractions.

TICKET TO SUCCESS

Keep these questions in mind as you read through the section. Then respond in your own words and in complete sentences.

1. Explain how you would use the order of operations when simplifying the expression $8 - \left(\frac{6}{11}\right)\left(1\frac{5}{6}\right)$.
2. What is a complex fraction?
3. Rewrite $\dfrac{\frac{5}{6}}{\frac{1}{3}}$ as a multiplication problem.
4. How does the fraction bar that separates the numerator of the complex fraction from its denominator in Example 5 work like parentheses?

Suppose your hometown hires you to paint a mural on the side of a local building. The available wall space is a large rectangle that measures $12\frac{1}{2}$ feet high by $14\frac{3}{8}$ feet wide. You begin by drawing a diagonal line from one corner of the rectangle to the opposite corner. The line creates two equal triangles. How would you calculate the area of each triangle? To answer this problem, we need to set up the following expression:

$$\frac{\left(12\frac{1}{2}\right)\left(14\frac{3}{8}\right)}{2}$$

This expression contains both fractions and mixed numbers. In this section, we will learn how to simplify similar expressions, such as the one above.

A Simplifying Expressions Involving Fractions and Mixed Numbers

Let's use what we've learned about fractions in this chapter to work the following examples.

PRACTICE PROBLEMS

1. Simplify the expression:

$$4 + \left(1\frac{1}{2}\right)\left(2\frac{3}{4}\right)$$

EXAMPLE 1 Simplify the expression: $5 + \left(2\frac{1}{2}\right)\left(3\frac{2}{3}\right)$.

SOLUTION The rule for order of operations indicates that we should multiply $2\frac{1}{2}$ times $3\frac{2}{3}$ and then add 5 to the result.

$$5 + \left(2\frac{1}{2}\right)\left(3\frac{2}{3}\right) = 5 + \left(\frac{5}{2}\right)\left(\frac{11}{3}\right)$$ Change the mixed numbers to improper fractions.

$$= 5 + \frac{55}{6}$$ Multiply the improper fractions.

$$= \frac{30}{6} + \frac{55}{6}$$ Write 5 as $\frac{30}{6}$ so both numbers have the same denominator.

$$= \frac{85}{6}$$ Add fractions by adding their numerators.

$$= 14\frac{1}{6}$$ Write the answer as a mixed number.

2. Simplify:

$$\left(\frac{2}{3} + \frac{1}{6}\right)\left(2\frac{5}{6} + 1\frac{1}{3}\right)$$

EXAMPLE 2 Simplify: $\left(\frac{3}{4} + \frac{5}{8}\right)\left(2\frac{3}{8} + 1\frac{1}{4}\right)$.

SOLUTION We begin by combining the numbers inside the parentheses.

$$\frac{3}{4} + \frac{5}{8} = \frac{3 \cdot 2}{4 \cdot 2} + \frac{5}{8} \quad \text{and} \quad 2\frac{3}{8} = \quad 2\frac{3}{8} = \quad 2\frac{3}{8}$$

$$= \frac{6}{8} + \frac{5}{8} \quad\quad\quad\quad\quad + 1\frac{1}{4} = + 1\frac{1 \cdot 2}{4 \cdot 2} = + 1\frac{2}{8}$$

$$= \frac{11}{8} \quad\quad\quad\quad\quad\quad\quad\quad\quad\quad\quad\quad 3\frac{5}{8}$$

Now that we have combined the expressions inside the parentheses, we can complete the problem by multiplying the results.

$$\left(\frac{3}{4} + \frac{5}{8}\right)\left(2\frac{3}{8} + 1\frac{1}{4}\right) = \left(\frac{11}{8}\right)\left(3\frac{5}{8}\right)$$

$$= \frac{11}{8} \cdot \frac{29}{8}$$ Change $3\frac{5}{8}$ to an improper fraction.

$$= \frac{319}{64}$$ Multiply fractions.

$$= 4\frac{63}{64}$$ Write the answer as a mixed number.

3. Simplify:

$$\frac{3}{7} + \frac{1}{3}\left(1\frac{1}{2} + 4\frac{1}{2}\right)^2$$

EXAMPLE 3 Simplify: $\frac{3}{5} + \frac{1}{2}\left(3\frac{2}{3} + 4\frac{1}{3}\right)^2$.

SOLUTION We begin by combining the expressions inside the parentheses.

$$\frac{3}{5} + \frac{1}{2}\left(3\frac{2}{3} + 4\frac{1}{3}\right)^2 = \frac{3}{5} + \frac{1}{2}(8)^2$$ The sum inside the parentheses is 8.

$$= \frac{3}{5} + \frac{1}{2}(64)$$ The square of 8 is 64.

$$= \frac{3}{5} + 32$$ $\frac{1}{2}$ of 64 is 32.

$$= 32\frac{3}{5}$$ The result is a mixed number.

Answers
1. $8\frac{1}{8}$
2. $3\frac{17}{36}$
3. $12\frac{3}{7}$

B Complex Fractions

We will continue to use properties we have developed previously as we encounter a new situation: simplifying a *complex fraction*.

> **Definition**
>
> A **complex fraction** is a fraction in which the numerator and/or the denominator are themselves fractions or combinations of fractions.

Each of the following is a complex fraction:

$$\frac{\frac{3}{4}}{\frac{5}{6}}, \quad \frac{3 + \frac{1}{2}}{2 - \frac{3}{4}}, \quad \frac{\frac{1}{2} + \frac{2}{3}}{\frac{3}{4} - \frac{1}{6}}$$

EXAMPLE 4 Simplify: $\dfrac{\frac{3}{4}}{\frac{5}{6}}$.

SOLUTION This is actually the same as the problem $\frac{3}{4} \div \frac{5}{6}$, because the bar between $\frac{3}{4}$ and $\frac{5}{6}$ indicates division. Therefore, it must be true that

$$\frac{\frac{3}{4}}{\frac{5}{6}} = \frac{3}{4} \div \frac{5}{6}$$

$$= \frac{3}{4} \cdot \frac{6}{5}$$

$$= \frac{18}{20}$$

$$= \frac{9}{10}$$

In Example 4, we use the fact that division by a number and multiplication by its reciprocal produce the same result. We are taking a new problem, simplifying a complex fraction, and thinking of it in terms of a problem we have done previously: division by a fraction.

EXAMPLE 5 Simplify: $\dfrac{\frac{1}{2} + \frac{2}{3}}{\frac{3}{4} - \frac{1}{6}}$.

SOLUTION Let's decide to call the numerator of this complex fraction the *top* of the fraction and its denominator the *bottom* of the complex fraction. It will be less confusing if we name them this way. The LCD for all the denominators on the top and bottom is 12, so we can multiply the top and bottom of this complex fraction by 12 and be sure all the denominators will divide it exactly. This will leave us with only whole numbers on the top and bottom.

$$\frac{\frac{1}{2} + \frac{2}{3}}{\frac{3}{4} - \frac{1}{6}} = \frac{12\left(\frac{1}{2} + \frac{2}{3}\right)}{12\left(\frac{3}{4} - \frac{1}{6}\right)} \quad \text{Multiply the top and bottom by the LCD.}$$

$$= \frac{12 \cdot \frac{1}{2} + 12 \cdot \frac{2}{3}}{12 \cdot \frac{3}{4} - 12 \cdot \frac{1}{6}} \quad \text{Distributive property}$$

4. Simplify:

$$\dfrac{\frac{2}{3}}{\frac{5}{9}}$$

5. Simplify: $\dfrac{\frac{1}{2} + \frac{3}{4}}{\frac{2}{3} - \frac{1}{4}}$

> **NOTE**
> We are going to simplify this complex fraction by two different methods. This is the first method.

Answers
4. $1\frac{1}{5}$
5. 3

$$= \frac{6+8}{9-2}$$ Multiply each fraction by 12.

$$= \frac{14}{7}$$ Add on top and subtract on bottom.

$$= 2$$ Reduce to lowest terms.

The problem can be worked in another way also. We can simplify the top and bottom of the complex fraction separately. Simplifying the top, we have

$$\frac{1}{2} + \frac{2}{3} = \frac{1 \cdot \mathbf{3}}{2 \cdot \mathbf{3}} + \frac{2 \cdot \mathbf{2}}{3 \cdot \mathbf{2}} = \frac{3}{6} + \frac{4}{6} = \frac{7}{6}$$

Simplifying the bottom, we have

$$\frac{3}{4} - \frac{1}{6} = \frac{3 \cdot \mathbf{3}}{4 \cdot \mathbf{3}} - \frac{1 \cdot \mathbf{2}}{6 \cdot \mathbf{2}} = \frac{9}{12} - \frac{2}{12} = \frac{7}{12}$$

We now write the original complex fraction again using the simplified expressions for the top and bottom. Then we proceed as we did in Example 4.

$$\frac{\frac{1}{2} + \frac{2}{3}}{\frac{3}{4} - \frac{1}{6}} = \frac{\frac{7}{6}}{\frac{7}{12}}$$

$$= \frac{7}{6} \div \frac{7}{12}$$ The divisor is $\frac{7}{12}$.

$$= \frac{7}{6} \cdot \frac{12}{7}$$ Replace $\frac{7}{12}$ by its reciprocal and multiply.

$$= \frac{\cancel{7} \cdot 2 \cdot \cancel{6}}{\cancel{6} \cdot \cancel{7}}$$ Divide out common factors.

$$= 2$$ ∎

EXAMPLE 6 Simplify: $\dfrac{3 + \frac{1}{2}}{2 - \frac{3}{4}}$.

SOLUTION The simplest approach here is to multiply both the top and bottom by the LCD for all fractions, which is 4.

$$\frac{3 + \frac{1}{2}}{2 - \frac{3}{4}} = \frac{\mathbf{4}\left(3 + \frac{1}{2}\right)}{\mathbf{4}\left(2 - \frac{3}{4}\right)}$$ Multiply the top and bottom by 4.

$$= \frac{\mathbf{4} \cdot 3 + \mathbf{4} \cdot \frac{1}{2}}{\mathbf{4} \cdot 2 - \mathbf{4} \cdot \frac{3}{4}}$$ Distributive property

$$= \frac{12 + 2}{8 - 3}$$ Multiply each number by 4.

$$= \frac{14}{5}$$ Add on top and subtract on bottom.

$$= 2\frac{4}{5}$$ ∎

NOTE
The fraction bar that separates the numerator of the complex fraction from its denominator works like parentheses. If we were to rewrite this problem without it, we would write it like this:

$$\left(\frac{1}{2} + \frac{2}{3}\right) \div \left(\frac{3}{4} - \frac{1}{6}\right)$$

That is why we simplify the top and bottom of the complex fraction separately and then divide.

6. Simplify: $\dfrac{4 + \frac{2}{3}}{3 - \frac{1}{4}}$

Answer
6. $1\frac{23}{33}$

Problem Set 2.9

Moving Toward Success

"The only place where success comes before work is in the dictionary."
—Donald Kendall, 1921–present, American businessman

1. If you have completed an assignment and have time left over, what should you do to further your success in this course?

2. If you have extra time, one thing you could do is read the next section in the book and work more problems. How would this be helpful?

A Use the rule for order of operations to simplify each of the following. [Examples 1–3]

1. $3 + \left(1\frac{1}{2}\right)\left(2\frac{2}{3}\right)$

2. $7 - \left(1\frac{3}{5}\right)\left(2\frac{1}{2}\right)$

3. $8 - \left(\frac{6}{11}\right)\left(1\frac{5}{6}\right)$

4. $10 + \left(2\frac{4}{5}\right)\left(\frac{5}{7}\right)$

5. $\frac{2}{3}\left(1\frac{1}{2}\right) + \frac{3}{4}\left(1\frac{1}{3}\right)$

6. $\frac{2}{5}\left(2\frac{1}{2}\right) + \frac{5}{8}\left(3\frac{1}{5}\right)$

7. $2\left(1\frac{1}{2}\right) + 5\left(6\frac{2}{5}\right)$

8. $4\left(5\frac{3}{4}\right) + 6\left(3\frac{5}{6}\right)$

9. $\left(\frac{3}{5} + \frac{1}{10}\right)\left(\frac{1}{2} + \frac{3}{4}\right)$

10. $\left(\frac{2}{9} + \frac{1}{3}\right)\left(\frac{1}{5} + \frac{1}{10}\right)$

11. $\left(2 + \frac{2}{3}\right)\left(3 + \frac{1}{8}\right)$

12. $\left(3 - \frac{3}{4}\right)\left(3 + \frac{1}{3}\right)$

13. $\left(1 + \frac{5}{6}\right)\left(1 - \frac{5}{6}\right)$

14. $\left(2 - \frac{1}{4}\right)\left(2 + \frac{1}{4}\right)$

15. $\frac{2}{3} + \frac{1}{3}\left(2\frac{1}{2} + \frac{1}{2}\right)^2$

16. $\frac{3}{5} + \frac{1}{4}\left(2\frac{1}{2} - \frac{1}{2}\right)^3$

17. $2\frac{3}{8} + \frac{1}{2}\left(\frac{1}{3} + \frac{5}{3}\right)^3$

18. $8\frac{2}{3} + \frac{1}{3}\left(\frac{8}{5} + \frac{7}{5}\right)^2$

19. $2\left(\frac{1}{2} + \frac{1}{3}\right) + 3\left(\frac{2}{3} + \frac{1}{4}\right)$

20. $5\left(\frac{1}{5} + \frac{3}{10}\right) + 2\left(\frac{1}{10} + \frac{1}{2}\right)$

B Simplify each complex fraction as much as possible. [Examples 4–6]

21. $\dfrac{\frac{2}{3}}{\frac{3}{4}}$

22. $\dfrac{\frac{5}{6}}{\frac{3}{12}}$

23. $\dfrac{\frac{2}{3}}{\frac{4}{3}}$

24. $\dfrac{\frac{7}{9}}{\frac{5}{9}}$

25. $\dfrac{\frac{11}{20}}{\frac{5}{10}}$

26. $\dfrac{\frac{9}{16}}{\frac{3}{4}}$

27. $\dfrac{\frac{1}{2}+\frac{1}{3}}{\frac{1}{2}-\frac{1}{3}}$

28. $\dfrac{\frac{1}{4}+\frac{1}{5}}{\frac{1}{4}-\frac{1}{5}}$

29. $\dfrac{\frac{5}{8}-\frac{1}{4}}{\frac{1}{8}+\frac{1}{2}}$

30. $\dfrac{\frac{3}{4}+\frac{1}{3}}{\frac{2}{3}+\frac{1}{6}}$

31. $\dfrac{\frac{9}{20}-\frac{1}{10}}{\frac{1}{10}+\frac{9}{20}}$

32. $\dfrac{\frac{1}{2}+\frac{2}{3}}{\frac{3}{4}+\frac{5}{6}}$

33. $\dfrac{1+\frac{2}{3}}{1-\frac{2}{3}}$

34. $\dfrac{5-\frac{3}{4}}{2+\frac{3}{4}}$

35. $\dfrac{2+\frac{5}{6}}{5-\frac{1}{3}}$

36. $\dfrac{9-\frac{11}{5}}{3+\frac{13}{10}}$

37. $\dfrac{3+\frac{5}{6}}{1+\frac{5}{3}}$

38. $\dfrac{10+\frac{9}{10}}{5+\frac{4}{5}}$

39. $\dfrac{\frac{1}{3}+\frac{3}{4}}{2-\frac{1}{6}}$

40. $\dfrac{3+\frac{5}{2}}{\frac{5}{6}+\frac{1}{4}}$

41. $\dfrac{\frac{5}{6}}{3+\frac{2}{3}}$

42. $\dfrac{9-\frac{3}{2}}{\frac{7}{4}}$

B Simplify each of the following complex fractions. [Examples 5–6]

43. $\dfrac{2\frac{1}{2} + \frac{1}{2}}{3\frac{3}{5} - \frac{2}{5}}$

44. $\dfrac{5\frac{3}{8} + \frac{5}{8}}{4\frac{1}{4} + 1\frac{3}{4}}$

45. $\dfrac{2 + 1\frac{2}{3}}{3\frac{5}{6} - 1}$

46. $\dfrac{5 + 8\frac{3}{5}}{2\frac{3}{10} + 4}$

47. $\dfrac{3\frac{1}{4} - 2\frac{1}{2}}{5\frac{3}{4} + 1\frac{1}{2}}$

48. $\dfrac{9\frac{3}{8} + 2\frac{5}{8}}{6\frac{1}{2} + 7\frac{1}{2}}$

49. $\dfrac{3\frac{1}{4} + 5\frac{1}{6}}{2\frac{1}{3} + 3\frac{1}{4}}$

50. $\dfrac{8\frac{5}{6} + 1\frac{2}{3}}{7\frac{1}{3} + 2\frac{1}{4}}$

51. $\dfrac{6\frac{2}{3} + 7\frac{3}{4}}{8\frac{1}{2} + 9\frac{7}{8}}$

52. $\dfrac{3\frac{4}{5} - 1\frac{9}{10}}{6\frac{5}{6} - 2\frac{3}{4}}$

53. What is twice the sum of $2\frac{1}{5}$ and $\frac{3}{6}$?

54. Find 3 times the difference of $1\frac{7}{9}$ and $\frac{2}{9}$.

55. Add $5\frac{1}{4}$ to the sum of $\frac{3}{4}$ and 2.

56. Subtract $\frac{7}{8}$ from the product of 2 and $3\frac{1}{2}$.

Applying the Concepts

57. Tri-cities The Google Earth image shows a right triangle between three cities in Colorado. If the distance between Edgewater and Denver is 4 miles, and the distance between Denver and North Washington is $2\frac{1}{2}$ miles, what is the area of the triangle created by the three cities?

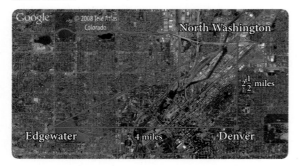

58. Tri-cities The Google Earth image shows a right triangle between three cities in California. If the distance between Pomona and Ontario is $5\frac{7}{10}$ miles, and the distance between Ontario and Upland is $3\frac{3}{5}$ miles, what is the area of the triangle created by the three cities?

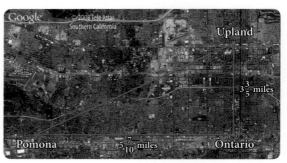

59. Remember our discussion of the mural at the beginning of the section. The expression

$$\frac{\left(12\frac{1}{2}\right)\left(14\frac{3}{8}\right)}{2}$$

was given to represent the area of two triangles that make up the mural. How would we simplify this expression?

60. Body Temperature Suppose your normal body temperature is $98\frac{3}{5}°$ Fahrenheit. If your temperature goes up $3\frac{1}{5}°$ on Monday and then down $1\frac{4}{5}°$ on Tuesday, what is your temperature on Tuesday?

61. Suppose you were commissioned to paint a rectangular mural that is $11\frac{3}{8}$ feet high and $21\frac{3}{4}$ feet long. If you draw a diagonal line from one corner of the rectangle to the other, dividing the area into two equal triangles, what is the area of one of the triangles?

62. Suppose you were commissioned to paint a rectangular mural that is $12\frac{1}{2}$ feet high and $16\frac{5}{8}$ feet long. If you paint a border around the edge of the mural, what is the outer perimeter of the border?

Maintaining Your Skills

These problems review the four basic operations with fractions from this chapter.

Perform the indicated operations.

63. $\frac{3}{4} \cdot \frac{8}{9}$

64. $8 \cdot \frac{5}{6}$

65. $\frac{2}{3} \div 4$

66. $\frac{7}{8} \div \frac{14}{24}$

67. $\frac{3}{7} - \frac{2}{7}$

68. $\frac{6}{7} + \frac{9}{14}$

69. $10 - \frac{2}{9}$

70. $\frac{2}{3} - \frac{3}{5}$

Chapter 2 Summary

EXAMPLES

1. Each of the following is a fraction:
$$\frac{1}{2}, \frac{3}{4}, \frac{8}{1}, \frac{7}{3}$$

■ Definition of Fractions [2.1]

A fraction is any number that can be written in the form $\frac{a}{b}$, where a and b are numbers and b is not 0. The number a is called the *numerator*, and the number b is called the *denominator*.

---

2. Change $\frac{3}{4}$ to an equivalent fraction with denominator 12.
$$\frac{3}{4} = \frac{3 \cdot 3}{4 \cdot 3} = \frac{9}{12}$$

■ Properties of Fractions [2.1]

Multiplying the numerator and the denominator of a fraction by the same nonzero number will produce an equivalent fraction. The same is true for dividing the numerator and denominator by the same nonzero number. In symbols the properties look like this: If a, b, and c are numbers and b and c are not 0, then

Property 1 $\quad \dfrac{a}{b} = \dfrac{a \cdot c}{b \cdot c} \qquad$ **Property 2** $\quad \dfrac{a}{b} = \dfrac{a \div c}{b \div c}$

---

3. $\dfrac{5}{1} = 5, \quad \dfrac{5}{5} = 1$

■ Fractions and the Number 1 [2.1]

If a represents any number, then
$$\frac{a}{1} = a \quad \text{and} \quad \frac{a}{a} = 1 \quad (\text{where } a \text{ is not } 0)$$

---

4. $\dfrac{90}{588} = \dfrac{\cancel{2} \cdot \cancel{3} \cdot 3 \cdot 5}{\cancel{2} \cdot 2 \cdot \cancel{3} \cdot 7 \cdot 7}$
$= \dfrac{3 \cdot 5}{2 \cdot 7 \cdot 7}$
$= \dfrac{15}{98}$

■ Reducing Fractions to Lowest Terms [2.2]

To reduce a fraction to lowest terms, factor the numerator and the denominator, and then divide both the numerator and denominator by any factors they have in common.

---

5. $\dfrac{3}{5} \cdot \dfrac{4}{7} = \dfrac{3 \cdot 4}{5 \cdot 7} = \dfrac{12}{35}$

■ Multiplying Fractions [2.3]

To multiply fractions, multiply numerators and multiply denominators.

---

6. If the base of a triangle is 10 inches and the height is 7 inches, then the area is
$A = \dfrac{1}{2}bh$
$= \dfrac{1}{2} \cdot 10 \cdot 7$
$= 5 \cdot 7$
$= 35$ square inches

■ The Area of a Triangle [2.3]

The formula for the area of a triangle with base b and height h is

$$A = \frac{1}{2}bh$$

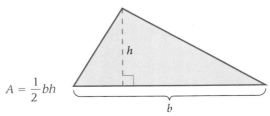

Reciprocals [2.4]

Any two numbers whose product is 1 are called *reciprocals*. The numbers $\frac{2}{3}$ and $\frac{3}{2}$ are reciprocals, because their product is 1.

Division with Fractions [2.4]

7. $\dfrac{3}{8} \div \dfrac{1}{3} = \dfrac{3}{8} \cdot \dfrac{3}{1} = \dfrac{9}{8}$

To divide by a fraction, multiply by its reciprocal. That is, the quotient of two fractions is defined to be the product of the first fraction with the reciprocal of the second fraction (the divisor).

Least Common Denominator (LCD) [2.5]

The *least common denominator* (LCD) for a set of denominators is the smallest number that is exactly divisible by each denominator.

Addition and Subtraction of Fractions [2.5]

8. $\dfrac{1}{8} + \dfrac{3}{8} = \dfrac{1+3}{8}$
$= \dfrac{4}{8}$
$= \dfrac{1}{2}$

To add (or subtract) two fractions with a common denominator, add (or subtract) numerators and use the common denominator. In symbols: If a, b, and c are numbers with c not equal to 0, then

$$\frac{a}{c} + \frac{b}{c} = \frac{a+b}{c} \quad \text{and} \quad \frac{a}{c} - \frac{b}{c} = \frac{a-b}{c}$$

Additional Facts about Fractions

1. In some books fractions are called *rational numbers*.
2. Every whole number can be written as a fraction with a denominator of 1.
3. The commutative, associative, and distributive properties are true for fractions.
4. The word *of* as used in the expression "$\frac{2}{3}$ of 12" indicates that we are to multiply $\frac{2}{3}$ and 12.
5. Two fractions with the same value are called *equivalent fractions*.

Mixed-Number Notation [2.6]

A mixed number is the sum of a whole number and a fraction. The + sign is not shown when we write mixed numbers; it is implied. The mixed number $4\frac{2}{3}$ is actually the sum $4 + \frac{2}{3}$.

Changing Mixed Numbers to Improper Fractions [2.6]

9. $\underset{\substack{\uparrow \\ \text{Mixed} \\ \text{number}}}{4\dfrac{2}{3}} = \dfrac{3 \cdot 4 + 2}{3} = \underset{\substack{\uparrow \\ \text{Improper} \\ \text{fraction}}}{\dfrac{14}{3}}$

To change a mixed number to an improper fraction, we write the mixed number showing the + sign and add as usual. The result is the same if we multiply the denominator of the fraction by the whole number and add what we get to the numerator of the fraction, putting this result over the denominator of the fraction.

Changing an Improper Fraction to a Mixed Number [2.6]

10. Change $\frac{14}{3}$ to a mixed number.

$$3\overline{)14} \qquad \frac{14}{3} = 4\frac{2}{3}$$
$$\underline{12}$$
$$2$$

Quotient, Divisor, Remainder

To change an improper fraction to a mixed number, divide the denominator into the numerator. The quotient is the whole-number part of the mixed number. The fraction part is the remainder over the divisor.

Multiplication and Division with Mixed Numbers [2.7]

11. $2\frac{1}{3} \cdot 1\frac{3}{4} = \frac{7}{3} \cdot \frac{7}{4} = \frac{49}{12} = 4\frac{1}{12}$

To multiply or divide two mixed numbers, change each to an improper fraction and multiply or divide as usual.

Addition and Subtraction with Mixed Numbers [2.8]

12.
$$3\frac{4}{9} = 3\frac{4}{9} = 3\frac{4}{9}$$
$$+2\frac{2}{3} = +2\frac{2 \cdot 3}{3 \cdot 3} = +2\frac{6}{9}$$
$$\qquad\qquad\qquad\qquad 5\frac{10}{9} = 6\frac{1}{9}$$

Common denominator. Add whole numbers. Add fractions.

To add or subtract two mixed numbers, add or subtract the whole-number parts and the fraction parts separately. This is best done with the numbers written in columns.

Borrowing with Mixed Numbers [2.8]

13.
$$4\frac{1}{3} = 4\frac{2}{6} = 3\frac{8}{6}$$
$$-1\frac{5}{6} = -1\frac{5}{6} = -1\frac{5}{6}$$
$$\qquad\qquad\qquad\qquad 2\frac{3}{6} = 2\frac{1}{2}$$

It is sometimes necessary to borrow when doing subtraction with mixed numbers. We always change to a common denominator before we actually borrow.

Complex Fractions [2.9]

14.
$$\frac{4 + \frac{1}{3}}{2 - \frac{5}{6}} = \frac{6\left(4 + \frac{1}{3}\right)}{6\left(2 - \frac{5}{6}\right)}$$
$$= \frac{6 \cdot 4 + 6 \cdot \frac{1}{3}}{6 \cdot 2 - 6 \cdot \frac{5}{6}}$$
$$= \frac{24 + 2}{12 - 5}$$
$$= \frac{26}{7} = 3\frac{5}{7}$$

A fraction that contains a fraction in its numerator or denominator is called a *complex fraction*.

COMMON MISTAKES

1. The most common mistake when working with fractions occurs when we try to add two fractions without using a common denominator. For example,

$$\frac{2}{3} + \frac{4}{5} \neq \frac{2+4}{3+5}$$

   If the two fractions we are trying to add don't have the same denominators, then we must rewrite each one as an equivalent fraction with a common denominator. *We never add denominators when adding fractions.*

   Note: We do *not* need a common denominator when multiplying fractions.

2. A common mistake made with division of fractions occurs when we multiply by the reciprocal of the first fraction instead of the reciprocal of the divisor. For example,

$$\frac{2}{3} \div \frac{5}{6} \neq \frac{3}{2} \cdot \frac{5}{6}$$

   Remember, we perform division by multiplying by the reciprocal of the divisor (the fraction to the right of the division symbol).

3. If the answer to a problem turns out to be a fraction, that fraction should always be written in lowest terms. It is a mistake not to reduce to lowest terms.

4. A common mistake when working with mixed numbers is to confuse mixed-number notation for multiplication of fractions. The notation $3\frac{2}{5}$ does *not* mean 3 *times* $\frac{2}{5}$. It means 3 *plus* $\frac{2}{5}$.

5. Another mistake occurs when multiplying mixed numbers. The mistake occurs when we don't change the mixed number to an improper fraction before multiplying and instead try to multiply the whole numbers and fractions separately. Like this:

$$2\frac{1}{2} \cdot 3\frac{1}{3} = (2 \cdot 3) + \left(\frac{1}{2} \cdot \frac{1}{3}\right) \quad \text{Mistake}$$
$$= 6 + \frac{1}{6}$$
$$= 6\frac{1}{6}$$

   Remember, the correct way to multiply mixed numbers is to first change to improper fractions and then multiply numerators and multiply denominators. This is correct:

$$2\frac{1}{2} \cdot 3\frac{1}{3} = \frac{5}{2} \cdot \frac{10}{3} = \frac{50}{6} = 8\frac{2}{6} = 8\frac{1}{3} \quad \text{Correct}$$

Chapter 2 Review

Reduce each of the following fractions to lowest terms. [2.2]

1. $\dfrac{6}{8}$
2. $\dfrac{12}{36}$
3. $\dfrac{110}{70}$
4. $\dfrac{45}{75}$

Multiply the following fractions. (That is, find the product in each case, and reduce to lowest terms.) [2.3]

5. $\dfrac{1}{5}(5)$
6. $\dfrac{80}{27}\left(\dfrac{3}{20}\right)$
7. $\dfrac{96}{25} \cdot \dfrac{15}{98} \cdot \dfrac{35}{54}$
8. $\dfrac{3}{5} \cdot 75 \cdot \dfrac{2}{3}$

Find the following quotients. (That is, divide and reduce to lowest terms.) [2.4]

9. $\dfrac{8}{9} \div \dfrac{4}{3}$
10. $\dfrac{9}{10} \div 3$
11. $\dfrac{15}{36} \div \dfrac{10}{9}$
12. $\dfrac{18}{49} \div \dfrac{36}{28}$

Perform the indicated operations. Reduce all answers to lowest terms. [2.5]

13. $\dfrac{6}{8} - \dfrac{2}{8}$
14. $\dfrac{9}{10} + \dfrac{11}{10}$
15. $3 + \dfrac{1}{2}$
16. $\dfrac{9}{52} + \dfrac{5}{78}$
17. $\dfrac{11}{126} - \dfrac{5}{84}$
18. $\dfrac{3}{10} + \dfrac{7}{25} + \dfrac{3}{4}$

Change to improper fractions. [2.6]

19. $3\dfrac{5}{8}$
20. $7\dfrac{2}{3}$

Change to mixed numbers. [2.6]

21. $\dfrac{15}{4}$
22. $\dfrac{110}{8}$

Perform the indicated operations. [2.7, 2.8]

23. $2 \div 3\dfrac{1}{4}$
24. $4\dfrac{7}{8} \div 2\dfrac{3}{5}$
25. $6 \cdot 2\dfrac{1}{2} \cdot \dfrac{4}{5}$
26. $3\dfrac{1}{5} + 4\dfrac{2}{5}$
27. $8\dfrac{2}{3} + 9\dfrac{1}{4}$
28. $5\dfrac{1}{3} - 2\dfrac{8}{9}$

Simplify each of the following as much as possible. [2.9]

29. $3 + 2\left(4\dfrac{1}{3}\right)$
30. $\left(2\dfrac{1}{2} + \dfrac{3}{4}\right)\left(2\dfrac{1}{2} - \dfrac{3}{4}\right)$

Simplify each complex fraction as much as possible. [2.9]

31. $\dfrac{1 + \dfrac{2}{3}}{1 - \dfrac{2}{3}}$
32. $\dfrac{3 - \dfrac{3}{4}}{3 + \dfrac{3}{4}}$
33. $\dfrac{\dfrac{7}{8} - \dfrac{1}{2}}{\dfrac{1}{4} + \dfrac{1}{2}}$
34. $\dfrac{2\dfrac{1}{8} + 3\dfrac{1}{3}}{1\dfrac{1}{6} + 5\dfrac{1}{4}}$

35. **Defective Parts** If $\dfrac{1}{10}$ of the items in a shipment of 200 items are defective, how many are defective? [2.3]

36. **Number of Students** If 80 students took a math test and $\dfrac{3}{4}$ of them passed, then how many students passed the test? [2.3]

37. **Translating** What is 3 times the sum of $2\dfrac{1}{4}$ and $\dfrac{3}{4}$? [2.9]

38. **Translating** Subtract $\dfrac{5}{6}$ from the product of $1\dfrac{1}{2}$ and $\dfrac{2}{3}$. [2.9]

39. **Cooking** If a recipe that calls for $2\dfrac{1}{2}$ cups of flour will make 48 cookies, how much flour is needed to make 36 cookies? [2.7]

40. **Length of Wood** A piece of wood $10\dfrac{3}{4}$ inches long is divided into 6 equal pieces. How long is each piece? [2.7]

41. **Cooking** A recipe that calls for $3\dfrac{1}{2}$ tablespoons of oil is tripled. How much oil must be used in the tripled recipe? [2.7]

42. **Sheep Feed** A rancher fed his sheep $10\dfrac{1}{2}$ pounds of feed on Monday, $9\dfrac{3}{4}$ pounds on Tuesday, and $12\dfrac{1}{4}$ pounds on Wednesday. How many pounds of feed did he use on these 3 days? [2.8]

43. Find the area and the perimeter of the triangle below. [2.7, 2.8]

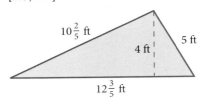

Chapter 2 Cumulative Review

Simplify.

1. 172
    6,050
    + 95

2. 4,320
    × 17

3. $\left(\dfrac{9}{13}\right)^2$

4. $17 - (11 - 4)$

5. $\dfrac{5}{8} - \dfrac{3}{8}$

6. $17\dfrac{13}{16} - 9\dfrac{5}{12}$

7. $\dfrac{2}{3} \cdot \dfrac{6}{11}$

8. $4^3 - 2^2$

9. $4 \div \dfrac{1}{6}$

10. $196 \div 14 \div 7$

11. $\dfrac{92}{21}$

12. $\dfrac{180}{252}$

13. 25^2

14. $(7 + 4) - (8 - 3)$

15. $\dfrac{2}{3} + \dfrac{1}{4}$

16. $1,030(604)$

17. $\dfrac{3}{12} \div \dfrac{9}{16}$

18. $4 + 5^2 - 2 \cdot 6 + 3$

19. $12 \cdot \dfrac{3}{4}$

20. $\left(\dfrac{2}{3}\right)^2 \cdot \left(\dfrac{1}{2}\right)^3$

21. $2{,}639 - 365$

22. $\dfrac{7\dfrac{1}{9}}{2\dfrac{2}{3}}$

23. $\left(\dfrac{3}{5} + \dfrac{2}{25}\right) - \dfrac{4}{125}$

24. $6 + \dfrac{3}{11} \div \dfrac{1}{2}$

25. Round the following numbers to the nearest ten, then add.
    747
    116
    + 222

26. Find the sum of $\dfrac{2}{3}$, $\dfrac{1}{9}$, and $\dfrac{3}{4}$.

27. Write the fraction $\dfrac{3}{13}$ as an equivalent fraction with a denominator of 39.

28. Reduce $\dfrac{14}{49}$ to lowest terms.

29. Add $\dfrac{4}{5}$ to half of $\dfrac{4}{9}$.

30. **Neptune's Diameter** The planet Neptune has an equatorial diameter of about 30,760 miles. Write out Neptune's diameter in words and expanded form.

The chart shows the number of babies born to different age groups of women.

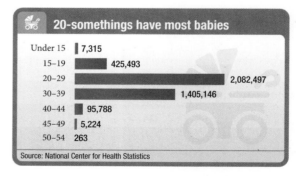

31. **Babies** How many babies were born to mothers between the ages of 30 and 54?

Chapter 2 Test

Perform the indicated operations. Reduce all answers to lowest terms. [2.5, 2.8]

1. $\dfrac{3}{10} + \dfrac{1}{10}$

2. $\dfrac{5}{8} - \dfrac{2}{8}$

3. $4 + \dfrac{3}{5}$

4. $\dfrac{3}{10} + \dfrac{2}{5}$

5. $\dfrac{5}{6} + \dfrac{2}{9} + \dfrac{1}{4}$

6. $7\dfrac{1}{3} + 2\dfrac{3}{8}$

7. $5\dfrac{1}{6} - 1\dfrac{1}{2}$

8. $6 - \dfrac{3}{5}$

9. $\dfrac{3}{5} - \dfrac{4}{15}$

10. $\dfrac{4}{9} + \dfrac{2}{3} - 1$

11. $\dfrac{9}{16} - \dfrac{1}{4}$

12. $2\dfrac{2}{3} - 1\dfrac{5}{6}$

13. $3\dfrac{1}{5} - \dfrac{7}{10}$

14. $\dfrac{7}{12} - \dfrac{1}{3}$

15. $2\dfrac{5}{8} - 2\dfrac{1}{4}$

16. $3\dfrac{2}{3} + \dfrac{5}{6}$

17. $4\dfrac{3}{5} + 2\dfrac{1}{15}$

18. $1\dfrac{8}{9} + 1\dfrac{7}{12} - \dfrac{3}{4}$

Simplify each of the following as much as possible. [2.9]

19. $4 + 3\left(4\dfrac{1}{4}\right)$

20. $\left(2\dfrac{1}{3} + \dfrac{1}{2}\right)\left(3\dfrac{2}{3} - \dfrac{1}{6}\right)$

21. $\dfrac{\dfrac{11}{12} - \dfrac{2}{3}}{\dfrac{1}{6} + \dfrac{1}{3}}$

22. $5 - 2\left(1\dfrac{2}{3}\right)$

23. $\left(5\dfrac{1}{3} - \dfrac{5}{6}\right)\left(2\dfrac{1}{2} + \dfrac{1}{4}\right)$

24. $\left(\dfrac{9}{2}\right)^2 - \dfrac{7}{2}$

25. $\left(\dfrac{1}{3}\right)\left(2\dfrac{1}{2}\right) - \dfrac{2}{3}$

26. $\dfrac{4\dfrac{1}{6}}{1\dfrac{1}{9}}$

27. $\dfrac{\dfrac{8}{15}}{5\dfrac{1}{3}}$

28. $\dfrac{\left(1\dfrac{7}{8} + \dfrac{3}{4}\right)}{\left(4\dfrac{1}{4} - \dfrac{1}{2}\right)}$

29. **Sewing** A dress that is $31\dfrac{1}{6}$ inches long is shortened by $3\dfrac{2}{3}$ inches. What is the new length of the dress? [2.8]

30. Find the perimeter of the triangle below. [2.8]

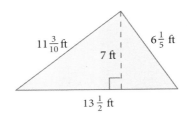

Suppose the table shows the monthly average rainfall in Hawaii for the first 6 months of the year.

Month	Rainfall (inches)
January	$3\dfrac{1}{3}$
February	$2\dfrac{2}{5}$
March	$2\dfrac{7}{10}$
April	$1\dfrac{1}{3}$
May	1
June	$\dfrac{2}{5}$

31. How much rain does Hawaii get during the first three months of the year? [2.8]

32. How much more rain falls in Hawaii in April than in June? [2.8]

Chapter 2 Projects

FRACTIONS AND MIXED NUMBERS

Group Project

Recipe

Number of People 2

Time Needed 5 minutes

Equipment Pencil and paper

Background Here are the ingredients for Martha Stewart's recipe for chocolate chip cookies.

Chocolate Chip Cookies

Makes 2 dozen

1 cup unsalted butter, room temperature
1 1/2 cups packed light-brown sugar
1/2 cup granulated sugar
1 teaspoon pure vanilla extract
1 large egg, room temperature
2 cups all-purpose flour
1/2 teaspoon baking soda
1/2 teaspoon salt
12 ounces semisweet chocolate, coarsely chopped, or one 12-ounce bag semisweet chocolate chips

Procedure Rewrite the recipe to make 3 dozen cookies by multiplying the quantities by $1\frac{1}{2}$.

____ cups (____ sticks) unsalted butter, room temperature

____ cups packed light-brown sugar

____ cups granulated sugar

____ teaspoons pure vanilla extract

____ large eggs, room temperature

____ cups all-purpose flour

____ teaspoons baking soda

____ teaspoons salt

____ ounces semisweet chocolate, coarsely chopped, or ____ 12-ounce bags semisweet chocolate chips

Research Project

Sophie Germain

The photograph at the right shows the street sign in Paris named for the French mathematician Sophie Germain (1776-1831). Among her contributions to mathematics is her work with prime numbers. In this chapter, we had an introductory look at some of the classifications for numbers, including the prime numbers. Within the prime numbers themselves, there are still further classifications. In fact, a Sophie Germain prime is a prime number P, for which both P and $2P + 1$ are primes. For example, the prime number 2 is the first Sophie Germain prime because both 2 and $2 \cdot 2 + 1 = 5$ are prime numbers. The next Germain prime is 3 because both 3 and $2 \cdot 3 + 1 = 7$ are primes.

Sophie Germain was born on April 1, 1776, in Paris, France. She taught herself mathematics by reading the books in her father's library at home. Today she is recognized most for her work in number theory, which includes her work with prime numbers. Research the life of Sophie Germain. Write a short essay that includes information on her work with prime numbers and how her results contributed to solving Fermat's Theorem almost 200 years later.

A Glimpse of Algebra

In algebra, we add and subtract fractions in the same way that we have added and subtracted fractions in this chapter. For example, consider the expression

$$\frac{x}{3} + \frac{2}{3}$$

The two fractions have the same denominator. So to add these fractions, all we have to do is add the numerators to get $x + 2$. The denominator of the sum is the common denominator 3.

$$\frac{x}{3} + \frac{2}{3} = \frac{x+2}{3}$$

Here are some further examples.

EXAMPLE 1 Add or subtract as indicated.

a. $\frac{x}{5} + \frac{4}{5}$ b. $\frac{x+5}{8} - \frac{3}{8}$ c. $\frac{4x}{10} + \frac{3x}{10}$ d. $\frac{5}{x} + \frac{3}{x}$

SOLUTION In each case the denominators are the same. We add or subtract the numerators and write the sum or difference over the common denominator.

a. $\frac{x}{5} + \frac{4}{5} = \frac{x+4}{5}$

b. $\frac{x+5}{8} - \frac{3}{8} = \frac{x+5-3}{8} = \frac{x+2}{8}$

c. $\frac{4x}{10} + \frac{3x}{10} = \frac{4x+3x}{10} = \frac{7x}{10}$ $\quad 4x + 3x = (x + x + x + x) + (x + x + x)$
$\quad\quad\quad\quad\quad\quad\quad\quad\quad\quad = 7x$

d. $\frac{5}{x} + \frac{3}{x} = \frac{5+3}{x} = \frac{8}{x}$ ∎

To add or subtract fractions that do not have the same denominator, we must first find the LCD. We then change each fraction to an equivalent fraction that has the LCD for its denominator. Finally, when all that has been done, we add or subtract the numerators and put the result over the common denominator.

EXAMPLE 2 Add: $\frac{x}{3} + \frac{1}{2}$.

SOLUTION The LCD for 3 and 2 is 6. We multiply the numerator and the denominator of the first fraction by 2, and the numerator and the denominator of the second fraction by 3, to change each fraction to an equivalent fraction with the LCD for a denominator. We then add the numerators as usual. Here is how it looks.

$\frac{x}{3} + \frac{1}{2} = \frac{x \cdot \mathbf{2}}{3 \cdot \mathbf{2}} + \frac{1 \cdot \mathbf{3}}{2 \cdot \mathbf{3}}$ Change to equivalent fractions. Also, $x \cdot 2$ is the same as $2x$, because multiplication is commutative.

$= \frac{2x}{6} + \frac{3}{6}$

$= \frac{2x + 3}{6}$ Add the numerators. ∎

PRACTICE PROBLEMS

1. Add or subtract as indicated.
   a. $\frac{x}{6} + \frac{5}{6}$
   b. $\frac{x+3}{7} - \frac{1}{7}$
   c. $\frac{9x}{4} + \frac{4x}{4}$
   d. $\frac{9}{x} + \frac{3}{x}$

NOTE Remember, the LCD is the least common denominator. It is the smallest expression that is divisible by each of the denominators.

2. Add: $\frac{x}{3} + \frac{1}{5}$.

Answers
1. a. $\frac{x+5}{6}$ b. $\frac{x+2}{7}$ c. $\frac{13x}{4}$ d. $\frac{12}{x}$
2. $\frac{5x+3}{15}$

3. Add: $\dfrac{5}{x} + \dfrac{2}{3}$.

NOTE
In Examples 3 and 4, it is understood that x cannot be 0. Do you know why?

4. Add: $\dfrac{1}{2} + \dfrac{3}{x} + \dfrac{1}{5}$.

5. Add or subtract as indicated.
   a. $3 + \dfrac{x}{8}$
   b. $1 - \dfrac{a}{4}$
   c. $6 + \dfrac{2}{x}$
   d. $x + \dfrac{2}{3}$

Answers
3. $\dfrac{15 + 2x}{3x}$
4. $\dfrac{7x + 30}{10x}$
5. a. $\dfrac{24 + x}{8}$ b. $\dfrac{4 - a}{4}$
   c. $\dfrac{6x + 2}{x}$ d. $\dfrac{3x + 2}{3}$

Chapter 2 Fractions and Mixed Numbers

EXAMPLE 3 Add: $\dfrac{4}{x} + \dfrac{2}{3}$.

SOLUTION The LCD for x and 3 is $3x$. We multiply the numerator and the denominator of the first fraction by 3, and the numerator and the denominator of the second fraction by x, to get two fractions with the same denominator. We then add the numerators.

$$\dfrac{4}{x} + \dfrac{2}{3} = \dfrac{4 \cdot 3}{x \cdot 3} + \dfrac{2 \cdot x}{3 \cdot x} \qquad \text{Change to equivalent fractions.}$$

$$= \dfrac{12}{3x} + \dfrac{2x}{3x}$$

$$= \dfrac{12 + 2x}{3x} \qquad \text{Add the numerators.}$$

EXAMPLE 4 Add: $\dfrac{1}{2} + \dfrac{5}{x} + \dfrac{1}{3}$.

SOLUTION The LCD for 2, x, and 3 is $6x$.

$$\dfrac{1}{2} + \dfrac{5}{x} + \dfrac{1}{3} = \dfrac{1 \cdot 3x}{2 \cdot 3x} + \dfrac{5 \cdot 6}{x \cdot 6} + \dfrac{1 \cdot 2x}{3 \cdot 2x} \qquad \text{Change to equivalent fractions.}$$

$$= \dfrac{3x}{6x} + \dfrac{30}{6x} + \dfrac{2x}{6x}$$

$$= \dfrac{5x + 30}{6x} \qquad \text{Add the numerators.}$$

In this chapter we changed mixed numbers to improper fractions. For example, the mixed number $3\dfrac{4}{5}$ can be changed to an improper fraction as follows:

$$3\dfrac{4}{5} = 3 + \dfrac{4}{5} = \dfrac{3 \cdot 5}{1 \cdot 5} + \dfrac{4}{5} = \dfrac{15}{5} + \dfrac{4}{5} = \dfrac{19}{5}$$

A similar kind of problem in algebra would be to add 2 and $\dfrac{x}{8}$.

EXAMPLE 5 Add or subtract as indicated.
   a. $2 + \dfrac{x}{8}$ b. $1 - \dfrac{a}{2}$ c. $5 + \dfrac{3}{x}$ d. $x + \dfrac{3}{5}$

SOLUTION We can think of each whole number and the letter x in part (d) as a fraction with denominator 1. In each case we multiply the numerator and the denominator of the first number by the denominator of the fraction.

a. $2 + \dfrac{x}{8} = \dfrac{2 \cdot 8}{1 \cdot 8} + \dfrac{x}{8} = \dfrac{16}{8} + \dfrac{x}{8} = \dfrac{16 + x}{8}$

b. $1 - \dfrac{a}{2} = \dfrac{1 \cdot 2}{1 \cdot 2} - \dfrac{a}{2} = \dfrac{2}{2} - \dfrac{a}{2} = \dfrac{2 - a}{2}$

c. $5 + \dfrac{3}{x} = \dfrac{5 \cdot x}{1 \cdot x} + \dfrac{3}{x} = \dfrac{5x}{x} + \dfrac{3}{x} = \dfrac{5x + 3}{x}$

d. $x + \dfrac{3}{5} = \dfrac{x \cdot 5}{x \cdot 5} + \dfrac{3}{5} = \dfrac{5x}{5} + \dfrac{3}{5} = \dfrac{5x + 3}{5}$

A Glimpse of Algebra Problems

Add or subtract as indicated.

1. $\dfrac{x}{4} + \dfrac{3}{4}$
2. $\dfrac{x}{8} - \dfrac{5}{8}$
3. $\dfrac{x+6}{5} - \dfrac{4}{5}$
4. $\dfrac{x+1}{3} + \dfrac{2}{3}$

5. $\dfrac{5x}{8} + \dfrac{2x}{8}$
6. $\dfrac{9x}{7} - \dfrac{3x}{7}$
7. $\dfrac{6}{x} - \dfrac{4}{x}$
8. $\dfrac{3}{x} + \dfrac{7}{x}$

For each sum or difference, find the LCD, change to equivalent fractions, and then add or subtract numerators as indicated.

9. $\dfrac{x}{2} + \dfrac{1}{3}$
10. $\dfrac{x}{6} - \dfrac{3}{4}$
11. $\dfrac{x}{2} - \dfrac{1}{4}$
12. $\dfrac{x}{3} - \dfrac{1}{6}$

13. $\dfrac{3}{x} + \dfrac{3}{4}$
14. $\dfrac{2}{x} - \dfrac{1}{3}$
15. $\dfrac{4}{5} - \dfrac{1}{x}$
16. $\dfrac{3}{4} - \dfrac{1}{x}$

17. $\dfrac{1}{3} + \dfrac{2}{x} + \dfrac{1}{4}$
18. $\dfrac{1}{5} + \dfrac{1}{x} + \dfrac{1}{3}$
19. $\dfrac{1}{2} + \dfrac{1}{x} + \dfrac{1}{4}$
20. $\dfrac{1}{3} + \dfrac{1}{x} + \dfrac{1}{6}$

Add or subtract as indicated.

21. $3 + \dfrac{x}{4}$

22. $2 + \dfrac{x}{7}$

23. $5 - \dfrac{x}{2}$

24. $6 - \dfrac{x}{8}$

25. $4 - \dfrac{a}{7}$

26. $6 - \dfrac{a}{4}$

27. $1 + \dfrac{2a}{5}$

28. $3 + \dfrac{4a}{9}$

29. $8 + \dfrac{3}{x}$

30. $7 + \dfrac{9}{x}$

31. $2 - \dfrac{5}{x}$

32. $3 - \dfrac{1}{x}$

33. $x + \dfrac{3}{4}$

34. $x + \dfrac{5}{6}$

35. $x + \dfrac{2x}{9}$

36. $x + \dfrac{3x}{5}$

37. $a - \dfrac{4a}{7}$

38. $a - \dfrac{3a}{5}$

39. $2x - \dfrac{3x}{4}$

40. $3x - \dfrac{2x}{5}$

Decimals

3

Chapter Outline

3.1 Decimal Notation and Place Value

3.2 Addition and Subtraction with Decimals

3.3 Multiplication with Decimals; Circumference and Area of a Circle

3.4 Division with Decimals

3.5 Fractions and Decimals, and the Volume of a Sphere

3.6 Square Roots and the Pythagorean Theorem

The People's Republic of China hosted the 2008 Summer Olympics in their capital city of Beijing. Extensive efforts were made to ensure the more than 11,000 international athletes had top-of-the-line facilities in which to practice and compete. The Beijing National Stadium, also known as the Bird's Nest, was one of the games most recognizable structures. Construction on the stadium took five years and cost more than $400 million.

MEN'S 400-METER RACE, 2008 OLYMPICS		
Medal	Runner	Time (seconds)
Gold	LaShawn Merritt	43.75
Silver	Jeremy Wariner	44.74
Bronze	David Neville	44.80

Source: ESPN

The Bird's Nest hosted the opening and closing ceremonies of the 2008 Olympic Games, as well as all the track and field events. The above table shows the medal-winning times for the men's 400-meter race. The times in the chart are in seconds, accurate to the nearest hundredth of a second. In this chapter, we will study numbers like these to obtain a good working knowledge of the decimal numbers we see everywhere around us.

Preview

Key Words	Definition
Decimal Point	The symbol used to separate the integer and fractional parts
Radius	The length of a straight line that stretches from the center of the circle to any point on its edge
Diameter	The length of a straight line that stretches from one side of a circle, through the center, to the other side
Circumference	The distance around a circle
Sphere	A round, three-dimensional geometric object
Hemisphere	A half of a sphere
Radical	A symbol that indicates taking the square root of a number
Square Root	The number we square to get the original number
Irrational Number	A number that cannot be exactly written in the form of a fraction or decimal
Right Triangle	A triangle that contains a 90° angle
Hypotenuse	The longest side in a right triangle

Chapter Outline

3.1 Decimal Notation and Place Value
- A Understand place value for decimal numbers.
- B Write decimal numbers in words and with digits.
- C Convert decimals to fractions and fractions to decimals.
- D Solve applications involving decimals.

3.2 Addition and Subtraction with Decimals
- A Add and subtract decimals.
- B Solve applications involving addition and subtraction of decimals.

3.3 Multiplication with Decimals; Circumference and Area of a Circle
- A Multiply decimal numbers.
- B Solve application problems involving decimals.
- C Find the circumference of a circle.

3.4 Division with Decimals
- A Divide decimal numbers.
- B Solve application problems involving decimals.

3.5 Fractions and Decimals, and the Volume of a Sphere
- A Convert fractions to decimals.
- B Convert decimals to fractions.
- C Simplify expressions containing fractions and decimals.
- D Solve applications involving fractions and decimals.

3.6 Square Roots and the Pythagorean Theorem
- A Find square roots of numbers.
- B Use decimals to approximate square roots.
- C Solve problems with the Pythagorean Theorem.

Decimal Notation and Place Value

3.1

OBJECTIVES

A Understand place value for decimal numbers.

B Write decimal number in words and with digits.

C Convert decimals to fractions and fractions to decimals.

D Solve applications involving decimals.

TICKET TO SUCCESS

Keep these questions in mind as you read through the section. Then respond in your own words and in complete sentences.

1. Why do we use a decimal point?
2. Name the first three place values to the right of the decimal.
3. How do you convert between a fraction and a decimal?
4. What is the rule for rounding decimal numbers?

In this chapter, we will focus our attention on *decimals*. Anyone who has used money in the United States has worked with decimals already. For example, suppose you decide to eat out at your local fast food restaurant. Once you have ordered your food, the cashier says your total bill is "6 dollars and 25 cents." However, you see

Decimal point

on the cash register's screen. The register has translated your total bill into a number that uses a decimal point. What is interesting and useful about decimals is their relationship to fractions and to powers of ten. The work we have done up to now—especially our work with fractions—can be used to develop the properties of decimal numbers.

3.1 Decimal Notation and Place Value 205

A Place Value

In Chapter 1, we developed the idea of place value for the digits in a whole number. At that time, we gave the name and the place value of each of the first seven columns in our number system, as follows:

Millions Column	Hundred Thousands Column	Ten Thousands Column	Thousands Column	Hundreds Column	Tens Column	Ones Column
1,000,000	100,000	10,000	1,000	100	10	1

As we move from right to left, we multiply by 10 each time. The value of each column is 10 times the value of the column on its right, with the right most column being 1. Up until now, we have always looked at place value as increasing by a factor of 10 each time we move one column to the left.

Ten Thousands	Thousands	Hundreds	Tens	Ones
10,000 ←	1,000 ←	100 ←	10 ←	1
	Multiply by 10.	Multiply by 10.	Multiply by 10.	Multiply by 10.

To understand the idea behind decimal numbers, we notice that moving in the opposite direction, from left to right, we *divide* by 10 each time.

Ten Thousands	Thousands	Hundreds	Tens	Ones
10,000 →	1,000 →	100 →	10 →	1
	Divide by 10.	Divide by 10.	Divide by 10.	Divide by 10.

If we keep going to the right, the next column will have to be

$$1 \div 10 = \frac{1}{10} \quad \text{Tenths}$$

The next one after that will be

$$\frac{1}{10} \div 10 = \frac{1}{10} \cdot \frac{1}{10} = \frac{1}{100} \quad \text{Hundredths}$$

After that, we have

$$\frac{1}{100} \div 10 = \frac{1}{100} \cdot \frac{1}{10} = \frac{1}{1,000} \quad \text{Thousandths}$$

We could continue this pattern as long as we wanted. We simply divide by 10 to move one column to the right. (And remember, dividing by 10 gives the same result as multiplying by $\frac{1}{10}$.)

To show where the ones column is, we use a *decimal point* between the ones column and the tenths column.

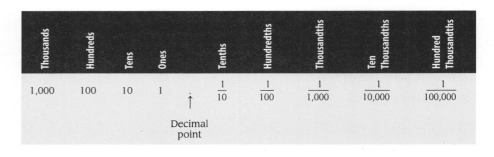

> **NOTE**
> Because the digits to the right of the decimal point have fractional place values, numbers with digits to the right of the decimal point are called decimal fractions. In this book, we will also call them decimal numbers, or simply decimals for short.

The ones column can be thought of as the middle column, with columns larger than 1 to the left and columns smaller than 1 to the right. The first column to the right of the ones column is the tenths column, the next column to the right is the hundredths column, the next is the thousandths column, and so on. The decimal point is always written between the ones column and the tenths column.

We can use the place value of decimal fractions to write them in expanded form.

EXAMPLE 1 Write 423.576 in expanded form.

SOLUTION $423.576 = 400 + 20 + 3 + \dfrac{5}{10} + \dfrac{7}{100} + \dfrac{6}{1,000}$

B Writing Decimals with Words

We can also use the place value of decimal fractions to write them in words.

EXAMPLE 2 Write each number in words.
- **a.** 0.4
- **b.** 0.04
- **c.** 0.004

SOLUTION
- **a.** 0.4 is "four tenths."
- **b.** 0.04 is "four hundredths."
- **c.** 0.004 is "four thousandths."

When a decimal fraction contains digits to the left of the decimal point, we use the word "and" to indicate where the decimal point is when writing the number in words.

EXAMPLE 3 Write each number in words.
- **a.** 5.4
- **b.** 5.04
- **c.** 5.004

SOLUTION
- **a.** 5.4 is "five and four tenths."
- **b.** 5.04 is "five and four hundredths."
- **c.** 5.004 is "five and four thousandths."

EXAMPLE 4 Write 3.64 in words.

SOLUTION The number 3.64 is read "three and sixty-four hundredths." The place values of the digits are as follows:

3 . 6 4
↑   ↑   ↖
3 ones   6 tenths   4 hundredths

We read the decimal part as "sixty-four hundredths" because

$$6 \text{ tenths} + 4 \text{ hundredths} = \dfrac{6}{10} + \dfrac{4}{100} = \dfrac{60}{100} + \dfrac{4}{100} = \dfrac{64}{100}$$

EXAMPLE 5 Write 25.4936 in words.

SOLUTION Using the idea given in Example 4, we write 25.4936 in words as "twenty-five and four thousand, nine hundred thirty-six ten thousandths."

PRACTICE PROBLEMS

1. Write 785.462 in expanded form.

2. Write in words.
   - **a.** 0.06
   - **b.** 0.7
   - **c.** 0.008

3. Write in words.
   - **a.** 5.06
   - **b.** 4.7
   - **c.** 3.008

> **NOTE** Sometimes we name decimal fractions by simply reading the digits from left to right and using the word "point" to indicate where the decimal point is. For example, using this method the number 5.04 is read "five point zero four."

4. Write in words.
   - **a.** 5.98
   - **b.** 5.098

5. Write 305.406 in words.

Answers
1–5. See Solutions Section.

6. Write each number as a fraction or a mixed number. Do not reduce to lowest terms.
   a. 0.06
   b. 5.98
   c. 305.406

C Converting Between Fractions and Decimals

In order to understand addition and subtraction of decimals in the next section, we need to be able to convert decimal numbers to fractions or mixed numbers.

EXAMPLE 6 Write each number as a fraction or a mixed number. Do not reduce to lowest terms.

   a. 0.004   b. 3.64   c. 25.4936

SOLUTION

a. Because 0.004 is 4 thousandths, we write

$$0.004 = \frac{4}{1,000}$$

↑ Three digits after the decimal point   ↖ Three zeros

b. Looking over the work in Example 4, we can write

$$3.64 = 3\frac{64}{100}$$

↑ Two digits after the decimal point   ↖ Two zeros

c. From the way in which we wrote 25.4936 in words in Example 5, we have

$$25.4936 = 25\frac{4936}{10,000}$$

↑ Four digits after the decimal point   ↖ Four zeros

D Applications with Decimals

The rule for rounding decimal numbers is similar to the rule for rounding whole numbers. If the digit in the column to the right of the one we are rounding to is 5 or more, we add 1 to the digit in the column we are rounding to; otherwise, we leave it alone. We then replace all digits to the right of the column we are rounding to with zeros if they are to the left of the decimal point.

Let's apply what we have learned about decimals to the following real-life problem.

7. Round each number in the bar chart from Example 9 to the nearest tenth of a dollar.

EXAMPLE 7 The bar chart below shows some average ticket prices at different ball parks for a recent major league baseball season. Round each ticket price to the nearest dollar.

SOLUTION Using our rule for rounding decimal numbers, we have the following results:
   Least expensive: $14.31 rounds to $14
   League average: $26.74 rounds to $27
   Most expensive: $52.32 rounds to $52

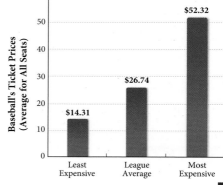

Answers
6. a. $\frac{6}{100}$ b. $5\frac{98}{100}$ c. $305\frac{406}{1,000}$
7. Least expensive, $14.30;
   League average, $26.70;
   Most expensive, $52.30

Problem Set 3.1

Moving Toward Success

"Sit down before fact as a little child, be prepared to give up every conceived notion, follow humbly wherever and whatever abysses nature leads, or you will learn nothing."

—Thomas Huxley, 1825–1895, English biologist

1. What will you do if you find you need more than the recommended two hours for every one hour of class to master the material in this course?

2. Which of the following techniques will you use to help you memorize a definition?
   a. Read the definition.
   b. Reread the definition.
   c. Say the definition out loud and/or explain it to another person.
   d. Write the definition down on a separate sheet of notes.
   e. Review the definition and analyze how it applies to your homework problems.
   f. All the above

A Give the place value of the 5 in each of the following numbers. [Examples 2–5]

1. 458.327
2. 327.458
3. 29.52
4. 25.92

5. 0.00375
6. 0.00532
7. 275.01
8. 0.356

9. 539.76
10. 0.123456

B Write out the name of each number in words. [Examples 2–5]

11. 0.3
12. 0.03

13. 0.015
14. 0.0015

15. 3.4
16. 2.04

17. 52.7
18. 46.8

B Write each of the following as a decimal number.

19. Fifty-five hundredths

20. Two hundred thirty-five ten thousandths

21. Six and nine tenths

22. Forty-five thousand and six hundred twenty-one thousandths

23. Eleven and eleven hundredths

24. Twenty-six thousand, two hundred forty-five and sixteen hundredths

25. One hundred and two hundredths

26. Seventy-five and seventy-five hundred thousandths

27. Three thousand and three thousandths

28. One thousand, one hundred eleven and one hundred eleven thousandths

C Write each number as a fraction or a mixed number. Do not reduce your answers. [Example 6]

29. 405.36
30. 362.78
31. 9.009
32. 60.06

33. 1.234
34. 12.045
35. 0.00305
36. 2.00106

For each pair of numbers, place the correct symbol, < or >, between the numbers.

37. a. 0.02   0.2
    b. 0.3   0.032

38. a. 0.45   0.5
    b. 0.5   0.56

39. Write the following numbers in order from smallest to largest.
    0.02 0.05 0.025 0.052 0.005 0.002

40. Write the following numbers in order from smallest to largest.
    0.2 0.02 0.4 0.04 0.42 0.24

41. Which of the following numbers will round to 7.5?
    7.451 7.449 7.54 7.56

42. Which of the following numbers will round to 3.2?
    3.14999 3.24999 3.279 3.16111

C Change each decimal to a fraction, and then reduce to lowest terms. [Example 6]

43. 0.25 **44.** 0.75 **45.** 0.125 **46.** 0.375

47. 0.625 **48.** 0.0625 **49.** 0.875 **50.** 0.1875

Estimating For each pair of numbers, choose the number that is closest to 10.

51. 9.9 and 9.99 **52.** 8.5 and 8.05 **53.** 10.5 and 10.05 **54.** 10.9 and 10.99

Estimating For each pair of numbers, choose the number that is closest to 0.

55. 0.5 and 0.05 **56.** 0.10 and 0.05 **57.** 0.01 and 0.02 **58.** 0.1 and 0.01

Applying the Concepts

100 Meters World record times in the 100-meter sprint have gone down consistently over the years. The table here shows some examples. Use the information to answer the following questions.

59. What is the place value of the 2 in Carl Lewis's time in 1988?

60. Write Usain Bolt's time using words.

Year	Athlete	Time (Seconds)
1964	Bob Hayes	10.06
1988	Carl Lewis	9.92
1999	Maurice Green	9.79
2009	Usain Bolt	9.58

Source: International Association of Athletics Federations

61. Penny Weight If you have a penny dated anytime from 1959 through 1982, its original weight was 3.11 grams. If the penny has a date of 1983 or later, the original weight was 2.5 grams. Write the two weights in words.

1959–1982   1983–present

62. Halley's Comet Halley's comet was seen from the earth during 1986. It will be another 76.1 years before it returns. Write 76.1 in words.

63. Nutrition A 50-gram egg contains 0.15 milligram of riboflavin. Write 0.15 in words.

64. Nutrition One medium banana contains 0.64 milligrams of B_6. Write 0.64 in words.

65. Gasoline Prices The bar chart below was created from a survey by the U.S. Department of Energy's Energy Information Administration during the month of April 2011. It gives the average price of regular gasoline for the state of California on each Monday of the month. Use the information in the chart to fill in the table.

66. Speed and Time The bar chart below was created from data given by Car and Driver magazine. It gives the minimum time in seconds for a Toyota Echo to reach various speeds from a complete stop. Use the information in the chart to fill in the table.

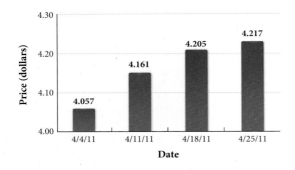

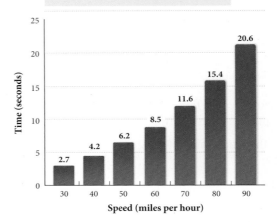

Getting Ready for the Next Section

In the next section we will do addition and subtraction with decimals. To understand that process, we need to understand the process of addition and subtraction with mixed numbers.

Find each of the following sums and differences. (Add or subtract.)

67. $4\dfrac{3}{10} + 2\dfrac{1}{100}$

68. $5\dfrac{35}{100} + 2\dfrac{3}{10}$

69. $8\dfrac{5}{10} - 2\dfrac{4}{100}$

70. $6\dfrac{3}{100} - 2\dfrac{125}{1{,}000}$

71. $5\dfrac{1}{10} + 6\dfrac{2}{100} + 7\dfrac{3}{1{,}000}$

72. $4\dfrac{3}{1{,}000} + 6\dfrac{3}{10} + 7\dfrac{123}{1{,}000}$

Maintaining Your Skills

Write the fractions in order from smallest to largest.

73. $\dfrac{3}{8}\quad \dfrac{3}{16}\quad \dfrac{3}{4}\quad \dfrac{3}{10}$

74. $\dfrac{3}{4}\quad \dfrac{1}{4}\quad \dfrac{5}{4}\quad \dfrac{1}{2}$

Place the correct inequality symbol, < or > between each pair of numbers.

75. $\dfrac{3}{8}\quad \dfrac{5}{6}$

76. $\dfrac{9}{10}\quad \dfrac{10}{11}$

77. $\dfrac{1}{12}\quad \dfrac{1}{13}$

78. $\dfrac{3}{4}\quad \dfrac{5}{8}$

Addition and Subtraction with Decimals

3.2

OBJECTIVES

A Add and subtract decimals.

B Solve applications involving addition and subtraction of decimals.

TICKET TO SUCCESS

Keep these questions in mind as you read through the section. Then respond in your own words and in complete sentences.

1. What is the rule to add or subtract decimal numbers?
2. When adding numbers with decimals, why is it important to line up the decimal points?
3. When working an addition problem that involves decimals, why might we first change each decimal to a mixed number?
4. Write an application problem that involves subtracting decimals.

The chart shows the top finishing times (in seconds) for the men's 500-meter speed skating race during the 2010 Vancouver Olympics.

VANCOUVER OLYMPICS		
Skater	**Country**	**Time (seconds)**
Charles Hamelin	Canada	40.770
Si-Bak Sung	Korea	40.821
Francois-Louis Tremblay	Canada	41.326
Yoo-Gy Kwak	Korea	41.620

Source: ESPN

In order to analyze the different finishing times, it is important that you are able to add and subtract decimals, and that is what we will cover in this section.

A Combining Decimals

Suppose you are earning $8.50 an hour and you receive a raise of $1.25 an hour. To add the two rates of pay, we align the decimal points, and then add the columns.

$8.50
+ $1.25
―――
$9.75

Your new hourly rate of pay is $9.75.

To see why this is true in general, we can use mixed-number notation.

$$8.50 = 8\frac{50}{100}$$
$$+\ 1.25 = 1\frac{25}{100}$$
$$9.75 = 9\frac{75}{100}$$

We can visualize the mathematics above by thinking in terms of dollar bills and some change.

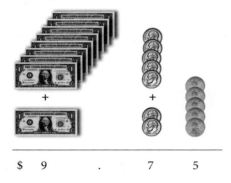

Here is another example:

EXAMPLE 1 Add by first changing to fractions: 25.43 + 2.897 + 379.6.

SOLUTION We first change each decimal to a mixed number. We then write each fraction using the least common denominator and add as usual.

$$25.43 = 25\frac{43}{100} = 25\frac{430}{1,000}$$
$$2.897 = 2\frac{897}{1,000} = 2\frac{897}{1,000}$$
$$+\ 379.6 = 379\frac{6}{10} = 379\frac{600}{1,000}$$
$$406\frac{1,927}{1,000} = 407\frac{927}{1,000} = 407.927$$

Again, the result is the same if we just line up the decimal points and add as if we were adding whole numbers.

```
   25.430    Notice that we can fill in zeros on the right to help
    2.897    keep the numbers in the correct columns. Doing this
+ 379.600    does not change the value of any of the numbers.
  407.927
```

Notice that the decimal point in the answer is directly below the decimal points in the problem. ∎

The same thing would happen if we were to subtract two decimal numbers. We can use these facts to write a rule for addition and subtraction of decimal numbers.

PRACTICE PROBLEMS

1. Change each decimal to a fraction, and then add. Write your answer as a decimal.
   a. 38.45 + 456.073
   b. 38.045 + 456.73

Answers
1. a. 494.523   b. 494.775

3.2 Addition and Subtraction with Decimals

> **Rule** Addition (or Subtraction) of Decimal Numbers
> To add (or subtract) decimal numbers, we line up the decimal points and add (or subtract) as usual. The decimal point in the result is written directly below the decimal points in the problem.

We will use this rule for the rest of the examples in this section.

EXAMPLE 2   Subtract: $39.812 - 14.236$.

SOLUTION   We write the numbers vertically, with the decimal points lined up, and subtract as usual.

$$\begin{array}{r} 39.812 \\ -14.236 \\ \hline 25.576 \end{array}$$

2. Subtract: $78.674 - 23.431$.

EXAMPLE 3   Add: $8 + 0.002 + 3.1 + 0.04$.

SOLUTION   To make sure we keep the digits in the correct columns, we can write zeros to the right of the rightmost digits.

$$8 = 8.000$$
$$3.1 = 3.100$$
$$0.04 = 0.040$$

Writing the extra zeros here is really equivalent to finding a common denominator for the fractional parts of the original number—now we have a thousandths column in all the numbers.

This doesn't change the value of any of the numbers, and it makes our task easier. Now we have

$$\begin{array}{r} 8.000 \\ 0.002 \\ 3.100 \\ +\ 0.040 \\ \hline 11.142 \end{array}$$

3. Add: $16 + 0.033 + 4.6 + 0.08$.

EXAMPLE 4   Subtract: $5.9 - 3.0814$.

SOLUTION   In this case, it is very helpful to write 5.9 as 5.9000, since we will have to borrow in order to subtract.

$$\begin{array}{r} \overset{8\,9\,9\,10}{5.9000} \\ -\ 3.0814 \\ \hline 2.8186 \end{array}$$

4. Subtract:
   a. $6.7 - 2.05$
   b. $6.7 - 2.0563$

EXAMPLE 5   Subtract 3.09 from the sum of 9 and 5.472.

SOLUTION   Writing the problem in symbols, we have

$$(9 + 5.472) - 3.09 = 14.472 - 3.09$$
$$= 11.382$$

5. Subtract 5.89 from the sum of 7 and 3.567.

Answers
2. 55.243
3. 20.713
4. a. 4.65   **b.** 4.6437
5. 4.677

B Applications

Here are more real-life examples of combining decimals:

EXAMPLE 6 While I was writing this section of the book, I stopped to have lunch with a friend at a coffee shop near my office. The bill for lunch was $15.64. I gave the person at the cash register a $20 bill. For change, I received four $1 bills, a quarter, a nickel, and a penny. Was my change correct?

SOLUTION To find the total amount of money I received in change, we add

```
Four $1 bills  = $4.00
One quarter   =  0.25
One nickel    =  0.05
One penny     =  0.01
Total         = $4.31
```

To find out if this is the correct amount, we subtract the amount of the bill from $20.00.

```
  $20.00
−  15.64
  $ 4.36
```

The change was not correct. It is off by 5 cents. Instead of the nickel, I should have been given a dime. ∎

EXAMPLE 7 Find the perimeter of each of the following stamps. Write your answer as a decimal, rounded to the nearest tenth, if necessary.

a.

Each side is 3.5 centimeters.

b.

Base = 2.625 inches
Other two sides = 1.875 inches

SOLUTION To find the perimeter, we add the lengths of all the sides together.
  a. $P = 3.5 + 3.5 + 3.5 + 3.5 = 14.0$ cm
  b. $P = 2.625 + 1.875 + 1.875 = 6.375$ in. ≈ 6.4 in. ∎

---

6. If you pay for a purchase of $9.56 with a $10 bill, how much money should you receive in change? What will you do if the change that is given to you is one quarter, two dimes, and four pennies?

7. Find the perimeter of each stamp in Example 7 from the dimensions given below.
  a. Each side is 1.38 inches
  b. Base = 6.6 centimeters, other two sides = 4.7 centimeters

Answers
6. $0.44; Tell the clerk that you have been given too much change. Instead of two dimes, you should have received one dime and one nickel.
7. a. 5.52 in. **b.** 16.0 cm

Problem Set 3.2

Moving Toward Success

"You can't depend on your eyes when your imagination is out of focus."

—Mark Twain, 1835–1910, American humorist and writer

1. What does it mean to increase the effectiveness of your study time?
2. What things may prevent efficient study time? What should you do to avoid these things?

A Find each of the following sums. (Add.) [Examples 1, 3]

1. 2.91 + 3.28
2. 8.97 + 2.04
3. 0.04 + 0.31 + 0.78
4. 0.06 + 0.92 + 0.65

5. 3.89 + 2.4
6. 7.65 + 3.8
7. 4.532 + 1.81 + 2.7
8. 9.679 + 3.49 + 6.5

9. 0.081 + 5 + 2.94
10. 0.396 + 7 + 3.96
11. 5.0003 + 6.78 + 0.004
12. 27.0179 + 7.89 + 0.009

13. 7.123
     8.12
   + 9.1

14. 5.432
     4.32
   + 3.2

15. 9.001
     8.01
   + 7.1

16. 6.003
     5.02
   + 4.1

17. 89.7854
      3.4
     65.35
   + 100.006

18. 57.4698
      9.89
     32.032
   + 572.0079

19. 543.21
   + 123.45

20. 987.654
   + 456.789

A Find each of the following differences. (Subtract.) [Examples 2, 4]

21. 99.34 − 88.23
22. 47.69 − 36.58
23. 5.97 − 2.4
24. 9.87 − 1.04

25. 6.3 − 2.08
26. 7.5 − 3.04
27. 149.37 − 28.96
28. 796.45 − 32.68

29. 45 − 0.067 **30.** 48 − 0.075 **31.** 8 − 0.327 **32.** 12 − 0.962

33. 765.432 − 234.567 **34.** 654.321 − 123.456

A Subtract. [Example 4]

35. 34.07   **36.** 25.008   **37.** 40.04   **38.** 50.05   **39.** 768.436   **40.** 495.237
       − 6.18   − 3.119   − 4.4   − 5.5   −356.998   − 247.668

A Add and subtract as indicated. [Examples 1–5]

41. (7.8 − 4.3) + 2.5   **42.** (8.3 − 1.2) + 3.4   **43.** 7.8 − (4.3 + 2.5)   **44.** 8.3 − (1.2 + 3.4)

45. (9.7 − 5.2) − 1.4   **46.** (7.8 − 3.2) − 1.5   **47.** 9.7 − (5.2 − 1.4)   **48.** 7.8 − (3.2 − 1.5)

49. Subtract 5 from the sum of 8.2 and 0.072.   **50.** Subtract 8 from the sum of 9.37 and 2.5.

51. What number is added to 0.035 to obtain 4.036?   **52.** What number is added to 0.043 to obtain 6.054?

Applying the Concepts

53. Beijing Olympics The chart from the chapter's introduction shows the medal-winning times for the men's 400 meters in the 2008 Olympics. How much faster was LaShawn Merritt's time than Jeremy Wariner's time?

MEN'S 400-METER RACE, 2008 OLYMPICS		
Medal	Runner	Time (seconds)
Gold	LaShawn Merritt	43.75
Silver	Jeremy Wariner	44.74
Bronze	David Neville	44.80

Source: ESPN

54. Computers The chart shows how many internet users can be found in the countries listed. What is the total number of internet users that can be found in these three countries?

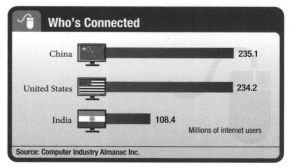

55. Take-Home Pay A college professor making $2,105.96 per month has $311.93 deducted from her check for federal income tax, $158.21 for retirement, and $64.72 for state income tax. How much does the professor take home after the deductions have been taken from her monthly income?

56. Take-Home Pay A cook making $1,504.75 a month has deductions of $157.32 for federal income tax, $58.52 for Social Security, and $45.12 for state income tax. How much does the cook take home after the deductions have been taken from his check?

57. Perimeter of a Stamp This stamp shows the Mexican artist Frida Kahlo. The stamp is the first U.S. stamp to honor a Hispanic woman. The image area of the stamp has a width of 0.84 inches and a length of 1.41 inches. Find the perimeter of the image.

58. Perimeter of a Stamp This stamp honors the Italian scientist Enrico Fermi. The image area of the stamp has a width of 21.4 millimeters and a length of 35.8 millimeters. Find the perimeter of the image.

59. Change A person buys $4.57 worth of candy. If he pays for the candy with a $10 bill, how much change should he receive?

60. Checking Account A checking account contains $342.38. If checks are written for $25.04, $36.71, and $210, how much money is left in the account?

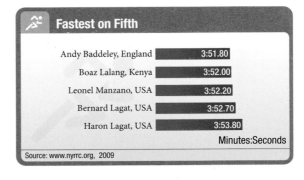

The chart shows times for several finishers of the Fifth Avenue Mile. Use the information to answer the following questions.

61. How much faster was Andy Baddeley than Leonel Manzano?

62. How much faster was Boaz Lalang than Haron Legat?

Fastest on Fifth

Andy Baddeley, England	3:51.80
Boaz Lalang, Kenya	3:52.00
Leonel Manzano, USA	3:52.20
Bernard Lagat, USA	3:52.70
Haron Lagat, USA	3:53.80

Minutes:Seconds

Source: www.nyrrc.org, 2009

63. Geometry A rectangle has a perimeter of 9.5 inches. If the length is 2.75 inches, find the width.

64. Geometry A rectangle has a perimeter of 11 inches. If the width is 2.5 inches, find the length.

65. Change Suppose you eat dinner in a restaurant and the bill comes to $16.76. If you give the cashier a $20 bill and a penny, how much change should you receive? List the bills and coins you should receive for change.

66. Change Suppose you buy some tools at the hardware store and the bill comes to $37.87. If you give the cashier two $20 bills and 2 pennies, how much change should you receive? List the bills and coins you should receive for change.

Sequences Find the next number in each sequence.

67. 2.5, 2.75, 3, . . .

68. 3.125, 3.375, 3.625, . . .

Getting Ready for the Next Section

To understand how to multiply decimals, we need to understand multiplication with whole numbers, fractions, and mixed numbers. The following problems review these concepts.

69. $\dfrac{1}{10} \cdot \dfrac{3}{10}$

70. $\dfrac{5}{10} \cdot \dfrac{6}{10}$

71. $\dfrac{3}{100} \cdot \dfrac{17}{100}$

72. $\dfrac{7}{100} \cdot \dfrac{31}{100}$

73. $5\left(\dfrac{3}{10}\right)$

74. $7 \cdot \dfrac{7}{10}$

75. $56 \cdot 25$

76. $39(48)$

77. $\dfrac{5}{10} \times \dfrac{3}{10}$

78. $\dfrac{5}{100} \times \dfrac{3}{1,000}$

79. $2\dfrac{1}{10} \times \dfrac{7}{100}$

80. $3\dfrac{5}{10} \times \dfrac{4}{100}$

81. $305(436)$

82. $403(522)$

83. $5(420 + 3)$

84. $3(550 + 2)$

Maintaining Your Skills

Use the rule for order of operations to simplify each expression.

85. $30 \div 5 \cdot 2$

86. $60 \div 3 \cdot 10$

87. $22 - 2 \cdot 3$

88. $37 - 7 \cdot 2$

89. $12 + 18 \div 2 - 1$

90. $15 + 10 \div 5 - 4$

91. $3 \cdot 5^2 - 75 \div 5 + 2^3$

92. $2 \cdot 3^2 - 18 \div 3 + 2^4$

Multiplication with Decimals; Circumference and Area of a Circle

3.3

OBJECTIVES

A Multiply decimal numbers.

B Solve application problems involving decimals.

C Find the circumference of a circle.

TICKET TO SUCCESS

Keep these questions in mind as you read through the section. Then respond in your own words and in complete sentences.

1. Briefly explain the rule for multiplying with decimals.
2. How might estimation help us solve a multiplication problem that involves decimals?
3. Give the formulas for finding the following measurements of a circle:
   a. diameter
   b. circumference (using the radius)
   c. area
4. Explain how you would find the volume of a cylinder with a radius 2 inches and a height of 4 inches.

The distance around a circle is called the circumference. If you know the circumference of a bicycle wheel, and you ride the bicycle for one mile, you can calculate how many times the wheel has turned through one complete revolution. In this section, we learn how to multiply decimal numbers, and this gives us the information we need to work with circles and their circumferences.

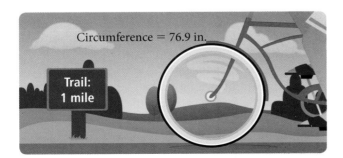

A Multiplying with Decimals

Before we introduce circumference, we need to back up and discuss multiplication with decimals. Suppose that during a half-price sale a calendar that usually sells for $6.42 is priced at $3.21. Therefore, it must be true that

$$\frac{1}{2} \text{ of } 6.42 \text{ is } 3.21$$

But, because $\frac{1}{2}$ can be written as 0.5 and *of* translates to *multiply*, we can also write this problem as

$$0.5 \times 6.42 = 3.21$$

If we were to ignore the decimal points in this problem and simply multiply 5 and 642, the result would be 3,210. So, multiplication with decimal numbers is similar to multiplication with whole numbers. The difference lies in deciding where to place the decimal point in the answer. To find out how this is done, we can use fraction notation.

NOTE
To indicate multiplication we are using a × sign here instead of a dot so we won't confuse the decimal points with the multiplication symbol.

PRACTICE PROBLEMS

1. Change each decimal to a fraction and multiply. Write your answer as a decimal.
   a. 0.4×0.6
   b. 0.04×0.06

EXAMPLE 1 Change each decimal to a fraction and multiply.

$$0.5 \times 0.3$$

SOLUTION Changing each decimal to a fraction and multiplying, we have

$$0.5 \times 0.3 = \frac{5}{10} \times \frac{3}{10} \qquad \text{Change to fractions.}$$

$$= \frac{15}{100} \qquad \text{Multiply numerators and multiply denominators.}$$

$$= 0.15 \qquad \text{Write the answer in decimal form.}$$

The result is 0.15, which has two digits to the right of the decimal point. ∎

What we want to do now is find a shortcut that will allow us to multiply decimals without first having to change each decimal number to a fraction. Let's look at another example.

2. Change each decimal to a fraction and multiply. Write your answer as a decimal.
   a. 0.5×0.007
   b. 0.05×0.07

EXAMPLE 2 Change each decimal to a fraction and multiply: 0.05×0.003.

SOLUTION $0.05 \times 0.003 = \frac{5}{100} \times \frac{3}{1,000} \qquad \text{Change to fractions.}$

$$= \frac{15}{100,000} \qquad \text{Multiply numerators and multiply denominators.}$$

$$= 0.00015 \qquad \text{Write the answer in decimal form.}$$

The result is 0.00015, which has a total of five digits to the right of the decimal point. ∎

Looking over these first two examples, we can see that the digits in the result are just what we would get if we simply forgot about the decimal points and multiplied; that is, $3 \times 5 = 15$. Then the decimal point in the result is placed so that the total number of digits to its right is the same as the total number of digits to the right of both decimal points in the original two numbers. The reason this is true becomes clear when we look at the denominators after we have changed from decimals to fractions. Let's look at a third example.

Answers
1. a. 0.24   b. 0.0024
2. Both are 0.0035.

EXAMPLE 3
Multiply: 2.1×0.07.

SOLUTION

$2.1 \times 0.07 = 2\dfrac{1}{10} \times \dfrac{7}{100}$ Change to fractions.

$ = \dfrac{21}{10} \times \dfrac{7}{100}$

$ = \dfrac{147}{1,000}$ Multiply numerators and multiply denominators.

$ = 0.147$ Write the answer as a decimal.

Again, the digits in the answer come from multiplying $21 \times 7 = 147$. The decimal point is placed so that there are three digits to its right, because that is the total number of digits to the right of the decimal points in 2.1 and 0.07. ∎

We summarize this discussion with the following rule.

> **Rule Multiplication with Decimals**
> To multiply two decimal numbers, follow these steps:
> 1. Multiply as you would if the decimal points were not there.
> 2. Place the decimal point in the answer so that the number of digits to its right is equal to the total number of digits to the right of the decimal points in the original two numbers in the problem.

Let's practice a couple more.

EXAMPLE 4
How many digits will be to the right of the decimal point in the following product?

$$2.987 \times 24.82$$

SOLUTION There are three digits to the right of the decimal point in 2.987 and two digits to the right in 24.82. Therefore, there will be $3 + 2 = 5$ digits to the right of the decimal point in their product. ∎

EXAMPLE 5
Multiply: 3.05×4.36.

SOLUTION We can set this up as if it were a multiplication problem with whole numbers. We multiply and then place the decimal point in the correct position in the answer.

```
    3.05   ← Move decimal 2 digits to the right.
  × 4.36   ← Move decimal 2 digits to the right.
   1830
    915
  12 20
  13.2980
      ↑
      └── The decimal point is placed so that there are 2 + 2 = 4
          digits to its right.
```
∎

---

3. Change each decimal to a fraction and multiply. Write your answer as a decimal.
   a. 3.5×0.04
   b. 0.35×0.4

4. How many digits will be to the right of the decimal point in the following products?
   a. 3.706×55.88
   b. 37.06×0.5588

5. Multiply.
   a. 4.03×5.22
   b. 40.3×0.522

Answers
3. Both are 0.14.
4. a. 5 b. 6
5. Both are 21.0366.

As you can see, multiplying decimal numbers is just like multiplying whole numbers, except that we must place the decimal point in the result in the correct position.

Estimating

Look back to Example 5. We could have placed the decimal point in the answer by rounding the two numbers to the nearest whole number and then multiplying them. Because 3.05 rounds to 3 and 4.36 rounds to 4, and the product of 3 and 4 is 12, we estimate that the answer to 3.05×4.36 will be close to 12. By placing the decimal point in the actual product 132980 between the 3 and the 2, we see that the answer is in fact a number close to 12.

EXAMPLE 6 Estimate the answer to each of the following products:
  a. 29.4×8.2 **b.** 68.5×172 **c.** $(6.32)^2$

SOLUTION
  a. Because 29.4 is approximately 30 and 8.2 is approximately 8, we estimate this product to be about $30 \times 8 = 240$. (If we were to multiply 29.4 and 8.2, we would find the product to be exactly 241.08.)
  b. Rounding 68.5 to 70 and 172 to 170, we estimate this product to be $70 \times 170 = 11,900$. (The exact answer is 11,782.) Note here that we do not always round the numbers to the nearest whole number when making estimates. The idea is to round to numbers that will be easy to multiply.
  c. Because 6.32 is approximately 6 and $6^2 = 36$, we estimate our answer to be close to 36. (The actual answer is 39.9424.) ∎

Combined Operations

We can use the rule for order of operations to simplify expressions involving decimal numbers and addition, subtraction, and multiplication.

EXAMPLE 7 Simplify: $0.05(4.2 + 0.03)$.

SOLUTION We begin by adding inside the parentheses.

$$0.05(4.2 + 0.03) = 0.05(4.23) \quad \text{Add.}$$
$$= 0.2115 \quad \text{Multiply.}$$

Notice that we could also have used the distributive property first, and the result would be unchanged.

$$0.05(4.2 + 0.03) = 0.05(4.2) + 0.05(0.03) \quad \text{Distributive property}$$
$$= 0.210 + 0.0015 \quad \text{Multiply.}$$
$$= 0.2115 \quad \text{Add.} \quad ∎$$

EXAMPLE 8 Simplify: $4.8 + 12(3.2)^2$.

SOLUTION According to the rule for order of operations, we must first evaluate the number with an exponent, then multiply, and finally add.

$$4.8 + 12(3.2)^2 = 4.8 + 12(10.24) \quad (3.2)^2 = 10.24$$
$$= 4.8 + 122.88 \quad \text{Multiply.}$$
$$= 127.68 \quad \text{Add.} \quad ∎$$

---

6. Estimate the answer to each product.
  a. 82.3×5.8
  b. 37.5×178
  c. $(8.21)^2$

7. Simplify:
  a. $0.03(5.5 + 0.02)$
  b. $0.03(0.55 + 0.002)$

8. Simplify.
  a. $5.7 + 14(2.4)^2$
  b. $0.57 + 1.4(2.4)^2$

Answers
6. a. 480 **b.** 7,200 **c.** 64
7. a. 0.1656 **b.** 0.01656
8. a. 86.34 **b.** 8.634

B Applications

Let's use what we have learned to work some application problems.

EXAMPLE 9 Find the area of each of the following stamps. Round to the nearest hundredth.

a.  Each side is 35.0 millimeters.

b.  Length = 1.56 inches
Width = 0.99 inches

SOLUTION Applying our formulas for area, we have

a. $A = s^2 = (35 \text{ mm})^2 = 1{,}225 \text{ mm}^2$

b. $A = lw = (1.56 \text{ in.})(0.99 \text{ in.}) = 1.54 \text{ in}^2$

EXAMPLE 10 Sally earns $6.82 for each of the first 36 hours she works in one week, and $10.23 in overtime pay for each additional hour she works in the same week. How much money will she make if she works 42 hours in one week?

SOLUTION The difference between 42 and 36 is 6 hours of overtime pay. The total amount of money she will make is

$$\underbrace{6.82(36)}_{\text{Pay for the first 36 hours}} + \underbrace{10.23(6)}_{\text{Pay for the next 6 hours}} = 245.52 + 61.38$$
$$= 306.90$$

She will make $306.90 for working 42 hours in one week.

C Circumference

Since we now have a working knowledge of decimals, we can begin our discussion on circumference. We start with two definitions.

Definition

The **radius** of a circle is the length of a straight line that stretches from the center of the circle to any point on its edge. The radius is denoted by the letter r.

---

9. Find the area of each stamp in Example 9 from the dimensions given below. Round answers to the nearest hundredth.
   a. Each side is 1.38 inches.

   b. Length = 39.6 millimeters, width = 25.1 millimeters

10. How much will Sally make if she works 50 hours in one week?

NOTE
To estimate the answer to Example 10 before doing the actual calculations, we would do the following:
$6(40) + 10(6) = 240 + 60 = 300$

Answers
9. a. 1.90 in^2 b. 993.96 mm^2
10. $388.74

> **Definition**
> The **diameter** of a circle is the distance from one side of the circle to the other, crossing through the center. The diameter is denoted by the letter d.

We can use these definitions to understand circumference.

> **FACTS FROM GEOMETRY**   **The Circumference of a Circle**
>
> The *circumference* of a circle is the distance around the outside of the circle, just as the perimeter of a polygon is the distance around the outside. The circumference of a circle can be found by measuring its radius or diameter and then using the appropriate formula. In Figure 1, we can see that the diameter is twice the radius, or
>
> $$d = 2r$$
>
> The relationship between the circumference and the diameter or radius is not as obvious. As a matter of fact, it takes some fairly complicated mathematics to show just what the relationship between the circumference and the diameter is.
>
>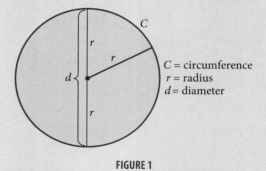
>
> C = circumference
> r = radius
> d = diameter
>
> **FIGURE 1**
>
> If you took a string and actually measured the circumference of a circle by wrapping the string around the circle and then measured the diameter of the same circle, you would find that the ratio of the circumference to the diameter, $\frac{C}{d}$, would be approximately equal to 3.14. The actual ratio of C to d in any circle is an irrational number, which is a number that can't be written in decimal form. We use the symbol π (Greek pi) to represent this ratio. In symbols, the relationship between the circumference and the diameter in any circle is
>
> $$\frac{C}{d} = \pi$$
>
> Knowing what we do about the relationship between division and multiplication, we can rewrite this formula as
>
> $$C = \pi d$$
>
> This is the formula for the circumference of a circle. For now, when we do the actual calculations, we will use the approximation 3.14 for π.
>
> Because $d = 2r$, the same formula written in terms of the radius is
>
> $$C = 2\pi r$$

3.3 Multiplication with Decimals; Circumference and Area of a Circle

Here are some examples that show how we use the formulas given above to find the circumference of a circle.

EXAMPLE 11 Find the circumference of a circle with a diameter of 5 feet.

SOLUTION Substituting 5 for d in the formula $C = \pi d$, and using 3.14 for π, we have

$C \approx 3.14(5)$
$ = 15.7$ feet

11. Find the circumference of a circle with a diameter of 3 centimeters.

EXAMPLE 12 Find the circumference of each coin.

a. 1 Euro coin (Round to the nearest whole number.)

Diameter = 23.25 millimeters

b. Susan B. Anthony dollar (Round to the nearest hundredth.)

Radius = 0.52 inch

SOLUTION Applying our formulas for circumference, we have

a. $C = \pi d \approx (3.14)(23.25) \approx 73$ mm
b. $C = 2\pi r \approx 2(3.14)(0.52) \approx 3.27$ in.

12. Find the circumference for each coin in Example 12 from the dimensions given below. Round answers to the nearest hundredth.
    a. Diameter = 0.92 inches
    b. Radius = 13.20 millimeters

FACTS FROM GEOMETRY **Other Formulas Involving π**

Two figures are presented here, along with some important formulas that are associated with each figure. As you can see, each of the formulas contains the number π. For now, when we do the actual calculations, we will use the approximation 3.14 for π.

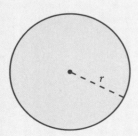

Area = π(radius)^2
$A = \pi r^2$

FIGURE 2 Circle

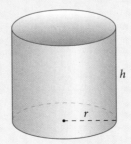

Volume = π(radius)^2(height)
$V = \pi r^2 h$

FIGURE 3 Right Circular Cylinder

Answers
11. 9.42 cm
12. a. 2.89 in. **b.** 82.90 mm

You will see how to use the formulas for the area of a circle and volume of a cylinder in the following examples.

EXAMPLE 13 Find the area of a circle with a diameter of 10 feet.

SOLUTION The formula for the area of a circle is $A = \pi r^2$. Because the radius r is half the diameter and the diameter is 10 feet, the radius is 5 feet. Therefore,

$$A \approx \pi r^2 = (3.14)(5)^2 = (3.14)(25) = 78.5 \text{ ft}^2$$
∎

13. Find the area of a circle with a diameter of 20 feet.

EXAMPLE 14 The drinking straw shown has a radius of 0.125 inch and a length of 6 inches. To the nearest thousandth, find the volume of liquid that it will hold.

14. Find the volume of the straw in Example 14, if the radius is doubled. Round your answer to the nearest thousandth.

SOLUTION The total volume is found from the formula for the volume of a right circular cylinder. In this case, the radius is $r = 0.125$, and the height is $h = 6$. We approximate π with 3.14.

$$\begin{aligned}V &= \pi r^2 h \\ &\approx (3.14)(0.125)^2(6) \\ &\approx (3.14)(0.015625)(6) \\ &\approx 0.294 \text{ in}^3 \text{ to the nearest thousandth}\end{aligned}$$
∎

Answers
13. 314 ft²
14. 1.178 in³

Problem Set 3.3

Moving Toward Success

"It was character that got us out of bed, commitment that moved us into action, and discipline that enabled us to follow through."

—Zig Ziglar, 1926–present, American motivational speaker and author

1. How will intending to succeed in this course affect your actual success?
2. Is it enough to just intend to succeed in this course or do you have to follow through with actions? What are those actions?

A Find each of the following products. (Multiply.) [Examples 1–3, 5]

1. 0.7 × 0.4
2. 0.8 × 0.3
3. 0.07 × 0.4
4. 0.8 × 0.03
5. 0.03 × 0.09
6. 0.07 × 0.002

7. 2.6(0.3)
8. 8.9(0.2)
9. 0.9 × 0.88
10. 0.8 × 0.99
11. 3.12 × 0.005
12. 4.69 × 0.006

13. 4.003 × 6.07
14. 7.0001 × 3.04
15. 5(0.006)
16. 7(0.005)
17. 75.14 × 2.5
18. 963.8 × 0.24

19. 0.1 × 0.02
20. 0.3 × 0.02
21. 2.796(10)
22. 97.531(100)
23. 0.0043 × 100
24. 12.345 × 1,000

25. 49.94 × 1,000
26. 157.02 × 10,000
27. 987.654 × 10,000
28. 1.23 × 100,000

A Perform the following operations according to the rule for order of operations. [Examples 7, 8]

29. 2.1(3.5 − 2.6)
30. 5.4(9.9 − 6.6)
31. 0.05(0.02 + 0.03)
32. 0.04(0.07 + 0.09)

33. 2.02(0.03 + 2.5)
34. 4.04(0.05 + 6.6)
35. (2.1 + 0.03)(3.4 + 0.05)
36. (9.2 + 0.01)(3.5 + 0.03)

37. (2.1 − 0.1)(2.1 + 0.1)
38. (9.6 − 0.5)(9.6 + 0.5)
39. 3.08 − 0.2(5 + 0.03)
40. 4.09 + 0.5(6 + 0.02)

41. $4.23 - 5(0.04 + 0.09)$ **42.** $7.89 - 2(0.31 + 0.76)$ **43.** $2.5 + 10(4.3)^2$ **44.** $3.6 + 15(2.1)^2$

45. $100(1 + 0.08)^2$ **46.** $500(1 + 0.12)^2$ **47.** $(1.5)^2 + (2.5)^2 + (3.5)^2$ **48.** $(1.1)^2 + (2.1)^2 + (3.1)^2$

Applying the Concepts

Solve each of the following word problems. Note that not all of the problems are solved by simply multiplying the numbers in the problems. Many of the problems involve addition and subtraction as well as multiplication.

49. Google Earth This Google Earth image shows an aerial view of a crop circle found near Wroughton, England. If the crop circle has a radius of 59.13 meters, what is its circumference? Use the approximation 3.14 for π. Round to the nearest hundredth.

50. Google Earth This is a Google Earth image of the Louvre Museum in Paris, France. The pyramid that dominates the Napoleon Courtyard has a height of 21.65 meters and a square base with sides of 35.50 meters. What is the volume of the pyramid to the nearest whole number? Use the formula for the volume of a pyramid $V = \left(\frac{1}{3}\right)$ (area of the base)(height).

51. Number Problem What is the product of 6 and the sum of 0.001 and 0.02?

52. Number Problem Find the product of 8 and the sum of 0.03 and 0.002.

53. Number Problem What does multiplying a decimal number by 100 do to the decimal point?

54. Number Problem What does multiplying a decimal number by 1,000 do to the decimal point?

55. Home Mortgage On a certain home mortgage, there is a monthly payment of $9.66 for every $1,000 that is borrowed. What is the monthly payment on this type of loan if $143,000 is borrowed?

56. Caffeine Content If 1 cup of regular coffee contains 105 milligrams of caffeine, how much caffeine is contained in 3.5 cups of coffee?

57. Geometry of a Coin The $1 coin shown here depicts Sacagawea and her infant son. The diameter of the coin is 26.5 mm, and the thickness is 2.00 mm. Find the following, rounding your answers to the nearest hundredth. Use 3.14 for π.

a. the circumference of the coin

b. the area of one face of the coin

c. the volume of the coin

58. Geometry of a Coin The Susan B. Anthony dollar shown here has a radius of 0.52 inches and a thickness of 0.0079 inches. Find the following, rounding your answers to the nearest ten thousandth, if necessary. Use 3.14 for π.

a. the circumference of the coin

b. the area of one face of the coin

c. the volume of the coin

59. Area of a Stamp This stamp shows the Mexican artist Frida Kahlo. The image area of the stamp has a width of 0.84 inches and a length of 1.41 inches. Find the area of the image. Round to the nearest hundredth.

60. Area of a Stamp This stamp honors the Italian scientist Enrico Fermi. The image area of the stamp has a width of 21.4 millimeters and a length of 35.8 millimeters. Find the area of the image. Round to the nearest whole number.

C Circumference Find the circumference and the area of each circle. Use 3.14 for π. [Examples 11–13]

61.

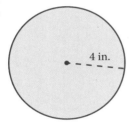

62.

63. Circumference The radius of the earth is approximately 3,900 miles. Find the circumference of the earth at the equator. (The equator is a circle around the earth that divides the earth into two equal halves.)

64. Circumference The radius of the moon is approximately 1,100 miles. Find the circumference of the moon around its equator.

65. Bicycle Wheel The wheel on a bicycle is such that the distance from the center of the wheel to the outside of the tire is 26.75 inches. If you walk the bicycle so that the wheel turns through one complete revolution, how many inches did you walk? Round to the nearest inch.

66. Model Plane A model plane is flying in a circle with a radius of 40 feet. To the nearest foot, how far does it fly in one complete trip around the circle?

Find the volume of each right circular cylinder. Use 3.14 for π. [Examples 14]

67.

68.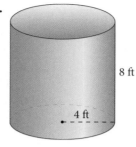

69.

70.

Getting Ready for the Next Section

To get ready for the next section, which covers division with decimals, we will review division with whole numbers and fractions.

Simplify each of the following.

71. 3,758 ÷ 2

72. 9,900 ÷ 22

73. 50,032 ÷ 33

74. 90,902 ÷ 5

75. 20)5,960

76. 30)4,620

77. 4 × 8.7

78. 5 × 6.7

79. 27 × 1.848

80. 35 × 32.54

81. 38)31,350

82. 25)377,800

Maintaining Your Skills

83. Write the numbers in order from smallest to largest.

$1\frac{5}{6}$ $\frac{3}{2}$ $1\frac{2}{3}$ $\frac{25}{12}$

84. Write the numbers in order from smallest to largest.

$1\frac{11}{12}$ $\frac{19}{12}$ $\frac{4}{3}$ $1\frac{1}{6}$

Extending the Concepts

85. Containment System Holding tanks for hazardous liquids are often surrounded by containment tanks that will hold the hazardous liquid if the main tank begins to leak. We see that the center tank has a height of 16 feet and a radius of 6 feet. The outside containment tank has a height of 4 feet and a radius of 8 feet. If the center tank is full of heating fuel and develops a leak at the bottom, will the containment tank be able to hold all the heating fuel that leaks out?

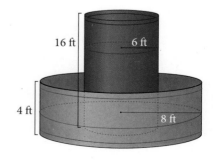

Division with Decimals

3.4

OBJECTIVES

A Divide decimal numbers.

B Solve application problems involving decimals.

TICKET TO SUCCESS

Keep these questions in mind as you read through the section. Then respond in your own words and in complete sentences.

1. Briefly explain the rule for dividing with decimals.
2. In a division problem, how many zeros can we write after the rightmost digit in a decimal number without changing the value of the number?
3. If the divisor in a division problem is a decimal, how would we change it to a whole number and not change the value of the problem?
4. Why is a student's GPA considered a weighted average?

During the 2008 Beijing Olympics, American swimmer Michael Phelps won 8 gold medals, breaking the record for most gold medals won in a single Olympics.

BEIJING OLYMPICS		
Swimmer	Country	Time (seconds)
Michael Phelps	USA	1:42.96
Taehwan Park	Korea	1:44.85
Peter Vanderkaay	USA	1:45.14
Jean Basson	Russia	1:45.97

Source: ESPN

The chart shows the top finishing times for the men's 200-meter freestyle swim during the 2008 Olympics. The times are shown in minutes and seconds. For example, Phelps' time is one minute, forty-two and ninety-six one hundredths seconds. An Olympic pool is 50 meters long, so each swimmer will have to complete 4 lengths during a 200-meter race.

During the race, each swimmer keeps track of how long it takes him to complete each length. To find the time of a swimmer's average lap, we need to be able to divide with decimal numbers, which we will learn in this section.

A Dividing with Decimals

To begin our discussion on dividing with decimal numbers, lets work the long division problem below.

EXAMPLE 1 Divide: $5{,}974 \div 20$.

SOLUTION

```
        298
    20)5974
        40
        ---
        197
        180
        ---
         174
         160
         ---
          14
```

In the past, we have written this answer as $298\frac{14}{20}$ or, after reducing the fraction, $298\frac{7}{10}$. Because $\frac{7}{10}$ can be written as 0.7, we could also write our answer as 298.7. This last form of our answer is exactly the same result we obtain if we write 5,974 as 5,974.0 and continue the division until we have no remainder. Here is how it looks:

```
         298.7
    20) 5974.0
        40
        ---
        197
        180
        ---
         174
         160
         ---
          14 0
          14 0
          ----
             0
```

Notice that we place the decimal point in the answer directly above the decimal point in the problem.

Let's try another division problem. This time, one of the numbers in the problem will be a decimal.

EXAMPLE 2 Divide: $34.8 \div 4$.

SOLUTION We can use the ideas from Example 1 and divide as usual. The decimal point in the answer will be placed directly above the decimal point in the problem.

```
       8.7         Check:    8.7
    4)34.8                 ×   4
      32                   ----
      --                   34.8
       2 8
       2 8
       ---
         0
```

The answer is 8.7.

We can use these facts to write a rule for dividing decimal numbers.

> **Rule Division of Decimal Numbers**
>
> To divide a decimal by a whole number, we do the usual long division as if there were no decimal point involved. The decimal point in the answer is placed directly above the decimal point in the problem.

PRACTICE PROBLEMS

1. Divide: $4{,}626 \div 30$.

NOTE
We can estimate the answer to Example 1 by rounding 5,974 to 6,000 and dividing by 20:

$$\frac{6{,}000}{20} = 300$$

NOTE
We can avoid making mistakes with division because we can always check our results with multiplication.

2. Divide.
   a. $33.5 \div 5$
   b. $34.5 \div 5$
   c. $35.5 \div 5$

Answers
1. 154.2
2. a. 6.7 b. 6.9 c. 7.1

3.4 Division with Decimals

Here are some more examples to illustrate the procedure.

EXAMPLE 3   Divide: $49.896 \div 27$.

SOLUTION

```
         1.848
    27)49.896
       27
       ---
       22 8
       21 6
       ----
        1 29
        1 08
        ----
          216
          216
          ---
            0
```

Check:
```
    1.848
  ×    27
  ------
   12 936
   36 96
   ------
   49.896
```

We can write as many zeros as we choose after the rightmost digit in a decimal number without changing the value of the number. For example,

$$6.91 = 6.910 = 6.9100 = 6.91000$$

There are times when this can be very useful, as Example 4 shows.

EXAMPLE 4   Divide: $1{,}138.9 \div 35$.

SOLUTION

```
         32.54
    35)1138.90
       105
       ---
        88
        70
        --
        18 9
        17 5
        ----
         1 40
         1 40
         ----
            0
```

Write 0 after the 9. It doesn't change the original number, but it gives us another digit to bring down.

Check:
```
    32.54
  ×    35
  ------
   162 70
   976 2
   -------
   1,138.90
```

Until now we have considered only division of a decimal number by a whole number. Extending division to include division of a decimal number by another decimal number is a matter of knowing what to do about the decimal point in the divisor.

EXAMPLE 5   Divide: $31.35 \div 3.8$.

SOLUTION   In fraction form, this problem is equivalent to

$$\frac{31.35}{3.8}$$

If we want to write the divisor as a whole number, we can multiply the numerator and the denominator of this fraction by 10:

$$\frac{31.35 \times \mathbf{10}}{3.8 \times \mathbf{10}} = \frac{313.5}{38}$$

3. Divide.
  a. $47.448 \div 18$
  b. $474.48 \div 18$

4. Divide.
  a. $1{,}138.5 \div 25$
  b. $113.85 \div 25$

5. Divide.
  a. $13.23 \div 4.2$
  b. $13.23 \div 0.42$

Answers
3. a. 2.636   **b.** 26.36
4. a. 45.54   **b.** 4.554
5. a. 3.15   **b.** 31.5

NOTE
We do not always use the rules for rounding numbers to make estimates. For example, to estimate the answer to Example 5, 31.35 ÷ 3.8, we can get a rough estimate of the answer by reasoning that 3.8 is close to 4 and 31.35 is close to 32. Therefore, our answer will be approximately 32 ÷ 4 = 8.

So, since this fraction is equivalent to the original fraction, our original division problem is equivalent to

$$\begin{array}{r} 8.25 \\ 38{\overline{\smash{\big)}\,313.50}} \\ \underline{304} \\ 95 \\ \underline{76} \\ 190 \\ \underline{190} \\ 0 \end{array}$$ Put 0 after the last digit.

We can summarize division with decimal numbers by listing the following points, as illustrated by the first five examples.

Summary of Division with Decimals
1. We divide decimal numbers by the same process used in Chapter 1 to divide whole numbers. The decimal point in the answer is placed directly above the decimal point in the dividend.
2. We are free to write as many zeros after the last digit in a decimal number as we need.
3. If the divisor is a decimal, we can change it to a whole number by moving the decimal point to the right as many places as necessary so long as we move the decimal point in the dividend the same number of places.

EXAMPLE 6 Divide and round the answer to the nearest hundredth.

$$0.3778 \div 0.25$$

SOLUTION First, we move the decimal point two places to the right.

$$0.25.\overline{)\,.37.78}$$

Then we divide, using long division.

$$\begin{array}{r} 1.5112 \\ 25{\overline{\smash{\big)}\,37.7800}} \\ \underline{25} \\ 12\,7 \\ \underline{12\,5} \\ 28 \\ \underline{25} \\ 30 \\ \underline{25} \\ 50 \\ \underline{50} \\ 0 \end{array}$$

Rounding to the nearest hundredth, we have 1.51. We actually did not need to have this many digits to round to the hundredths column. We could have stopped at the thousandths column and rounded off.

6. Divide and round your answer to the nearest hundredth.
$$0.4553 \div 0.32$$

NOTE
Moving the decimal point two places in both the divisor and the dividend is justified like this:

$$\frac{0.3778 \times \mathbf{100}}{0.25 \times \mathbf{100}} = \frac{37.78}{25}$$

Answer
6. 1.42

EXAMPLE 7 Divide and round to the nearest tenth: $17 \div 0.03$.

SOLUTION Because we are rounding to the nearest tenth, we will continue dividing until we have a digit in the hundredths column. We don't have to go any further to round to the tenths column.

```
         566.66
0.03.)17.00.00
      15
      ──
       2 0
       1 8
       ──
         20
         18
         ──
          2 0
          1 8
          ──
            20
            18
            ──
             2
```

Rounding to the nearest tenth, we have 566.7. ■

7. Divide and round to the nearest tenth.
   a. $19 \div 0.06$
   b. $1.9 \div 0.06$

B Applications

EXAMPLE 8 If a man earning $7.26 an hour receives a paycheck for $235.95, how many hours did he work?

SOLUTION To find the number of hours the man worked, we divide $235.95 by $7.26.

```
          32.5
7.26.)235.95.0
      217 8
      ─────
       18 15
       14 52
       ─────
        3 63 0
        3 63 0
        ──────
             0
```

The man worked 32.5 hours. ■

8. A woman earning $6.54 an hour receives a paycheck for $186.39. How many hours did the woman work?

EXAMPLE 9 A telephone company charges $0.43 for the first minute and then $0.33 for each additional minute for a long-distance call. If a long-distance call costs $3.07, how many minutes was the call?

SOLUTION To solve this problem we need to find the number of additional minutes for the call. To do so, we first subtract the cost of the first minute from the total cost, and then we divide the result by the cost of each additional minute. Without showing the actual arithmetic involved, the solution looks like this:

$$\text{The number of additional minutes} = \frac{\overset{\text{Total cost of the call}}{3.07} - \overset{\text{Cost of the first minute}}{0.43}}{\underset{\text{Cost of each additional minute}}{0.33}} = \frac{2.64}{0.33} = 8$$

The call was 9 minutes long. (The number 8 is the number of additional minutes past the first minute.) ■

9. If the phone company in Example 9 charged $4.39 for a call, how long was the call?

Answers
7. a. 316.7   b. 31.7
8. 28.5 hours
9. 13 minutes

Weighted Average

Suppose Alana earned the grades shown in the table during her first semester in college.

Class	Units	Grade
Algebra	5	B
Chemistry	4	C
English	3	A
History	3	B

When her grades arrived in the mail, she believed she had a 3.0 grade point average, because the A and C grades averaged to a B. Actually, her GPA was a little less than a 3.0. Can you calculate her GPA?

When you calculate your grade point average (GPA), you are calculating what is called a *weighted average*. To calculate your grade point average, you must first calculate the number of grade points you have earned in each class that you have completed. The number of grade points for a class is the product of the number of units the class is worth times the value of the grade received. The table below shows the value that is assigned to each grade.

Grade	Value
A	4
B	3
C	2
D	1
F	0

If you earn a B in a 4-unit class, you earn $4 \times 3 = 12$ grade points. A grade of C in the same class gives you $4 \times 2 = 8$ grade points. To find your grade point average for one term (a semester or quarter), you must add your grade points and divide that total by the number of units. Round your answer to the nearest hundredth.

EXAMPLE 10 Calculate Alana's grade point average using the information in the table above.

SOLUTION We begin by writing in two more columns, one for the value of each grade (4 for an A, 3 for a B, 2 for a C, 1 for a D, and 0 for an F), and another for the grade points earned for each class. To fill in the grade points column, we multiply the number of units by the value of the grade.

Class	Units	Grade	Value	Grade Points
Algebra	5	B	3	$5 \times 3 = 15$
Chemistry	4	C	2	$4 \times 2 = 8$
English	3	A	4	$3 \times 4 = 12$
History	3	B	3	$3 \times 3 = 9$
Total Units:	15			Total Grade Points: 44

To find her grade point average, we divide 44 by 15 and round (if necessary) to the nearest hundredth.

$$\text{Grade point average} = \frac{44}{15} = 2.93$$

10. If Alana had earned a B in chemistry, instead of a C, what grade point average would she have?

Answer
10. 3.20

Problem Set 3.4

Moving Toward Success

"We all learn best in our own ways. Some people do better studying one subject at a time, while some do better studying three things at once. Some people do best studying in structured, linear way, while others do best jumping around, 'surrounding' a subject rather than traversing it. Some people prefer to learn by manipulating models, and others by reading."

—Bill Gates, 1955–present, American entrepreneur and founder of Microsoft Co.

1. Each student learns in a different way. Go online and research learning styles (e.g., visual, auditory, kinesthetic). How do you prefer to learn?
2. How might you apply your preferred learning style(s) to this class?

A Perform each of the following divisions. [Examples 1–5]

1. $394 \div 20$
2. $486 \div 30$
3. $248 \div 40$
4. $372 \div 80$

5. $5\overline{)26}$
6. $8\overline{)36}$
7. $25\overline{)276}$
8. $50\overline{)276}$

9. $28.8 \div 6$
10. $15.5 \div 5$
11. $77.6 \div 8$
12. $31.48 \div 4$

13. 35)92.05 **14.** 26)146.38 **15.** 45)1900.8 **16.** 55)342.1

17. 86.7 ÷ 34 **18.** 411.4 ÷ 44 **19.** 29.7 ÷ 22 **20.** 488.4 ÷ 88

21. 4.5)29.25 **22.** 3.3)21.978 **23.** 0.11)1.089 **24.** 0.75)2.40

25. 2.3)0.115 **26.** 6.6)0.198 **27.** 0.012)1.068 **28.** 0.052)0.23712

29. 1.1)2.42 **30.** 2.2)7.26

Carry out each of the following divisions only so far as needed to round the results to the nearest hundredth. [Examples 6, 7]

31. $26\overline{)35}$

32. $18\overline{)47}$

33. $3.3\overline{)56}$

34. $4.4\overline{)75}$

35. $0.1234 \div 0.5$

36. $0.543 \div 2.1$

37. $19 \div 7$

38. $16 \div 6$

39. $0.059\overline{)0.69}$

40. $0.048\overline{)0.49}$

41. $1.99 \div 0.5$

42. $0.99 \div 0.5$

43. $2.99 \div 0.5$

44. $3.99 \div 0.5$

Calculator Problems Work each of the following problems on your calculator. If rounding is necessary, round to the nearest hundred thousandth.

45. $7 \div 9$

46. $11 \div 13$

47. $243 \div 0.791$

48. $67.8 \div 37.92$

49. $0.0503 \div 0.0709$

50. $429.87 \div 16.925$

Applying the Concepts

51. Google Earth The Google Earth map shows Yellowstone National Park. There is an average of 2.3 moose per square mile. If there are about 7,986 moose in Yellowstone, how many square miles does Yellowstone cover? Round to the nearest square mile.

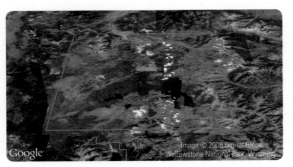

52. Google Earth The Google Earth image shows a corn field. A farmer harvests 29,952 bushels of corn. If the farmer harvested 130 bushels per acre, how many acres does the field cover?

53. Hot Air Balloon Since the pilot of a hot air balloon can only control the balloon's altitude, he relies on the winds for travel. To ride on the jet streams, a hot air balloon must rise as high as 12 kilometers. Convert this to miles by dividing by 1.61. Round your answer to the nearest tenth of a mile.

54. Hot Air Balloon December and January are the best times for traveling in a hot-air balloon because the jet streams in the Northern Hemisphere are the strongest. They reach speeds of 400 kilometers per hour. Convert this to miles per hour by dividing by 1.61. Round to the nearest whole number.

55. The swimmers in the table completed four laps during the 200-meter race. Use the information to find Michael Phelps' average lap time.

56. Use the table to find Peter Vanderkaay's average lap time.

BEIJING OLYMPICS		
Swimmer	Country	Time (seconds)
Michael Phelps	USA	1:42.96
Taehwan Park	Korea	1:44.85
Peter Vanderkaay	USA	1:45.14
Jean Basson	Russia	1:45.97

Source: ESPN

57. Gas Mileage If a car travels 336 miles on 15 gallons of gas, how far will the car travel on 1 gallon of gas?

58. Wages How many hours does a person making $6.78 per hour have to work in order to earn $257.64?

59. Wages Suppose a woman earns $6.78 an hour for the first 36 hours she works in a week and then $10.17 an hour in overtime pay for each additional hour she works in the same week. If she makes $294.93 in one week, how many hours did she work overtime?

60. Wages Suppose a woman makes $286.08 in one week. If she is paid $5.96 an hour for the first 36 hours she works and then $8.94 an hour in overtime pay for each additional hour she works in the same week, how many hours did she work overtime that week?

61. Phone Bill Suppose a telephone company charges $0.41 for the first minute and then $0.32 for each additional minute for a long-distance call. If a long-distance call costs $2.33, how many minutes was the call?

62. Phone Bill Suppose a telephone company charges a total of $0.45 for the first three minutes and then $0.29 for each additional minute for a long-distance call. If a long-distance call costs $2.77, how many minutes was the call?

63. Women's Golf The table gives the top five money earners for the Ladies' Professional Golf Association (LPGA) in 2010. Fill in the last column of the table by finding the average earnings per event for each golfer. Round your answers to the nearest dollar.

Rank	Name	Number of Events	Total Earnings	Average per Event
1.	Na Yeon Choi	23	$1,871,165.50	
2.	Jiyai Shin	19	$1,783,127.00	
3.	Cristie Kerr	21	$1,601,551.75	
4.	Yani Tseng	19	$1,573,529.00	
5.	Suzann Pettersen	19	$1,557,174.50	

Source: Ladies' Professional Golf Association

64. Men's Golf The table gives the top five money earners for the men's Professional Golf Association (PGA) in 2010. Fill in the last column of the table by finding the average earnings per event for each golfer. Round your answers to the nearest dollar.

Rank	Name	Number of Events	Total Earnings	Average per Event
1.	Matt Kuchar	26	$4,910,477	
2.	Jim Furyk	21	$4,809,622	
3.	Ernie Els	20	$4,558,861	
4.	Dustin Johnson	23	$4,473,122	
5.	Steve Stricker	19	$4,190,235	

Source: Professional Golf Association

Getting Ready for the Next Section

In the next section we will consider the relationship between fractions and decimals in more detail. The problems below review some of the material that is necessary to make a successful start in the next section.

Reduce to lowest terms.

65. $\dfrac{75}{100}$ **66.** $\dfrac{220}{1,000}$ **67.** $\dfrac{12}{18}$ **68.** $\dfrac{15}{30}$

69. $\dfrac{75}{200}$ **70.** $\dfrac{220}{2,000}$ **71.** $\dfrac{38}{100}$ **72.** $\dfrac{75}{1,000}$

Write each fraction as an equivalent fraction with denominator 10.

73. $\dfrac{3}{5}$ **74.** $\dfrac{1}{2}$

Write each fraction as an equivalent fraction with denominator 100.

75. $\dfrac{3}{5}$ **76.** $\dfrac{17}{20}$

Write each fraction as an equivalent fraction with denominator 15.

77. $\dfrac{4}{5}$ **78.** $\dfrac{2}{3}$ **79.** $\dfrac{4}{1}$ **80.** $\dfrac{2}{1}$ **81.** $\dfrac{6}{5}$ **82.** $\dfrac{7}{3}$

Divide.

83. $3 \div 4$ **84.** $3 \div 5$ **85.** $7 \div 8$ **86.** $3 \div 8$

Maintaining Your Skills

Simplify.

87. $15\left(\dfrac{2}{3} + \dfrac{3}{5}\right)$ **88.** $15\left(\dfrac{4}{5} - \dfrac{1}{3}\right)$ **89.** $4\left(\dfrac{1}{2} + \dfrac{1}{4}\right)$ **90.** $6\left(\dfrac{1}{3} + \dfrac{1}{2}\right)$

Fractions and Decimals, and the Volume of a Sphere

3.5

OBJECTIVES

A Convert fractions to decimals.

B Convert decimals to fractions.

C Simplify expressions containing fractions and decimals.

D Solve applications involving fractions and decimals.

TICKET TO SUCCESS

Keep these questions in mind as you read through the section. Then respond in your own words and in complete sentences.

1. How do you convert a fraction to a decimal?
2. Explain how you would convert a decimal to a fraction.
3. Write 36 thousandths in decimal form and in fraction form.
4. What formula would you use to find the volume of a sphere?

If you are shopping for clothes and a store has a sale advertising $\frac{1}{3}$ off the regular price, how much can you expect to pay for a pair of pants that normally sells for $31.95? If the sale price of the pants is $22.30, have they really been marked down by $\frac{1}{3}$? To answer questions like these, we need to know how to solve problems that involve fractions and decimals together.

We begin this section by showing how to convert back and forth between fractions and decimals.

A Converting Fractions to Decimals

You may recall that the notation we use for fractions can be interpreted as implying division. That is, the fraction $\frac{3}{4}$ can be thought of as meaning "3 divided by 4." We can use this idea to convert fractions to decimals.

EXAMPLE 1 Write $\frac{3}{4}$ as a decimal.

SOLUTION Dividing 3 by 4, we have

$$\begin{array}{r} .75 \\ 4\overline{)3.00} \\ \underline{2\,8}\downarrow \\ 20 \\ \underline{20} \\ 0 \end{array}$$

The fraction $\frac{3}{4}$ is equal to the decimal 0.75. ∎

PRACTICE PROBLEMS

1. Write as a decimal.
   a. $\frac{2}{5}$
   b. $\frac{3}{5}$
   c. $\frac{4}{5}$

Answers
1. a. 0.4 b. 0.6 c. 0.8

2. Write as a decimal correct to the thousandths column.

   a. $\dfrac{11}{12}$

   b. $\dfrac{12}{13}$

EXAMPLE 2 Write $\dfrac{7}{12}$ as a decimal correct to the thousandths column.

SOLUTION Because we want the decimal to be rounded to the thousandths column, we divide to the ten thousandths column and round off to the thousandths column.

$$\begin{array}{r} .5833 \\ 12\overline{)7.0000} \\ \underline{6\,0} \\ 1\,00 \\ \underline{96} \\ 40 \\ \underline{36} \\ 40 \\ \underline{36} \\ 4 \end{array}$$

Rounding off to the thousandths column, we have 0.583. Because $\dfrac{7}{12}$ is not exactly the same as 0.583, we write

$$\dfrac{7}{12} \approx 0.583$$

where the symbol ≈ is read "is approximately."

If we wrote more zeros after 7.0000 in Example 2, the pattern of 3s would continue for as many places as we could want. When we get a sequence of digits that repeat like this, 0.58333 . . . , we can indicate the repetition by writing

$0.58\overline{3}$ The bar over the 3 indicates that the 3 repeats from there on. ■

3. Write $\dfrac{5}{11}$ as a decimal.

EXAMPLE 3 Write $\dfrac{3}{11}$ as a decimal.

SOLUTION Dividing 3 by 11, we have

$$\begin{array}{r} .272727 \\ 11\overline{)3.000000} \\ \underline{2\,2} \\ 80 \\ \underline{77} \\ 30 \\ \underline{22} \\ 80 \\ \underline{77} \\ 30 \\ \underline{22} \\ 80 \\ \underline{77} \\ 3 \end{array}$$

NOTE
The bar over the 2 and the 7 in $0.\overline{27}$ is used to indicate that the pattern repeats itself indefinitely.

No matter how long we continue the division, the remainder will never be 0, and the pattern will continue. We write the decimal form of $\dfrac{3}{11}$ as $0.\overline{27}$, where

$0.\overline{27} = 0.272727\ldots$ The dots mean "and so on." ■

Answers
2. a. 0.917   b. 0.923
3. $0.\overline{45}$

B Converting Decimals to Fractions

To convert decimals to fractions, we take advantage of the place values we assigned to the digits to the right of the decimal point.

EXAMPLE 4   Write 0.38 as a fraction in lowest terms.

SOLUTION   0.38 is 38 hundredths, or

$$0.38 = \frac{38}{100}$$

$$= \frac{19}{50} \qquad \text{Divide the numerator and the denominator by 2 to reduce to lowest terms.}$$

The decimal 0.38 is equal to the fraction $\frac{19}{50}$.

We could check our work here by converting $\frac{19}{50}$ back to a decimal. We do this by dividing 19 by 50. That is,

```
       .38
   50)19.00
      15 0
       4 00
       4 00
          0
```

4. Write as a fraction in lowest terms.
  a. 0.48
  b. 0.048

EXAMPLE 5   Convert 0.075 to a fraction.

SOLUTION   We have 75 thousandths, or

$$0.075 = \frac{75}{1,000}$$

$$= \frac{3}{40} \qquad \text{Divide the numerator and the denominator by 25 to reduce to lowest terms.}$$

5. Convert 0.025 to a fraction.

EXAMPLE 6   Write 15.6 as a mixed number.

SOLUTION   Converting 0.6 to a fraction, we have

$$0.6 = \frac{6}{10} = \frac{3}{5} \qquad \text{Reduce to lowest terms.}$$

Since $0.6 = \frac{3}{5}$, we have $15.6 = 15\frac{3}{5}$.

6. Write 12.8 as a mixed number.

C Problems Containing Both Fractions and Decimals

We continue this section by working some problems that involve both fractions and decimals.

EXAMPLE 7   Simplify: $\frac{19}{50}(1.32 + 0.48)$.

SOLUTION   In Example 4, we found that $0.38 = \frac{19}{50}$. Therefore we can rewrite the problem as

$$\frac{19}{50}(1.32 + 0.48) = 0.38(1.32 + 0.48) \qquad \text{Convert all numbers to decimals.}$$

$$= 0.38(1.80) \qquad \text{Add: } 1.32 + 0.48.$$

$$= 0.684 \qquad \text{Multiply: } 0.38 \times 1.80.$$

7. Simplify: $\frac{14}{25}(2.43 + 0.27)$.

Answers

4. a. $\frac{12}{25}$ **b.** $\frac{6}{125}$

5. $\frac{1}{40}$

6. $12\frac{4}{5}$

7. 1.512

Chapter 3 Decimals

8. Simplify: $\dfrac{1}{4} + 0.25\left(\dfrac{3}{5}\right)$.

EXAMPLE 8 Simplify: $\dfrac{1}{2} + (0.75)\left(\dfrac{2}{5}\right)$.

SOLUTION We could do this problem one of two different ways. First, we could convert all fractions to decimals and then simplify.

$$\dfrac{1}{2} + (0.75)\left(\dfrac{2}{5}\right) = 0.5 + 0.75(0.4) \quad \text{Convert to decimals.}$$
$$= 0.5 + 0.300 \quad \text{Multiply: } 0.75 \times 0.4.$$
$$= 0.8 \quad \text{Add.}$$

Or we could convert 0.75 to $\dfrac{3}{4}$ and then simplify.

$$\dfrac{1}{2} + 0.75\left(\dfrac{2}{5}\right) = \dfrac{1}{2} + \dfrac{3}{4}\left(\dfrac{2}{5}\right) \quad \text{Convert decimals to fractions.}$$
$$= \dfrac{1}{2} + \dfrac{3}{10} \quad \text{Multiply: } \dfrac{3}{4} \times \dfrac{2}{5}.$$
$$= \dfrac{5}{10} + \dfrac{3}{10} \quad \text{The common denominator is 10.}$$
$$= \dfrac{8}{10} \quad \text{Add numerators.}$$
$$= \dfrac{4}{5} \quad \text{Reduce to lowest terms.}$$

The answers are equivalent. That is, $0.8 = \dfrac{8}{10} = \dfrac{4}{5}$. Either method can be used with problems of this type. ∎

9. Simplify: $\left(\dfrac{1}{3}\right)^{3}(5.4) + \left(\dfrac{1}{5}\right)^{2}(2.5)$.

EXAMPLE 9 Simplify: $\left(\dfrac{1}{2}\right)^{3}(2.4) + \left(\dfrac{1}{4}\right)^{2}(3.2)$.

SOLUTION This expression can be simplified without any conversions between fractions and decimals. To begin, we evaluate all numbers that contain exponents. Then we multiply. After that, we add.

$$\left(\dfrac{1}{2}\right)^{3}(2.4) + \left(\dfrac{1}{4}\right)^{2}(3.2) = \dfrac{1}{8}(2.4) + \dfrac{1}{16}(3.2) \quad \text{Evaluate exponents.}$$
$$= 0.3 + 0.2 \quad \text{Multiply by } \dfrac{1}{8} \text{ and } \dfrac{1}{16}.$$
$$= 0.5 \quad \text{Add.} \quad ∎$$

D Applications

10. A shirt that normally sells for $35.50 is on sale for $\dfrac{1}{4}$ off. What is the sale price of the shirt? (Round to the nearest cent.)

EXAMPLE 10 If a shirt that normally sells for $27.99 is on sale for $\dfrac{1}{3}$ off, what is the sale price of the shirt?

SOLUTION To find out how much the shirt is marked down, we must find $\dfrac{1}{3}$ of 27.99. That is, we multiply $\dfrac{1}{3}$ and 27.99, which is the same as dividing 27.99 by 3.

$$\dfrac{1}{3}(27.99) = \dfrac{27.99}{3} = 9.33$$

The shirt is marked down $9.33. The sale price is the original price less the amount it is marked down:

$$\text{Sale price} = 27.99 - 9.33 = 18.66$$

The sale price is $18.66. We also could have solved this problem by simply multiplying the original price by $\dfrac{2}{3}$, since, if the shirt is marked $\dfrac{1}{3}$ off, then the sale price must be $\dfrac{2}{3}$ of the original price. Multiplying by $\dfrac{2}{3}$ is the same as dividing by 3 and then multiplying by 2. The answer would be the same. ∎

Answers
8. $\dfrac{2}{5}$, or 0.4
9. 0.3
10. $26.63

EXAMPLE 11
Find the area of the stamp. Write your answer as a decimal, rounded to the nearest hundredth.

Base = $2\frac{5}{8}$ inches

Height = $1\frac{1}{4}$ inches

SOLUTION We can work the problem using fractions and then convert the answer to a decimal.

$$A = \frac{1}{2}bh = \frac{1}{2}\left(2\frac{5}{8}\right)\left(1\frac{1}{4}\right) = \frac{1}{2} \cdot \frac{21}{8} \cdot \frac{5}{4} = \frac{105}{64} \approx 1.64 \text{ in}^2$$

Or, we can convert the fractions to decimals and then work the problem.

$$A = \frac{1}{2}bh = 0.5(2.625)(1.25) \approx 1.64 \text{ in}^2 \quad \blacksquare$$

11. Find the area of the stamp in Example 11 if
Base = 6.6 centimeters,
Height = 3.3 centimeters

FACTS FROM GEOMETRY **The Volume of a Sphere**

Figure 1 shows a sphere and the formula for its volume. Because the formula contains both the fraction $\frac{4}{3}$ and the number π, and we have been using 3.14 for π, we can think of the formula as containing both a fraction and a decimal.

Volume = $\frac{4}{3}\pi(\text{radius})^3$
$= \frac{4}{3}\pi r^3$

FIGURE 1

EXAMPLE 12
Figure 2 is composed of a right circular cylinder with half a sphere on top. (A half-sphere is called a *hemisphere*.) To the nearest tenth, find the total volume enclosed by the figure.

10 in.

5 in.

FIGURE 2

12. If the radius in Figure 2 is doubled so that it becomes 10 inches instead of 5 inches, what is the new volume of the figure? Round your answer to the nearest tenth.

Answers
11. 10.89 cm²
12. 5,233.3 in³

SOLUTION The total volume is found by adding the volume of the cylinder to the volume of the hemisphere.

$$V = \text{volume of cylinder} + \text{volume of hemisphere}$$

$$= \pi r^2 h + \frac{1}{2} \cdot \frac{4}{3}\pi r^3$$

$$\approx (3.14)(5)^2(10) + \frac{1}{2} \cdot \frac{4}{3}(3.14)(5)^3$$

$$\approx (3.14)(25)(10) + \frac{1}{2} \cdot \frac{4}{3}(3.14)(125)$$

$$\approx 785 + \frac{2}{3}(392.5) \qquad \text{Multiply: } \tfrac{1}{2} \cdot \tfrac{4}{3} = \tfrac{4}{6} = \tfrac{2}{3}.$$

$$\approx 785 + \frac{785}{3} \qquad \text{Multiply: } 2(392.5) = 785.$$

$$\approx 785 + 261.7 \qquad \text{Divide 785 by 3, and round to the nearest tenth.}$$

$$\approx 1{,}046.7 \text{ in}^3 \qquad \blacksquare$$

Problem Set 3.5

Moving Toward Success

"All that we are is the result of what we have thought."
—Buddha, 563–483 BC, Hindu Prince Gautama Siddharta and founder of Buddhism

1. Why do you think it is important to create pictures in your mind as you learn mathematics?
2. Why can it be helpful to sometimes read this book or your notes out loud as you study?

A Each circle below is divided into 8 equal parts. The number below each circle indicates what fraction of the circle is shaded. Convert each fraction to a decimal. [Examples 1–3]

1. $\frac{1}{8}$
2. $\frac{3}{8}$
3. $\frac{5}{8}$
4. $\frac{7}{8}$

A Complete the following tables by converting each fraction to a decimal. [Examples 1–3]

5.

Fraction	$\frac{1}{4}$	$\frac{2}{4}$	$\frac{3}{4}$	$\frac{4}{4}$
Decimal				

6.

Fraction	$\frac{1}{5}$	$\frac{2}{5}$	$\frac{3}{5}$	$\frac{4}{5}$	$\frac{5}{5}$
Decimal					

7.

Fraction	$\frac{1}{6}$	$\frac{2}{6}$	$\frac{3}{6}$	$\frac{4}{6}$	$\frac{5}{6}$	$\frac{6}{6}$
Decimal						

A Convert each of the following fractions to a decimal. [Examples 1–3]

8. $\frac{1}{2}$
9. $\frac{12}{25}$
10. $\frac{14}{25}$
11. $\frac{14}{32}$
12. $\frac{18}{32}$

A Write each fraction as a decimal correct to the hundredths column. [Examples 1–3]

13. $\frac{12}{13}$
14. $\frac{17}{19}$
15. $\frac{3}{11}$
16. $\frac{5}{11}$

17. $\frac{2}{23}$
18. $\frac{3}{28}$
19. $\frac{12}{43}$
20. $\frac{15}{51}$

Chapter 3 Decimals

B Complete the following table by converting each decimal to a fraction. [Examples 4–6]

21.
Decimal	0.125	0.250	0.375	0.500	0.625	0.750	0.875
Fraction							

22.
Decimal	0.1	0.2	0.3	0.4	0.5	0.6	0.7	0.8	0.9
Fraction									

B Write each decimal as a fraction in lowest terms. [Examples 4–6]

23. 0.15   **24.** 0.45   **25.** 0.08   **26.** 0.06   **27.** 0.375   **28.** 0.475

B Write each decimal as a mixed number. [Example 6]

29. 5.6   **30.** 8.4   **31.** 5.06   **32.** 8.04   **33.** 1.22   **34.** 2.11

C Simplify each of the following as much as possible, and write all answers as decimals. [Examples 7–9]

35. $\frac{1}{2}(2.3 + 2.5)$ **36.** $\frac{3}{4}(1.8 + 7.6)$ **37.** $\frac{1.99}{\frac{1}{2}}$ **38.** $\frac{2.99}{\frac{1}{2}}$

39. $3.4 - \frac{1}{2}(0.76)$ **40.** $6.7 - \frac{1}{5}(0.45)$ **41.** $\frac{2}{5}(0.3) + \frac{3}{5}(0.3)$ **42.** $\frac{1}{8}(0.7) + \frac{3}{8}(0.7)$

43. $6\left(\frac{3}{5}\right)(0.02)$ **44.** $8\left(\frac{4}{5}\right)(0.03)$ **45.** $\frac{5}{8} + 0.35\left(\frac{1}{2}\right)$ **46.** $\frac{7}{8} + 0.45\left(\frac{3}{4}\right)$

47. $\left(\frac{1}{3}\right)^2(5.4) + \left(\frac{1}{2}\right)^3(3.2)$ **48.** $\left(\frac{1}{5}\right)^2(7.5) + \left(\frac{1}{4}\right)^2(6.4)$ **49.** $(0.25)^2 + \left(\frac{1}{4}\right)^2(3)$ **50.** $(0.75)^2 + \left(\frac{1}{4}\right)^2(7)$

Applying the Concepts

Hiking The chart shows the lengths of some of Yosemite's most popular trails.

51. Change the length of the Upper Yosemite Fall Trail to a mixed number.

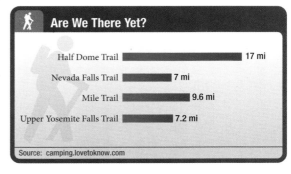

52. Change the length of the Mile Trail to a mixed number.

53. Price of Beef If each pound of beef costs $4.99, how much does $3\frac{1}{4}$ pounds cost?

54. Price of Gasoline What does it cost to fill a $15\frac{1}{2}$-gallon gas tank if the gasoline is priced at 429.9¢ per gallon?

55. Sale Price A dress that costs $78.99 is on sale for $\frac{1}{3}$ off. What is the sale price of the dress?

56. Sale Price A suit that normally sells for $221 is on sale for $\frac{1}{4}$ off. What is the sale price of the suit?

57. Perimeter of the Sierpinski Triangle The diagram below shows one stage of what is known as the Sierpinski triangle. Each triangle in the diagram has three equal sides. The large triangle is made up of 4 smaller triangles. If each side of the large triangle is 2 inches, and each side of the smaller triangles is 1 inch, what is the perimeter of the shaded region?

58. Perimeter of the Sierpinski Triangle The diagram below shows another stage of the Sierpinski triangle. Each triangle in the diagram has three equal sides. The largest triangle is made up of a number of smaller triangles. If each side of the large triangle is 2 inches, and each side of the smallest triangles is 0.5 inch, what is the perimeter of the shaded region?

59. Average Gain in Stock Price The table below shows the amount of change each day of one week in 2011 for the price of Microsoft Stock. Complete the table by converting each fraction to a decimal, rounding to the nearest hundredth if necessary.

CHANGE IN STOCK PRICE		
Date	Gain ($)	As a Decimal ($) (to the nearest hundredth)
April 25, 2011	$\frac{1}{20}$	
April 26, 2011	$\frac{9}{20}$	
April 27, 2011	$\frac{2}{25}$	
April 28, 2011	$\frac{1}{4}$	
April 29, 2011	$\frac{63}{100}$	

Source: Google Finance

60. Average Gain in Stock Price The table below shows the approximate amount of change each day of one week in 2011 for the stock price of Amazon.com. Complete the table by converting each fraction to a decimal, rounding to the nearest hundredth, if necessary.

CHANGE IN STOCK PRICE		
Date	Gain ($)	As a Decimal ($) (to the nearest hundredth)
April 25, 2011	$\frac{23}{100}$	
April 26, 2011	$3\frac{9}{10}$	
April 27, 2011	$13\frac{2}{5}$	
April 28, 2011	$\frac{89}{100}$	
April 29, 2011	$1\frac{2}{5}$	

Source: Google Finance

61. Nutrition If 1 ounce of ground beef contains 50.75 calories and 1 ounce of halibut contains 27.5 calories, what is the difference in calories between a $4\frac{1}{2}$-ounce serving of ground beef and a $4\frac{1}{2}$-ounce serving of halibut?

62. Nutrition If a 1-ounce serving of baked potato contains 48.3 calories and a 1-ounce serving of chicken contains 24.6 calories, how many calories are in a meal of $5\frac{1}{4}$ ounces of chicken and a $3\frac{1}{3}$-ounce baked potato?

Taxi Ride Recently, the Texas Junior College Teachers Association annual conference was held in Austin. At that time a taxi ride in Austin was $1.25 for the first $\frac{1}{5}$ of a mile and $0.25 for each additional $\frac{1}{5}$ of a mile. The charge for a taxi to wait is $12.00 per hour. Use this information for Problems 63 through 66.

63. If the distance from one of the convention hotels to the airport is 7.5 miles, how much will it cost to take a taxi from that hotel to the airport?

64. If you were to tip the driver of the taxi in Problem 63 $1.50, how much would it cost to take a taxi from the hotel to the airport?

65. Suppose the distance from one of the hotels to one of the western dance clubs in Austin is 12.4 miles. If the fare meter in the taxi gives the charge for that trip as $16.50, is the meter working correctly?

66. Suppose that the distance from a hotel to the airport is 8.2 miles, and the ride takes 20 minutes. Is it more expensive to take a taxi to the airport or to just sit in the taxi?

Volume Find the volume of each sphere. Round to the nearest hundredth. Use 3.14 for π. [Example 12]

67.
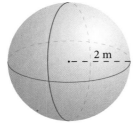

68.
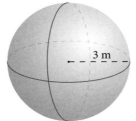

Volume Find the volume of each figure. Round to the nearest tenth. Use 3.14 for π. [Example 12]

69.
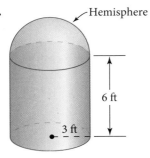

70.
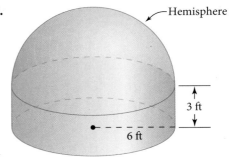

Area Find the total area enclosed by each figure below. Use 3.14 for π.

71.
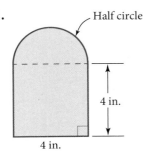

72.
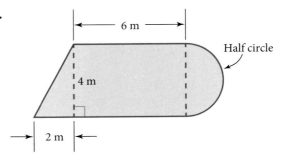

73. 4,256 ÷ 2

74. 7,700 ÷ 11

75. 41,056 ÷ 22

76. 94,256 ÷ 5

77. 15 × 2.463

78. 22 × 45.26

79. $42\overline{)23646}$

80. $32\overline{)457120}$

81. Write the fractions in order from smallest to largest.

$\dfrac{2}{5} \quad \dfrac{4}{5} \quad \dfrac{3}{10} \quad \dfrac{1}{2}$

82. Write the fractions in order from smallest to largest.

$\dfrac{4}{5} \quad \dfrac{1}{4} \quad \dfrac{1}{10} \quad \dfrac{17}{100}$

Maintaining Your Skills

83. Find the sum of 827 and 25.

84. Find the difference of 827 and 25.

85. Find the product of 827 and 25.

86. Find the quotient of 827 and 25.

Square Roots and the Pythagorean Theorem

3.6

OBJECTIVES

A Find square roots of numbers.

B Use decimals to approximate square roots.

C Solve problems with the Pythagorean theorem.

TICKET TO SUCCESS

Keep these questions in mind as you read through the section. Then respond in your own words and in complete sentences.

1. What is the square root of a number?
2. What are perfect squares?
3. What two numbers will the square root of 20 fall between?
4. What is the Pythagorean theorem?

Figure 1 shows the front view of a tool shed. How do we find the length d of the diagonal part of the roof? (Imagine that you are drawing the plans for the shed. Since the shed hasn't been built yet, you can't just measure the diagonal, but you need to know how long it will be so you can buy the correct amount of material to build the shed.)

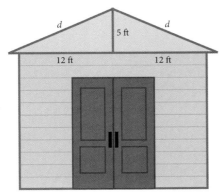

FIGURE 1

There is a formula from geometry that gives the length d:

$$d = \sqrt{12^2 + 5^2}$$

where $\sqrt{}$ is called the *square root symbol*, or the *radical*. If we simplify what is under the square root symbol, we have this:

$$d = \sqrt{144 + 25}$$
$$= \sqrt{169}$$

3.6 Square Roots and the Pythagorean Theorem

The expression $\sqrt{169}$ stands for the number we square to get 169. Because $13 \cdot 13 = 169$, that number is 13. Therefore, the length d in our original diagram is 13 feet. In this section, we'll work more with square roots, such as the one above.

A Square Roots

Here is a more detailed discussion of square roots. In Chapter 1, we did some work with exponents. In particular, we spent some time finding squares of numbers. For example, we considered expressions like this:

$$5^2 = 5 \cdot 5 = 25$$
$$7^2 = 7 \cdot 7 = 49$$
$$x^2 = x \cdot x$$

We say that "the square of 5 is 25" and "the square of 7 is 49." To square a number, we multiply it by itself. When we ask for the *square root* of a given number, we want to know what number we square in order to obtain the given number. We say that the square root of 49 is 7, because 7 is the number we square to get 49. Likewise, the square root of 25 is 5, because $5^2 = 25$. Remember, the symbol we use to denote square root is $\sqrt{}$, which is also called a *radical sign*. Here is the precise definition of square root:

> **Definition**
>
> The **square root** of a positive number a, written $\sqrt{a}$, is the number we square to get a.
>
> If $\sqrt{a} = b$, then $b^2 = a$.

NOTE
The square root we are describing here is actually the principal square root. There is another square root that is a negative number. We won't see it in this book, but, if you go on to take an algebra course, you will see it there.

We list some common square roots in Table 1.

TABLE 1		
Statement	In Words	Reason
$\sqrt{0} = 0$	The square root of 0 is 0	Because $0^2 = 0$
$\sqrt{1} = 1$	The square root of 1 is 1	Because $1^2 = 1$
$\sqrt{4} = 2$	The square root of 4 is 2	Because $2^2 = 4$
$\sqrt{9} = 3$	The square root of 9 is 3	Because $3^2 = 9$
$\sqrt{16} = 4$	The square root of 16 is 4	Because $4^2 = 16$
$\sqrt{25} = 5$	The square root of 25 is 5	Because $5^2 = 25$

Numbers like 1, 9, and 25, whose square roots are whole numbers, are called *perfect squares*. To find the square root of a perfect square, we look for the whole number that is squared to get the perfect square. The following examples involve square roots of perfect squares.

PRACTICE PROBLEMS

1. Simplify: $4\sqrt{25}$.

EXAMPLE 1 Simplify: $7\sqrt{64}$.

SOLUTION The expression $7\sqrt{64}$ means 7 times $\sqrt{64}$. To simplify this expression, we write $\sqrt{64}$ as 8 and multiply.

$$7\sqrt{64} = 7 \cdot 8 = 56$$

We know $\sqrt{64} = 8$, because $8^2 = 64$. ∎

Answer
1. 20

EXAMPLE 2 Simplify: $\sqrt{9} + \sqrt{16}$.

SOLUTION We write $\sqrt{9}$ as 3 and $\sqrt{16}$ as 4. Then we add.

$$\sqrt{9} + \sqrt{16} = 3 + 4 = 7$$

2. Simplify: $\sqrt{36} + \sqrt{4}$.

EXAMPLE 3 Simplify: $\sqrt{\dfrac{25}{81}}$.

SOLUTION We are looking for the number we square (multiply times itself) to get $\dfrac{25}{81}$. We know that when we multiply two fractions, we multiply the numerators and multiply the denominators. Because $5 \cdot 5 = 25$ and $9 \cdot 9 = 81$, the square root of $\dfrac{25}{81}$ must be $\dfrac{5}{9}$.

$$\sqrt{\dfrac{25}{81}} = \dfrac{5}{9} \text{ because } \left(\dfrac{5}{9}\right)^2 = \dfrac{5}{9} \cdot \dfrac{5}{9} = \dfrac{25}{81}$$

3. Simplify: $\sqrt{\dfrac{36}{100}}$.

In Examples 4–6, we simplify each expression as much as possible.

EXAMPLE 4 Simplify: $12\sqrt{25} = 12 \cdot 5 = 60$.

Simplify each expression as much as possible.

4. $14\sqrt{36}$

EXAMPLE 5 Simplify: $\sqrt{100} - \sqrt{36} = 10 - 6 = 4$.

5. $\sqrt{81} - \sqrt{25}$

EXAMPLE 6 Simplify: $\sqrt{\dfrac{49}{121}} = \dfrac{7}{11}$ because $\left(\dfrac{7}{11}\right)^2 = \dfrac{7}{11} \cdot \dfrac{7}{11} = \dfrac{49}{121}$.

6. $\sqrt{\dfrac{64}{121}}$

B Approximating Square Roots

So far in this section we have been concerned only with square roots of perfect squares. The next question is, "What about square roots of numbers that are not perfect squares, like $\sqrt{7}$, for example?" We know that

$$\sqrt{4} = 2 \quad \text{and} \quad \sqrt{9} = 3$$

And because 7 is between 4 and 9, $\sqrt{7}$ should be between $\sqrt{4}$ and $\sqrt{9}$. That is, $\sqrt{7}$ should be between 2 and 3. But what is it exactly? The answer is, we cannot write it exactly in decimal or fraction form. Because of this, it is called an *irrational number*. We can approximate it with a decimal, but we can never write it exactly with a decimal. Table 2 gives some decimal approximations for $\sqrt{7}$. The decimal approximations were obtained by using a calculator. We could continue the list to any accuracy we desired. However, we would never reach a number in decimal form whose square was exactly 7.

TABLE 2

APPROXIMATIONS FOR THE SQUARE ROOT OF 7

Accurate to the Nearest	The Square Root of 7 Is	Check by Squaring
Tenth	$\sqrt{7} = 2.6$	$(2.6)^2 = 6.76$
Hundredth	$\sqrt{7} = 2.65$	$(2.65)^2 = 7.0225$
Thousandth	$\sqrt{7} = 2.646$	$(2.646)^2 = 7.001316$
Ten thousandth	$\sqrt{7} = 2.6458$	$(2.6458)^2 = 7.00025764$

Answers
2. 8
3. $\dfrac{3}{5}$
4. 84
5. 4
6. $\dfrac{8}{11}$

7. Give a decimal approximation for the expression $5\sqrt{14}$ that is accurate to the nearest ten thousandth.

EXAMPLE 7 Give a decimal approximation for the expression $5\sqrt{12}$ that is accurate to the nearest ten thousandth.

SOLUTION Let's agree not to round to the nearest ten thousandth until we have first done all the calculations. Using a calculator, we find $\sqrt{12} \approx 3.4641016$. Therefore,

$$5\sqrt{12} \approx 5(3.4641016) \qquad \sqrt{12} \text{ on calculator}$$
$$= 17.320508 \qquad \text{Multiply.}$$
$$= 17.3205 \qquad \text{Round to the nearest ten thousandth.}$$

8. Approximate $\sqrt{405} + \sqrt{147}$ to the nearest hundredth.

EXAMPLE 8 Approximate $\sqrt{301} + \sqrt{137}$ to the nearest hundredth.

SOLUTION Using a calculator to approximate the square roots, we have

$$\sqrt{301} + \sqrt{137} \approx 17.349352 + 11.704700 = 29.054052$$

To the nearest hundredth, the answer is 29.05.

9. Approximate $\sqrt{\dfrac{7}{12}}$ to the nearest thousandth.

EXAMPLE 9 Approximate $\sqrt{\dfrac{7}{11}}$ to the nearest thousandth.

SOLUTION Because we are using calculators, we first change $\dfrac{7}{11}$ to a decimal and then find the square root.

$$\sqrt{\dfrac{7}{11}} \approx \sqrt{0.6363636} \approx 0.7977240$$

To the nearest thousandth, the answer is 0.798.

C The Pythagorean Theorem

Now we'll see how square roots can help us calculate the longest side of a triangle using the *Pythagorean theorem*.

> **FACTS FROM GEOMETRY   Pythagorean Theorem**
>
> A *right triangle* is a triangle that contains a 90° (or right) angle. The longest side in a right triangle is called the *hypotenuse*, and we use the letter c to denote it. The two shorter sides are denoted by the letters a and b. The Pythagorean theorem states that the hypotenuse is the square root of the sum of the squares of the two shorter sides. Here is the previous statement in symbols:
>
> $$c = \sqrt{a^2 + b^2}$$
>
>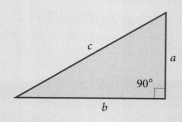

Answers
7. 18.7083
8. 32.25
9. 0.764

EXAMPLE 10
Find the length of the hypotenuse in each right triangle.

a.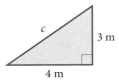
b.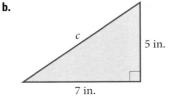

SOLUTION We apply the formula given above.

a. When $a = 3$ and $b = 4$,
$$c = \sqrt{3^2 + 4^2}$$
$$= \sqrt{9 + 16}$$
$$= \sqrt{25}$$
$$= 5 \text{ meters}$$

b. When $a = 5$ and $b = 7$,
$$c = \sqrt{5^2 + 7^2}$$
$$= \sqrt{25 + 49}$$
$$= \sqrt{74}$$
$$\approx 8.60 \text{ inches}$$

In part a, the solution is a whole number, whereas in part b, we must use a calculator to get 8.60 as an approximation to $\sqrt{74}$. ■

EXAMPLE 11
A ladder is leaning against the top of a 6-foot wall. If the bottom of the ladder is 8 feet from the wall, how long is the ladder?

SOLUTION A picture of the situation is shown in Figure 2. We let c denote the length of the ladder. Applying the Pythagorean theorem, we have

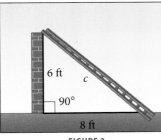

FIGURE 2

$$c = \sqrt{6^2 + 8^2}$$
$$= \sqrt{36 + 64}$$
$$= \sqrt{100}$$
$$= 10 \text{ feet}$$

The ladder is 10 feet long. ■

---

10. Find the length of the hypotenuse in each right triangle.

a.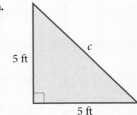

b.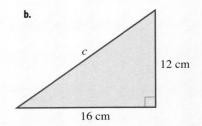

11. A wire from the top of a 12-foot pole is fastened to the ground by a stake that is 5 feet from the bottom of the pole. What is the length of the wire?

Answers
10. a. Approx. 7.07 ft   b. 20 cm
11. 13 ft

Problem Set 3.6

Moving Toward Success

"Concentrate all your thoughts upon the work at hand. The sun's rays do not burn until brought to a focus."

—Alexander Graham Bell, 1847–1922, Scottish born American inventor and educator

1. A mnemonic device is a mental tool that uses an acronym or a short verse to help a person remember. Why do you think a mnemonic device is a helpful learning tool?
2. Have you used a mnemonic device to help you study mathematics? If so, what was it? If not, create one for this section.

A Find each of the following square roots without using a calculator.

1. $\sqrt{64}$
2. $\sqrt{100}$
3. $\sqrt{81}$
4. $\sqrt{49}$
5. $\sqrt{36}$
6. $\sqrt{144}$
7. $\sqrt{25}$
8. $\sqrt{169}$

A Simplify each of the following expressions without using a calculator. [Examples 1–6]

9. $3\sqrt{25}$
10. $9\sqrt{49}$
11. $6\sqrt{64}$
12. $11\sqrt{100}$
13. $15\sqrt{9}$
14. $8\sqrt{36}$
15. $16\sqrt{9}$
16. $9\sqrt{16}$
17. $\sqrt{49} + \sqrt{64}$
18. $\sqrt{1} + \sqrt{0}$
19. $\sqrt{16} - \sqrt{9}$
20. $\sqrt{25} - \sqrt{4}$
21. $3\sqrt{25} + 9\sqrt{49}$
22. $6\sqrt{64} + 11\sqrt{100}$
23. $15\sqrt{9} - 9\sqrt{16}$
24. $7\sqrt{49} - 2\sqrt{4}$
25. $\sqrt{\frac{16}{49}}$
26. $\sqrt{\frac{100}{121}}$
27. $\sqrt{\frac{36}{64}}$
28. $\sqrt{\frac{81}{144}}$

Indicate whether each of the statements in Problems 29–32 is *True* or *False*.

29. $\sqrt{4} + \sqrt{9} = \sqrt{4+9}$

30. $\sqrt{\dfrac{16}{25}} = \dfrac{\sqrt{16}}{\sqrt{25}}$

31. $\sqrt{25 \cdot 9} = \sqrt{25} \cdot \sqrt{9}$

32. $\sqrt{100} - \sqrt{36} = \sqrt{100-36}$

Calculator Problems

Use a calculator to work problems 33 through 52.

Approximate each of the following square roots to the nearest ten thousandth.

33. $\sqrt{1.25}$ **34.** $\sqrt{12.5}$ **35.** $\sqrt{125}$ **36.** $\sqrt{1250}$

Approximate each of the following expressions to the nearest hundredth.

37. $2\sqrt{3}$ **38.** $3\sqrt{2}$ **39.** $5\sqrt{5}$ **40.** $5\sqrt{3}$

41. $\dfrac{\sqrt{3}}{3}$ **42.** $\dfrac{\sqrt{2}}{2}$ **43.** $\sqrt{\dfrac{1}{3}}$ **44.** $\sqrt{\dfrac{1}{2}}$

Approximate each of the following expressions to the nearest thousandth.

45. $\sqrt{12} + \sqrt{75}$ **46.** $\sqrt{18} + \sqrt{50}$ **47.** $\sqrt{87}$ **48.** $\sqrt{68}$

49. $2\sqrt{3} + 5\sqrt{3}$ **50.** $3\sqrt{2} + 5\sqrt{2}$ **51.** $7\sqrt{3}$ **52.** $8\sqrt{2}$

C Find the length of the hypotenuse in each right triangle. Round to the nearest hundredth, if necessary. [Examples 10, 11]

53.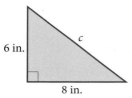

54.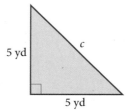

55.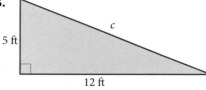

56.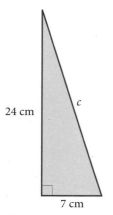

57.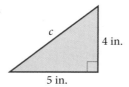

58.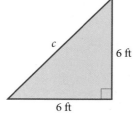

Applying the Concepts

59. Google Earth The Google Earth image shows a right triangle between three cities in the Los Angeles area. If the distance between Pomona and Ontario is 5.7 miles, and the distance between Ontario and Upland is 3.6 miles, what is the distance between Pomona and Upland? Round to the nearest tenth of a mile.

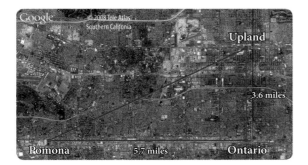

60. Google Earth The Google Earth image shows three cities in Colorado. If the distance between Denver and North Washington is 2.5 miles, and the distance between Edgewater and Denver is 4 miles, what is the distance between North Washington and Edgewater? Round to the nearest tenth.

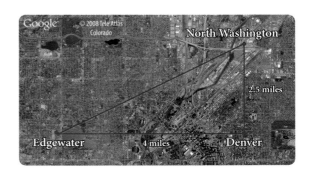

61. Geometry One end of a wire is attached to the top of a 24-foot pole; the other end of the wire is anchored to the ground 18 feet from the bottom of the pole. If the pole makes an angle of 90° with the ground, find the length of the wire.

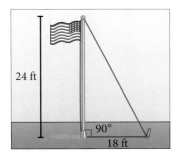

62. Geometry Two children are trying to cross a stream. They want to use a log that goes from one bank to the other. If the left bank is 5 feet higher than the right bank and the stream is 12 feet wide, how long must a log be to just barely reach?

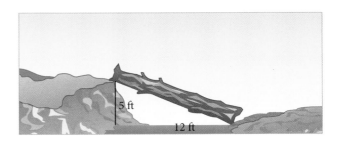

63. Geometry A ladder is leaning against the top of a 15-foot wall. If the bottom of the ladder is 20 feet from the wall, how long is the ladder?

64. Geometry A wire from the top of a 24-foot pole is fastened to the ground by a stake that is 10 feet from the bottom of the pole. How long is the wire?

65. Surveying A surveying team wants to calculate the length of a straight tunnel through a mountain. They form a right angle by connecting lines from each end of the proposed tunnel. One of the connecting lines is 3 miles, and the other is 4 miles. What is the length of the proposed tunnel?

66. Surveying A surveying team wants to calculate the length of a straight tunnel through a mountain. They form a right angle by connecting lines from each end of the proposed tunnel. One of the connecting lines is 6 miles, and the other is 8 miles. What is the length of the proposed tunnel?

Maintaining Your Skills

Factor each of the following numbers into the product of two numbers, one of which is a perfect square.

67. 32   **68.** 200   **69.** 75   **70.** 12

71. 50   **72.** 20   **73.** 40   **74.** 18

75. 16   **76.** 27   **77.** 98   **78.** 72

79. 48   **80.** 121

The problems below review material involving fractions and mixed numbers. Perform the indicated operations. Write your answers as whole numbers, proper fractions, or mixed numbers.

81. $\dfrac{5}{7} \cdot \dfrac{14}{25}$ **82.** $1\dfrac{1}{4} \div 2\dfrac{1}{8}$ **83.** $4\dfrac{3}{10} + 5\dfrac{2}{100}$ **84.** $8\dfrac{1}{5} + 1\dfrac{1}{10}$

85. $3\dfrac{2}{10} \cdot 2\dfrac{5}{10}$ **86.** $6\dfrac{9}{10} \div 2\dfrac{3}{10}$ **87.** $7\dfrac{1}{10} - 4\dfrac{3}{10}$ **88.** $3\dfrac{7}{10} - 1\dfrac{97}{100}$

89. $\dfrac{\dfrac{3}{8}}{\dfrac{6}{7}}$ **90.** $\dfrac{\dfrac{3}{4}}{\dfrac{1}{2} + \dfrac{1}{4}}$ **91.** $\dfrac{\dfrac{2}{3} + \dfrac{3}{5}}{\dfrac{2}{3} - \dfrac{3}{5}}$ **92.** $\dfrac{\dfrac{4}{5} - \dfrac{1}{3}}{\dfrac{4}{5} + \dfrac{1}{3}}$

Chapter 3 Summary

EXAMPLES

1. The number 4.123 in words is "four and one hundred twenty-three thousandths."

■ Place Value [3.1]

The place values for the first five places to the right of the decimal point are

Decimal Point	Tenths	Hundredths	Thousandths	Ten Thousandths	Hundred Thousandths
.	$\frac{1}{10}$	$\frac{1}{100}$	$\frac{1}{1,000}$	$\frac{1}{10,000}$	$\frac{1}{100,000}$

■ Rounding Decimals [3.1]

2. 357.753 rounded to the nearest
   Tenth: 357.8
   Ten: 360

If the digit in the column to the right of the one we are rounding to is 5 or more, we add 1 to the digit in the column we are rounding to; otherwise, we leave it alone. We then replace all digits to the right of the column we are rounding to with zeros if they are to the left of the decimal point; otherwise, we simply delete them.

■ Addition and Subtraction with Decimals [3.2]

3.
```
   3.400
  25.060
 + 0.347
 ───────
  28.807
```

To add (or subtract) decimal numbers, we align the decimal points and add (or subtract) as if we were adding (or subtracting) whole numbers. The decimal point in the answer goes directly below the decimal points in the problem.

■ Multiplication with Decimals [3.3]

4. If we multiply 3.49 × 5.863, there will be a total of 2 + 3 = 5 digits to the right of the decimal point in the answer.

To multiply two decimal numbers, we multiply as if the decimal points were not there. The decimal point in the product has as many digits to the right as there are total digits to the right of the decimal points in the two original numbers.

■ Division with Decimals [3.4]

5.
```
         1.39
   2.5.)3.4.75
        2 5
        ───
          97
          75
          ──
          2 25
          2 25
          ────
             0
```

To begin a division problem with decimals, we make sure that the divisor is a whole number. If it is not, we move the decimal point in the divisor to the right as many places as it takes to make it a whole number. We must then be sure to move the decimal point in the dividend the same number of places to the right. Once the divisor is a whole number, we divide as usual. The decimal point in the answer is placed directly above the decimal point in the dividend.

Changing Fractions to Decimals [3.5]

6. $\frac{4}{15} = 0.2\overline{6}$ because

$$\begin{array}{r} .266 \\ 15\overline{)4.000} \\ \underline{3\,0} \\ 1\,00 \\ \underline{90} \\ 100 \\ \underline{90} \\ 10 \end{array}$$

To change a fraction to a decimal, we divide the numerator by the denominator.

Changing Decimals to Fractions [3.5]

7. $0.781 = \frac{781}{1,000}$

To change a decimal to a fraction, we write the digits to the right of the decimal point over the appropriate power of 10.

Square Roots [3.6]

8. $\sqrt{49} = 7$ because $7^2 = 7 \cdot 7 = 49$

The *square root* of a positive number a, written $\sqrt{a}$, is the number we square to get a.

Pythagorean Theorem [3.6]

In any right triangle, the length of the longest side (the hypotenuse) is equal to the square root of the sum of the squares of the two shorter sides. This is called the Pythagorean theorem.

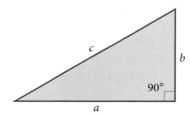

$c = \sqrt{a^2 + b^2}$

Chapter 3 Review

Give the place value of the 7 in each of the following numbers. [3.1]

1. 36.007
2. 121.379

Write each of the following as a decimal number. [3.1]

3. Thirty-seven and forty-two ten thousandths
4. One hundred and two hundred two hundred thousandths

Round 98.7654 to the nearest: [3.1]

5. hundredth
6. hundred

Perform the following operations. [3.2, 3.3, 3.4]

7. $3.78 + 2.036$
8. $11.076 - 3.297$
9. 6.7×5.43
10. $0.89(24.24)$
11. $29.07 \div 3.8$
12. $0.7134 \div 0.58$
13. $65\overline{)460.85}$
14. $(0.25)^3$
15. $13.27 - 7.541$
16. $8.52 + 5.4$
17. $0.24(4.2)$
18. $42.5 \div 3.4$
19. $6.5\overline{)221.65}$
20. $(0.42)^2$
21. 2.6×1.6
22. $5.7 \div 0.15$

Write as a decimal. [3.5]

23. $\dfrac{7}{8}$
24. $\dfrac{3}{16}$

Write as a fraction in lowest terms. [3.5]

25. 0.705
26. 0.246

Write as a mixed number. [3.5]

27. 14.125
28. 5.05

Simplify each of the following expressions as much as possible. [3.5, 3.6]

29. $3.3 - 4(0.22)$
30. $54.987 - 2(3.05 + 0.151)$
31. $125\left(\dfrac{3}{5}\right) + 4$
32. $\dfrac{3}{5}(0.9) + \dfrac{2}{5}(0.4)$
33. $0.3(1.7)(2.4)$
34. $3.6(1.4 - 0.5)$
35. $4(1.2 - 0.7) - 1.6$
36. $\sqrt{169}$
37. $\sqrt{121}$

38. Find the length of the hypotenuse of the following right triangle. Round to the nearest hundredth. [3.6]

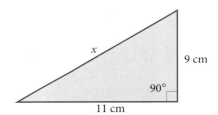

Chapter 3 Cumulative Review

Simplify.

1. 3,781 + 298
2. 903 − 576
3. 56(287)
4. 2.106 − 1.79
5. 24)̅1̅4̅9̅.̅2̅8̅
6. $\frac{2}{7} + \frac{3}{5}$
7. 4.3(12.96)
8. 1,292 ÷ 17
9. $\frac{5}{14} \div \frac{15}{21}$

10. Round 463,612 to the nearest thousand.
11. Change $\frac{63}{4}$ to a mixed number.
12. Change $2\frac{1}{5}$ to an improper fraction.
13. Find the product of $2\frac{1}{2}$ and 8.
14. Change each decimal into a fraction.

Decimal	Fraction
0.125	
0.250	
0.375	
0.500	
0.625	
0.750	
0.875	
1	

15. Give the quotient of 72 and 8.
16. Identify the property or properties used in the following: 2 · (x · 3) = (2 · 3) · x
17. Translate into symbols, then simplify: Three times the sum of 13 and 4 is 51.
18. Reduce $\frac{120}{70}$.

19. True or False? Adding the same number to the numerator and denominator of a fraction produces an equivalent fraction.

Simplify.

20. $6(4)^2 - 8(2)^3$
21. $\frac{6 + 2(4)}{8 + 10}$
22. $10\left(\frac{1}{2}\right) + 6\left(\frac{2}{3}\right)$
23. $\frac{2}{3}(0.45) + \frac{4}{5}(0.8)$
24. $\left(\frac{1}{2}\right)^3 + \left(\frac{1}{4}\right)^2$
25. $\left(3\frac{1}{3} - \frac{1}{2}\right)\left(4\frac{1}{2} + \frac{3}{4}\right)$

26. **Average Score** Lorena has scores of 83, 85, 79, 93, and 80 on her first five math tests. What is her average score for these five tests?

27. **Recipe** A muffin recipe calls for $2\frac{3}{4}$ cups of flour. If the recipe is tripled, how many cups of flour will be needed?

28. **Hourly Wage** If you earn $384 for working 40 hours, what is your hourly wage?

The illustration shows the longest courses in U.S. Open history. If each course is 18 holes, use the information to answer the following questions. Round to the nearest hundredth if necessary.

Longest Courses in U.S. Open History
- Torrey Pines, South Course (2008) 7,643 yds
- Bethpage State Park, Black Course (2009) 7,426 yds
- Winged Foot Golf Course, West Course (2006) 7,264 yds
- Oakmont Country Club (2007) 7,230 yds
- Pinehurst, No. 2 Course (2005) 7,214 yds

Source: http://www.usopen.com

29. How long is the average hole for Oakmont Country Club?
30. How long is the average hole at Torrey Pines, South Course?

Chapter 3 Test

1. Write the decimal number 5.053 in words. [3.1]

2. Give the place value of the 4 in the number 53.0543. [3.1]

3. Write seventeen and four hundred six ten thousandths as a decimal number. [3.1]

4. Round 46.7549 to the nearest hundredth. [3.1]

Perform the following operations. [3.2, 3.3, 3.4, 3.5]

5. $7 + 0.6 + 0.58$

6. $12.032 - 5.976$

7. $5.7(6.24)$

8. $22.672 \div (2.6)$

9. $4.3 + 7.06$

10. $14.32 - 5.413$

11. $0.4(2.8)(1.4)$

12. $10.224 \div 42.6$

13. Write $\dfrac{23}{25}$ as a decimal. [3.5]

14. Write 0.56 as a fraction in lowest terms. [3.5]

Simplify each of the following expressions as much as possible. [3.5, 3.6]

15. $5.2(2.8 + 0.02)$

16. $5.2 - 3(0.17)$

17. $23.852 - 3(2.01 + 0.231)$

18. $\dfrac{3}{5}(0.6) - \dfrac{2}{3}(0.15)$

19. $14.2 - 4(4.03 - 2.1)$

20. $\dfrac{4}{5}(0.23) + 5(4.02 - 1.9)$

21. $\sqrt{36}$

22. $\sqrt{81}$

23. A person purchases $8.47 worth of goods at a drugstore. If a $20 bill is used to pay for the purchases, how much change is received?

24. If coffee sells for $6.99 per pound, how much will 3.5 pounds of coffee cost?

25. If a person earns $262 for working 40 hours, what is the person's hourly wage?

The following illustration shows the heights of some of the tallest buildings in the United States. Use the information to answer the following questions. Round to the nearest tenth if necessary.

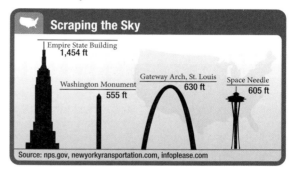

Scraping the Sky
Empire State Building 1,454 ft
Washington Monument 555 ft
Gateway Arch, St. Louis 630 ft
Space Needle 605 ft
Source: nps.gov, newyorkyransportation.com, infoplease.com

26. If the Empire State Building is 102 floors, what is the average height of each floor.

27. If the Washington Monument is 55 floors, what is the average height of each floor.

28. Find the length of the hypotenuse of the following right triangle. Round to the nearest hundredth. [3.6]

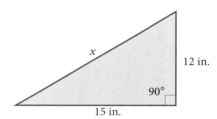

Chapter 3 Projects

DECIMALS

Group Project

Unwinding the Spiral of Roots

Number of People 2–3

Time Needed 8–12 minutes

Equipment Pencil, ruler, graph paper, scissors, and tape

Background The diagram below is called the Spiral of Roots. We can use the Spiral of Roots to visualize square roots of whole numbers.

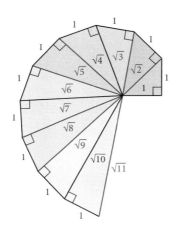

Procedure

1. Carefully cut out each triangle from the Spiral of Roots above.

2. Line up the triangles horizontally on the coordinate system shown here so that the side of length 1 is on the bottom and the hypotenuse is on the left. Note that the first triangle is shown in place, and the outline of the second triangle is next to it. The 1-unit side of each triangle should fit in each of the 1-unit spaces on the bottom.

3. On the graph, plot a point at the tip of each triangle. Then, connect these points to create a line graph. Each vertical line has a length that is represented by the square root of one of the first 10 counting numbers.

Research Project

The Wizard of Oz

In the 1939 movie *The Wizard of Oz*, the Scarecrow (played by Ray Bolger) sings "If I Only Had a Brain." Upon receiving a diploma from the great Oz, he rapidly recites a math theorem in an attempt to display his new knowledge. Unfortunately, the Scarecrow's inability to recite the Pythagorean theorem might lead one to doubt the effectiveness of his diploma. Watch this scene in the movie. Write down the Scarecrow's speech and explain the errors.

A Glimpse of Algebra

In the beginning of this chapter and in Chapter 1, we wrote numbers in expanded form. If we were to write the number 345 in expanded form and then in terms of powers of 10, it would look like this:

$$345 = 300 + 40 + 5$$
$$= 3 \cdot 100 + 4 \cdot 10 + 5$$
$$= 3 \cdot 10^2 + 4 \cdot 10 + 5$$

If we replace the 10s with x's in this last expression, we get what is called a *polynomial*. It looks like this:

$$3x^2 + 4x + 5$$

Polynomials are to algebra what whole numbers written in expanded form in terms of powers of 10 are to arithmetic. As in other expressions in algebra, we can use any variable we choose. Here are some other examples of polynomials:

$$4x - 5 \qquad a^2 + 5a + 6 \qquad y^3 + 3y^2 + 3y + 1$$

When we add two whole numbers, we add in columns. That is, if we add 345 and 234, we write one number under the other and add the numbers in the ones column, then the numbers in the tens column, and finally the numbers in the hundreds column. Here is how it looks:

$$\begin{array}{r} 345 \\ +\,234 \\ \hline 579 \end{array} \quad \text{or} \quad \begin{array}{r} 3 \cdot 10^2 + 4 \cdot 10 + 5 \\ +\,2 \cdot 10^2 + 3 \cdot 10 + 4 \\ \hline 5 \cdot 10^2 + 7 \cdot 10 + 9 \end{array}$$

(Hundreds, Tens, Ones)

We add polynomials in the same manner. If we want to add $3x^2 + 4x + 5$ and $2x^2 + 3x + 4$, we write one polynomial under the other, and then add in columns.

$$\begin{array}{r} 3x^2 + 4x + 5 \\ +\,2x^2 + 3x + 4 \\ \hline 5x^2 + 7x + 9 \end{array}$$

The sum of the two polynomials is the polynomial $5x^2 + 7x + 9$. We add only the digits. Notice that the variable parts (the letters) stay the same (just as the powers of 10 did when we added 345 and 234).

Here are some more examples.

EXAMPLE 1 Add $3x^2 + 2x + 6$ and $4x^2 + 7x + 3$.

SOLUTION We write one polynomial under the other and add in columns.

$$\begin{array}{r} 3x^2 + 2x + 6 \\ +\,4x^2 + 7x + 3 \\ \hline 7x^2 + 9x + 9 \end{array}$$

The sum of the two polynomials is $7x^2 + 9x + 9$. ∎

EXAMPLE 2 Add $4a + 2$ and $5a + 9$.

SOLUTION We write one polynomial under the other and add in columns.

$$\begin{array}{r} 4a + 2 \\ +\,5a + 9 \\ \hline 9a + 11 \end{array}$$

∎

PRACTICE PROBLEMS

1. Add $2x^2 + 4x + 2$ and $4x^2 + 3x + 5$.

2. Add $3a + 7$ and $2a + 6$.

Answers
1. $6x^2 + 7x + 7$
2. $5a + 13$

Chapter 3 Decimals

3. Add $2x^3 + 5x^2 + 3x + 6$, $3x^3 + 4x^2 + 9x + 8$, and $4x^3 + 2x^2 + 3x + 2$.

EXAMPLE 3 Add $4x^3 + 2x^2 + 4x + 1$, $2x^3 + 3x^2 + 9x + 6$, and $2x^3 + 2x^2 + 2x + 2$.

SOLUTION We add three polynomials the same way we add two of them. We write them one under the other and add in columns.

$$\begin{array}{r} 4x^3 + 2x^2 + 4x + 1 \\ 2x^3 + 3x^2 + 9x + 6 \\ +\ 2x^3 + 2x^2 + 2x + 2 \\ \hline 8x^3 + 7x^2 + 15x + 9 \end{array}$$

∎

4. Add $3y^2 + 4y + 6$ and $6y^2 + 2$.

EXAMPLE 4 Add $5y^2 + 3y + 6$ and $2y^2 + 3$.

SOLUTION We write one polynomial under the other, so that the terms with y^2 line up, and the terms without any y's line up.

$$\begin{array}{r} 5y^2 + 3y + 6 \\ +\ 2y^2\ \ \ \ \ \ \ + 3 \\ \hline 7y^2 + 3y + 9 \end{array}$$

∎

5. Add $5x^3 + 7x^2 + 3x + 1$ and $2x^2 + 4x + 1$.

EXAMPLE 5 Add $2x^3 + 4x^2 + 2x + 6$ and $3x^2 + 2x + 1$.

SOLUTION Again, we line up terms with the same variable part and add.

$$\begin{array}{r} 2x^3 + 4x^2 + 2x + 6 \\ +\ \ \ \ \ \ \ \ \ 3x^2 + 2x + 1 \\ \hline 2x^3 + 7x^2 + 4x + 7 \end{array}$$

∎

Answers
3. $9x^3 + 11x^2 + 15x + 16$
4. $9y^2 + 4y + 8$
5. $5x^3 + 9x^2 + 7x + 2$

A Glimpse of Algebra Problems

In each case, add the polynomials.

1. $4x^2 + 2x + 3$
    $\underline{+ 2x^2 + 7x + 5}$

2. $3x^2 + 4x + 5$
    $\underline{+ 5x^2 + 4x + 3}$

3. $2a + 3$
    $\underline{+ 3a + 5}$

4. $5a + 2$
    $\underline{+ 2a + 1}$

5. $3x + 4$
    $2x + 1$
    $\underline{+ 4x + 1}$

6. $2x + 1$
    $3x + 2$
    $\underline{+ 4x + 3}$

7. $2y^3 + 3y^2 + 4y + 5$
    $\underline{+ 3y^3 + 2y^2 + 5y + 2}$

8. $4y^3 + 2y^2 + 6y + 7$
    $\underline{+ 5y^3 + 6y^2 + 2y + 8}$

9. Add $3x^2 + 4x + 3$ and $3x^2 + 2$.

10. Add $5x^2 + 6x + 7$ and $4x^2 + 2$.

11. Add $3a^2 + 4$ and $7a + 2$.

12. Add $2a^2 + 5$ and $4a + 3$.

13. Add $5x^3 + 4x^2 + 7x + 3$ and $3x^2 + 9x + 10$.

14. Add $2x^3 + 7x^2 + 3x + 1$ and $4x^2 + 3x + 8$.

Ratio and Proportion

4

Chapter Outline

4.1 Ratios
4.2 Rates and Unit Pricing
4.3 Solving Equations by Division
4.4 Proportions
4.5 Applications of Proportions
4.6 Similar Figures

When construction on the Paris Eiffel Tower was completed in the late 19th century, many Parisians found it ugly. A noted French novelist, Guy de Maupassant, was reported to eat lunch in one of the restaurants in the tower every day, despite claiming to detest the tower. He said it was because it was the only place in town where he could not see the hated structure. Originally built with a 20-year lease, the city intended to dismantle the tower when the lease expired. However, its height made it useful for telegraphy, and France used it to dispatch resources to the front line during the World War I Battle of Marne. Now it is the most visited paid monument in the world, having hosted over 200 million visitors to date.

The Eiffel Tower is also one of the most replicated structures in the world. Below is a table showing the locations and heights of some of the replicas.

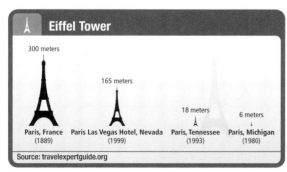

As you can see, these replicas are all different sizes. In mathematics, we can use *ratios* to compare the heights of these towers. For instance, we say that the height of the tower in Paris, France compared to the height of the tower in Paris, Michigan are in a ratio of 50 to 1. In this chapter, we study ratios like this one. We will also observe how ratios are very closely related to fractions and decimals, which we have already studied.

Preview

Key Words	Definition
Ratio	A comparison between two numbers
Rate	A ratio between two quantities that have different units, neither of which can be converted to the other
Unit Pricing	The ratio of price to quantity
Proportion	A statement that two ratios are equal

Chapter Outline

4.1 Ratios
- A Express ratios as fractions in lowest terms.
- B Use ratios to solve application problems.

4.2 Rates and Unit Pricing
- A Express rates as ratios.
- B Use ratios to write a unit price.

4.3 Solving Equations by Division
- A Divide expressions containing a variable.
- B Solve equations using division.

4.4 Proportions
- A Name the terms in a proportion.
- B Use the fundamental property of proportions to solve a proportion.

4.5 Applications of Proportions
- A Use proportions to solve application problems.

4.6 Similar Figures
- A Use proportions to find the lengths of sides of similar triangles.
- B Use proportions to find the lengths of sides of similar figures.
- C Draw a figure similar to a given figure, given the length of one side.
- D Use similar figures to solve application problems.

Ratios

4.1

OBJECTIVES

A Express ratios as fractions in lowest terms.

B Use ratios to solve application problems.

TICKET TO SUCCESS

Keep these questions in mind as you read through the section. Then respond in your own words and in complete sentences.

1. What is a ratio?
2. How would you express a ratio as a fraction in lowest terms?
3. When will the ratio of two numbers be a complex fraction?
4. Write an application problem that uses a ratio.

Rido/Shutterstock.com

The *ratio* of two numbers is a way of comparing them. If we say that the ratio of two numbers is 2 to 1, then the first number is twice as large as the second number. For example, if there are 10 men and 5 women enrolled in a math class, then the ratio of men to women is 10 to 5. Because 10 is twice as large as 5, we can also say that the ratio of men to women is 2 to 1.

We define the ratio of two numbers in terms of fractions below.

A Ratios as Fractions in Lowest Terms

> **Definition**
>
> A **ratio** is a comparison between two numbers and is represented as a fraction, where the first number in the ratio is the numerator and the second number in the ratio is the denominator.
>
> If a and b are any two numbers, then the ratio of a to b is $\dfrac{a}{b}$.   ($b \neq 0$)

We handle ratios the same way we handle fractions. For example, when we said that the ratio of 10 men to 5 women was the same as the ratio 2 to 1, we were actually saying

$$\frac{10}{5} = \frac{2}{1} \qquad \text{Reduce to lowest terms.}$$

4.1 Ratios   **281**

PRACTICE PROBLEMS

1. a. Express the ratio of 32 to 48 as a fraction in lowest terms.
   b. Express the ratio of 3.2 to 4.8 as a fraction in lowest terms.
   c. Express the ratio of 0.32 to 0.48 as a fraction in lowest terms.

2. a. Give the ratio of $\frac{3}{5}$ to $\frac{9}{10}$ as a fraction in lowest terms.
   b. Give the ratio of 0.6 to 0.9 as a fraction in lowest terms.

3. a. Write the ratio of 0.06 to 0.12 as a fraction in lowest terms.
   b. Write the ratio of 600 to 1200 as a fraction in lowest terms.

NOTE Another symbol used to denote ratio is the colon (:). The ratio of, say, 5 to 4 can be written as 5:4. Although we will not use it here, this notation is fairly common.

Answers
1. All are $\frac{2}{3}$.
2. Both are $\frac{2}{3}$.
3. Both are $\frac{1}{2}$.

Because we have already studied fractions in detail, much of the introductory material on ratios will seem like review.

EXAMPLE 1 Express the ratio of 16 to 48 as a fraction in lowest terms.

SOLUTION Because the ratio is 16 to 48, the numerator of the fraction is 16 and the denominator is 48.

$$\frac{16}{48} = \frac{1}{3}$$

Notice that the first number in the ratio becomes the numerator of the fraction, and the second number in the ratio becomes the denominator. ∎

EXAMPLE 2 Give the ratio of $\frac{2}{3}$ to $\frac{4}{9}$ as a fraction in lowest terms.

SOLUTION We begin by writing the ratio of $\frac{2}{3}$ to $\frac{4}{9}$ as a complex fraction. The numerator is $\frac{2}{3}$, and the denominator is $\frac{4}{9}$. Then we simplify.

$$\frac{\frac{2}{3}}{\frac{4}{9}} = \frac{2}{3} \cdot \frac{9}{4} \quad \text{Division by } \tfrac{4}{9} \text{ is the same as multiplication by } \tfrac{9}{4}.$$

$$= \frac{18}{12} \quad \text{Multiply.}$$

$$= \frac{3}{2} \quad \text{Reduce to lowest terms.} \quad \blacksquare$$

EXAMPLE 3 Write the ratio of 0.08 to 0.12 as a fraction in lowest terms.

SOLUTION When the ratio is in reduced form, it is customary to write it with whole numbers and not decimals. For this reason, we multiply the numerator and the denominator of the ratio by 100 to clear it of decimals. Then we reduce to lowest terms.

$$\frac{0.08}{0.12} = \frac{0.08 \times 100}{0.12 \times 100} \quad \text{Multiply the numerator and the denominator by 100 to clear the ratio of decimals.}$$

$$= \frac{8}{12} \quad \text{Multiply.}$$

$$= \frac{2}{3} \quad \text{Reduce to lowest terms.} \quad \blacksquare$$

Table 1 shows several more ratios and their fractional equivalents. Notice that in each case the fraction has been reduced to lowest terms. Also, the ratio that contains decimals has been rewritten as a fraction that does not contain decimals.

TABLE 1

Ratio	Fraction	Fraction In Lowest Terms	
25 to 35	$\frac{25}{35}$	$\frac{5}{7}$	
35 to 25	$\frac{35}{25}$	$\frac{7}{5}$	
8 to 2	$\frac{8}{2}$	$\frac{4}{1}$	We can also write this as just 4.
$\frac{1}{4}$ to $\frac{3}{4}$	$\frac{\frac{1}{4}}{\frac{3}{4}}$	$\frac{1}{3}$	because $\frac{\frac{1}{4}}{\frac{3}{4}} = \frac{1}{4} \cdot \frac{4}{3} = \frac{1}{3}$
0.6 to 1.7	$\frac{0.6}{1.7}$	$\frac{6}{17}$	because $\frac{0.6 \times 10}{1.7 \times 10} = \frac{6}{17}$

B Applications of Ratios

Quite often, ratios are applied to real-life situations. Let's practice a couple now.

EXAMPLE 4   During a game, a basketball player makes 12 out of the 18 free throws he attempts. Write the ratio of the number of free throws he makes to the number of free throws he attempts as a fraction in lowest terms.

SOLUTION   Because he makes 12 out of 18, we want the ratio 12 to 18, or

$$\frac{12}{18} = \frac{2}{3}$$

Because the ratio is 2 to 3, we can say that, in this particular game, he made 2 out of every 3 free throws he attempted. ∎

EXAMPLE 5   A solution of alcohol and water contains 15 milliliters of water and 5 milliliters of alcohol. Find the ratio of alcohol to water, water to alcohol, water to total solution, and alcohol to total solution. Write each ratio as a fraction and reduce to lowest terms.

SOLUTION   There are 5 milliliters of alcohol and 15 milliliters of water, so there are 20 milliliters of solution (alcohol + water). The ratios are as follows:

The ratio of alcohol to water is 5 to 15, or

$$\frac{5}{15} = \frac{1}{3} \quad \text{In lowest terms}$$

The ratio of water to alcohol is 15 to 5, or

$$\frac{15}{5} = \frac{3}{1} \quad \text{In lowest terms}$$

The ratio of water to total solution is 15 to 20, or

$$\frac{15}{20} = \frac{3}{4} \quad \text{In lowest terms}$$

The ratio of alcohol to total solution is 5 to 20, or

$$\frac{5}{20} = \frac{1}{4} \quad \text{In lowest terms}$$

∎

4. Suppose the basketball player in Example 4 makes 12 out of 16 free throws. Write the ratio again using these new numbers.

5. A solution of alcohol and water contains 12 milliliters of water and 4 milliliters of alcohol. Find the ratio of alcohol to water, water to alcohol, and water to total solution. Write each ratio as a fraction and reduce to lowest terms.

Answers

4. $\frac{3}{4}$

5. $\frac{1}{3}, \frac{3}{1}, \frac{3}{4}$

Problem Set 4.1

Moving Toward Success

"Three Rules of Work: Out of clutter find simplicity; From discord find harmony; In the middle of difficulty lies opportunity."
—Albert Einstein, 1879–1955, German-born American physicist

1. Why is focus important when studying? What are some factors that may distract your focus?
2. Does self-doubt or questioning the purpose of this course affect your focus? Why or why not?

A Write each of the following ratios as a fraction in lowest terms. None of the answers should contain decimals.
[Examples 1–3]

1. 8 to 6
2. 6 to 8
3. 64 to 12
4. 12 to 64

5. 100 to 250
6. 250 to 100
7. 13 to 26
8. 36 to 18

9. $\frac{3}{4}$ to $\frac{1}{4}$
10. $\frac{5}{8}$ to $\frac{3}{8}$
11. $\frac{7}{3}$ to $\frac{6}{3}$
12. $\frac{9}{5}$ to $\frac{11}{5}$

13. $\frac{6}{5}$ to $\frac{6}{7}$
14. $\frac{5}{3}$ to $\frac{1}{3}$
15. $2\frac{1}{2}$ to $3\frac{1}{2}$
16. $5\frac{1}{4}$ to $1\frac{3}{4}$

17. $2\frac{2}{3}$ to $\frac{5}{3}$
18. $\frac{1}{2}$ to $3\frac{1}{2}$
19. 0.05 to 0.15
20. 0.21 to 0.03

21. 0.3 to 3
22. 0.5 to 10
23. 1.2 to 10
24. 6.4 to 0.8

25.
a. What is the ratio of shaded squares to nonshaded squares?

b. What is the ratio of shaded squares to total squares?

c. What is the ratio of nonshaded squares to total squares?

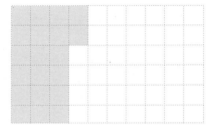

26.
a. What is the ratio of shaded squares to nonshaded squares?

b. What is the ratio of shaded squares to total squares?

c. What is the ratio of nonshaded squares to total squares?

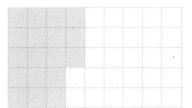

Applying the Concepts

27. Biggest Hits The chart shows the artists that had the most songs in the number one spot on Billboard's Hot 100. Use the information to find the ratio of hits The Beatles had to the hits The Supremes had.

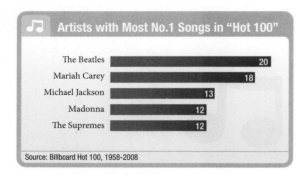

28. Google Earth The Google Earth image shows Crater Lake National Park in Oregon. The park covers 266 square miles and the lake covers 20 square miles. What is the ratio of the park's area to the lake's area?

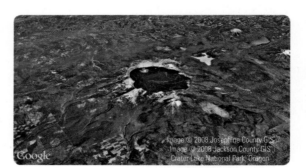

Write each of the following ratios as a fraction in lowest terms. None of the answers should contain decimals.

29. 100 mg to 5 mg

30. 25 g to 1 g

31. 375 mg to 10 mg

32. 450 mg to 20 mg

33. Family Budget
A family of four budgeted the following amounts for some of their monthly bills:

Food bill $400
Gas bill $100
Utilities $150
Rent $650

a. What is the ratio of the rent to the food bill?

b. What is the ratio of the gas bill to the food bill?

c. What is the ratio of the utilities to the food bill?

d. What is the ratio of the rent to the utilities?

34. Nutrition
One cup of breakfast cereal was found to contain the following:

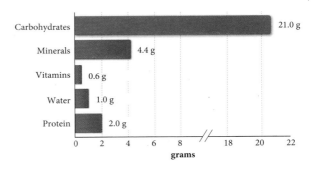

a. Find the ratio of water to protein.

b. Find the ratio of carbohydrates to protein.

c. Find the ratio of vitamins to minerals.

d. Find the ratio of protein to vitamins and minerals.

35. Profit and Revenue
The following bar chart shows the profit and revenue of the Baby Steps Shoe Company each quarter for one year.

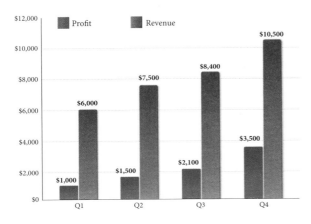

Find the ratio of revenue to profit for each of the following quarters. Write your answer in lowest terms.

a. Q1   b. Q2   c. Q3   d. Q4

e. Find the ratio of revenue to profit for the entire year.

36. Geometry
Regarding the diagram below, AC represents the length of the line segment that starts at A and ends at C. From the diagram we see that AC = 8.

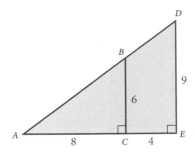

a. Find the ratio of BC to AC.

b. What is the length AE?

c. Find the ratio of DE to AE.

37. Major League Baseball The following table shows the number of games won during the 2010 baseball season by several National League teams.

Team	Number of Wins
New York Mets	79
Atlanta Braves	91
Washington Nationals	69
St. Louis Cardinals	86
Houston Astros	76
Arizona Diamondbacks	65
Cincinnati Reds	91
San Francisco Giants	92

Source: mlb.com

a. What is the ratio of wins of the St. Louis Cardinals to the Houston Astros?

b. What is the ratio of wins of the Atlanta Braves to the Cincinnati Reds?

c. What is the ratio of wins of the Houston Astros to the San Francisco Giants?

38. Buying an iPod The hard drive of an Apple iPod determines how many songs you will be able to store and carry around with you. The table below compares the size of the hard drive, song capacity, and cost of three iPods.

iPod Type	Hard-Drive Size	Number of Songs	Cost
Shuffle	2 GB	500 songs	$49.00
Nano	8 GB	2000 songs	$149.00
Classic	160 GB	40,000 songs	$249.00

Source: apple.com/ipod

a. What is the ratio of hard drive size between the Shuffle and the Nano? Between the Shuffle and the Classic?

b. What is the ratio of number of songs between the Shuffle and the Nano? Between the Shuffle and the Classic?

Getting Ready for the Next Section

The following problems review material from a previous section. Reviewing these problems will help you with the next section.

Write as a decimal.

39. $\dfrac{90}{5}$ **40.** $\dfrac{120}{3}$ **41.** $\dfrac{125}{2}$ **42.** $\dfrac{2}{10}$ **43.** $\dfrac{1.23}{2}$

44. $\dfrac{1.39}{2}$ **45.** $\dfrac{88}{0.5}$ **46.** $\dfrac{1.99}{0.5}$ **47.** $\dfrac{46}{0.25}$ **48.** $\dfrac{92}{0.25}$

Divide. Round answers to the nearest thousandth.

49. $0.48 \div 5.5$ **50.** $0.75 \div 11.5$ **51.** $2.19 \div 46$ **52.** $1.25 \div 50$

Maintaining Your Skills

Multiply and divide as indicated.

53. $\dfrac{3}{4} \cdot \dfrac{5}{6}$ **54.** $\dfrac{7}{8} \cdot 32$ **55.** $\dfrac{11}{16} \div \dfrac{1}{8}$ **56.** $13 \div \dfrac{1}{3}$

57. $\dfrac{65}{72} \cdot \dfrac{108}{273}$ **58.** $\dfrac{165}{84} \cdot \dfrac{24}{195}$ **59.** $\dfrac{3}{4} \div \dfrac{1}{8} \cdot 16$ **60.** $\dfrac{1}{4} \div \dfrac{1}{12} \cdot 6$

Rates and Unit Pricing

4.2

OBJECTIVES

A Express rates as ratios.

B Use ratios to write a unit price.

TICKET TO SUCCESS

Keep these questions in mind as you read through the section. Then respond in your own words and in complete sentences.

1. What is a rate?
2. When is a rate written in simplest terms?
3. What is unit pricing?
4. Give an example of how you would use a ratio to write a unit price.

We will now apply our discussion from the last section to two specific ratios called *rate* and *unit pricing*. Let's begin with rate.

A Rates

Whenever a ratio compares two quantities that have different units (and neither unit can be converted to the other), then the ratio is called a *rate*. For example, if we were to travel 120 miles in 3 hours, then our average rate of speed expressed as the ratio of miles to hours would be

$$\frac{120 \text{ miles}}{3 \text{ hours}} = \frac{40 \text{ miles}}{1 \text{ hour}}$$ Divide the numerator and the denominator by 3 to reduce to lowest terms.

The ratio $\frac{40 \text{ miles}}{1 \text{ hour}}$ can be expressed as

$$40 \frac{\text{miles}}{\text{hour}} \quad \text{or} \quad 40 \text{ miles/hour} \quad \text{or} \quad 40 \text{ miles per hour}$$

A rate is expressed in simplest form when the numerical part of the denominator is 1. To accomplish this we use division.

4.2 Rates and Unit Pricing

PRACTICE PROBLEMS

1. A car travels 107 miles in 2 hours. What is the car's rate in miles per hour?

2. A car travels 192 miles on 6 gallons of gas. Give the ratio of miles to gallons as a rate in miles per gallon.

3. A supermarket sells vegetable juice in three different containers at the following prices:
   5.5 ounces, 48¢
   11.5 ounces, 75¢
   46 ounces, $2.19
   Give the unit price in cents per ounce for each one. Round to the nearest tenth of a cent, if necessary.

Answers
1. 53.5 miles/hour
2. 32 miles/gallon
3. 8.7¢/ounce, 6.5¢/ounce, 4.8¢/ounce

EXAMPLE 1 A train travels 125 miles in 2 hours. What is the train's rate in miles per hour?

SOLUTION The ratio of miles to hours is

$$\frac{125 \text{ miles}}{2 \text{ hours}} = 62.5 \frac{\text{miles}}{\text{hour}} \quad \text{Divide 125 by 2.}$$

$$= 62.5 \text{ miles per hour}$$

If the train travels 125 miles in 2 hours, then its average rate of speed is 62.5 miles per hour.

EXAMPLE 2 A car travels 90 miles on 5 gallons of gas. Give the ratio of miles to gallons as a rate in miles per gallon.

SOLUTION The ratio of miles to gallons is

$$\frac{90 \text{ miles}}{5 \text{ gallons}} = 18 \frac{\text{miles}}{\text{gallon}} \quad \text{Divide 90 by 5.}$$

$$= 18 \text{ miles/gallon}$$

The gas mileage of the car is 18 miles per gallon.

B Unit Pricing

One kind of rate that is very common is *unit pricing*. Unit pricing is the ratio of price to quantity when the quantity is one unit. Suppose a 1-liter bottle of a certain soft drink costs $1.19, whereas a 2-liter bottle of the same drink costs $1.39. Which is the better buy? That is, which has the lower price per liter?

$$\frac{\$1.19}{1 \text{ liter}} = \$1.19 \text{ per liter}$$

$$\frac{\$1.39}{2 \text{ liter}} = \$0.695 \text{ per liter}$$

The unit price for the 1-liter bottle is $1.19 per liter, whereas the unit price for the 2-liter bottle is 69.5¢ per liter. The 2-liter bottle is a better buy.

EXAMPLE 3 A supermarket sells low-fat milk in three different containers at the following prices:

1 gallon $3.59
$\frac{1}{2}$ gallon $1.99
1 quart $1.29   (1 quart = $\frac{1}{4}$ gallon)

Give the unit price in dollars per gallon for each one.

SOLUTION Because 1 quart = $\frac{1}{4}$ gallon, we have

1-gallon container $\frac{\$3.59}{1 \text{ gallon}} = \frac{\$3.59}{1 \text{ gallon}} = \$3.59$ per gallon

$\frac{1}{2}$-gallon container $\frac{\$1.99}{\frac{1}{2} \text{ gallon}} = \frac{\$1.99}{0.5 \text{ gallon}} = \$3.98$ per gallon

1-quart container $\frac{\$1.29}{1 \text{ quart}} = \frac{\$1.29}{0.25 \text{ gallon}} = \$5.16$ per gallon

The 1-gallon container has the lowest unit price, whereas the 1-quart container has the highest unit price.

Problem Set 4.2

Moving Toward Success

"When walking, walk. When eating, eat."
—Zen Proverb

1. What type of study environment is conducive to success?
2. Describe the place you typically study. How can you improve this space to help foster success in this course?

A Express each of the following rates as a ratio with the given units. [Examples 1, 2]

1. **Miles/Hour** A car travels 220 miles in 4 hours. What is the rate of the car in miles per hour?

2. **Miles/Hour** A train travels 360 miles in 5 hours. What is the rate of the train in miles per hour?

3. **Kilometers/Hour** It takes a car 3 hours to travel 252 kilometers. What is the rate in kilometers per hour?

4. **Kilometers/Hour** In 6 hours an airplane travels 4,200 kilometers. What is the rate of the airplane in kilometers per hour?

5. **Gallons/Second** The flow of water from a water faucet can fill a 3-gallon container in 15 seconds. Give the ratio of gallons to seconds as a rate in gallons per second.

6. **Gallons/Minute** A 225-gallon drum is filled in 3 minutes. What is the rate in gallons per minute?

7. **Liters/Minute** It takes 4 minutes to fill a 56-liter gas tank. What is the rate in liters per minute?

8. **Liters/Hour** The gas tank on a car holds 60 liters of gas. At the beginning of a 6-hour trip, the tank is full. At the end of the trip, it contains only 12 liters. What is the rate at which the car uses gas in liters per hour?

9. **Miles/Gallon** A car travels 95 miles on 5 gallons of gas. Give the ratio of miles to gallons as a rate in miles per gallon.

10. **Miles/Gallon** On a 384-mile trip, an economy car uses 8 gallons of gas. Give this as a rate in miles per gallon.

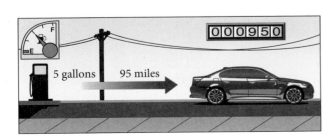

11. **Miles/Liter** The gas tank on a car has a capacity of 75 liters. On a full tank of gas, the car travels 325 miles. What is the gas mileage in miles per liter?

12. **Miles/Liter** A car pulling a trailer can travel 105 miles on 70 liters of gas. What is the gas mileage in miles per liter?

13. **Gas Prices** The snapshot shows the gas prices for the different regions of the United States. If a man bought 12 gallons of gas for $44.40, where might he live?

14. **Pitchers** The chart shows the major league pitchers with the most career strikeouts. If Pedro Martinez pitched 2,827 innings, how many strikeouts does he throw per inning? Round to the nearest hundredth.

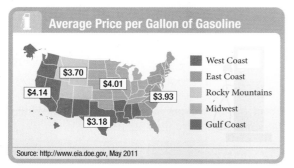

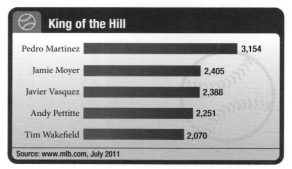

Nursing Intravenous (IV) infusions are often ordered in either milliliters per hour or milliliters per minute.

15. What was the infusion rate in milliliters per hour if it took 5 hours to administer 2,400 mL?

16. What was the infusion rate in milliliters per minute if 42 milliliters were administered in 6 minutes?

Unit Pricing

17. **Cents/Ounce** A 20-ounce package of frozen peas is priced at 99¢. Give the unit price in cents per ounce.

18. **Dollars/Pound** A 4-pound bag of cat food costs $8.12. Give the unit price in dollars per pound.

19. **Best Buy** Find the unit price in cents per diaper for each of the packages shown here. Which is the better buy? Round to the nearest tenth of a cent.

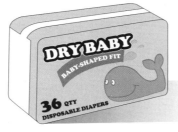

36 disposable diapers
$12.49

38 disposable diapers
$11.99

20. Best Buy Find the unit price in cents per pill for each of the packages shown here. Which is the better buy? Round to the nearest tenth of a cent.

100 pills
$5.99

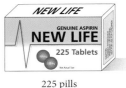
225 pills
$13.96

Currency Conversions There are a number of online calculators that will show what the money in one country is worth in another country. One such converter, the XE Universal Currency Converter®, uses live, up-to-the-minute currency rates. Use the information shown here to determine what the equivalent to one U.S. dollar is for each of the following denominations. Round to the nearest thousandth, where necessary.

21. $100.00 U.S. dollars are equivalent to 67.438 euros.

22. $50.00 U.S. dollars are equivalent to $47.911 Canadian dollars.

23. $40.00 U.S. dollars are equivalent to 24.252 British pounds.

24. $25.00 U.S. dollars are equivalent to 2,016.69 Japanese yen.

25. Food Prices Using unit rates is a way to compare prices of different sized packages to see which price is really the best deal. Suppose we compare the cost of a box of Cheerios sold at three different stores for the following prices:

Store	Size	Cost
A	11.3 ounce box	$4.00
B	18 ounce box	$4.99
C	180 ounce case	$52.90

Which size is the best buy? Give the cost per ounce for that size.

26. Cell Phone Plans All cell phone plans are not created equal. The number of minutes and the monthly charges can vary greatly. Suppose the table shows four plans presented by four different cell phone providers.

Carriers	Plan Name	Monthly Minutes	Monthly Cost	Plan cost per Minute
AT&T	Nation 450	450	$39.99	
Sprint	Basic	200	$29.99	
T-Mobile	Individual Basic	500	$39.99	
Verizon	Nationwide Basic	450	$39.99	

Find the cost per minute for each plan. Based on your results, which plan should you go with?

27. Miles/Hour A car travels 675.4 miles in $12\frac{1}{2}$ hours. Give the rate in miles per hour to the nearest hundredth.

28. Miles/Hour At the beginning of a trip, the odometer on a car read 32,567.2 miles. At the end of the trip, it read 32,741.8 miles. If the trip took $4\frac{1}{4}$ hours, what was the rate of the car in miles per hour to the nearest tenth?

29. Miles/Gallon If a truck travels 128.4 miles on 13.8 gallons of gas, what is the gas mileage in miles per gallon? (Round to the nearest tenth.)

30. Cents/Day If a 15-day supply of vitamins costs $1.62, what is the price in cents per day?

Hourly Wages Jane has a job at the local Marcy's department store. The graph shows how much Jane earns for working 8 hours per day for 5 days.

31. What is her daily rate of pay? (Assume she works 8 hours per day.)

32. What is her weekly rate of pay? (Assume she works 5 days per week.)

33. What is her annual rate of pay? (Assume she works 50 weeks per year.)

34. What is her hourly rate of pay? (Assume she works 8 hours per day.)

Getting Ready for the Next Section

Solve each equation by finding a number to replace n that will make the equation a true statement.

35. $2 \cdot n = 12$ 36. $3 \cdot n = 27$ 37. $6 \cdot n = 24$ 38. $8 \cdot n = 16$

39. $20 = 5 \cdot n$ 40. $35 = 7 \cdot n$ 41. $650 = 10 \cdot n$ 42. $630 = 7 \cdot n$

Maintaining Your Skills

Add and subtract as indicated.

43. $\dfrac{1}{2} + \dfrac{3}{8}$ 44. $\dfrac{7}{6} - \dfrac{1}{3}$ 45. $\dfrac{2}{5} - \dfrac{3}{8}$ 46. $\dfrac{5}{8} + \dfrac{3}{4}$

47. $\dfrac{11}{12} - \dfrac{9}{10}$ 48. $\dfrac{13}{15} + \dfrac{1}{10}$ 49. $\dfrac{5}{6} - \dfrac{1}{3} + \dfrac{4}{3}$ 50. $\dfrac{7}{8} + \dfrac{1}{8} - \dfrac{1}{16}$

Extending the Concepts

51. **Unit Pricing** Suppose the makers of Wisk liquid detergent cut the size of its popular midsize jug from 100 ounces (3.125 quarts) to 80 ounces (2.5 quarts). At the same time it lowered the price from $6.99 to $5.75. Fill in the table and use your results to decide which of the two sizes is the better buy.

WISK LAUNDRY DETERGENT	Old	New
Size	100 ounces	80 ounces
Container cost	$6.99	$5.75
Price per quart		

Solving Equations by Division

4.3

OBJECTIVES

A Divide expressions containing a variable.

B Solve equations using division.

TICKET TO SUCCESS

Keep these questions in mind as you read through the section. Then respond in your own words and in complete sentences.

1. In your own words, explain what a solution to an equation is.
2. When you divide one side of an equation by a number, why must you divide the other side by the same number?
3. What is the result of dividing $7 \cdot y$ by 7?
4. Explain how division is used to solve the equation $30 = 5 \cdot a$.

Suppose you and two friends went to a carnival. You bought 3 tickets, one for each person, to ride the Ferris wheel. The ticket seller asked you for a total of $12. How much did each ticket cost? We can use this information to set up the following equation:

$$3 \cdot n = 12$$

In Chapter 1, we solved equations like $3 \cdot n = 12$ by finding a number with which to replace n that would make the equation a true statement. The solution for the equation $3 \cdot n = 12$ is $n = 4$, because

when → $n = 4$
the equation → $3 \cdot n = 12$
becomes → $3 \cdot 4 = 12$
                     $12 = 12$ A true statement

The problem with this method of solving equations is that we have to guess at the solution and then check it in the equation to see if it works. In this section, we will develop a method of solving equations like $3 \cdot n = 12$ that does not require any guessing.

In Chapter 2, we simplified expressions such as

$$\frac{2 \cdot 2 \cdot 3 \cdot 5 \cdot 7}{2 \cdot 5}$$

by dividing out any factors common to the numerator and the denominator. For example,

$$\frac{2 \cdot \cancel{2} \cdot 3 \cdot \cancel{5} \cdot 7}{\cancel{2} \cdot \cancel{5}} = 2 \cdot 3 \cdot 7 = 42$$

4.3 Solving Equations by Division

The same process works with expressions that have variables for some of their factors. For example, the expression

$$\frac{2 \cdot n \cdot 7 \cdot 11}{n \cdot 11}$$

can be simplified by dividing out the factors common to the numerator and the denominator—namely, n and 11.

$$\frac{2 \cdot \cancel{n} \cdot 7 \cdot \cancel{11}}{\cancel{n} \cdot \cancel{11}} = 2 \cdot 7 = 14$$

Therefore, we can solve for n in the equation $3 \cdot n = 12$ from the Ferris wheel problem like this:

$$\frac{\cancel{3} \cdot n}{\cancel{3}} = \frac{12}{3}$$
$$n = 4$$

A Dividing Expressions

Let's practice some more examples of solving equations by division.

EXAMPLE 1 Divide the expression $5 \cdot n$ by 5.

SOLUTION Applying the method above, we have

$$5 \cdot n \text{ divided by 5 is } \frac{\cancel{5} \cdot n}{\cancel{5}} = n$$

If you are having trouble understanding this process because there is a variable involved, consider what happens when we divide 6 by 2 and when we divide 6 by 3. Because $6 = 2 \cdot 3$, when we divide by 2 we get 3.

$$\frac{6}{2} = \frac{\cancel{2} \cdot 3}{\cancel{2}} = 3$$

When we divide by 3, we get 2.

$$\frac{6}{3} = \frac{2 \cdot \cancel{3}}{\cancel{3}} = 2$$

EXAMPLE 2 Divide $7 \cdot y$ by 7.

SOLUTION Dividing by 7, we have

$$7 \cdot y \text{ divided by 7 is } \frac{\cancel{7} \cdot y}{\cancel{7}} = y$$

B Solving for a Variable

Based on the previous examples, we can use division to solve equations such as $3 \cdot n = 12$. Notice that the left side of the equation is $3 \cdot n$. The equation is solved when we have just n, instead of $3 \cdot n$, on the left side and a number on the right side. That is, we have solved the equation when we have rewritten it as

$$n = \text{a number}$$

We can accomplish this by dividing *both* sides of the equation by 3.

$$\frac{\cancel{3} \cdot n}{\cancel{3}} = \frac{12}{3} \quad \text{Divide both sides by 3.}$$
$$n = 4$$

PRACTICE PROBLEMS

1. Divide the expression $8 \cdot n$ by 8.

2. Divide $3 \cdot y$ by 3.

NOTE
The choice of the letter we use for the variable is not important. The process works just as well with y as it does with n. The letters used for variables in equations are most often the letters a, n, x, y, or z.

NOTE
In the last chapter of this book, we will devote a lot of time to solving equations. For now, we are concerned only with equations that can be solved by division.

Answers
1. n
2. y

Because 12 divided by 3 is 4, the solution to the equation is $n = 4$, which we know to be correct from our discussion at the beginning of this section. Notice that it would be incorrect to divide just the left side by 3 and not the right side also. It is important to remember that whenever we divide one side of an equation by a number, we must also divide the other side by the same number.

EXAMPLE 3 Solve the equation $7 \cdot y = 42$ for y by dividing both sides by 7.

SOLUTION Dividing both sides by 7, we have

$$\frac{7 \cdot y}{7} = \frac{42}{7}$$
$$y = 6$$

We can check our solution by replacing y with 6 in the original equation.

When → $y = 6$
the equation → $7 \cdot y = 42$
becomes → $7 \cdot 6 = 42$
$42 = 42$ A true statement

3. Solve the equation $8 \cdot n = 40$ by dividing both sides by 8.

EXAMPLE 4 Solve for a: $30 = 5 \cdot a$.

SOLUTION Our method of solving equations by division works regardless of which side the variable is on. In this case, the right side is $5 \cdot a$, and we would like it to be just a. Dividing both sides by 5, we have

$$\frac{30}{5} = \frac{5 \cdot a}{5}$$
$$6 = a$$

The solution is $a = 6$. (If 6 is a, then a is 6.)

4. Solve for a: $35 = 7 \cdot a$.

We can write our solutions as improper fractions, mixed numbers, or decimals. Let's agree to write our answers as either whole numbers, proper fractions, or mixed numbers unless otherwise stated.

Answers
3. 5
4. 5

Problem Set 4.3

Moving Toward Success

"When you change the way you look at things, the things you look at change."
— Dr. Wayne W. Dyer, 1940–present, American motivational speaker and author

1. An inner dialogue of negative statements, such as "I'll never be able to learn this material" or "I'm never going to use this stuff," distracts you from achieving success in this course. What are other examples of negative statements you may have thought or heard from other people?

2. Write three positive statements you can say when you notice your mind participating in a negative inner dialogue.

A Simplify each of the following expressions by dividing out any factors common to the numerator and the denominator and then simplifying the result. [Examples 1, 2]

1. $\dfrac{3 \cdot 5 \cdot 5 \cdot 7}{3 \cdot 5}$
2. $\dfrac{2 \cdot 2 \cdot 3 \cdot 5 \cdot 7}{2 \cdot 5 \cdot 7}$
3. $\dfrac{2 \cdot n \cdot 3 \cdot 3 \cdot 5}{n \cdot 5}$
4. $\dfrac{3 \cdot 5 \cdot n \cdot 7 \cdot 7}{3 \cdot n \cdot 7}$

5. $\dfrac{2 \cdot 2 \cdot n \cdot 7 \cdot 11}{2 \cdot n \cdot 11}$
6. $\dfrac{3 \cdot n \cdot 7 \cdot 13 \cdot 17}{n \cdot 13 \cdot 17}$
7. $\dfrac{9 \cdot n}{9}$
8. $\dfrac{8 \cdot a}{8}$

9. $\dfrac{4 \cdot y}{4}$
10. $\dfrac{7 \cdot x}{7}$

B Solve each of the following equations by dividing both sides by the appropriate number. Be sure to show the division in each case. [Examples 3, 4]

11. $4 \cdot n = 8$
12. $2 \cdot n = 8$
13. $5 \cdot x = 35$
14. $7 \cdot x = 35$

15. $3 \cdot y = 21$
16. $7 \cdot y = 21$
17. $6 \cdot n = 48$
18. $16 \cdot n = 48$

19. $5 \cdot a = 40$
20. $10 \cdot a = 40$
21. $3 \cdot x = 6$
22. $8 \cdot x = 40$

23. $2 \cdot y = 2$
24. $2 \cdot y = 12$
25. $3 \cdot a = 18$
26. $4 \cdot a = 4$

27. $5 \cdot n = 25$
28. $9 \cdot n = 18$
29. $6 = 2 \cdot x$
30. $56 = 7 \cdot x$

31. $42 = 6 \cdot n$
32. $30 = 5 \cdot n$
33. $4 = 4 \cdot y$
34. $90 = 9 \cdot y$

35. $63 = 7 \cdot y$
36. $3 = 3 \cdot y$
37. $2 \cdot n = 7$
38. $4 \cdot n = 10$

39. $6 \cdot x = 21$ **40.** $7 \cdot x = 8$ **41.** $5 \cdot a = 12$ **42.** $8 \cdot a = 13$

43. $4 = 7 \cdot y$ **44.** $3 = 9 \cdot y$ **45.** $10 = 13 \cdot y$ **46.** $9 = 11 \cdot y$

47. $12 \cdot x = 30$ **48.** $16 \cdot x = 56$ **49.** $21 = 14 \cdot n$ **50.** $48 = 20 \cdot n$

51. Suppose you and two of your friends bought tickets to ride a bungee jump ride at the fair. If the total cost of the tickets was $27, how much did each ticket cost?

52. Suppose you and five of your friends bought tickets to ride on the log ride at the fair. If the total cost of the tickets was $24, how much did each ticket cost?

Getting Ready for the Next Section

Reduce.

53. $\dfrac{6}{8}$ **54.** $\dfrac{17}{34}$

Multiply.

55. $3(0.4)$ **56.** $\dfrac{2}{3} \cdot 6$

Divide.

57. $65 \div 10$ **58.** $1.2 \div 8$

Maintaining Your Skills

Write each fraction or mixed number as an equivalent decimal number.

59. $\dfrac{3}{4}$ **60.** $\dfrac{2}{5}$ **61.** $5\dfrac{1}{2}$ **62.** $8\dfrac{1}{4}$

63. $\dfrac{3}{100}$ **64.** $\dfrac{2}{50}$ **65.** $\dfrac{3}{8}$ **66.** $\dfrac{5}{8}$

Write each decimal as an equivalent proper fraction or mixed number.

67. 0.34   **68.** 0.08   **69.** 2.4   **70.** 5.05

71. 1.75   **72.** 3.125   **73.** 0.875   **74.** 0.375

4.4 Proportions

OBJECTIVES

A Name the terms in a proportion.

B Use the fundamental property of proportions to solve a proportion.

TICKET TO SUCCESS

Keep these questions in mind as you read through the section. Then respond in your own words and in complete sentences.

1. What is a proportion?
2. Label the terms in the following proportion: $\frac{a}{b} = \frac{c}{d}$.
3. What is the fundamental property of proportions?
4. For the proportion $\frac{2}{5} = \frac{4}{x}$, find the product of the means and the product of the extremes.

Millions of people turn to the Internet to view music videos of their favorite musician. Many websites offer different sizes of video based on the speed of a user's internet connection. Even though the figures below are not the same size, their sides are proportional. Later in this chapter, we will use proportions to find the unknown height h in the larger figure.

In this section, we will solve problems using proportions. As you will see, proportions can model a number of everyday applications.

Definition

A statement that two ratios are equal is called a **proportion**. If $\frac{a}{b}$ and $\frac{c}{d}$ are two equal ratios, then the statement

$$\frac{a}{b} = \frac{c}{d}$$

is called a proportion.

A Terms of a Proportion

Each of the four numbers in a proportion is called a *term* of the proportion. We number the terms of a proportion as follows:

$$\text{First term} \longrightarrow \frac{a}{b} = \frac{c}{d} \longleftarrow \text{Third term}$$
$$\text{Second term} \longrightarrow \quad\quad \longleftarrow \text{Fourth term}$$

The first and fourth terms of a proportion are called the *extremes*, and the second and third terms of a proportion are called the *means*.

$$\text{Means} \longrightarrow \frac{a}{b} = \frac{c}{d} \longleftarrow \text{Extremes}$$

EXAMPLE 1 In the proportion $\frac{3}{4} = \frac{6}{8}$, name the four terms, the means, and the extremes.

SOLUTION The terms are numbered as follows:

First term = 3   Third term = 6
Second term = 4   Fourth term = 8

The means are 4 and 6; the extremes are 3 and 8. ∎

The final thing we need to know about proportions is expressed in the following property.

B The Fundamental Property of Proportions

Fundamental Property of Proportions

In any proportion, the product of the extremes is equal to the product of the means. This property is also referred to as the means/extremes property, and in symbols, it looks like this:

$$\text{If } \frac{a}{b} = \frac{c}{d} \text{ then } ad = bc$$

EXAMPLE 2 Verify the fundamental property of proportions for the following proportions.

a. $\frac{3}{4} = \frac{6}{8}$ b. $\frac{17}{34} = \frac{1}{2}$

SOLUTION We verify the fundamental property by finding the product of the means and the product of the extremes in each case.

PRACTICE PROBLEMS

1. In the proportion $\frac{2}{3} = \frac{6}{9}$, name the four terms, the means, and the extremes.

2. Verify the fundamental property of proportions for the following proportions.
   a. $\frac{5}{6} = \frac{15}{18}$
   b. $\frac{13}{39} = \frac{1}{3}$
   c. $\frac{\frac{2}{3}}{\frac{5}{3}} = \frac{2}{5}$
   d. $\frac{0.12}{0.18} = \frac{2}{3}$

Answers
1. See Solutions Section.
2. See Solutions Section.

Proportion	Product of the Means	Product of the Extremes
a. $\frac{3}{4} = \frac{6}{8}$	$4 \cdot 6 = 24$	$3 \cdot 8 = 24$
b. $\frac{17}{34} = \frac{1}{2}$	$34 \cdot 1 = 34$	$17 \cdot 2 = 34$

For each proportion the product of the means is equal to the product of the extremes. ∎

We can use the fundamental property of proportions to solve an equation that has the form of a proportion.

EXAMPLE 3 Solve for x.

$$\frac{2}{3} = \frac{4}{x}$$

SOLUTION Applying the fundamental property of proportions, we have

If $\quad \frac{2}{3} = \frac{4}{x}$

then $\quad 2 \cdot x = 3 \cdot 4 \quad$ The product of the extremes equals the product of the means.

$\quad\quad 2x = 12 \quad$ Multiply.

The result is an equation. We know from Section 4.3 that we can divide both sides of an equation by the same nonzero number without changing the solution to the equation. In this case, we divide both sides by 2 to solve for x.

$$2x = 12$$

$$\frac{2x}{2} = \frac{12}{2} \quad \text{Divide both sides by 2.}$$

$$x = 6 \quad \text{Simplify each side.}$$

The solution is 6. We can check our work by using the fundamental property of proportions.

$\quad\quad 12 \quad\quad\quad 12$
Product of Product of
the means the extremes

Because the product of the means and the product of the extremes are equal, our work is correct. ∎

EXAMPLE 4 Solve for y: $\frac{5}{y} = \frac{10}{13}$.

SOLUTION We apply the fundamental property and solve as we did in Example 3.

If $\quad \frac{5}{y} = \frac{10}{13}$

then $\quad 5 \cdot 13 = y \cdot 10 \quad$ The product of the extremes equals the product of the means.

$\quad\quad 65 = 10y \quad$ Multiply $5 \cdot 13$.

$\quad\quad \frac{65}{10} = \frac{10y}{10} \quad$ Divide both sides by 10.

$\quad\quad 6.5 = y \quad\quad 65 \div 10 = 6.5$

The solution is 6.5. We could check our result by substituting 6.5 for y in the original proportion and then finding the product of the means and the product of the extremes. ∎

3. Find the missing term.
   a. $\frac{3}{4} = \frac{9}{x}$
   b. $\frac{5}{8} = \frac{3}{x}$

NOTE
In some of these problems you will be able to see what the solution is just by looking the problem over. In those cases it is still best to show all the work involved in solving the proportion. It is good practice for the more difficult problems.

4. Solve for y: $\frac{2}{y} = \frac{8}{19}$.

Answers
3. a. 12 b. 4.8
4. 4.75

5. Find n.
   a. $\dfrac{n}{6} = \dfrac{0.3}{15}$
   b. $\dfrac{0.35}{n} = \dfrac{7}{100}$

6. Solve for x:
   a. $\dfrac{\frac{3}{4}}{7} = \dfrac{x}{8}$
   b. $\dfrac{6}{\frac{3}{5}} = \dfrac{15}{x}$

7. Solve $\dfrac{b}{18} = 0.5$.

EXAMPLE 5 Find n if $\dfrac{n}{3} = \dfrac{0.4}{8}$.

SOLUTION We proceed as we did in the previous two examples.

If $\dfrac{n}{3} = \dfrac{0.4}{8}$

then $n \cdot 8 = 3(0.4)$ The product of the extremes equals the product of the means.

$8n = 1.2$ $3(0.4) = 1.2$

$\dfrac{8n}{8} = \dfrac{1.2}{8}$ Divide both sides by 8.

$n = 0.15$ $1.2 \div 8 = 0.15$

The missing term is 0.15. ∎

EXAMPLE 6 Solve for x: $\dfrac{\frac{2}{3}}{5} = \dfrac{x}{6}$.

SOLUTION We begin by multiplying the means and multiplying the extremes.

If $\dfrac{\frac{2}{3}}{5} = \dfrac{x}{6}$

then $\dfrac{2}{3} \cdot 6 = 5 \cdot x$ The product of the extremes equals the product of the means.

$4 = 5 \cdot x$ $\dfrac{2}{3} \cdot 6 = 4$

$\dfrac{4}{5} = \dfrac{5 \cdot x}{5}$ Divide both sides by 5.

$\dfrac{4}{5} = x$

The missing term is $\dfrac{4}{5}$, or 0.8. ∎

EXAMPLE 7 Solve $\dfrac{b}{15} = 2$.

SOLUTION Since the number 2 can be written as the ratio of 2 to 1, we can write this equation as a proportion, and then solve as we have in the examples above.

$\dfrac{b}{15} = 2$

$\dfrac{b}{15} = \dfrac{2}{1}$ Write 2 as a ratio.

$b \cdot 1 = 15 \cdot 2$ Product of the extremes equals the product of the means.

$b = 30$ ∎

The procedure for finding a missing term in a proportion is always the same. We first apply the fundamental property of proportions to find the product of the extremes and the product of the means. Then we solve the resulting equation.

Answers
5. a. 0.12 b. 5
6. a. $\dfrac{6}{7}$ b. $\dfrac{3}{2}$
7. 9

Problem Set 4.4

Moving Toward Success

"Work joyfully and peacefully, knowing that right thoughts and right efforts will inevitably bring about right results."

—James Allen, 1855–1942, New Zealand statesman and Minister of Defense

1. How would you react if you received a poor grade on a homework assignment or test?
   a. Dwell or feel bad about yourself.
   b. Review the mistakes you made.
   c. Discuss with your instructor how to do better.
   d. Revise your study skills to ensure greater success.
   e. Answers b, c, and d
2. What steps will you take to better your grade on the next assignment or test?

A For each of the following proportions, name the means, name the extremes, and show that the product of the means is equal to the product of the extremes. [Examples 1, 2]

1. $\dfrac{1}{3} = \dfrac{5}{15}$
2. $\dfrac{6}{12} = \dfrac{1}{2}$
3. $\dfrac{10}{25} = \dfrac{2}{5}$
4. $\dfrac{5}{8} = \dfrac{10}{16}$

5. $\dfrac{\frac{1}{3}}{\frac{1}{2}} = \dfrac{4}{6}$
6. $\dfrac{2}{\frac{1}{4}} = \dfrac{4}{\frac{1}{2}}$
7. $\dfrac{0.5}{5} = \dfrac{1}{10}$
8. $\dfrac{0.3}{1.2} = \dfrac{1}{4}$

B Find the missing term in each of the following proportions. Set up each problem like the examples in this section. Write your answers as fractions in lowest terms. [Examples 3–7]

9. $\dfrac{2}{5} = \dfrac{4}{x}$
10. $\dfrac{3}{8} = \dfrac{9}{x}$
11. $\dfrac{1}{y} = \dfrac{5}{12}$
12. $\dfrac{2}{y} = \dfrac{6}{10}$
13. $\dfrac{x}{4} = \dfrac{3}{8}$
14. $\dfrac{x}{5} = \dfrac{7}{10}$

15. $\dfrac{5}{9} = \dfrac{x}{2}$
16. $\dfrac{3}{7} = \dfrac{x}{3}$
17. $\dfrac{3}{7} = \dfrac{3}{x}$
18. $\dfrac{2}{9} = \dfrac{2}{x}$
19. $\dfrac{x}{2} = 7$
20. $\dfrac{x}{3} = 10$

21. $\dfrac{\frac{1}{2}}{y} = \dfrac{\frac{1}{3}}{12}$
22. $\dfrac{\frac{2}{3}}{y} = \dfrac{\frac{1}{3}}{5}$
23. $\dfrac{n}{12} = \dfrac{\frac{1}{4}}{\frac{1}{2}}$
24. $\dfrac{n}{10} = \dfrac{\frac{3}{5}}{\frac{3}{8}}$
25. $\dfrac{10}{20} = \dfrac{20}{n}$
26. $\dfrac{8}{4} = \dfrac{4}{n}$

27. $\dfrac{x}{10} = \dfrac{10}{2}$ 28. $\dfrac{x}{12} = \dfrac{12}{48}$ 29. $\dfrac{y}{12} = 9$ 30. $\dfrac{y}{16} = 0.75$ 31. $\dfrac{0.4}{1.2} = \dfrac{1}{x}$ 32. $\dfrac{5}{0.5} = \dfrac{20}{x}$

33. $\dfrac{0.3}{0.12} = \dfrac{n}{0.16}$ 34. $\dfrac{0.01}{0.1} = \dfrac{n}{10}$ 35. $\dfrac{0.5}{x} = \dfrac{1.4}{0.7}$ 36. $\dfrac{0.3}{x} = \dfrac{2.4}{0.8}$

37. $\dfrac{168}{324} = \dfrac{56}{x}$ 38. $\dfrac{280}{530} = \dfrac{112}{x}$ 39. $\dfrac{429}{y} = \dfrac{858}{130}$ 40. $\dfrac{573}{y} = \dfrac{2{,}292}{316}$

41. $\dfrac{n}{39} = \dfrac{533}{507}$ 42. $\dfrac{n}{47} = \dfrac{1{,}003}{799}$ 43. $\dfrac{756}{903} = \dfrac{x}{129}$ 44. $\dfrac{321}{1{,}128} = \dfrac{x}{376}$

Computer Monitor The resolution of a computer monitor is the the number of pixels in the length multiplied by the number of pixels in the height. The ratio between the two measurements is the aspect ratio. Most newer computer monitors have an aspect ratio of 16:9. Fill in the following blanks for screen resolutions, assuming they have an aspect ratio of 16:9.

45. _____ by 1440

46. 1920 by _____

Getting Ready for the Next Section

Divide.

47. 360 ÷ 18

48. 2,700 ÷ 6

Multiply.

49. 3.5(85)

50. 4.75(105)

Solve each equation.

51. $\dfrac{x}{10} = \dfrac{270}{6}$ 52. $\dfrac{x}{45} = \dfrac{8}{18}$ 53. $\dfrac{x}{25} = \dfrac{4}{20}$ 54. $\dfrac{x}{3.5} = \dfrac{85}{1}$

Maintaining Your Skills

Give the place value of the 5 in each number.

55. 250.14

56. 2.5014

Add or subtract as indicated.

57. 2.3 + 0.18 + 24.036

58. 5 + 0.03 + 1.9

59. 3.18 − 2.79

60. 3.4 − 1.975

Applications of Proportions

4.5

OBJECTIVES

A Use proportions to solve application problems.

TICKET TO SUCCESS

Keep these questions in mind as you read through the section. Then respond in your own words and in complete sentences.

1. What does it mean to translate a word problem into a proportion?
2. Write a word problem for the proportion $\frac{2}{5} = \frac{4}{x}$.
3. Explain how you would solve an application problem that involves two ratios that compare the same quantities.
4. Name some jobs that may frequently require solving proportion problems.

Proportions can be used to solve a variety of word problems. The examples that follow show some of these word problems. In each case, we will translate the word problem into a proportion and then solve the proportion using the method developed in this chapter.

Let's begin by solving a proportion that would help us predict how many hits a baseball player will get in a specific number of games.

A Applications

EXAMPLE 1 A baseball player gets 8 hits in the first 18 games of the season. If he continues at the same rate, how many hits will he get in 45 games?

SOLUTION We let x represent the number of hits he will get in 45 games. Then

$$x \text{ is to } 45 \text{ as } 8 \text{ is to } 18$$

$$\text{Hits} \longrightarrow \frac{x}{45} = \frac{8}{18} \longleftarrow \text{Hits}$$
$$\text{Games} \longrightarrow \phantom{\frac{x}{45}} \phantom{\frac{8}{18}} \longleftarrow \text{Games}$$

Notice that the two ratios are comparing the same quantities, hits to games. We solve the proportion as follows:

PRACTICE PROBLEMS

1. A softball player gets 10 hits in the first 18 games of the season. If she continues at the same rate, how many hits will she get in:
   a. 54 games
   b. 27 games

Answers
1. a. 30 hits b. 15 hits

4.5 Applications of Proportions

307

$$18x = 360 \qquad 45 \cdot 8 = 360$$
$$\frac{\cancel{18}x}{\cancel{18}} = \frac{360}{18} \qquad \text{Divide both sides by 18.}$$
$$x = 20 \qquad 360 \div 18 = 20$$

If he continues to hit at the rate of 8 hits in 18 games, he will get 20 hits in 45 games. ■

EXAMPLE 2 A woman drives her car 270 miles in 6 hours. If she continues at the same rate, how far will she travel in 10 hours?

SOLUTION We let x represent the distance traveled in 10 hours. Using x, we translate the problem into the following proportion:

$$\begin{array}{c}\text{Miles} \longrightarrow \\ \text{Hours} \longrightarrow \end{array} \frac{x}{10} = \frac{270}{6} \begin{array}{c} \longleftarrow \text{Miles} \\ \longleftarrow \text{Hours}\end{array}$$

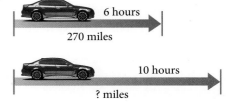

Notice again that the two ratios in the proportion compare the same quantities; that is, both ratios compare miles to hours. In words, this proportion says,

x miles is to 10 hours as 270 miles is to 6 hours

$$\frac{x}{10} = \frac{270}{6}$$

Next, we solve the proportion.

$$x \cdot 6 = 10 \cdot 270$$
$$x \cdot 6 = 2{,}700$$
$$\frac{x \cdot \cancel{6}}{\cancel{6}} = \frac{2{,}700}{6}$$
$$x = 450 \text{ miles}$$

If the woman continues at the same rate, she will travel 450 miles in 10 hours. ■

EXAMPLE 3 The scale on a map indicates that 1 inch on the map corresponds to an actual distance of 85 miles. Two cities are 3.5 inches apart on the map. What is the actual distance between the two cities?

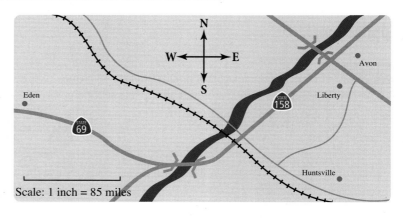

SOLUTION We let x represent the actual distance between the two cities. The proportion is

$$\begin{array}{c}\text{Miles} \longrightarrow \\ \text{Inches} \longrightarrow \end{array} \frac{x}{3.5} = \frac{85}{1} \begin{array}{c} \longleftarrow \text{Miles} \\ \longleftarrow \text{Inches}\end{array}$$

$$x \cdot 1 = 3.5(85)$$
$$x = 297.5 \text{ miles}$$ ■

---

2. A man drives his car 288 miles in 6 hours. If he continues at the same rate, how far will he travel in
   a. 10 hours?
   b. 11 hours?

3. The scale on a map indicates that 1 inch on the map corresponds to an actual distance of 105 miles. Two cities are 4.75 inches apart on the map. What is the actual distance between the two cities?

Answers
2. a. 480 miles **b.** 528 miles
3. 498.75 mi

Problem Set 4.5

Moving Toward Success

"Desire is the key to motivation, but it's determination and commitment to an unrelenting pursuit of your goal—a commitment to excellence—that will enable you to attain the success you seek."

—Mario Andretti, 1940–present, Italian-born American race car driver

1. If you have the intention and confidence that you will master this material, then you are more likely to accomplish your goals. Give a real-life example that supports this statement.
2. Why is it important to be resilient when taking this course?

A Solve each of the following word problems by translating the statement into a proportion. Be sure to show the proportion used in each case. [Examples 1–3]

1. **Distance** A woman drives her car 235 miles in 5 hours. At this rate how far will she travel in 7 hours?

2. **Distance** An airplane flies 1,260 miles in 3 hours. How far will it fly in 5 hours?

3. **Basketball** A basketball player scores 162 points in 9 games. At this rate how many points will he score in 20 games?

4. **Football** In the first 4 games of the season, a football team scores a total of 68 points. At this rate how many points will the team score in 11 games?

5. **Mixture** A solution contains 8 pints of antifreeze and 5 pints of water. How many pints of water must be added to 24 pints of antifreeze to get a solution with the same concentration?

6. **Nutrition** If 10 ounces of a certain breakfast cereal contain 3 ounces of sugar, how many ounces of sugar do 25 ounces of the same cereal contain?

7. **Map Reading** The scale on a map indicates that 1 inch corresponds to an actual distance of 95 miles. Two cities are 4.5 inches apart on the map. What is the actual distance between the two cities?

8. **Map Reading** A map is drawn so that every 2.5 inches on the map corresponds to an actual distance of 100 miles. If the actual distance between two cities is 350 miles, how far apart are they on the map?

9. **Farming** A farmer knows that of every 50 eggs his chickens lay, only 45 will be marketable. If his chickens lay 1,000 eggs in a week, how many of them will be marketable?

10. **Manufacturing** Of every 17 parts manufactured by a certain machine, only 1 will be defective. How many parts were manufactured by the machine if 8 defective parts were found?

11. **Medicine** A patient is given a prescription of 10 pills. The total prescription contains 355 milligrams. How many milligrams is contained in each pill?

12. **Medicine** A child is given a prescription for 9 mg of a drug. If she has to take 3 chewable tablets, what is the strength of each tablet?

13. **Medicine** An oral medication has a dosage strength of 275 mg/5 mL. If a patient takes a dosage of 300 mg, how many milliliters does he take? Round to the nearest tenth.

14. **Medicine** An atropine sulfate injection has a dosage strength of 0.1 mg/mL. If 4.5 mL was given to the patient, how many milligrams did she receive?

15. **Medicine** A tablet has a strength of 45 mg. If a patient is prescribed a dose of 112.5 mg, how many tablets does he take?

16. **Medicine** A tablet has a dosage strength of 35 mg. What was the prescribed dosage if the patient was told to take 1.5 tablets?

Model Trains The size of a model train relative to an actual train is referred to as its scale. Each scale is associated with a ratio as shown in the table. For example, an HO model train has a ratio of 1 to 87, meaning it is $\frac{1}{87}$ as large as an actual train.

17. **Length of a Boxcar** How long is an actual boxcar that has an HO scale model 5 inches long? Give your answer in inches, then divide by 12 to give the answer in feet.

18. **Length of a Flatcar** How long is an actual flatcar that has an LGB scale model 24 inches long? Give your answer in feet.

Scale	Ratio
LGB	1 to 22.5
#1	1 to 32
O	1 to 43.5
S	1 to 64
HO	1 to 87
TT	1 to 120

19. **Travel Expenses** A traveling salesman figures it costs 55¢ for every mile he drives his car. How much does it cost him a week to drive his car if he travels 570 miles a week?

20. **Travel Expenses** A family plans to drive their car during their annual vacation. The car can go 350 miles on a tank of gas, which is 18 gallons of gas. The vacation they have planned will cover 1,785 miles. How many gallons of gas will that take?

21. **Nutrition** A 9-ounce serving of pasta contains 159 grams of carbohydrates. How many grams of carbohydrates do 15 ounces of this pasta contain?

22. **Nutrition** If 100 grams of ice cream contains 13 grams of fat, how much fat is in 250 grams of ice cream?

23. **Travel Expenses** If a car travels 378.9 miles on 50 liters of gas, how many liters of gas will it take to go 692 miles if the car travels at the same rate? Round to the nearest tenth.

24. **Nutrition** If 125 grams of peas contain 26 grams of carbohydrates, how many grams of carbohydrates do 375 grams of peas contain?

25. Elections During a recent election, 47 of every 100 registered voters in a certain city voted. If there were 127,900 registered voters in that city, how many people voted?

26. Map Reading The scale on a map is drawn so that 4.5 inches corresponds to an actual distance of 250 miles. If two cities are 7.25 inches apart on the map, how many miles apart are they? Round to the nearest tenth.

27. Students to Teachers The chart shows the student to teacher ratio for elementary and secondary schools in the United States from 1975 to 2010. If a school had 1,400 students in 1985, how many teachers did the school have? Round to the nearest teacher.

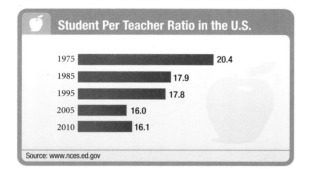

28. Bridges The chart shows the lengths of the longest bridges in the United States. The ratio of feet to meters is given by $\frac{3.28}{1}$. Using this information, convert the length of the Golden Gate Bridge to meters. Round to the nearest hundredth.

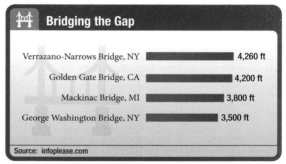

29. Google Earth The Google Earth image shows the western side of The Mall in Washington, D.C. If the scale indicates that one inch is 800 meters and the distance between the Lincoln Memorial and the World War II memorial is $\frac{17}{16}$ inches, what is the actual distance between the two landmarks?

30. Google Earth The Google Earth image shows Disney World in Florida. A scale indicates that one inch is 200 meters. If the distance between Splash Mountain and the Jungle Cruise is 190 meters, what is the distance on the map in inches?

Medicine Liquid medication is usually given in milligrams per milliliter. Use the information to find the amount a patient should take for a prescribed dosage.

31. A patient is prescribed a dosage of Ceclor® of 561 mg. The dosage strength is 187 mg per 5 mL. How many milliliters should he take?

32. A particular brand of amoxicillin has a dosage strength of 125 mg/5 mL. If a patient is prescribed a dosage of 25 mg, how many milliliters should she take?

Nursing For children, the amount of medicine prescribed is often determined by the child's mass. Usually it is calculated from the milligrams per kilogram per day listed on the medication's box.

33. How much should an 18 kg child be given a day if the dosage is 50 mg/kg/day?

34. How much should a 13 kg child be given a day if the dosage is 24 mg/kg/day?

Getting Ready for the Next Section

Simplify.

35. $\dfrac{320}{160}$

36. $21 \cdot 105$

37. $2{,}205 \div 15$

38. $\dfrac{48}{24}$

Solve each equation.

39. $\dfrac{x}{5} = \dfrac{28}{7}$

40. $\dfrac{x}{4} = \dfrac{6}{3}$

41. $\dfrac{x}{21} = \dfrac{105}{15}$

42. $\dfrac{b}{15} = 2$

Maintaining Your Skills

The problems below are a review of some of the concepts we covered previously.

Find the following products. (Multiply.)

43. 2.7×0.5

44. $(0.7)^2$

45. 3.18×1.2

46. $(0.3)^4$

Find the following quotients. (Divide.)

47. $2.8 \div 0.7$

48. $0.042 \div 0.21$

49. $24 \div 0.15$

50. $6.99 \div 2.33$

Divide and round answers to the nearest hundredth.

51. $5{,}679 \div 30.9$

52. $4{,}070 \div 64.2$

Similar Figures

4.6

OBJECTIVES

A Use proportions to find the lengths of sides of similar triangles.

B Use proportions to find the lengths of sides of other similar figures.

C Draw a figure similar to a given figure, given the length of one side.

D Use similar figures to solve application problems.

TICKET TO SUCCESS

Keep these questions in mind as you read through the section. Then respond in your own words and in complete sentences.

1. What are similar figures?
2. How do we know if corresponding sides of two triangles are proportional?
3. When labeling a triangle *ABC*, how do we label the sides?
4. Write an application problem that uses similar figures.

This 8-foot-high bronze sculpture "Cellarman" in Napa, California, is an exact replica of the smaller, 12-inch sculpture. Both pieces are the product of artist Tim Lloyd of Arroyo Grande, California.

In mathematics, when two or more objects have the same shape, but are different sizes, we say they are similar. If two figures are similar, then their corresponding sides are proportional.

In order to give more details on what we mean by corresponding sides of similar figures, it will be helpful to introduce a simple way to label the parts of a triangle.

4.6 Similar Figures

A Similar Triangles

Two triangles that have the same shape are similar when their corresponding sides are proportional, or have the same ratio. The triangles below are similar.

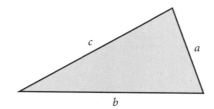
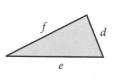

Corresponding Sides	Ratio
Side a corresponds with side d.	$\dfrac{a}{d}$
Side b corresponds with side e.	$\dfrac{b}{e}$
Side c corresponds with side f.	$\dfrac{c}{f}$

Because their corresponding sides are proportional, we write

$$\frac{a}{d} = \frac{b}{e} = \frac{c}{f}$$

EXAMPLE 1 The two triangles below are similar. Find side x.

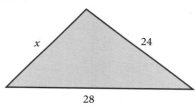
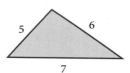

SOLUTION To find the length x, we set up a proportion of equal ratios. The ratio of x to 5 is equal to the ratio of 24 to 6 and to the ratio of 28 to 7. Algebraically, we have

$$\frac{x}{5} = \frac{24}{6} \quad \text{and} \quad \frac{x}{5} = \frac{28}{7}$$

We can solve either proportion to get our answer. The first gives us

$$\frac{x}{5} = 4 \qquad \frac{24}{6} = 4$$

$x = 4 \cdot 5$ Multiply both sides by 5.

$x = 20$ Simplify. ∎

B Other Similar Figures

When one shape or figure is either a reduced or enlarged copy of the same shape or figure, we consider them similar. For example, video viewed over the Internet was once confined to a small "postage stamp" size. Now it is common to see larger video over the Internet. Although the width and height have increased, the shape of the video has not changed.

NOTE
One way to label the important parts of a triangle is to label the vertices with capital letters and the sides with lower-case letters.

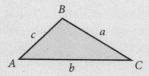

Notice that side a is opposite vertex A, side b is opposite vertex B, and side c is opposite vertex C. Also, because each vertex is the vertex of one of the angles of the triangle, we refer to the three interior angles as A, B, and C.

PRACTICE PROBLEMS

1. The two triangles below are similar. Find the missing side, x.

Answer
1. 35

EXAMPLE 2 The width and height of the two video clips are proportional. Find the height, h, in pixels of the larger video window.

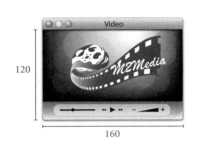

SOLUTION We write our proportion as the ratio of the height of the new video to the height of the old video is equal to the ratio of the width of the new video to the width of the old video.

$$\frac{h}{120} = \frac{320}{160}$$

$$\frac{h}{120} = 2$$

$$h = 2 \cdot 120$$

$$h = 240$$

The height of the larger video is 240 pixels. ∎

C Drawing Similar Figures

Graph paper can help us draw similar figures.

EXAMPLE 3 Draw a triangle similar to triangle ABC, if AC is proportional to DF. Make E the third vertex of the new triangle.

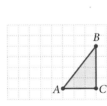
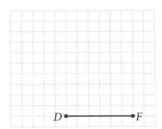

SOLUTION We see that AC is 3 units in length and BC has a length of 4 units. Since AC is proportional to DF, which has a length of 6 units, we set up a proportion to find the length EF.

$$\frac{EF}{BC} = \frac{DF}{AC}$$

$$\frac{EF}{4} = \frac{6}{3}$$

$$\frac{EF}{4} = 2$$

$$EF = 8$$

2. Find the height, h, in pixels of a video clip proportional to those in Example 2 with a width of 360 pixels.

NOTE
A pixel is the smallest dot made on a computer monitor. Many computer monitors have a width of 800 pixels and a height of 600 pixels.

3. Draw a triangle similar to triangle ABC, if AC is proportional to GI.

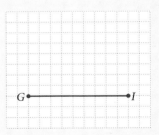

Answers
2. 270 pixels
3. See Solutions Section.

Now we can draw EF with a length of 8 units, then complete the triangle by drawing line DE

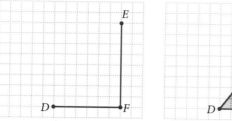

We have drawn triangle *DEF* similar to triangle *ABC*.

D Applications

The next two problems show how similar figures may appear in real-life.

EXAMPLE 4 A building casts a shadow of 105 feet while a 21-foot flagpole casts a shadow that is 15 feet. Find the height of the building.

SOLUTION The figure shows both the building and the flagpole, along with their respective shadows. From the figure it is apparent that we have two similar triangles. Letting x = the height of the building, we have

$$\frac{x}{21} = \frac{105}{15}$$

$15x = 2205$ Extremes/means property

$x = 147$ Divide both sides by 15.

The height of the building is 147 feet.

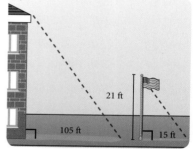

EXAMPLE 5 The instruments in the violin family include the bass, cello, viola, and violin. These instruments can be considered similar figures because the entire length of each instrument is proportional to its body length.

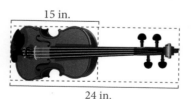

The entire length of a violin is 24 inches, while the body length is 15 inches. Find the body length of a cello if the entire length is 48 inches.

SOLUTION Let b equal the body length of the cello, and set up the proportion.

$$\frac{b}{15} = \frac{48}{24}$$

$$\frac{b}{15} = 2$$

$$b = 2 \cdot 15$$

$$b = 30$$

The body length of a cello is 30 inches.

---

4. A building casts a shadow of 42 feet, while an 18-foot flagpole casts a shadow that is 12 feet. Find the height of the building.

5. Find the body length of an instrument proportional to the violin family that has a total length of 32 inches.

NOTE
These numbers are whole number approximations used to simplify our calculations.

Answer
4. 63 ft
5. 20 in.

Problem Set 4.6

Moving Toward Success

"Avoiding occasions of expense by cultivating peace, we should remember also that timely disbursements to prepare for danger frequently prevent much greater disbursements to repel it."

—George Washington, 1732–1799, First U.S. President

1. What does it mean to come to class prepared?
2. How does coming to class prepared help to reduce stress?

A In problems 1–4, for each pair of similar triangles, set up a proportion and find the unknown. [Example 1]

1.
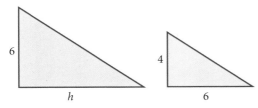

2.
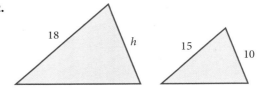

3.
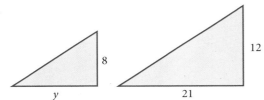

4.
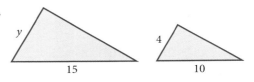

B In problems 5–10, for each pair of similar figures, set up a proportion and find the unknown. [Example 2]

5.
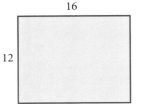

6.
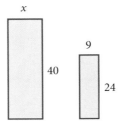

7.

8.
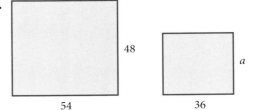

9.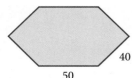

10.

C For each problem, draw a figure on the grid on the right that is similar to the given figure. [Example 3]

11. *AC* is proportional to *DF*.

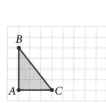

12. *AB* is proportional to *DE*.

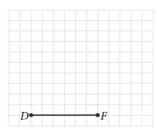 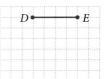

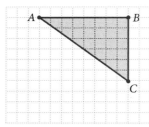

13. *BC* is proportional to *EF*.

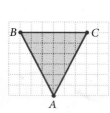

14. *AC* is proportional to *DF*.

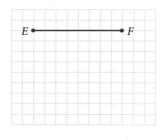

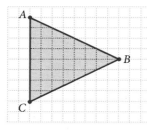

15. *DC* is proportional to *HG*.

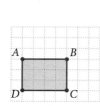

16. *AD* is proportional to *EH*.

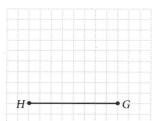 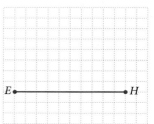

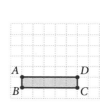

17. *AB* is proportional to *FG*.

18. *BC* is proportional to *FG*.

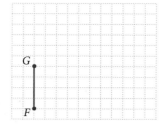

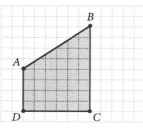

Applying the Concepts

19. Length of a Bass The entire length of a violin is 24 inches, while its body length is 15 inches. The bass is an instrument proportional to the violin. If the total length of a bass is 72 inches, find its body length.

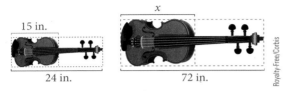

20. Length of an Instrument The entire length of a violin is 24 inches, while the body length is 15 inches. Another instrument proportional to the violin has a body length of 25 inches. What is the total length of this instrument?

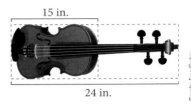

21. Video Resolution A new graphics card can increase the resolution of a computer's monitor. Suppose a monitor has a horizontal resolution of 800 pixels and a vertical resolution of 600 pixels. By adding a new graphics card, the resolutions remain in the same proportions, but the horizontal resolution increases to 1,280 pixels. What is the new vertical resolution?

22. Video Resolution Suppose a monitor has a horizontal resolution of 640 pixels and a vertical resolution of 480 pixels. By adding a new graphics card, the resolutions remain in the same proportions, but the vertical resolution increases to 786 pixels. What is the new horizontal resolution?

23. Screen Resolution The display of a 20" computer monitor is proportional to that of a 23" monitor. A 20" monitor has a horizontal resolution of 1,680 pixels and a vertical resolution of 1,050 pixels. If a 23" monitor has a horizontal resolution of 1,920 pixels, what is its vertical resolution?

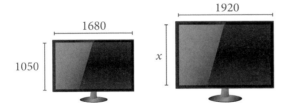

24. Screen Resolution The display of a 20" computer monitor is proportional to that of a 17" monitor. A 20" monitor has a horizontal resolution of 1,680 pixels and a vertical resolution of 1,050 pixels. If a 17" monitor has a vertical resolution of 900 pixels, what is its horizontal resolution?

25. Height of a Tree A tree casts a shadow 38 feet long, while a 6-foot man casts a shadow 4 feet long. How tall is the tree?

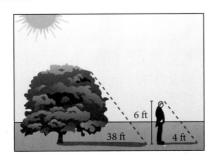

26. Height of a Building A building casts a shadow 128 feet long, while a 24-foot flagpole casts a shadow 32 feet long. How tall is the building?

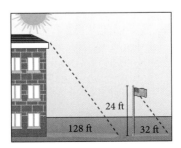

27. Eiffel Tower Recall our discussion at the beginning of this chapter about Eiffel Tower replicas. At the Paris Las Vegas Hotel is a replica of the Eiffel Tower in France. The heights of the tower in Las Vegas and the tower in France are 460 feet and 1,063 feet respectively. The base of the Eiffel Tower in France is 410 feet wide. What is the width of the base of the tower in Las Vegas? Round to the nearest foot.

28. Pyramids The Luxor Hotel in Las Vegas is almost an exact model of the pyramid of Khafre, the second largest Egyptian pyramid. The heights of the Luxor hotel and the pyramid of Khafre are 350 feet and 470 feet respectively. If the base of the pyramid in Khafre was 705 feet wide, what is the width of the base of the Luxor Hotel?

Maintaining Your Skills

The problems below are a review of the four basic operations with fractions and decimals.

Add.

29. 2.03 + 11.958 + 0.002

30. $\frac{3}{4} + \frac{1}{6} + \frac{5}{8}$

Subtract.

31. 65.002 − 24.003

32. $5\frac{1}{8} - 2\frac{5}{8}$

Multiply.

33. 42.18 × 0.0025

34. $7\frac{1}{7} \times 2\frac{1}{3}$

Divide.

35. 378.9 ÷ 21.05

36. $12.25 \div \frac{3}{4}$

37. Find the sum of $2\frac{2}{3}$ and $1\frac{1}{2}$.

38. Find the difference of $2\frac{2}{3}$ and $1\frac{1}{2}$.

39. Find the product of $2\frac{2}{3}$ and $1\frac{1}{2}$.

40. Find the quotient of $2\frac{2}{3}$ and $1\frac{1}{2}$.

Extending the Concepts

41. The rectangles shown here are similar, with similar rectangles within.

a. In the smaller figure, what is the ratio of the shaded to nonshaded rectangles?

b. Shade the larger rectangle such that the ratio of shaded to nonshaded rectangles is $\frac{1}{2}$.

c. For each of the figures, what is the ratio of the shaded rectangles to total rectangles?

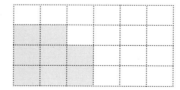
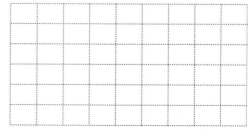

Chapter 4 Summary

EXAMPLES

■ Ratio [4.1]

1. The ratio of 6 to 8 is
$$\frac{6}{8}$$
which can be reduced to
$$\frac{3}{4}$$

The ratio of a to b is $\frac{a}{b}$. The *ratio* of two numbers is a way of comparing them using fraction notation.

■ Rates [4.2]

2. If a car travels 150 miles in 3 hours, then the ratio of miles to hours is considered a rate:
$$\frac{150 \text{ miles}}{3 \text{ hours}} = 50 \frac{\text{miles}}{\text{hour}}$$
$$= 50 \text{ miles per hour}$$

Whenever a ratio compares two quantities that have different units (and neither unit can be converted to the other), then the ratio is called a *rate*.

■ Unit Pricing [4.2]

3. If a 10-ounce package of frozen peas costs 69¢, then the price per ounce, or unit price, is
$$\frac{69 \text{ cents}}{10 \text{ ounces}} = 6.9 \frac{\text{cents}}{\text{ounce}}$$
$$= 6.9 \text{ cents per ounce}$$

The *unit price* of an item is the ratio of price to quantity when the quantity is one unit.

■ Solving Equations by Division [4.3]

4. Solve: $5 \cdot x = 40$
$$5 \cdot x = 40$$
$$\frac{5 \cdot x}{5} = \frac{40}{5} \quad \text{Divide both sides by 5.}$$
$$x = 8 \quad 40 \div 5 = 8$$

Dividing both sides of an equation by the same number will not change the solution to the equation. For example, the equation $5 \cdot x = 40$ can be solved by dividing both sides by 5.

■ Proportions [4.4]

5. The following is a proportion:
$$\frac{6}{8} = \frac{3}{4}$$

A *proportion* is an equation that indicates that two ratios are equal.

The numbers in a proportion are called *terms* and are numbered as follows:

$$\text{First term} \longrightarrow \frac{a}{b} = \frac{c}{d} \longleftarrow \text{Third term}$$
$$\text{Second term} \longrightarrow \phantom{\frac{a}{b} = \frac{c}{d}} \longleftarrow \text{Fourth term}$$

The first and fourth terms are called the *extremes*. The second and third terms are called the *means*.

$$\text{Means} \longrightarrow \frac{a}{b} = \frac{c}{d} \longleftarrow \text{Extremes}$$

Chapter 4 Summary

Fundamental Property of Proportions [4.4]

In any proportion the product of the extremes is equal to the product of the means. In symbols,

$$\text{if } \frac{a}{b} = \frac{c}{d} \text{ then } ad = bc$$

Finding an Unknown Term in a Proportion [4.4]

6. Find x: $\frac{2}{5} = \frac{8}{x}$.

$2 \cdot x = 5 \cdot 8$
$2 \cdot x = 40$
$\frac{2 \cdot x}{2} = \frac{40}{2}$
$x = 20$

To find the unknown term in a proportion, we apply the fundamental property of proportions and solve the equation that results by dividing both sides by the number that is multiplied by the unknown. For instance, if we want to find the unknown in the proportion

$$\frac{2}{5} = \frac{8}{x}$$

we use the fundamental property of proportions to set the product of the extremes equal to the product of the means.

Using Proportions to Find Unknown Length with Similar Figures [4.6]

7. Find x.

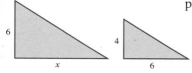

$\frac{4}{6} = \frac{6}{x}$
$36 = 4x$
$9 = x$

Two triangles that have the same shape are similar when their corresponding sides are proportional, or have the same ratio. The triangles below are similar.

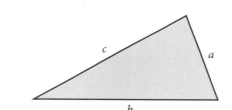 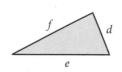

Corresponding Sides	**Ratio**
Side a corresponds with side d.	$\frac{a}{d}$
Side b corresponds with side e.	$\frac{b}{e}$
Side c corresponds with side f.	$\frac{c}{f}$

Because their corresponding sides are proportional, we write

$$\frac{a}{d} = \frac{b}{e} = \frac{c}{f}$$

> ⊘ **COMMON MISTAKES**
>
> A common mistake when working with ratios is to write the numbers in the wrong order when writing the ratio as a fraction. For example, the ratio 3 to 5 is equivalent to the fraction $\frac{3}{5}$. It cannot be written as $\frac{5}{3}$.

Chapter 4 Review

Write each of the following ratios as a fraction in lowest terms. [4.1]

42. 9 to 30

43. 30 to 9

44. $\frac{3}{7}$ to $\frac{4}{7}$

45. $\frac{8}{5}$ to $\frac{8}{9}$

46. $2\frac{1}{3}$ to $1\frac{2}{3}$

47. 3 to $2\frac{3}{4}$

48. 0.6 to 1.2

49. 0.03 to 0.24

50. $\frac{1}{5}$ to $\frac{3}{5}$

51. $\frac{2}{7}$ to $\frac{3}{7}$

Suppose the chart shows where each dollar spent on gasoline in the United States goes. Use the chart for problems 11–12. [4.1]

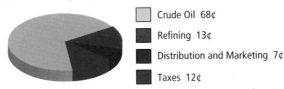

Crude Oil 68¢
Refining 13¢
Distribution and Marketing 7¢
Taxes 12¢

52. Ratio Find the ratio of money paid for taxes to money paid for crude oil.

53. Ratio Find the ratio of the cost of taxes to the combination of the cost of distribution, marketing, and refining.

54. Gas Mileage A car travels 285 miles on 15 gallons of gas. What is the rate of gas mileage in miles per gallon? [4.2]

55. Speed of Sound If it takes 2.5 seconds for sound to travel 2,750 feet, what is the speed of sound in feet per second? [4.2]

56. Unit Price A certain brand of ice cream comes in two different-sized cartons with prices marked as shown. Give the unit price for each carton, and indicate which is the better buy. [4.2]

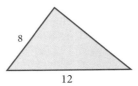

64-ounce carton
$5.79

32-ounce carton
$2.69

57. Unit Price A 6-pack of store-brand soda is $1.25, while a 24-pack of name-brand soda is $5.99. Find the price per soda for each, and determine which is less expensive. [4.2]

Solve each equation by division. [4.3]

58. $6 \cdot x = 12$

59. $5 \cdot x = 16$

Find the missing term in each of the following proportions. [4.4]

60. $\frac{5}{7} = \frac{35}{x}$

61. $\frac{n}{18} = \frac{18}{54}$

62. $\frac{\frac{1}{2}}{10} = \frac{y}{2}$

63. $\frac{x}{1.8} = \frac{5}{1.5}$

64. Chemistry Suppose every 2,000 milliliters of a solution contains 24 milliliters of a certain drug. How many milliliters of solution are required to obtain 18 milliliters of the drug? [4.5]

65. Medicine If $\frac{1}{2}$ cup of breakfast cereal contains 8 milligrams of calcium, how much calcium does $1\frac{1}{2}$ cups of the cereal contain? [4.5]

66. Medicine A patient received a dosage of 7.5 mg of a certain medication. How many tablets must he take if the tablet strength is 2.5 mg? [4.5]

67. Nursing A patient is told to take 300 mg of a certain medication daily. If he takes it in two sittings, how many milligrams is he taking each time he takes the medication? [4.5]

68. Similar Triangles The triangles below are similar figures. Find x. [4.6]

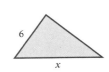

69. Find x if the two rectangles are similar.

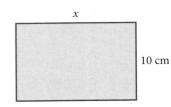

Chapter 4 Review 323

Chapter 4 Cumulative Review

Simplify.

1. 8359
    401
    +1762

2. 3011 − 1032

3. $\dfrac{378}{21}$

4. $(3 \cdot 8) \cdot 2$

5. $31\overline{)15{,}689}$

6. 5^3

7. $6 \cdot 2^3 - 1$

8. $135 \div 15$

9. $56 + 18$

10. $\dfrac{76}{4}$

11. $(11 - 2) + (403 - 102)$

12. $(3.6)(7.1)$

13. $83.6 - 12.12$

14. $6.4 + 3.12 + 5.07$

15. $30.6 \div 6.8$

16. $\left(\dfrac{1}{3}\right)^2\left(\dfrac{1}{2}\right)^3$

17. $5 \div \left(\dfrac{1}{4}\right)^2$

18. $\dfrac{1}{6} + \dfrac{2}{9}$

19. $5 \div \left(14 \div 1\dfrac{2}{3}\right)$

20. $12 - 5\dfrac{1}{4}$

Solve.

21. $\dfrac{x}{20} = \dfrac{5}{4}$

22. $\dfrac{9}{10} = \dfrac{18}{x}$

23. Find the perimeter and area of the figure below.

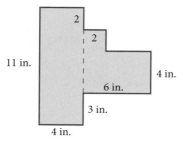

24. Find the perimeter and area of the figure below.

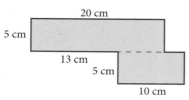

25. Find x if the two rectangles are similar.

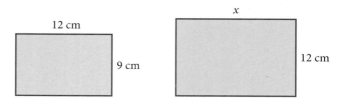

26. **Ratio** If the ratio of men to women in a self-defense class is 3 to 4, and there are 15 men in the class, how many women are in the class?

27. **Surfboard Length** A surfing company decides that a surfboard would be more efficient if its length were reduced by $3\dfrac{5}{8}$ inches. If the original length was 7 feet $\dfrac{3}{16}$ inches, what will be the new length of the board (in inches)?

28. **Average Distance** A bicyclist on a cross-country trip travels 72 miles the first day, 113 miles the second day, 108 miles the third day, and 95 miles the fourth day. What is her average distance traveled during the four days?

29. **Teaching** A teacher lectures on five sections in two class periods. If she continues at the same rate, on how many sections can the teacher lecture in 60 class periods?

Use the illustration to answer problems 30 and 31.

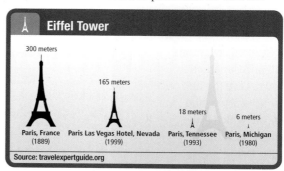

30. **Ratio** What is the ratio of the height of the Eiffel Tower in Las Vegas, Nevada to the height of the tower in Paris, France?

31. **Ratio** What is the ratio of the height of the Eiffel tower in Paris, Michigan to the height of the Eiffel tower in Paris, Tennessee?

Chapter 4 Test

Write each ratio as a fraction in lowest terms. [4.1]

1. 24 to 18
2. $\frac{3}{4}$ to $\frac{5}{6}$
3. 5 to $3\frac{1}{3}$
4. 0.18 to 0.6
5. $\frac{3}{11}$ to $\frac{5}{11}$

A family of three budgeted the following amounts for some of their monthly bills:

Family Budget

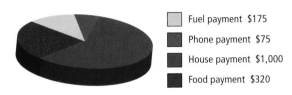

Fuel payment $175
Phone payment $75
House payment $1,000
Food payment $320

6. **Ratio** Find the ratio of house payment to fuel payment. [4.1]

7. **Ratio** Find the ratio of phone payment to food payment. [4.1]

8. **Gas Mileage** A car travels 414 miles on 18 gallons of gas. What is the rate of gas mileage in miles per gallon? [4.2]

9. **Unit Price** A certain brand of frozen orange juice comes in two different-sized cans with prices marked as shown. Give the unit price for each can, and indicate which is the better buy. [4.2]

 16-ounce can $4.16

 12-ounce can $3.36

Solve each equation by division. [4.3]

10. $5 \cdot n = 30$
11. $4 \cdot x = 25$

Find the unknown term in each proportion. [4.4]

12. $\frac{5}{6} = \frac{30}{x}$
13. $\frac{1.8}{6} = \frac{2.4}{x}$

14. **Baseball** A baseball player gets 9 hits in his first 21 at bats of the season. If he continues at the same rate, how many hits will he get in 56 at bats? [4.4]

15. **Map Reading** The scale on a map indicates that 1 inch on the map corresponds to an actual distance of 60 miles. Two cities are $2\frac{1}{4}$ inches apart on the map. What is the actual distance between the two cities? [4.4]

Nursing Sometimes body surface area is used to calculate the necessary dosage for a patient. [4.5]

16. The dosage for a drug is 15 mg/m². If an adult has a BSA of 1.8 m², what dosage should he take?

17. Find the dosage an adult should take if her BSA is 1.3 m² and the dosage strength is 25.5 mg/m².

18. The triangles below are similar figures. Find h. [4.6]

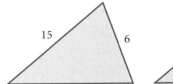

Use the illustration to answer problem 17.

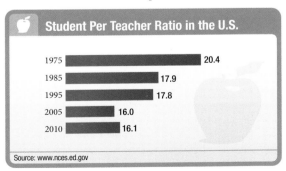

19. If a school in 2010 had 65 teachers, how many students are enrolled?

Chapter 4 Projects

RATIO AND PROPORTION

Group Project

Soil Texture

Number of People 2–3

Time Needed 8–12 minutes

Equipment Paper and pencil

Background Soil texture is defined as the relative proportions of sand, silt, and clay. The figure shows the relative sizes of each of these soil particles. People who study soil science, or work with soil, become very familiar with ratios.

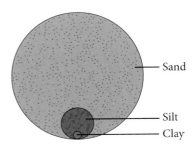

FIGURE 1 Relative sizes of sand, silt, and clay

Procedure A certain type of soil is one part silt, two parts clay, and three parts sand. Use your understanding of ratios and proportions to find the following ratios. Write these ratios as fractions.

1. Sand to total soil
2. Silt to total soil
3. Clay to total soil

4. What is the sum of the three fractions given in questions 1–3?

5. Let the 48 parts of the rectangle below each represent one cubic yard of the soil mixture above. Label each of the squares with either S (for sand), C (for clay), or T (for silt) based on the amount of each in 48 cubic yards of this soil.

Research Project

The Golden Ratio

For many people, the most visually pleasing rectangles are rectangles in which the ratio of length to width is a number called the *golden ratio*, which we have written below.

$$\text{Golden Ratio} = \frac{\sqrt{5} + 1}{2} \approx 1.6180339\ldots$$

Research the golden ratio in mathematics and give examples of where it is used in architecture and art, such as the Parthenon in Athens, Greece. Then measure the length and width of some rectangles around you (TV/computer monitor screen, picture frame, math book, calculator, a dollar, notebook paper, etc.). Calculate the ratio of length to width and indicate which are close to the golden ratio.

Challenge Project

SCRAPING THE SKY

Shanghai is a port city on the eastern coast of China. It is the most populous city in the country and has some of the tallest skyscrapers in the world. The city's buildings boast designs from a wide variety of styles. French, colonial, art deco, neoclassical, contemporary, and the traditional Jiangnan styles all make prominent appearances.

STEP 1 Use Google Earth to find Shanghai, China.

STEP 2 Click the **3D Buildings** layer and zoom in until you can view the city from the ground.

STEP 3 Locate four skyscrapers and use your cursor to measure the elevation of the top of each building.

STEP 4 Create a bar chart using your measurements. Write a ratio that represents the shortest building to the tallest building. What ratio represents the two tallest buildings?

A Glimpse of Algebra

In "A Glimpse of Algebra" in Chapter 3, we spent some time adding polynomials. Now we can use the formula for the area of a rectangle developed in Chapter 1, $A = lw$, to multiply some polynomials.

Suppose we have a rectangle with length $x + 3$ and width $x + 2$. Remember, the letter x is used to represent a number, so $x + 3$ and $x + 2$ are just numbers. Here is a diagram:

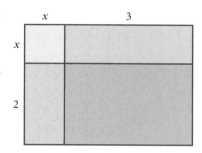

The area of the whole rectangle is the length times the width, or

Total area $= (x + 3)(x + 2)$

But we can also find the total area by first finding the area of each smaller rectangle, and then adding these smaller areas together. The area of each rectangle is its length times its width, as shown in the following diagram:

Because the total area $(x + 3)(x + 2)$ must be the same as the sum of the smaller areas, we have

$$(x + 3)(x + 2) = x^2 + 2x + 3x + 6$$
$$= x^2 + 5x + 6 \quad \text{Add } 2x \text{ and } 3x \text{ to get } 5x.$$

The polynomial $x^2 + 5x + 6$ is the product of the two polynomials $x + 3$ and $x + 2$. Here are some more examples.

EXAMPLE 1 Find the product of $x + 5$ and $x + 2$ by using the diagram.

PRACTICE PROBLEMS

1. Find the product of $x + 4$ and $x + 2$ by using the following diagram:

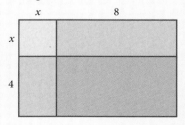

Answer
1. $x^2 + 6x + 8$

SOLUTION The total area is given by $(x + 5)(x + 2)$. We can fill in the smaller areas by multiplying length times width in each case.

	x	5
x	x^2	$5x$
2	$2x$	10

The product of $(x + 5)$ and $(x + 2)$ is

$$(x + 5)(x + 2) = x^2 + 2x + 5x + 10$$
$$= x^2 + 7x + 10$$

EXAMPLE 2 Find the product of $2x + 5$ and $3x + 2$ by using the diagram.

	$2x$	5
$3x$		
2		

SOLUTION We fill in each of the smaller rectangles by multiplying length times width in each case.

	$2x$	5
$3x$	$6x^2$	$15x$
2	$4x$	10

Using the information from the diagram, we have

$$(2x + 5)(3x + 2) = 6x^2 + 4x + 15x + 10$$
$$= 6x^2 + 19x + 10$$

2. Find the product of $3x + 7$ and $2x + 5$ by using the following diagram:

	$3x$	7
$2x$		
5		

Answer
2. $6x^2 + 29x + 35$

A Glimpse of Algebra Problems

Use the diagram in each problem to help multiply the polynomials.

1. $(x + 8)(x + 4)$

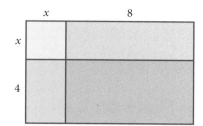

2. $(x + 1)(x + 3)$

3. $(2x + 3)(3x + 2)$

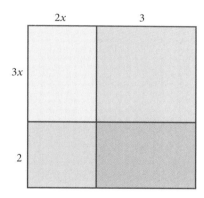

4. $(5x + 4)(6x + 1)$

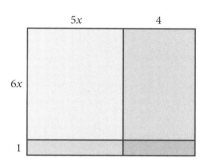

5. $(7x + 2)(3x + 4)$

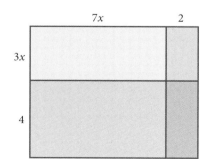

6. $(3x + 5)(2x + 5)$

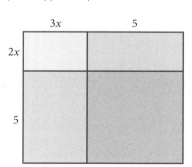

Multiply each of the following pairs of polynomials. You may draw a rectangle to assist you.

7. $(x + 2)(x + 5)$

8. $(x + 3)(x + 6)$

9. $(2x + 3)(x + 4)$

10. $(2x + 4)(x + 3)$

11. $(7x + 3)(2x + 5)$

12. $(5x + 4)(3x + 3)$

13. $(3x + 2)(3x + 2)$

14. $(2x + 3)(2x + 3)$

15. $(4a + 5)(a + 1)$

16. $(5a + 7)(a + 2)$

17. $(7y + 8)(6y + 9)$

18. $(9y + 3)(2y + 8)$

19. $(4 + 6x)(2 + 3x)$

20. $(5 + 3x)(2 + 5x)$

Percent

5

Chapter Outline

5.1 Percents, Decimals, and Fractions

5.2 Basic Percent Problems

5.3 General Applications of Percent

5.4 Sales Tax and Commission

5.5 Percent Increase or Decrease, and Discount

5.6 Interest

5.7 Pie Charts

In 1931, as the Great Depression held America in its grasp, construction began on what was to be the largest dam in the world. The Boulder Canyon project, now known officially as the Hoover Dam, was viewed as a way to control devastating downstream flooding from the Colorado River and provide water to Southern California and Arizona farmland. The first step in this huge venture was to divert the mighty Colorado through four tunnels cut through the canyon. Nearly five years later, two years ahead of schedule, the dam was completed. Based on the original plans, the dam's primary output was supposed to be water, but the generation of electricity has become the major product. The following table shows the current allocation of electricity produced by the dam's hydro-electric power generators.

POWER ALLOCATION	
Location	Percentage Allocated
Metropolitan Water District of Southern California	28.54%
State of Nevada	23.37%
State of Arizona	18.95%
Los Angeles, CA	15.42%
Southern California Edison	5.54%
Boulder City, NV	1.77%
Misc Southern California Cities	6.41%

Source: U.S. Department of the Interior, Bureau of Reclamation

Power Allocation

- Metropolitan Water District of Southern Califronia
- State of Nevada
- State of Arizona
- Los Angeles, CA
- Southern California Edison
- Boulder City, NV
- Misc. Southern California Cities

In this chapter, we will work more with percent and pie charts, like those in the above table.

335

Preview

Key Words	Definition
Percent	A method of comparing a number with the number 100
Commission	Earnings based on a percentage of sales
Interest	The amount of money paid for the use of money

Chapter Outline

5.1 Percents, Decimals, and Fractions
A   Understand the meaning of percent.
B   Change percents to decimals.
C   Change decimals to percents.
D   Change percents to fractions in lowest terms.
E   Change fractions to percents.

5.2 Basic Percent Problems
A   Solve percent problems using equations.
B   Solve percent problems using proportions.

5.3 General Applications of Percent
A   Solve application problems involving percent.

5.4 Sales Tax and Commission
A   Solve application problems involving sales tax.
B   Solve application problems involving commission.

5.5 Percent Increase or Decrease, and Discount
A   Find the percent increase.
B   Find the percent decrease.
C   Solve application problems involving the rate of discount.

5.6 Interest
A   Solve application problems involving annual interest.
B   Solve application problems involving simple interest.
C   Solve compound interest problems.

5.7 Pie Charts
A   Read a pie chart.
B   Construct a pie chart.

5.1 Percents, Decimals, and Fractions

OBJECTIVES

A Understand the meaning of percent.

B Change percents to decimals.

C Change decimals to percents.

D Change percents to fractions in lowest terms.

E Change fractions to percents.

TICKET TO SUCCESS

Keep these questions in mind as you read through the section. Then respond in your own words and in complete sentences.

1. What is a percent in terms of a fraction?
2. How do you change a percent to a decimal?
3. How do you change a decimal to a percent?
4. How would you change $\frac{5}{6}$ into a percent?

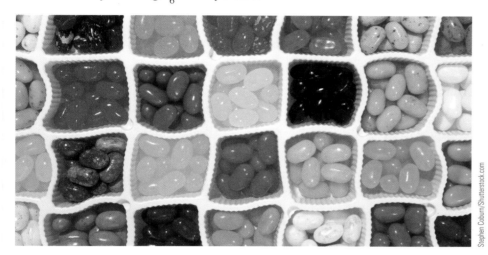

Suppose you went to a candy shop and filled a one-pound bag with assorted jelly beans. Later, you separated all the jelly beans by flavor to determine the following totals:

Flavor	Number of Beans	Percentage of Total Beans
Strawberry	120	30.00%
Cinnamon	25	6.25%
Grape	50	12.50%
Blueberry	50	12.50%
Piña Colada	25	6.25%
Popcorn	50	12.50%
Watermelon	80	20.00%
Total	400	100.00%

In this section, we will look at the meaning of percent and how it may appear in the form of a fraction. We will also learn to change decimals to percents and percents to decimals. After reading this section, you will be able to understand how we calculated the percentages in the jelly bean table based on each flavor's quantity.

PRACTICE PROBLEMS

Write each number as an equivalent fraction without the % symbol.

1. 40%

2. 80%

3. 15%

4. 37%

5. 8%

6. 150%

7. Change to a decimal.
   a. 25.2%
   b. 2.52%

A The Meaning of Percent

Percent means "per hundred." Writing a number as a percent is a way of comparing the number with the number 100. For example, the number 42% (the % symbol is read "percent") is the same as 42 one-hundredths. That is,

$$42\% = \frac{42}{100}$$

Percents are really fractions (or ratios) with denominator 100.

Here are some examples that show the meaning of percent.

EXAMPLE 1 $50\% = \frac{50}{100}$ ∎

EXAMPLE 2 $75\% = \frac{75}{100}$ ∎

EXAMPLE 3 $25\% = \frac{25}{100}$ ∎

EXAMPLE 4 $33\% = \frac{33}{100}$ ∎

EXAMPLE 5 $6\% = \frac{6}{100}$ ∎

EXAMPLE 6 $160\% = \frac{160}{100}$ ∎

B Changing Percents to Decimals

To change a percent to a decimal number, we simply use the meaning of percent that we just discussed.

EXAMPLE 7 Change 35.2% to a decimal.

SOLUTION We drop the % symbol and write 35.2 over 100.

$35.2\% = \frac{35.2}{100}$ Use the meaning of % to convert to a fraction with denominator 100.

$ = 0.352$ Divide 35.2 by 100. ∎

We see from Example 7 that 35.2% is the same as the decimal 0.352. The result is that the % symbol has been dropped and the decimal point has been moved two places to the *left*. Because % always means "per hundred," we will always end up moving the decimal point two places to the left when we change percents to decimals. Because of this, we can write the following rule:

> **Rule** Percent to Decimal
>
> To change a percent to a decimal, drop the % symbol and move the decimal point two places to the *left*, inserting zeros as placeholders if needed.

Here are some examples to illustrate how to use this rule.

Answers

1. $\frac{40}{100}$
2. $\frac{80}{100}$
3. $\frac{15}{100}$
4. $\frac{37}{100}$
5. $\frac{8}{100}$
6. $\frac{150}{100}$
7. a. 0.252   b. 0.0252

EXAMPLE 8 25% = 0.25

EXAMPLE 9 75% = 0.75   Notice that the results in Examples 8, 9, and 10 are consistent with the results in Examples 1, 2, and 3.

EXAMPLE 10 50% = 0.50

EXAMPLE 11 6.8% = 0.068   Notice here that we put a 0 in front of the 6 so we can move the decimal point two places to the left.

EXAMPLE 12 3.62% = 0.0362

EXAMPLE 13 0.4% = 0.004   This time we put two 0s in front of the 4 in order to be able to move the decimal point two places to the left.

EXAMPLE 14 Cortisone cream contains $\frac{1}{2}$% hydrocortisone. $\frac{1}{2}$% is equal to 0.5%. Writing this number as a decimal, we have

0.5% = 0.005

C Changing Decimals to Percents

Now we want to do the opposite of what we just did in Examples 7–14. We want to change decimals to percents. We know that 42% written as a decimal is 0.42, which means that in order to change 0.42 back to a percent, we must move the decimal point two places to the *right* and use the % symbol.

0.42 = 42%   Notice that we don't show the new decimal point if it is at the end of the number.

> **Rule Decimal to Percent**
> To change a decimal to a percent, we move the decimal point two places to the *right* and use the % symbol.

Examples 15–20 show how we use this rule.

EXAMPLE 15 0.27 = 27%

EXAMPLE 16 4.89 = 489%

EXAMPLE 17 0.2 = 0.20 = 20%   Notice here that we put a 0 after the 2 so we can move the decimal point two places to the right.

EXAMPLE 18 0.09 = 09% = 9%   Notice that we can drop the 0 at the left without changing the value of the number.

Change each percent to a decimal.
8. 40%

9. 80%

10. 15%

11. 5.6%

12. 4.86%

13. 0.6%

14. 0.58%

Write each decimal as a percent.
15. 0.35

16. 5.77

17. 0.4

18. 0.3

Answers
8. 0.40
9. 0.80
10. 0.15
11. 0.056
12. 0.0486
13. 0.006
14. 0.0058
15. 35%
16. 577%
17. 40%
18. 3%

Write each decimal as a percent.
19. 45

20. 0.69

Who Pays Health Care Bills

- Patient 19%
- Private insurance 36%
- Government 45%

21. Change 82% to a fraction in lowest terms.

22. Change 6.5% to a fraction in lowest terms.

Answers
19. 4,500%
20. 69%
21. $\frac{41}{50}$
22. $\frac{13}{200}$

Chapter 5 Percent

EXAMPLE 19 $25 = 25.00 = 2,500\%$ Here we put two 0s after the 5 so that we can move the decimal point two places to the right.

EXAMPLE 20 A softball player has a batting average of 0.650. As a percent, this number is $0.650 = 65.0\%$.

As you can see from the examples above, percent is just a way of comparing numbers to 100. To multiply decimals by 100, we move the decimal point two places to the right. To divide by 100, we move the decimal point two places to the left. Because of this, it is a fairly simple procedure to change percents to decimals and decimals to percents.

D Changing Percents to Fractions

Recall at the beginning of this section our discussion of percents and how they are really fractions (or ratios) with the denominator 100. We will now explore further how to change a percent to a fraction.

To change a percent to a fraction, drop the % symbol and write the original number over 100.

EXAMPLE 21 Suppose the pie chart in the margin shows who pays health care bills. Change each percent to a fraction.

SOLUTION In each case, we drop the percent symbol and write the number over 100. Then we reduce to lowest terms if possible.

$$19\% = \frac{19}{100} \qquad 45\% = \frac{45}{100} = \frac{9}{20} \qquad 36\% = \frac{36}{100} = \frac{9}{25}$$
$$\uparrow \qquad\qquad\qquad \uparrow$$
$$\text{Reduce.} \qquad\qquad \text{Reduce.}$$

EXAMPLE 22 Change 4.5% to a fraction in lowest terms.

SOLUTION We begin by writing 4.5 over 100:

$$4.5\% = \frac{4.5}{100}$$

We now multiply the numerator and the denominator by 10 so the numerator will be a whole number.

$$\frac{4.5}{100} = \frac{4.5 \times \mathbf{10}}{100 \times \mathbf{10}} \qquad \text{Multiply the numerator and the denominator by 10.}$$

$$= \frac{45}{1,000}$$

$$= \frac{9}{200} \qquad \text{Reduce to lowest terms.}$$

EXAMPLE 23 Change $32\frac{1}{2}$% to a fraction in lowest terms.

SOLUTION Writing $32\frac{1}{2}$% over 100 produces a complex fraction. We change $32\frac{1}{2}$ to an improper fraction and simplify.

$$32\frac{1}{2}\% = \frac{32\frac{1}{2}}{100}$$

$$= \frac{\frac{65}{2}}{100} \qquad \text{Change } 32\frac{1}{2} \text{ to the improper fraction } \frac{65}{2}.$$

$$= \frac{65}{2} \times \frac{1}{100} \qquad \text{Dividing by 100 is the same as multiplying by } \frac{1}{100}.$$

$$= \frac{\cancel{5} \cdot 13 \cdot 1}{2 \cdot \cancel{5} \cdot 20} \qquad \text{Multiply.}$$

$$= \frac{13}{40} \qquad \text{Reduce to lowest terms.} \blacksquare$$

Note that we could have changed our original mixed number to a decimal first and then changed to a fraction.

$$32\frac{1}{2}\% = 32.5\% = \frac{32.5}{100} = \frac{32.5 \times 10}{100 \times 10} = \frac{325}{1000} = \frac{\cancel{5} \cdot \cancel{5} \cdot 13}{\cancel{5} \cdot \cancel{5} \cdot 40} = \frac{13}{40}$$

The result is the same in both cases.

E Changing Fractions to Percents

To change a fraction to a percent, we can change the fraction to a decimal and then change the decimal to a percent.

EXAMPLE 24 Suppose the price your bookstore pays for your textbook is $\frac{7}{10}$ of the price you pay for your textbook. Write $\frac{7}{10}$ as a percent.

SOLUTION We can change $\frac{7}{10}$ to a decimal by dividing 7 by 10.

$$\begin{array}{r} 0.7 \\ 10\overline{)7.0} \\ \underline{7\ 0} \\ 0 \end{array}$$

We then change the decimal 0.7 to a percent by moving the decimal point two places to the *right* and using the % symbol:

$$0.7 = 70\% \qquad \blacksquare$$

You may have noticed that we could have saved some time in Example 24 by simply writing $\frac{7}{10}$ as an equivalent fraction with denominator 100. That is,

$$\frac{7}{10} = \frac{7 \cdot \mathbf{10}}{10 \cdot \mathbf{10}} = \frac{70}{100} = 70\%$$

This is a good way to convert fractions like $\frac{7}{10}$ to percents. It works well for fractions with denominators of 2, 4, 5, 10, 20, 25, and 50, because thtey are easy to change to fractions with denominators of 100.

---

23. Change $42\frac{1}{2}$% to a fraction in lowest terms.

24. Change to a percent.
   a. $\frac{9}{10}$
   b. $\frac{9}{20}$

Answers
23. $\frac{17}{40}$
24. a. 90% **b.** 45%

Chapter 5 Percent

25. Change to a percent.
   a. $\frac{5}{8}$
   b. $\frac{9}{8}$

EXAMPLE 25 Change $\frac{3}{8}$ to a percent.

SOLUTION We write $\frac{3}{8}$ as a decimal by dividing 3 by 8. We then change the decimal to a percent by moving the decimal point two places to the right and using the % symbol.

$$\frac{3}{8} = 0.375 = 37.5\%$$

```
      .375
   8)3.000
     24↓
      60
      56↓
      40
      40
       0
```

26. Change to a percent.
   a. $\frac{7}{12}$
   b. $\frac{13}{12}$

EXAMPLE 26 Change $\frac{5}{12}$ to a percent.

SOLUTION We begin by dividing 5 by 12.

```
       .4166
   12)5.0000
      4 8↓
        20
        12↓
        80
        72↓
         80
         72
```

NOTE
When rounding off, let's agree to round off to the nearest thousandth and then move the decimal point. Our answers in percent form will then be accurate to the nearest tenth of a percent, as in Example 26.

Because the 6s repeat indefinitely, we can use mixed number notation to write

$$\frac{5}{12} = 0.41\overline{6} = 41\frac{2}{3}\%$$

Or, rounding, we can write

$$\frac{5}{12} \approx 41.7\%$$ Round to the nearest tenth of a percent.

Table 1 lists some of the most commonly used fractions and decimals and their equivalent percents.

TABLE 1

Fraction	Decimal	Percent
$\frac{1}{2}$	0.5	50%
$\frac{1}{4}$	0.25	25%
$\frac{3}{4}$	0.75	75%
$\frac{1}{3}$	$0.3\overline{3}$	$33\frac{1}{3}\%$
$\frac{2}{3}$	$0.6\overline{6}$	$66\frac{2}{3}\%$
$\frac{1}{5}$	0.2	20%
$\frac{2}{5}$	0.4	40%
$\frac{3}{5}$	0.6	60%
$\frac{4}{5}$	0.8	80%

Answers
25. a. 62.5% **b.** 112.5%
26. a. $58\frac{1}{3}\% \approx 58.3\%$
    b. $108\frac{1}{3}\% \approx 108.3\%$

Problem Set 5.1

Moving Toward Success

"We throw all our attention on the utterly idle question whether A has done as well as B, when the only question is whether A has done as well as he could."

—William Graham Sumner, 1840–1910,
American academic and professor

1. Each chapter is a building block for the next. Briefly explain why this statement is true.
2. Why should you pay attention to the notes in the side columns of the section pages?

A Write each percent as a fraction with denominator 100. [Examples 1–6]

1. 20% 2. 40% 3. 60% 4. 80% 5. 24% 6. 48%

7. 65% 8. 35%

B Change each percent to a decimal. [Examples 7–14]

9. 23% 10. 34% 11. 92% 12. 87% 13. 9% 14. 7%

15. 3.4% 16. 5.8% 17. 6.34% 18. 7.25% 19. 0.9% 20. 0.6%

C Change each decimal to a percent. [Examples 15–20]

21. 0.23 22. 0.34 23. 0.92 24. 0.87 25. 0.45 26. 0.54

27. 0.03 28. 0.04 29. 0.6 30. 0.9 31. 0.8 32. 0.5

33. 0.27 34. 0.62 35. 1.23 36. 2.34

Chapter 5 Percent

D Change each percent to a fraction in lowest terms. [Examples 21–23]

37. 60%   **38.** 40%   **39.** 75%   **40.** 25%   **41.** 4%   **42.** 2%

43. 26.5%   **44.** 34.2%   **45.** 71.87% **46.** 63.6%   **47.** 0.75% **48.** 0.45%

49. $6\frac{1}{4}$%   **50.** $5\frac{1}{4}$%   **51.** $33\frac{1}{3}$%   **52.** $66\frac{2}{3}$%

E Change each fraction or mixed number to a percent. [Examples 24–26]

53. $\frac{1}{2}$ **54.** $\frac{1}{4}$ **55.** $\frac{3}{4}$ **56.** $\frac{2}{3}$ **57.** $\frac{1}{3}$ **58.** $\frac{1}{5}$

59. $\frac{4}{5}$ **60.** $\frac{1}{6}$ **61.** $\frac{7}{8}$ **62.** $\frac{1}{8}$ **63.** $\frac{7}{50}$ **64.** $\frac{9}{25}$

65. $3\frac{1}{4}$ **66.** $2\frac{1}{8}$ **67.** $1\frac{1}{2}$ **68.** $1\frac{3}{4}$

69. Change $\frac{21}{43}$ to the nearest tenth of a percent   **70.** Change $\frac{36}{49}$ to the nearest tenth of a percent

Applying the Concepts

71. Jelly Beans The table from the beginning of this section shows the quantities and percentages of jelly bean flavors in a one-pound bag.

Flavor	Number of Beans	Percentage of Total Beans
Strawberry	120	30%
Cinnamon	25	6.25%
Grape	50	12.5%
Blueberry	50	12.5%
Pina Colada	25	6.25%
Popcorn	50	12.5%
Watermelon	80	20%
Total	400	100%

Using the table, convert the percentage for the following jelly bean flavors to a decimal.

a. Strawberry

b. Cinnamon

c. Popcorn

72. U.S. Energy The pie chart shows where Americans get their energy.

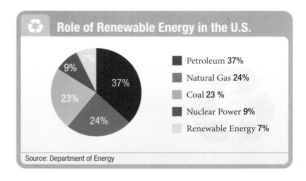

Using the chart, convert the percentage to a fraction for the following types of energy. Reduce to lowest terms.

a. Natural Gas

b. Nuclear Power

c. Petroleum

73. Paying Bills The pie chart below shows how households pay their bills.

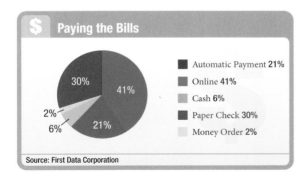

a. Convert each percent to a fraction.

b. Convert each percent to a decimal.

c. About how many times more likely are you to pay a bill online than by automatic payment?

74. Pizza Ingredients The pie chart below shows the decimal representation of each ingredient by weight that is used to make a sausage and mushroom pizza. We see that half of the pizza's weight comes from the crust. Change each decimal to a percent.

Mushroom and Sausage Pizza

Crust 0.5
Cheese 0.25
Sausage 0.075
Mushrooms 0.05
Tomato Sauce 0.125

Calculator Problems

Use a calculator to change each fraction to a decimal, and then change the decimal to a percent. Round all answers to the nearest tenth of a percent.

75. $\dfrac{29}{37}$ **76.** $\dfrac{18}{83}$ **77.** $\dfrac{6}{51}$ **78.** $\dfrac{8}{95}$ **79.** $\dfrac{236}{327}$ **80.** $\dfrac{568}{732}$

Getting Ready for the Next Section

Multiply.

81. 0.25(74) **82.** 0.15(63) **83.** 0.435(25) **84.** 0.635(45)

Divide. Round the answers to the nearest thousandth, if necessary.

85. $\dfrac{21}{42}$ **86.** $\dfrac{21}{84}$ **87.** $\dfrac{25}{0.4}$ **88.** $\dfrac{31.9}{78}$

Solve for n.

89. $42n = 21$ **90.** $25 = 0.40n$

Maintaining Your Skills

Write as a decimal.

91. $\dfrac{1}{8}$ **92.** $\dfrac{3}{8}$ **93.** $\dfrac{5}{8}$ **94.** $\dfrac{7}{8}$

95. $\dfrac{1}{16}$ **96.** $\dfrac{3}{16}$ **97.** $\dfrac{5}{16}$ **98.** $\dfrac{7}{16}$

Divide.

99. $\dfrac{1}{8} \div \dfrac{1}{16}$ **100.** $\dfrac{3}{8} \div \dfrac{3}{16}$ **101.** $\dfrac{5}{8} \div \dfrac{5}{16}$ **102.** $\dfrac{7}{8} \div \dfrac{7}{16}$

103. $0.125 \div 0.0625$ **104.** $0.375 \div 0.1875$ **105.** $0.625 \div 0.3125$ **106.** $0.875 \div 0.4375$

Basic Percent Problems

5.2

OBJECTIVES

A Solve percent problems using equations.

B Solve percent problems using proportions.

TICKET TO SUCCESS

Keep these questions in mind as you read through the section. Then respond in your own words and in complete sentences.

1. Give examples of the three types of percent problems.
2. Look at Example 1 in your text. The number 9.45 is what percent of 63?
3. Look at the food label in Example 7. What percent of the total fat is saturated fat?
4. Write an application problem that involves solving a percent problem using proportions.

The American Dietetic Association (ADA) recommends eating foods in which the number of calories from fat is less than 30% of the total number of calories. Foods that satisfy this requirement are considered healthy foods. Is the nutrition label shown below from a food that the ADA would consider healthy? This is the type of question we will be able to answer after we have worked through the examples in this section.

Nutrition Facts

Serving Size 1/2 cup (65g)
Servings Per Container: 8

Amount Per Serving

Calories 150   Calories from fat 90

	% Daily Value*
Total Fat 10g	16%
Saturated Fat 6g	32%
Cholesterol 35mg	12%
Sodium 30mg	1%
Total Carbohydrate 14g	5%
Dietary Fiber 0g	0%
Sugars 11g	
Protein 2g	

Vitamin A 6%	•	Vitamin C 0%	
Calcium 6%	•	Iron 0%	

*Percent Daily Values are based on a 2,000 calorie diet.

FIGURE 1 Nutrition label from vanilla ice cream

Chapter 5 Percent

A Solving Percent Problems Using Equations

This section is concerned with three kinds of word problems that are associated with percents. Here is an example of each type:

Type A: What number is 15% of 63?
Type B: What percent of 42 is 21?
Type C: 25 is 40% of what number?

The first method we use to solve all three types of problems involves translating the sentences into equations and then solving the equations. The following translations are used to write the sentences as equations:

English	Mathematics
is	=
of	· (multiply)
a number	n
what number	n
what percent	n

The word *is* always translates to an = sign. The word *of* almost always means multiply. The number we are looking for can be represented with a letter, such as n or x.

PRACTICE PROBLEMS

1. a. What number is 25% of 74?
   b. What number is 50% of 74?

EXAMPLE 1   What number is 15% of 63?

SOLUTION   We translate the sentence into an equation as follows:

What number is 15% of 63?
$$n = 0.15 \cdot 63$$

To do arithmetic with percents, we have to change to decimals. That is why 15% is rewritten as 0.15. Solving the equation, we have

$$n = 0.15 \cdot 63$$
$$n = 9.45$$

Therefore, 15% of 63 is 9.45.

2. a. What percent of 84 is 21?
   b. What percent of 84 is 42?

EXAMPLE 2   What percent of 42 is 21?

SOLUTION   We translate the sentence as follows:

What percent of 42 is 21?
$$n \cdot 42 = 21$$

We solve for n by dividing both sides by 42.

$$\frac{n \cdot 42}{42} = \frac{21}{42}$$
$$n = \frac{21}{42}$$
$$n = 0.50$$

Because the original problem asked for a percent, we change 0.50 to a percent:

$$n = 50\%$$

Therefore, 21 is 50% of 42.

Answers
1. a. 18.5   b. 37
2. a. 25%   b. 50%

5.2 Basic Percent Problems

EXAMPLE 3   25 is 40% of what number?

SOLUTION   Following the procedure from the first two examples, we have

$$25 \text{ is } 40\% \text{ of what number?}$$
$$25 = 0.40 \cdot n$$

Again, we changed 40% to 0.40 so we can do the arithmetic involved in the problem. Dividing both sides of the equation by 0.40, we have

$$\frac{25}{0.40} = \frac{0.40 \cdot n}{0.40}$$

$$\frac{25}{0.40} = n$$

$$62.5 = n$$

Therefore, 25 is 40% of 62.5. ■

3. **a.** 35 is 40% of what number?
   b. 70 is 40% of what number?

As you can see, all three types of percent problems are solved in a similar manner. We write *is* as =, *of* as ·, and *what number* as *n*. The resulting equation is then solved to obtain the answer to the original question. Here are some more examples.

EXAMPLE 4   What number is 43.5% of 25?

$$n = 0.435 \cdot 25$$
$$n = 10.9 \qquad \text{Rounded to the nearest tenth.}$$

Therefore, 10.9 is 43.5% of 25. ■

4. What number is 63.5% of 45? (Round to the nearest tenth.)

EXAMPLE 5   What percent of 78 is 31.9?

$$n \cdot 78 = 31.9$$

$$\frac{n \cdot 78}{78} = \frac{31.9}{78}$$

$$n = \frac{31.9}{78}$$

$$n = 0.409 \qquad \text{Rounded to the nearest thousandth.}$$

$$n = 40.9\%$$

Therefore, 40.9% of 78 is 31.9. ■

5. What percent of 85 is 11.9?

EXAMPLE 6   34 is 29% of what number?

$$34 = 0.29 \cdot n$$

$$\frac{34}{0.29} = \frac{0.29 \cdot n}{0.29}$$

$$\frac{34}{0.29} = n$$

$$117.2 = n \qquad \text{Rounded to the nearest tenth.}$$

Therefore, 34 is 29% of 117.2. ■

6. 62 is 39% of what number? (Round to the nearest tenth.)

Answers
3. **a.** 87.5 **b.** 175
4. 28.6
5. 14%
6. 159.0

7. The nutrition label below is from a package of vanilla frozen yogurt. What percent of the total number of calories is fat calories? Round your answer to the nearest tenth of a percent.

Nutrition Facts
Serving Size 1/2 cup (98g)
Servings Per Container: 4

Amount Per Serving	
Calories 160	Calories from fat 25

	% Daily Value*
Total Fat 2.5g	4%
Saturated Fat 1.5g	7%
Cholesterol 45mg	15%
Sodium 55mg	2%
Total Carbohydrate 26g	9%
Dietary Fiber 0g	0%
Sugars 19g	
Protein 8g	

Vitamin A 0%	•	Vitamin C 0%
Calcium 25%	•	Iron 0%

*Percent Daily Values are based on a 2,000 calorie diet.

Answer
7. 15.6% of the calories are from fat. (So far as fat content is concerned, the frozen yogurt is a healthier choice than the ice cream.)

Chapter 5 Percent

EXAMPLE 7 As we mentioned in the introduction to this section, the American Dietetic Association recommends eating foods in which the number of calories from fat is less than 30% of the total number of calories. According to the nutrition label below, what percent of the total number of calories is fat calories?

Nutrition Facts
Serving Size 1/2 cup (65g)
Servings Per Container: 8

Amount Per Serving	
Calories 150	Calories from fat 90

	% Daily Value*
Total Fat 10g	16%
Saturated Fat 6g	32%
Cholesterol 35mg	12%
Sodium 30mg	1%
Total Carbohydrate 14g	5%
Dietary Fiber 0g	0%
Sugars 11g	
Protein 2g	

Vitamin A 6%	•	Vitamin C 0%
Calcium 6%	•	Iron 0%

*Percent Daily Values are based on a 2,000 calorie diet.

Nutrition label from vanilla ice cream

SOLUTION To solve this problem, we must write the question in the form of one of the three basic percent problems shown in Examples 1–6. Because there are 90 calories from fat and a total of 150 calories, we can write the question this way: 90 is what percent of 150?

Now that we have written the question in the form of one of the basic percent problems, we simply translate it into an equation. Then we solve the equation.

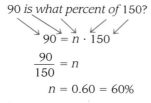

$$90 = n \cdot 150$$
$$\frac{90}{150} = n$$
$$n = 0.60 = 60\%$$

The number of calories from fat in this package of ice cream is 60% of the total number of calories. Thus the ADA would not consider this to be a healthy food.

B Solving Percent Problems Using Proportions

We can look at percent problems in terms of proportions also. For example, we know that 24% is the same as $\frac{24}{100}$, which reduces to $\frac{6}{25}$. That is,

$$\frac{24}{100} = \frac{6}{25}$$

↑ ↑ ↑
24 is to 100   as   6 is to 25

We can illustrate this visually with boxes of proportional lengths.

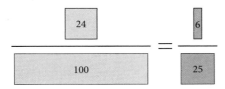

In general, we say

$$\frac{\text{Percent}}{100} = \frac{\text{Amount}}{\text{Base}}$$

Percent is to 100 as amount is to base.

EXAMPLE 8   What number is 15% of 63?

SOLUTION This is the same problem we worked in Example 1. We let n be the number in question. We reason that n will be smaller than 63 because it is only 15% of 63. The base is 63 and the amount is n. We compare n to 63 as we compare 15 to 100. Our proportion sets up as follows:

15 is to 100   as n is to 63

$$\frac{15}{100} = \frac{n}{63}$$

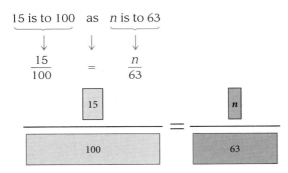

Solving the proportion, we have

$15 \cdot 63 = 100n$ Extremes/means property
$945 = 100n$ Simplify the left side.
$9.45 = n$ Divide each side by 100.

This gives us the same result we obtained in Example 1.

8. Rework Practice Problem 1 using proportions.

EXAMPLE 9   What percent of 42 is 21?

SOLUTION This is the same problem we worked in Example 2. We let n be the percent in question. The amount is 21 and the base is 42. We compare n to 100 as we compare 21 to 42. Here is our reasoning and proportion:

n is to 100   as   21 is to 42

$$\frac{n}{100} = \frac{21}{42}$$

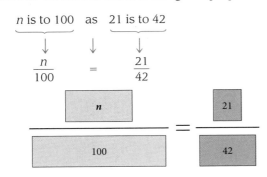

9. Rework Practice Problem 2 using proportions.

Answers
8. a. 18.5   **b.** 37
9. a. 25%   **b.** 50%

Solving the proportion, we have

$$42n = 21 \cdot 100 \quad \text{Extremes/means property}$$
$$42n = 2{,}100 \quad \text{Simplify the right side.}$$
$$n = 50 \quad \text{Divide each side by 42.}$$

Since n is a percent, our answer is 50%, giving us the same result we obtained in Example 2. ∎

EXAMPLE 10 25 is 40% of what number?

SOLUTION This is the same problem we worked in Example 3. We let n be the number in question. The base is n and the amount is 25. We compare 25 to n as we compare 40 to 100. Our proportion sets up as follows:

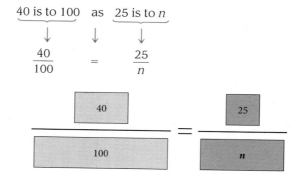

Solving the proportion, we have

$$40 \cdot n = 25 \cdot 100 \quad \text{Extremes/means property}$$
$$40 \cdot n = 2{,}500 \quad \text{Simplify the right side.}$$
$$n = 62.5 \quad \text{Divide each side by 40.}$$

So 25 is 40% of 62.5, which is the same result we obtained in Example 3. ∎

10. Rework Practice Problem 3 using proportions.

NOTE
When you work the problems in the problem set, use whichever method you like, unless your instructor indicates that you are to use one method instead of the other.

Answers
10. a. 87.5 b. 175

Problem Set 5.2

Moving Toward Success

"Determine what specific goal you want to achieve. Then dedicate yourself to its attainment with unswerving singleness of purpose, the trenchant zeal of a crusader."

—Paul J. Meyer, 1928–present, American businessman and publisher

1. Why is it better to attend all class sessions on time instead of just getting notes from a classmate?

2. When should you read the book?
   a. Before the material is covered in class
   b. During your homework time
   c. To prepare for an exam
   d. All the above

A B Solve each of the following problems. [Examples 1–6, 8–10]

1. What number is 25% of 32?
2. What number is 10% of 80?
3. What number is 20% of 120?
4. What number is 15% of 75?
5. What number is 54% of 38?
6. What number is 72% of 200?
7. What number is 11% of 67?
8. What number is 2% of 49?
9. What percent of 24 is 12?
10. What percent of 80 is 20?
11. What percent of 50 is 5?
12. What percent of 20 is 4?
13. What percent of 36 is 9?
14. What percent of 70 is 14?
15. What percent of 8 is 6?
16. What percent of 16 is 9?
17. 32 is 50% of what number?
18. 16 is 20% of what number?
19. 10 is 20% of what number?
20. 11 is 25% of what number?

21. 37 is 4% of what number?

22. 90 is 80% of what number?

23. 8 is 2% of what number?

24. 6 is 3% of what number?

25. What is 6.4% of 87?

26. What is 10% of 102?

27. 25% of what number is 30?

28. 10% of what number is 22?

29. 28% of 49 is what number?

30. 97% of 28 is what number?

31. 27 is 120% of what number?

32. 24 is 150% of what number?

33. 65 is what percent of 130?

34. 26 is what percent of 78?

35. What is 0.4% of 235,671?

36. What is 0.8% of 721,423?

37. 4.89% of 2,000 is what number?

38. 3.75% of 4,000 is what number?

39. Write a basic percent problem, the solution to which can be found by solving the equation $n = 0.25(350)$.

40. Write a basic percent problem, the solution to which can be found by solving the equation $n = 0.35(250)$.

41. Write a basic percent problem, the solution to which can be found by solving the equation $n \cdot 24 = 16$.

42. Write a basic percent problem, the solution to which can be found by solving the equation $n \cdot 16 = 24$.

43. Write a basic percent problem, the solution to which can be found by solving the equation $46 = 0.75 \cdot n$.

44. Write a basic percent problem, the solution to which can be found by solving the equation $75 = 0.46 \cdot n$.

Applying the Concepts

Nutrition For each nutrition label in Problems 45–48, find what percent of the total number of calories comes from fat calories. Then indicate whether the label is from a food considered healthy by the American Dietetic Association. Round to the nearest tenth of a percent if necessary.

45. Spaghetti

Nutrition Facts
Serving Size 2 oz. (56g per 1/8 of pkg) dry
Servings Per Container: 8

Amount Per Serving

Calories 210	Calories from fat 10
	% Daily Value*
Total Fat 1g	2%
Saturated Fat 0g	0%
Polyunsaturated Fat 0.5g	
Monounsaturated Fat 0g	
Cholesterol 0mg	0%
Sodium 0mg	0%
Total Carbohydrate 42g	14%
Dietary Fiber 2g	7%
Sugars 3g	
Protein 7g	
Vitamin A 0% •	Vitamin C 0%
Calcium 0% •	Iron 10%
Thiamin 30% •	Riboflavin 10%
Niacin 15% •	

*Percent Daily Values are based on a 2,000 calorie diet

46. Canned Italian tomatoes

Nutrition Facts
Serving Size 1/2 cup (121g)
Servings Per Container: about 3 1/2

Amount Per Serving

Calories 25	Calories from fat 0
	% Daily Value*
Total Fat 0g	0%
Saturated Fat 0g	0%
Cholesterol 0mg	0%
Sodium 300mg	12%
Potassium 145mg	4%
Total Carbohydrate 4g	2%
Dietary Fiber 1g	4%
Sugars 4g	
Protein 1g	
Vitamin A 20% •	Vitamin C 15%
Calcium 4% •	Iron 15%

*Percent Daily Values are based on a 2,000 calorie diet.

47. Shredded Romano cheese

Nutrition Facts
Serving Size 2 tsp (5g)
Servings Per Container: 34

Amount Per Serving

Calories 20	Calories from fat 10
	% Daily Value*
Total Fat 1.5g	2%
Saturated Fat 1g	5%
Cholesterol 5mg	2%
Sodium 70mg	3%
Total Carbohydrate 0g	0%
Fiber 0g	0%
Sugars 0g	
Protein 2g	
Vitamin A 0% •	Vitamin C 0%
Calcium 4% •	Iron 0%

*Percent Daily Values are based on a 2,000 calorie diet.

48. Tortilla chips

Nutrition Facts
Serving Size 1 oz (28g/About 12 chips)
Servings Per Container: about 2

Amount Per Serving

Calories 140	Calories from fat 60
	% Daily Value*
Total Fat 7g	1%
Saturated Fat 1g	6%
Cholesterol 0mg	0%
Sodium 170mg	7%
Total Carbohydrate 18g	6%
Dietary Fiber 1g	4%
Sugars less than 1g	
Protein 2g	
Vitamin A 0% •	Vitamin C 0%
Calcium 4% •	Iron 2%

*Percent Daily Values are based on a 2,000 calorie diet.

Getting Ready for the Next Section

Solve each equation.

49. $96 = n \cdot 120$

50. $2{,}400 = 0.48 \cdot n$

51. $114 = 150n$

52. $3{,}360 = 0.42n$

53. What number is 80% of 60?

54. What number is 25% of 300?

Maintaining Your Skills

Multiply.

55. 2×0.125

56. 3×0.125

57. 4×0.125

58. 5×0.125

59. The sequence below is an arithmetic sequence in which each term is found by adding $\frac{1}{8}$ to the previous term. Find the next three numbers in the sequence.

$$\frac{1}{4}, \frac{3}{8}, \frac{1}{2}, \ldots$$

60. The sequence below is an arithmetic sequence in which each term is found by adding $\frac{1}{16}$ to the previous term. Find the next three numbers in the sequence.

$$\frac{1}{8}, \frac{3}{16}, \frac{1}{4}, \ldots$$

Simplify.

61. $\frac{1}{4} - \frac{1}{8} + \frac{1}{2} - \frac{3}{8}$

62. $\frac{7}{8} - \frac{3}{4} + \frac{5}{8} - \frac{1}{2}$

Write as a decimal.

63. $\frac{2}{8}$

64. $\frac{4}{8}$

65. $\frac{6}{8}$

66. $\frac{8}{8}$

67. $\frac{2}{16}$

68. $\frac{4}{16}$

69. $\frac{6}{16}$

70. $\frac{8}{16}$

Write in order from smallest to largest.

71. $\frac{3}{8}, \frac{1}{4}, \frac{5}{8}, \frac{1}{8}, \frac{1}{2}, \frac{3}{4}, \frac{7}{8}$

72. $\frac{3}{16}, \frac{1}{8}, \frac{1}{4}, \frac{3}{8}, \frac{7}{16}, \frac{1}{16}, \frac{1}{2}, \frac{5}{16}$

General Applications of Percent

5.3

OBJECTIVES

A Solve application problems involving percent.

TICKET TO SUCCESS

Keep these questions in mind as you read through the section. Then respond in your own words and in complete sentences.

1. On the test mentioned in Example 1, how many questions would the student have answered correctly if she answered 40% of the questions correctly?
2. If the bottle in Example 2 contained 30 milliliters instead of 60, what would the answer be?
3. In Example 3, how many of the students were male?
4. How many of the students mentioned in Example 4 received a grade lower than an A?

ravl/Shutterstock.com

A recent online article by Reuters stated that Toys R Us reported a 5.4 percent rise in total sales during the peak holiday season of December 2010. This is a prime example of how we encounter percents in everyday life. As we progress through this chapter, we will work more with applications of percent. As a result, we will be better equipped to understand statements like the one above concerning Toys R Us.

In this section, we continue our study of percent by doing more of the translations that were introduced in Section 5.2. The better you are at working the problems in Section 5.2, the easier it will be for you to get started on the problems in this section.

A Applications Involving Percent

EXAMPLE 1 On a 120-question test, a student answered 96 correctly. What percent of the problems did the student work correctly?

SOLUTION We have 96 correct answers out of a possible 120. The problem can be restated as

96 *is what percent of* 120?

$$96 = n \cdot 120$$

$$\frac{96}{120} = \frac{n \cdot 120}{120} \quad \text{Divide both sides by 120.}$$

PRACTICE PROBLEMS

1. On a 150-question test, a student answered 114 correctly. What percent of the problems did the student work correctly?

Answer
1. 76%

$$n = \frac{96}{120}$$ Switch the left and right sides of the equation.

$n = 0.80$ Divide 96 by 120.

$ = 80\%$ Rewrite as a percent.

When we write a test score as a percent, we are comparing the original score to an equivalent score on a 100-question test. That is, 96 correct out of 120 is the same as 80 correct out of 100. ∎

EXAMPLE 2 How much HCl (hydrochloric acid) is in a 60-milliliter bottle that is marked 80% HCl?

SOLUTION If the bottle is marked 80% HCl, that means 80% of the solution is HCl and the rest is water. Because the bottle contains 60 milliliters, we can restate the question as:

What is 80% of 60?

$n = 0.80 \cdot 60$

$n = 48$

There are 48 milliliters of HCl in 60 milliliters of 80% HCl solution. ∎

EXAMPLE 3 If 48% of the students in a certain college are female and there are 2,400 female students, what is the total number of students in the college?

SOLUTION We restate the problem as:

2,400 is 48% of what number?

$2{,}400 = 0.48 \cdot n$

$$\frac{2{,}400}{0.48} = \frac{0.48 \cdot n}{0.48}$$ Divide both sides by 0.48.

$$n = \frac{2{,}400}{0.48}$$ Switch the left and right sides of the equation.

$n = 5{,}000$

There are 5,000 students. ∎

EXAMPLE 4 If 25% of the students in elementary algebra courses receive a grade of A, and there are 300 students enrolled in elementary algebra this year, how many students will receive As?

SOLUTION After reading the question a few times, we find that it is the same as this question.

What number is 25% of 300?

$n = 0.25 \cdot 300$

$n = 75$

Thus, 75 students will receive As in elementary algebra. ∎

Almost all application problems involving percents can be restated as one of the three basic percent problems we listed in Section 5.2. It takes some practice before the restating of application problems becomes automatic. You may have to review Section 5.2 and Examples 1–4 above several times before you can translate word problems into mathematical expressions yourself.

---

2. How much HCl is in a 40-milliliter bottle that is marked 75% HCl?

3. If 42% of the students in a certain college are female and there are 3,360 female students, what is the total number of students in the college?

4. Suppose in Example 4 that 35% of the students receive a grade of A. How many of the 300 students is that?

Answers
2. 30 milliliters
3. 8,000 students
4. 105 students

Problem Set 5.3

Moving Toward Success

"A painter told me that nobody could draw a tree without in some sort becoming a tree; or draw a child by studying the outlines of its form merely, but by watching for a time his motions and plays, the painter enters into his nature and can then draw."

—Ralph Waldo Emerson, 1803–1882, American poet and essayist

1. Will you learn the material better by studying other classmates' work or by working problems yourself? Explain.
2. Should you expect to master a new topic the first time you read it? Why or why not?

A Solve each of the following problems by first restating it as one of the three basic percent problems of Section 5.2. In each case, be sure to show the equation. [Examples 1–4]

1. **Test Scores** On a 120-question test a student answered 84 correctly. What percent of the problems did the student work correctly?

2. **Test Scores** An engineering student answered 81 questions correctly on a 90-question trigonometry test. What percent of the questions did she answer correctly? What percent were answered incorrectly?

3. **Basketball** A basketball player made 63 out of 75 free throws. What percent is this?

4. **Family Budget** A family spends $450 every month on food. If the family's income each month is $1,800, what percent of the family's income is spent on food?

5. **Chemistry** How much HCl (hydrochloric acid) is in a 60-milliliter bottle that is marked 75% HCl?

6. **Chemistry** How much acetic acid is in a 5-liter container of acetic acid and water that is marked 80% acetic acid? How much is water?

7. **Farming** A farmer owns 28 acres of land. Of the 28 acres, only 65% can be farmed. How many acres are available for farming? How many are not available for farming?

8. **Number of Students** Of the 420 students enrolled in a basic math class, only 30% are first-year students. How many are first-year students? How many are not?

9. **Determining a Tip** Servers and wait staff are often paid minimum wage and depend on tips for much of their income. It is common for tips to be 15% to 20% of the bill. After dinner at a local restaurant the total bill is $56.00. Since your service was above average you decide to give a 20% tip. Determine the amount of the tip you leave for your server.

10. **Determining a Tip** Suppose you decide to leave a 15% tip for services after your dinner out in the preceding problem. How much of a tip did you leave your server? How much smaller was the tip?

11. **Voting** In the 2008 presidential election, Barack Obama received 67.84% of the total electoral votes while John McCain received 32.16%. If there were 538 total votes cast by the electoral college, how many electoral votes did each candidate receive?

12. **Census Data** According to the U.S. Census Bureau's, national population estimates grouped by age and gender for July 2009, approximately 24.3% of the 307,006,550 people in the United States are under 18 years of age. How many people are in this age group? Round to the nearest person.

13. Census Data According to the U.S. Census Bureau's data for 2009, approximately 6.9% of the 307,006,550 people in the United States are under 5 years of age. How many people are in this age group? Round to the nearest person.

14. Census Data According to the U.S. Census Bureau's data for 2009, approximately 80.4% of the 307,006,550 people in the United States are high school graduates. How many people are high school graduates? Round to the nearest person.

15. Number of Students If 48% of the students in a certain college are female and there are 1,440 female students, what is the total number of students in the college?

16. Mixture Problem A solution of alcohol and water is 80% alcohol. The solution is found to contain 32 milliliters of alcohol. How many milliliters total (both alcohol and water) are in the solution?

17. Number of Graduates Suppose 60% of the graduating class in a certain high school goes on to college. If 240 students from this graduating class are going on to college, how many students are there in the graduating class?

18. Defective Parts In a shipment of airplane parts, 3% are known to be defective. If 15 parts are found to be defective, how many parts are in the shipment?

19. Number of Students There are 3,200 students at our school. If 52% of them are female, how many female students are there at our school?

20. Number of Students In a certain school, 75% of the students in first-year chemistry have had algebra. If there are 300 students in first-year chemistry, how many of them have had algebra?

21. Population In a city of 32,000 people, there are 10,000 people under 25 years of age. What percent of the population is under 25 years of age?

22. Electricity Below is the chart from this chapter's introduction that discusses the allocation of electricity produced by the Hoover Dam.

POWER ALLOCATION	
Location	Percentage Allocated
Metropolitan Water District of Southern California	28.54%
State of Nevada	23.37%
State of Arizona	18.95%
Los Angeles, CA	15.42%
Southern California Edison	5.54%
Boulder City, NV	1.77%
Misc Southern California Cities	6.41%

Source: U.S. Department of the Interior, Bureau of Reclamation

If the dam generates 4.0 billion kilowatt hours of electricity per year, how many kilowatt hours are allocated to

a. the state of Nevada?

b. Los Angeles, CA?

Calculator Problems

The following problems are similar to Problems 1–22. They should be set up the same way. Then the actual calculations should be done on a calculator.

23. Number of People Of 7,892 people attending an outdoor concert in Los Angeles, 3,972 are over 18 years of age. What percent is this? (Round to the nearest whole-number percent.)

24. Manufacturing A car manufacturer estimates that 25% of the new cars sold in one city have defective engine mounts. If 2,136 new cars are sold in that city, how many will have defective engine mounts?

25. Population The chart shows the most populated cities in the United States. If the population of New York City is about 42.5% of the state's population, what is the approximate population of the state?

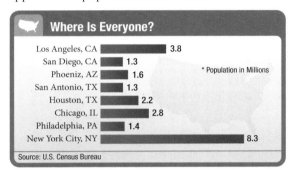

26. Employers The chart shows the importance employers place on different qualities of seasonal employees. If 14,253 employers were surveyed, how many think previous experience in the industry is important? Round to the nearest employer.

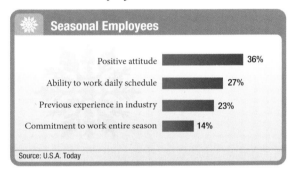

Getting Ready for the Next Section

Multiply.

27. 0.06(550)

28. 0.06(625)

29. 0.03(289,500)

30. 0.03(115,900)

Divide. Write your answers as decimals.

31. 5.44 ÷ 0.04

32. 4.35 ÷ 0.03

33. 19.80 ÷ 396

34. 11.82 ÷ 197

35. $\dfrac{1{,}836}{0.12}$

36. $\dfrac{115}{0.1}$

37. $\dfrac{90}{600}$

38. $\dfrac{105}{750}$

Maintaining Your Skills

The problems below review multiplication with fractions and mixed numbers.

Multiply.

39. $\dfrac{1}{2} \cdot \dfrac{2}{5}$

40. $\dfrac{3}{4} \cdot \dfrac{1}{3}$

41. $\dfrac{3}{4} \cdot \dfrac{5}{9}$

42. $\dfrac{5}{6} \cdot \dfrac{12}{13}$

43. $2 \cdot \dfrac{3}{8}$

44. $3 \cdot \dfrac{5}{12}$

45. $1\dfrac{1}{4} \cdot \dfrac{8}{15}$

46. $2\dfrac{1}{3} \cdot \dfrac{9}{10}$

Extending the Concepts

Batting averages in baseball are given as decimal numbers, rounded to the nearest thousandth. For example, at the end of April 2011, Matt Holiday had the highest batting average in the major leagues. At that time, he had 39 hits in 94 at bats, for a batting average of .415. This average is found by dividing the number of hits by the number of times he was at bat and then rounding to the nearest thousandth.

$$\text{Batting average} = \dfrac{\text{number of hits}}{\text{number of times at bat}} = \dfrac{39}{94} = 0.415$$

Because we can write any decimal number as a percent, we can convert batting averages to percents and use our knowledge of percent to solve problems. Looking at Matt Holiday's batting average as a percent, we can expect him to get a hit 41.5% of the times he is at bat.

Each of the following problems can be solved by converting batting averages to percents and translating the problem into one of our three basic percent problems. (All numbers are from the end of April 2011, according to espn.com.)

47. Jose Bautista had the highest batting average in the American League with 30 hits in 84 at bats. What percent of the time Bautista is at bat can we expect him to get a hit?

48. Andre Ethier had 44 hits in 119 at bats. What percent of the time can we expect Ethier to get a hit?

49. Adrian Gonzales was batting .314. If his batting average remains the same and he has 591 at bats in the 2011 season, how many hits will he have? Round to the nearest hit.

50. Miguel Cabrera was batting .342. If his batting average remains the same and he has 548 at bats in the 2011 season, how many hits will he have? Round to the nearest hit.

51. How many hits must Jose Bautista have in his next 50 times at bat to maintain a batting average of at least .357?

52. How many hits must Matt Holiday have in his next 50 at bats to maintain a batting average of at least .415?

5.4 Sales Tax and Commission

OBJECTIVES

A Solve application problems involving sales tax.

B Solve application problems involving commission.

TICKET TO SUCCESS

Keep these questions in mind as you read through the section. Then respond in your own words and in complete sentences.

1. How is the sales tax rate different from the sales tax?
2. Rework Example 1 using a sales tax rate of 7% instead of 6%.
3. How would you calculate a salesperson's commission based on a 4% commission rate?
4. Suppose the car salesperson in Example 5 receives a commission of $3,672. Assuming the same commission rate of 12%, how much did this car sell for?

Suppose you have decided to buy a new laptop that costs $549.99. Most likely, your total bill will be greater than the advertised purchase price for the computer because of the sales tax. If you purchase this computer in San Luis Obispo, CA where the sales tax rate is 8.25%, you would need to pay $45.37 in sales tax for a total of $595.36. In this section, we will work similar problems that involve sales tax. We will also solve application problems that involve commission.

A Sales Tax

Let's begin by calculating sales tax based on a purchase price and a given sales tax rate. We will solve these problems by first restating them in terms of the problems we have already learned how to solve.

EXAMPLE 1 Suppose the sales tax rate in Mississippi is 6% of the purchase price. If the price of a refrigerator is $550, how much sales tax must be paid?

SOLUTION Because the sales tax is 6% of the purchase price, and the purchase price is $550, the problem can be restated as

What is 6% of $550?

PRACTICE PROBLEMS

1. What is the sales tax on a new washing machine if the machine is purchased for $625 and the sales tax rate is 6%?

Answer
1. $37.50

NOTE

In Example 1, the sales tax rate is 6%, and the sales tax is $33. In most everyday communications, people say "The sales tax is 6%," which is incorrect. The 6% is the tax *rate*, and the $33 is the actual tax.

2. Suppose the sales tax rate is 3%. If the sales tax on a 10-speed bicycle is $4.35, what is the purchase price, and what is the total price of the bicycle?

3. Suppose the purchase price of two speakers is $197 and the sales tax is $11.82. What is the sales tax rate?

Answers
2. $145; $149.35
3. 6%

We solve this problem, as we did in Section 5.2, by translating it into an equation.

What is 6% of $550?
$$n = 0.06 \cdot 550$$
$$n = 33$$

The sales tax is $33. The total price of the refrigerator would be

$$\underset{\$550}{\text{Purchase price}} + \underset{\$33}{\text{Sales tax}} = \underset{\$583}{\text{Total price}}$$

∎

EXAMPLE 2 Suppose the sales tax rate is 4%. If the sales tax on a 10-speed bicycle is $5.44, what is the purchase price, and what is the total price of the bicycle?

SOLUTION We know that 4% of the purchase price is $5.44. We find the purchase price first by restating the problem as

$5.44 is 4% of what number?
$$5.44 = 0.04 \cdot n$$

We solve the equation by dividing both sides by 0.04.

$$\frac{5.44}{0.04} = \frac{0.04 \cdot n}{0.04} \quad \text{Divide both sides by 0.04.}$$

$$n = \frac{5.44}{0.04} \quad \text{Switch the left and right sides of the equation.}$$

$$n = 136 \quad \text{Divide.}$$

The purchase price is $136. The total price is the sum of the purchase price and the sales tax.

$$\begin{aligned}\text{Purchase price} &= \$136.00 \\ + \text{ Sales tax} &= \$5.44 \\ \hline \text{Total price} &= \$141.44\end{aligned}$$

∎

EXAMPLE 3 Suppose the purchase price of a stereo system is $396 and the sales tax is $19.80. What is the sales tax rate?

SOLUTION We restate the problem as

$19.80 is what percent of $396?
$$19.80 = n \cdot 396$$

To solve this equation, we divide both sides by 396.

$$\frac{19.80}{396} = \frac{n \cdot 396}{396} \quad \text{Divide both sides by 396.}$$

$$n = \frac{19.80}{396} \quad \text{Switch the left and right sides of the equation.}$$

$$n = 0.05 \quad \text{Divide.}$$
$$n = 5\% \quad 0.05 = 5\%$$

The sales tax rate is 5%.

∎

B Commission

Many salespeople work on a *commission* basis. That is, their earnings are a percentage of the amount they sell. The *commission rate* is a percent, and the actual commission they receive is a dollar amount.

EXAMPLE 4 A real estate agent gets 3% of the price of each house she sells. If she sells a house for $289,500, how much money does she earn?

SOLUTION The commission is 3% of the price of the house, which is $289,500. We restate the problem as

What is 3% of $289,500?

$$n = 0.03 \cdot 289{,}500$$
$$n = 8{,}685$$

The commission is $8,685.

4. A real estate agent gets 3% of the price of each house she sells. If she sells a house for $115,000, how much money does she earn?

EXAMPLE 5 Suppose a car salesperson's commission rate is 12%. If the commission on one of the cars is $1,836, what is the purchase price of the car?

SOLUTION 12% of the sales price is $1,836. The problem can be restated as

12% of what number is $1,836?

$$0.12 \cdot n = 1{,}836$$
$$\frac{0.12 \cdot n}{0.12} = \frac{1{,}836}{0.12} \quad \text{Divide both sides by 0.12.}$$
$$n = 15{,}300$$

The car sells for $15,300.

5. An appliance salesperson's commission rate is 10%. If the commission on one of the ovens is $115, what is the purchase price of the oven?

EXAMPLE 6 If the commission on a $600 dining room set is $90, what is the commission rate?

SOLUTION The commission rate is a percentage of the selling price. What we want to know is

$90 is what percent of $600?

$$90 = n \cdot 600$$
$$\frac{90}{600} = \frac{n \cdot 600}{600} \quad \text{Divide both sides by 600.}$$
$$n = \frac{90}{600} \quad \text{Switch the left and right sides of the equation.}$$
$$n = 0.15 \quad \text{Divide.}$$
$$n = 15\% \quad \text{Change to a percent.}$$

The commission rate is 15%.

6. If the commission on a $750 sofa is $105, what is the commission rate?

Answers
4. $3,450
5. $1,150
6. 14%

Problem Set 5.4

Moving Toward Success

"Real leaders are ordinary people with extraordinary determination."

—Unknown

1. Should you ask questions in class? Why or why not?
2. What do you plan to do if you have a question about a problem when you are outside of class?

A These problems should be solved by the method shown in this section. In each case show the equation needed to solve the problem. Write neatly, and show your work. [Examples 1–3]

1. **Sales Tax** Suppose the sales tax rate in Mississippi is 7% of the purchase price. If a new food processor sells for $750, how much is the sales tax?

2. **Sales Tax** If the sales tax rate is 5% of the purchase price, how much sales tax is paid on a television that sells for $980?

3. **Sales Tax and Purchase Price** Suppose the sales tax rate in Michigan is 6%. How much is the sales tax on a $45 concert ticket? What is the total price?

4. **Sales Tax and Purchase Price** Suppose the sales tax rate in Hawaii is 4%. How much tax is charged on a new car if the purchase price is $16,400? What is the total price?

5. **Total Price** The sales tax rate is 4%. If the sales tax on a 10-speed bicycle is $6, what is the purchase price? What is the total price?

6. **Total Price** The sales tax on a new microwave oven is $30. If the sales tax rate is 5%, what is the purchase price? What is the total price?

7. **Tax Rate** Suppose the purchase price of a dining room set is $450. If the sales tax is $22.50, what is the sales tax rate?

    ```
    ⊛FURNITURE PLUS⊛
    ----------------
         RECEIPT
    ----------------
    Dining Room Set   $450.00

    Tax Rate            ? %
    Tax               $22.50
    TOTAL            $472.50

    Receipt #1007  07/18/08  4:15PM
    ```

8. **Tax Rate** If the purchase price of a bottle of California wine is $24 and the sales tax is $1.50, what is the sales tax rate?

    ```
    ⊛ OAKS WINERY
    ----------------
         RECEIPT
    ----------------
    California wine   $24.00

    Tax Rate            ? %
    Tax                $1.50
    TOTAL             $25.50

    Receipt #128  07/30/08  2:32PM
    ```

9. **Energy** The chart shows the cost to install either solar panels or a wind turbine. A farmer is installing the equipment to generate energy from the wind. If he lives in a state that has a 6% sales tax rate, how much did the farmer pay in sales tax on the total equipment cost?

10. **Energy** Refer to the chart from the previous problem. If the sum of the tax on all items a farmer bought was $301.13 and the sales tax rate in his area is 7.5%, what was the price of the equipment before tax?

Solar Versus Wind Energy Costs

Solar Energy Equipment Cost:		Wind Energy Equipment Cost:	
Modules	$6200	Turbine	$3300
Fixed Rack	$1570	Tower	$3000
Charge Controller	$971	Cable	$715
Cable	$440		
TOTAL	**$9181**	**TOTAL**	**$7015**

Source: a Limited

B [Examples 4–6]

11. Commission A real estate agent has a commission rate of 3%. If a piece of property sells for $94,000, what is her commission?

12. Commission A tire salesperson has a 12% commission rate. If he sells a set of radial tires for $400, what is his commission?

13. Commission and Purchase Price Suppose a salesperson gets a commission rate of 12% on the lawnmowers she sells. If the commission on one of the mowers is $24, what is the purchase price of the lawnmower?

14. Commission and Purchase Price If an appliance salesperson gets 9% commission on all the appliances she sells, what is the price of a refrigerator if her commission is $67.50?

15. Commission Rate If the commission on an $800 washer is $112, what is the commission rate?

16. Commission Rate A realtor makes a commission of $11,400 on a $190,000 house he sells. What is his commission rate?

17. Phone Bill You recently received your monthly phone bill for service in your local area. The total of the bill was $53.35. You pay $14.36 in surcharges and federal and local taxes. What percent of your phone bill is made up of surcharges and taxes? Round your answer to the nearest tenth of a percent.

18. Wireless Phone Bill You recently received your Verizon wireless phone bill for the month. The total monthly bill is $70.52. Included in that total is $13.27 in surcharges and taxes. What percent of your wireless bill goes towards surcharges and taxes? Round your answer to the nearest tenth of a percent.

19. Gasoline Tax New York state has one of the highest gasoline taxes in the country. If gas is currently selling at $4.13 for a gallon of regular gas and the tax rate is 11.5%, how much of the price of a gallon of gas goes towards taxes?

20. Cigarette Tax In an effort to encourage people to quit smoking, many states place a high tax on a pack of cigarettes. Seventeen states place a tax of $2.00 or more on a pack of cigarettes, with New York being the highest at $4.35 per pack. If this is 42.9% of the cost of a pack of cigarettes in New York, how much does a pack of cigarettes cost before taxes?

21. Salary Plus Commission A computer salesperson earns a salary of $425 a week and a 6% commission on all sales over $4000 each week. Suppose she was able to sell $6,250 in computer parts and accessories one week. What was her salary for the week?

22. Salary Plus Bonus The manager for a computer store is paid a weekly salary of $650 plus a bonus amounting to 1.5% of the net earnings of the store each week. Find her total salary for the week when earnings for the store are $26,875.56. Round your answer to the nearest cent.

Calculator Problems

The following problems are similar to Problems 1–22. Set them up in the same way, but use a calculator for the calculations.

23. Sales Tax The sales tax rate on a certain item is 5.5%. If the purchase price is $216.95, how much is the sales tax? (Round to the nearest cent.)

24. Purchase Price If the sales tax rate is 4.75% and the sales tax is $18.95, what is the purchase price? What is the total price? (Both answers should be rounded to the nearest cent.)

25. Tax Rate The purchase price for a new suit is $229.50. If the sales tax is $10.33, what is the sales tax rate? (Round to the nearest tenth of a percent.)

26. Commission If the commission rate for a mobile home salesperson is 11%, what is the commission on the sale of a $15,794 mobile home?

27. Selling Price Suppose the commission rate on the sale of used cars is 13%. If the commission on one of the cars is $519.35, what did the car sell for?

28. Commission Rate If the commission on the sale of $79.40 worth of clothes is $14.29, what is the commission rate? (Round to the nearest percent.)

Getting Ready for the Next Section

Multiply.

29. 0.05(22,000)

30. 0.176(1,793,000)

31. 0.25(300)

32. 0.12(450)

Divide. Write your answers as decimals.

33. 4 ÷ 25

34. 7 ÷ 35

Subtract.

35. 25 − 21

36. 1,793,000 − 315,568

37. 450 − 54

38. 300 − 75

Add.

39. 396 + 19.8

40. 22,000 + 1,100

Maintaining Your Skills

The problems below review some basic concepts of division with fractions and mixed numbers.

Divide.

41. $\dfrac{1}{3} \div \dfrac{2}{3}$

42. $\dfrac{2}{3} \div \dfrac{1}{3}$

43. $2 \div \dfrac{3}{4}$

44. $3 \div \dfrac{1}{2}$

45. $\dfrac{3}{8} \div \dfrac{1}{4}$

46. $\dfrac{5}{9} \div \dfrac{2}{3}$

47. $2\dfrac{1}{4} \div \dfrac{1}{2}$

48. $1\dfrac{1}{4} \div 2\dfrac{1}{2}$

Percent Increase or Decrease, and Discount

5.5

OBJECTIVES

A Find the percent increase.

B Find the percent decrease.

C Solve application problems involving the rate of discount.

TICKET TO SUCCESS

Keep these questions in mind as you read through the section. Then respond in your own words and in complete sentences.

1. What is percent increase?
2. Write an application problem where you had to find the percent decrease of some specific statistics.
3. Suppose the shoes mentioned in Example 3 were on sale for $20, instead of $21. Calculate the new percent decrease in price.
4. What is discount rate?

Insurance companies gather statistics about stopping distances of cars. The following table and bar chart show some statistics for cars with different tires traveling 20 miles per hour on ice. The percent decrease column on the table shows how stopping distances of a car with special tires relate to the stopping distance of a car with regular tires.

	Stopping Distance	Percent Decrease
Regular tires	150 ft	0
Snow tires	151 ft	−1%
Studded snow tires	120 ft	20%
Reinforced tire chain	75 ft	50%

Source: Copyrighted table courtesy of The Casualty Adjuster's Guide

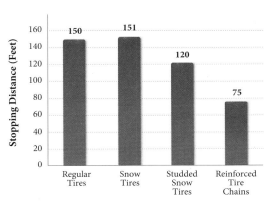

In this section, we will work problems involving percent increases or decreases, such as those for the snow tires. Many times it is more effective to state increases or decreases as percents, rather than the actual number, because with percent we are comparing everything to 100.

A Percent Increase

Let's begin with the following example that involves a percent increase.

EXAMPLE 1   If a person earns $22,000 a year and gets a 5% increase in salary, what is the new salary?

SOLUTION   We can find the dollar amount of the salary increase by finding 5% of $22,000.

$$0.05 \times 22{,}000 = 1{,}100$$

The increase in salary is $1,100. The new salary is the old salary plus the raise.

	$22,000	Old salary
+	1,100	Raise (5% of $22,000)
	$23,100	New salary

B Percent Decrease

Now let's work a couple of problems that involve percent decrease.

EXAMPLE 2   In 1997, there were approximately 1,477,000 arrests for driving under the influence of alcohol or drugs (DUI) in the United States. By 2007, the number of arrests for DUI had decreased 3.4% from the 1997 number. How many people were arrested for DUI in 2007? Round the answer to the nearest thousand.

SOLUTION   The decrease in the number of arrests is 3.4% of 1,477,000, or

$$0.034 \times 1{,}477{,}000 = 50{,}218$$

Subtracting this number from 1,477,000, we have the number of DUI arrests in 2007.

	1,477,000	Number of arrests in 1997
−	50,218	Decrease of 3.4%
	1,426,782	Number of arrests in 2007

To the nearest thousand, there were approximately 1,427,000 arrests for DUI in 2007.

EXAMPLE 3   Shoes that usually sell for $25 are on sale for $21. What is the percent decrease in price?

SOLUTION   We must first find the decrease in price. Subtracting the sale price from the original price, we have

$$\$25 - \$21 = \$4$$

The decrease is $4. To find the percent decrease (from the original price), we have

4 is what percent of 25?

$$4 = n \cdot 25$$

$$\frac{4}{25} = \frac{n \cdot 25}{25} \quad \text{Divide both sides by 25.}$$

$$n = \frac{4}{25} \quad \text{Switch the left and right sides of the equation.}$$

$$n = 0.16 \quad \text{Divide.}$$

$$n = 16\% \quad \text{Change to a percent.}$$

PRACTICE PROBLEMS

1. A person earning $18,000 a year gets a 7% increase in salary. What is the new salary?

2. In 1997, there were approximately 11,180 fatal DUI accidents. By 2007, that number had decreased by 4.4%. How many fatal DUI accidents were there in 2007? Round to the nearest hundred.

3. Shoes that usually sell for $35 are on sale for $28. What is the percent decrease in price?

Answers
1. $19,260
2. 10,700
3. 20%

The shoes that sold for $25 have been reduced by 16% to $21. In a problem like this, $25 is the *original* (or *marked*) price, $21 is the *sale price*, $4 is the *discount*, and 16% is the *rate of discount*.

C Discount Rate

In Example 3, $4 was the discount amount for the shoes on sale. Now we will work a couple of examples that deal directly with discount.

EXAMPLE 4   During a clearance sale, a suit that usually sells for $300 is marked "25% off." What is the discount? What is the sale price?

SOLUTION   To find the discount, we restate the problem as

$$\text{What is } 25\% \text{ of } 300?$$
$$n = 0.25 \cdot 300$$
$$n = 75$$

The discount is $75. The sale price is the original price less the discount.

$$\begin{array}{rl} \$300 & \text{Original price} \\ -\ 75 & \text{Less the discount (25\% of \$300)} \\ \hline \$225 & \text{Sale price} \end{array}$$

EXAMPLE 5   A man buys a washing machine on sale. The machine usually sells for $450, but it is on sale at 12% off. If the sales tax rate is 5%, how much is the total bill for the washer?

SOLUTION   First we have to find the sale price of the washing machine, and we begin by finding the discount.

$$\text{What is } 12\% \text{ of } \$450?$$
$$n = 0.12 \cdot 450$$
$$n = 54$$

The washing machine is marked down $54. The sale price is

$$\begin{array}{rl} \$450 & \text{Original price} \\ -\ 54 & \text{Discount (12\% of \$450)} \\ \hline \$396 & \text{Sale price} \end{array}$$

Because the sales tax rate is 5%, we find the sales tax as follows:

$$\text{What is } 5\% \text{ of } 396?$$
$$n = 0.05 \cdot 396$$
$$n = 19.80$$

The sales tax is $19.80. The total price the man pays for the washing machine is

$$\begin{array}{rl} \$396.00 & \text{Sale price} \\ +\ 19.80 & \text{Sales tax} \\ \hline \$415.80 & \text{Total price} \end{array}$$

---

4. During a sale, a microwave oven that usually sells for $550 is marked "15% off." What is the discount? What is the sale price?

5. A woman buys a new coat on sale. The coat usually sells for $45, but it is on sale at 15% off. If the sales tax rate is 5%, how much is the total bill for the coat?

Answers
4. $82.50; $467.50
5. $40.16

Problem Set 5.5

Moving Toward Success

"Study is the bane of childhood, the oil of youth, the indulgence of adulthood, and a restorative in old age."
— Walter Savage Landor, 1775–1864, English writer and poet

1. Before reading a section from the first to the last page, scan it briefly to review headers, graphics, and italicized or bold words. Why should you do this before reading the section?
2. Why should you pay attention to the italicized or bold words, and the information in colored boxes?

A **B** Solve each of these problems using the method developed in this section. [Examples 1–3]

1. **Salary Increase** If a person earns $23,000 a year and gets a 7% increase in salary, what is the new salary?

2. **DUI Accidents** In 1995, there were approximately 11,990 fatal DUI accidents. By 2008 that number had decreased by 10.9%. How many fatal DUI accidents were there in 2008? Round to the nearest hundred.

3. **Tuition Increase** The yearly tuition at a college is presently $6,000. Next year it is expected to increase by 17%. What will the tuition at this school be next year?

4. **Price Increase** A market increased the price of cheese selling for $4.98 per pound by 3%. What is the new price for a pound of cheese? (Round to the nearest cent.)

5. **Car Value** In one year a new car decreased in value by 20%. If it sold for $16,500 when it was new, what was it worth after 1 year?

6. **Calorie Content** A certain light beer has 20% fewer calories than the regular beer. If the regular beer has 120 calories per bottle, how many calories are in the same-sized bottle of the light beer?

7. **Salary Increase** A person earning $3,500 a month gets a raise of $350 per month. What is the percent increase in salary?

8. **Rate Increase** A student reader is making $6.50 per hour and gets a $0.70 raise. What is the percent increase? (Round to the nearest tenth of a percent.)

9. **Shoe Sale** Shoes that usually sell for $25 are on sale for $20. What is the percent decrease in price?

10. **Enrollment Decrease** The enrollment in a certain elementary school was 410 in 2007. In 2008, the enrollment in the same school was 328. Find the percent decrease in enrollment from 2007 to 2008.

11. **Golf Courses** The chart shows the lengths of some of the longest courses in U.S. Open History. What is the percent decrease from Torrey Pines, South Course to Pinehurst, No. 2 Course? Round to the nearest tenth of a percent.

12. **Health Care** The graph shows the rising cost of health care. What is the projected percent increase in health care costs from 2002 to 2014?

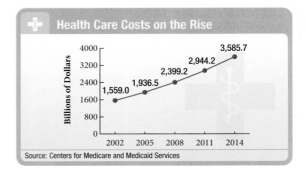

C [Examples 4, 5]

13. **Discount** During a clearance sale, a three-piece suit that usually sells for $300 is marked "15% off." What is the discount? What is the sale price?

14. **Sale Price** On opening day, a new music store offers a 12% discount on all electric guitars. If the regular price on a guitar is $550, what is the sale price?

15. **Stopping Distance** The graph below is from the section opener. Use the information to calculate the percent decrease in stopping distance between a car with snow tires versus a car with reinforced tire chains. Round to the nearest percent.

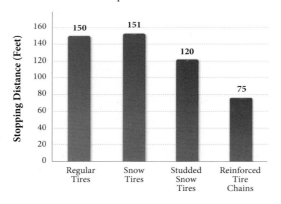

16. **Total Price** A bedroom set that normally sells for $1,450 is on sale for 10% off. If the sales tax rate is 5%, what is the total price of the bedroom set if it is bought while on sale?

17. **Real Estate Market** In 2006 the average price of a home began to fall in most real estate markets across the country. The median price of a single family home in the U.S. was $227,000 in 2006. Suppose the median price is now $195,500. By what percent did the median price of a single family home drop? Round your answer to the nearest tenth of a percent.

18. **Deep Discount** When buying some of today's newest electronic gadgets, good things come to those who wait. When Apple released its new iPhone in the summer of 2007, a 4GB model sold for $499. In June 2010, Apple released its new iPhone 4. The 16GB model sold for $199. What is the percent decrease in price for this new model? Round your answer to the nearest tenth of a percent.

19. **Losing Weight** According to the Centers for Disease Control and Prevention (CDC), more than 60% of U.S. adults are overweight, and about 15% of children and adolescents ages 6 to 19 are overweight. Your friend decides to go on a diet and goes from 155 pounds to 130 pounds over a 4 month period. What was her percentage weight loss? Round your answer to the nearest percent.

20. **Ordering Online** You are in the market for a new laptop. The model that you wish to purchase is $1,500 in a local store. However, you decide to buy the computer over the Internet for $1200, plus shipping charges. Taking into account shipping charges of $59, what percentage do you save by ordering it online? Round your answer to the nearest tenth of a percent.

21. **Product Error** When manufacturing a product, a certain amount of variation (or error) can occur in the process and still create a part or product that is useable. For one particular company, a 3% error is acceptable for their machine parts to be used safely. If the part they are manufacturing is 22.5 in. long, what is the range of measures that are acceptable for this part? Round your result to the nearest tenth of an inch.

22. **Home Remodeling** You have decided to update your house by laying a new wood floor in your living room. Your floor has an area of 440 sq ft. You decide to buy enough flooring to allow for a certain amount of waste so you purchase 470 sq ft of wood flooring materials. Express your waste allowance as a percent of the amount purchased. Round your result to the nearest tenth of a percent.

Calculator Problems

Set up the following problems the same way you set up Problems 1–22. Then use a calculator to do the calculations.

23. **Salary Increase** A teacher making $43,752 per year gets a 6.5% raise. What is the new salary?

24. **Utility Increase** A homeowner had a $95.90 electric bill in December. In January the bill was $107.40. Find the percent increase in the electric bill from December to January. (Round to the nearest whole number.)

25. **Soccer** The rules for soccer state that the playing field must be from 100 to 120 yards long and 55 to 75 yards wide. The 1999 Women's World Cup was played at the Rose Bowl on a playing field 116 yards long and 72 yards wide. The diagram below shows the smallest possible soccer field, the largest possible soccer field, and the soccer field at the Rose Bowl.

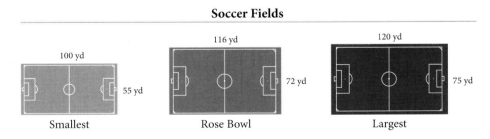

Soccer Fields

a. **Percent Increase** A team plays on the smallest field, then plays in the Rose Bowl. What is the percent increase in the area of the playing field from the smallest field to the Rose Bowl? Round to the nearest tenth of a percent.

b. **Percent Increase** A team plays a soccer game in the Rose Bowl. The next game is on a field with the largest dimensions. What is the percent increase in the area of the playing field from the Rose Bowl to the largest field? Round to the nearest tenth of a percent.

26. **Football** The diagrams below show the dimensions of playing fields for the National Football League (NFL), the Canadian Football League (CFL), and Arena Football.

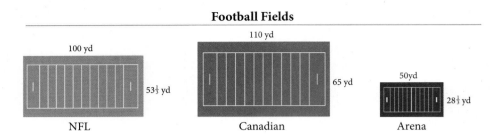

Football Fields

NFL: 100 yd × 53⅓ yd
Canadian: 110 yd × 65 yd
Arena: 50 yd × 28⅓ yd

a. **Percent Increase** Kurt Warner made a successful transition from Arena Football to the NFL, winning the Most Valuable Player award. What was the percent increase in the area of the fields he played on in moving from Arena Football to the NFL? Round to the nearest percent.

b. **Percent Decrease** Jeff Garcia played in the Canadian Football League before moving to the NFL. What was the percent decrease in the area of the fields he played on in moving from the CFL to the NFL? Round to the nearest tenth of a percent.

Getting Ready for the Next Section

Multiply. Round to nearest hundredth if necessary.

27. $0.07(2,000)$

28. $0.12(8,000)$

29. $600(0.04)\left(\dfrac{1}{6}\right)$

30. $900(0.06)\left(\dfrac{1}{4}\right)$

31. $10,150(0.06)\left(\dfrac{1}{4}\right)$

32. $10,302.25(0.06)\left(\dfrac{1}{4}\right)$

Add.

33. $3,210 + 224.7$

34. $900 + 13.50$

35. $10,000 + 150$

36. $10,150 + 152.25$

37. $10,302.25 + 154.53$

38. $10,456.78 + 156.85$

Simplify.

39. $2,000 + 0.07(2,000)$

40. $8,000 + 0.12(8,000)$

41. $3,000 + 0.07(3,000)$

42. $9,000 + 0.12(9,000)$

Maintaining Your Skills

The problems below review some basic concepts of addition of fractions and mixed numbers.
Add each of the following and reduce all answers to lowest terms.

43. $\dfrac{1}{3} + \dfrac{2}{3}$

44. $\dfrac{3}{8} + \dfrac{1}{8}$

45. $\dfrac{1}{2} + \dfrac{1}{4}$

46. $\dfrac{1}{5} + \dfrac{3}{10}$

47. $\dfrac{3}{4} + \dfrac{2}{3}$

48. $\dfrac{3}{8} + \dfrac{1}{6}$

49. $2\dfrac{1}{2} + 3\dfrac{1}{2}$

50. $3\dfrac{1}{4} + 2\dfrac{1}{8}$

Interest

5.6

OBJECTIVES

A Solve application problems involving annual interest.

B Solve application problems involving simple interest.

C Solve compound interest problems.

TICKET TO SUCCESS

Keep these questions in mind as you read through the section. Then respond in your own words and in complete sentences.

1. Explain the difference between interest and interest rate.
2. Define principal.
3. Define simple interest and give the formula used to calculate it.
4. Briefly explain and give an example of compound interest.

A Interest

Anyone who has borrowed money from a bank or other lending institution, or who has invested money in a savings account, is aware of *interest*. Interest is the amount of money paid for the use of money. If we put $500 in a savings account that pays 6% annually, the interest will be 6% of $500, or 0.06(500) = $30. The amount we invest ($500) is called the *principal*, the percent (6%) is the *interest rate*, and the money earned ($30) is the *interest*.

EXAMPLE 1 A man invests $2,000 in a savings plan that pays 7% per year. How much money will be in the account at the end of 1 year?

SOLUTION We first find the interest by taking 7% of the principal, $2,000.

$$\text{Interest} = 0.07(\$2,000)$$
$$= \$140$$

The interest earned in 1 year is $140. The total amount of money in the account at the end of a year is the original amount plus the $140 interest.

$$\begin{array}{rl} \$2,000 & \text{Original investment (principal)} \\ +\ \ \ 140 & \text{Interest (7\% of \$2,000)} \\ \hline \$2,140 & \text{Amount after 1 year} \end{array}$$

The amount in the account after 1 year is $2,140. ■

PRACTICE PROBLEMS

1. A man invests $3,000 in a savings plan that pays 8% per year. How much money will be in the account at the end of 1 year?

Answer
1. $3,240

5.6 Interest

379

2. If a woman borrows $7,500 from her local bank at 12% interest, how much does she pay back to the bank if she pays off the loan in 1 year?

EXAMPLE 2 A farmer borrows $8,000 from his local bank at 12%. How much does he pay back to the bank at the end of the year to pay off the loan?

SOLUTION The interest he pays on the $8,000 is

$$\text{Interest} = 0.12(\$8,000)$$
$$= \$960$$

At the end of the year, he must pay back the original amount he borrowed ($8,000) plus the interest at 12%.

	$8,000	Amount borrowed (principal)
+	960	Interest at 12%
	$8,960	Total amount to pay back

The total amount that the farmer pays back is $8,960. ∎

3. Suppose you carry an average daily balance of $5,243.56 on the credit card from Example 3. What will your finance charge be?

EXAMPLE 3 Suppose you carry an average daily balance of $1,523.65 on your credit card. If your credit card company charges 2.4% in finance charges per month based on the average daily balance, what will be your finance charge for the month?

SOLUTION Your finance charge based on your average daily balance will be

$$\text{Finance charge} = 0.024(1523.65)$$
$$= 36.57$$

You will pay $36.57 in finance charges based on your average daily balance of $1523.65. ∎

B Simple Interest

There are many situations in which interest on a loan is figured on other than a yearly basis. Many short-term loans are for only 30 or 60 days. In these cases we can use a formula to calculate the interest that has accumulated. This type of interest is called *simple interest*. The formula is

$$I = P \cdot R \cdot T$$

where

I = Interest
P = Principal
R = Interest rate (this is the percent)
T = Time (in years, 1 year = 360 days)

We could have used this formula to find the interest in Examples 1 and 2. In those two cases, T is 1. When the length of time is in days rather than years, it is common practice to use 360 days for 1 year, and we write T as a fraction. Examples 4 and 5 illustrate this procedure.

4. Another student takes out a loan like the one in Example 4. This loan is for $700 at 4%. How much interest does this student pay if the loan is paid back in 90 days?

EXAMPLE 4 A student takes out an emergency loan for tuition, books, and supplies. The loan is for $600 at an interest rate of 4%. How much interest does the student pay if the loan is paid back in 60 days?

Answers
2. $8,400
3. $125.85
4. $7

SOLUTION The principal P is \$600, the rate R is 4% = 0.04, and the time T is $\frac{60}{360}$. Notice that T must be given in years, and 60 days = $\frac{60}{360}$ year. Applying the formula, we have

$$I = P \cdot R \cdot T$$

$$I = 600 \cdot 0.04 \cdot \frac{60}{360}$$

$$I = 600 \cdot 0.04 \cdot \frac{1}{6} \qquad \frac{60}{360} = \frac{1}{6}$$

$$I = 4 \qquad \text{Multiply.}$$

The interest is \$4. ■

EXAMPLE 5 A woman deposits \$900 in an account that pays 6% annually. If she withdraws all the money in the account after 90 days, how much does she withdraw?

SOLUTION We have $P = \$900$, $R = 0.06$, and $T = 90$ days = $\frac{90}{360}$ year. Using these numbers in the formula, we have

$$I = P \cdot R \cdot T$$

$$I = 900 \times 0.06 \times \frac{90}{360}$$

$$I = 900 \times 0.06 \times \frac{1}{4} \qquad \frac{90}{360} = \frac{1}{4}$$

$$I = 13.5 \qquad \text{Multiply.}$$

The interest earned in 90 days is \$13.50. If the woman withdraws all the money in her account, she will withdraw

$$\begin{array}{rl} \$900.00 & \text{Original amount (principal)} \\ + \quad 13.50 & \text{Interest for 90 days} \\ \hline \$913.50 & \text{Total amount withdrawn} \end{array}$$

The woman will withdraw \$913.50. ■

5. Suppose \$1,200 is deposited in an account that pays 9.5% interest per year. If all the money is withdrawn after 120 days, how much money is withdrawn?

C Compound Interest

A second common kind of interest is *compound interest*. Compound interest includes interest paid on interest. We can use what we know about simple interest to help us solve problems involving compound interest.

EXAMPLE 6 A homemaker puts \$3,000 into a savings account that pays 7% compounded annually. How much money is in the account at the end of 2 years?

SOLUTION Because the account pays 7% annually, the simple interest at the end of 1 year is 7% of \$3,000.

$$\text{Interest after 1 year} = 0.07(\$3,000)$$
$$= \$210$$

Because the interest is paid annually, at the end of 1 year the total amount of money in the account is

$$\begin{array}{rl} \$3,000 & \text{Original amount} \\ + \quad 210 & \text{Interest for 1 year} \\ \hline \$3,210 & \text{Total in account after 1 year} \end{array}$$

6. If \$5,000 is put into an account that pays 6% compounded annually, how much money is in the account at the end of 2 years?

NOTE
If the interest earned in Example 6 were calculated using the formula for simple interest, $I = P \cdot R \cdot T$, the amount of money in the account at the end of two years would be \$3,420.00.

Answers
5. \$1,238
6. \$5,618

Chapter 5 Percent

The interest paid for the second year is 7% of this new total, or

$$\text{Interest paid the second year} = 0.07(\$3{,}210)$$
$$= \$224.70$$

At the end of 2 years, the total in the account is

$$\begin{array}{rl} \$3{,}210.00 & \text{Amount at the beginning of year 2} \\ +\ \ \ 224.70 & \text{Interest paid for year 2} \\ \hline \$3{,}434.70 & \text{Account after 2 years} \end{array}$$

At the end of 2 years, the account totals $3,434.70. The total interest earned during this 2-year period is $210 (first year) + $224.70 (second year) = $434.70. ∎

You may have heard of savings and loan companies that offer interest rates that are compounded quarterly. If the interest rate is, say, 6% and it is compounded quarterly, then after every 90 days ($\frac{1}{4}$ of a year) the interest is added to the account. If it is compounded semiannually, then the interest is added to the account every 6 months. Most accounts have interest rates that are compounded daily, which means the simple interest is computed daily and added to the account.

EXAMPLE 7 If $10,000 is invested in a savings account that pays 6% compounded quarterly, how much is in the account at the end of a year?

SOLUTION The interest for the first quarter ($\frac{1}{4}$ of a year) is calculated using the formula for simple interest.

$$I = P \cdot R \cdot T$$

$$I = \$10{,}000 \times 0.06 \times \frac{1}{4} \qquad \text{First quarter}$$

$$I = \$150$$

At the end of the first quarter, this interest is added to the original principal. The new principal is $10,000 + $150 = $10,150. Again, we apply the formula to calculate the interest for the second quarter.

$$I = \$10{,}150 \times 0.06 \times \frac{1}{4} \qquad \text{Second quarter}$$

$$I = \$152.25$$

The principal at the end of the second quarter is $10,150 + $152.25 = $10,302.25. The interest earned during the third quarter is

$$I = \$10{,}302.25 \times 0.06 \times \frac{1}{4} \qquad \text{Third quarter}$$

$$I = \$154.53 \qquad \text{Round to the nearest cent.}$$

The new principal is $10,302.25 + $154.53 = $10,456.78. Interest for the fourth quarter is

$$I = \$10{,}456.78 \times 0.06 \times \frac{1}{4} \qquad \text{Fourth quarter}$$

$$I = \$156.85 \qquad \text{Round to the nearest cent.}$$

The total amount of money in this account at the end of 1 year is

$$\$10{,}456.78 + \$156.85 = \$10{,}613.63 \qquad \blacksquare$$

---

7. If $20,000 is invested in an account that pays 8% compounded quarterly, how much is in the account at the end of a year?

Answer
7. $21,648.64

Problem Set 5.6

Moving Toward Success

"There is no such thing as a self-made man. You will reach your goals only with the help of others."

—George Shinn, 1941–present, Former owner of the Charlotte Hornets basketball team

1. Do you use a partner or group to study for this class? Why or why not?
2. How would a study group provide emotional support during this class?

A These problems are similar to the examples found in this section. They should be set up and solved in the same way. (Problems 1–12 involve simple interest.) [Examples 1–5]

1. **Savings Account** A man invests $2,000 in a savings plan that pays 8% per year. How much money will be in the account at the end of 1 year?

2. **Savings Account** How much simple interest is earned on $5,000 if it is invested for 1 year at 5%?

3. **Savings Account** A savings account pays 7% per year. How much interest will $9,500 invested in such an account earn in a year?

4. **Savings Account** A local bank pays 5.5% annual interest on all savings accounts. If $600 is invested in this type of account, how much will be in the account at the end of a year?

5. **Bank Loan** A farmer borrows $8,000 from his local bank at 7%. How much does he pay back to the bank at the end of the year when he pays off the loan?

6. **Bank Loan** If $400 is borrowed at a rate of 12% for 1 year, how much is the interest?

7. **Bank Loan** A bank lends one of its customers $3,000 at 10% for 1 year. If the customer pays the loan back at the end of the year, how much does he pay the bank?

8. **Bank Loan** If a loan of $2,000 at 20% for 1 year is to be paid back in one payment at the end of the year, how much does the borrower pay the bank?

9. **Student Loan** A student takes out an emergency loan for tuition, books, and supplies. The loan is for $600 with an annual interest rate of 5%. How much interest does the student pay if the loan is paid back in 60 days?

10. **Short-Term Loan** If a loan of $1,200 at 9% annual interest is paid off in 90 days, what is the interest?

11. Savings Account A woman deposits $800 in a savings account that pays 5% annual interest. If she withdraws all the money in the account after 120 days, how much does she withdraw?

12. Savings Account $1,800 is deposited in a savings account that pays 6% annual interest. If the money is withdrawn at the end of 30 days, how much interest is earned?

B The problems that follow involve compound interest. [Examples 6, 7]

13. Samuel invested $400 in a 6-month CD at a rate of 1.07%. If the interest is compounded quarterly, how much was in the CD at the end of 6 months? Round to the nearest cent.

14. If Alice deposited $200 in a $2\frac{1}{2}$ year CD at 1.51%, what will the CD make at the end of its term if interest is compounded quarterly. Use the compound interest formula and round to the nearest cent.

15. Compound Interest A woman puts $5,000 into a savings account that pays 6% compounded annually. How much money is in the account at the end of 2 years?

16. Compound Interest A savings account pays 5% compounded annually. If $10,000 is deposited in the account, how much is in the account after 2 years?

17. Compound Interest If $8,000 is invested in a savings account that pays 5% compounded quarterly, how much is in the account at the end of a year?

18. Compound Interest Suppose $1,200 is invested in a savings account that pays 6% compounded semiannually. How much is in the account at the end of $1\frac{1}{2}$ years?

Calculator Problems

The following problems should be set up in the same way in which Problems 1–18 have been set up. Then the calculations should be done on a calculator.

19. Savings Account A woman invests $917.26 in a savings account that pays 6.25% annually. How much is in the account at the end of a year?

20. Business Loan The owner of a clothing store borrows $6,210 for 1 year at 11.5% interest. If he pays the loan back at the end of the year, how much does he pay back?

21. Compound Interest Suppose $10,000 is invested in each account below. In each case find the amount of money in the account at the end of 5 years.
  a. Annual interest rate = 6%, compounded quarterly
  b. Annual interest rate = 6%, compounded monthly
  c. Annual interest rate = 5%, compounded quarterly
  d. Annual interest rate = 5%, compounded monthly

22. Compound Interest Suppose $5,000 is invested in each account below. In each case find the amount of money in the account at the end of 10 years.
  a. Annual interest rate = 5%, compounded quarterly
  b. Annual interest rate = 6%, compounded quarterly
  c. Annual interest rate = 7%, compounded quarterly
  d. Annual interest rate = 8%, compounded quarterly

Getting Ready for the Next Section

Change to a percent.

23. $\dfrac{75}{250}$
24. $\dfrac{150}{250}$
25. $\dfrac{400}{2,400}$
26. $\dfrac{200}{2,400}$

Multiply.

27. 0.3(360)
28. 0.4(360)
29. 0.45(360)
30. 0.15(360)

Divide.

31. 40 ÷ 5
32. 45 ÷ 5
33. 15 ÷ 5
34. 5 ÷ 5

Maintaining Your Skills

The problems below will allow you to review subtraction of fractions and mixed numbers.

35. $\dfrac{3}{4} - \dfrac{1}{4}$
36. $\dfrac{9}{10} - \dfrac{7}{10}$
37. $\dfrac{5}{8} - \dfrac{1}{4}$
38. $\dfrac{7}{10} - \dfrac{1}{5}$

39. $2 - \dfrac{4}{3}$
40. $2 + \dfrac{4}{3}$
41. $1 + \dfrac{1}{2}$
42. $1 - \dfrac{1}{2}$

43. $\dfrac{1}{3} - \dfrac{1}{4}$
44. $\dfrac{9}{12} - \dfrac{1}{5}$
45. $3\dfrac{1}{4} - 2$
46. $5\dfrac{1}{6} - 3\dfrac{1}{4}$

47. Find the sum of $\dfrac{8}{15}$ and $\dfrac{8}{35}$.

48. Find the difference of $\dfrac{8}{15}$ and $\dfrac{8}{35}$.

49. Find the product of $\dfrac{8}{15}$ and $\dfrac{8}{35}$.

50. Find the quotient of $\dfrac{8}{15}$ and $\dfrac{8}{35}$.

Extending the Concepts

The following problems are percent problems. Use any of the methods developed in this chapter to solve them.

51. Credit Card Debt Student credit-card debt is at an all-time high. Consolidated Credit Counseling Services Inc. reports that 20% of all college freshman got their first credit card in high school and nearly 40% sign up for one in their first year at college. Suppose your credit card company charges 1.3% in finance charges per month on the average daily balance in your credit card account. If your average daily balance for this month is $2,367.90 determine the finance charge for the month.

52. Finding Your Interest Rate In early January, your bank sent out a form called a 1099-INT, which summarizes the amount of interest you have received on a savings account for the previous year. If you received $72 interest for the year on an account in which you started with $1,200, determine the annual interest rate paid by your bank.

53. Movie Making The bar chart below shows the production costs for each of the six *Star Wars* movies. Find the percent increase or decrease in production costs from each *Star Wars* movie to the next. Round your results to the nearest tenth.

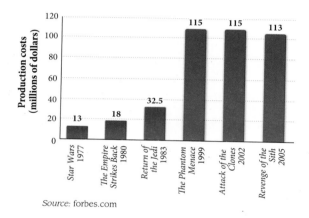

Source: forbes.com

54. Movie Making The table below shows the rounded amount of money each of the first four *Star Wars* movies brought in during the first weekend they were shown. Find the percent increase in opening weekend income from each *Star Wars* movie to the next. Round to the nearest percent.

Opening Weekend Income	
Star Wars (1977)	$1,554,000
The Empire Strikes Back (1980)	$6,415,000
Return of the Jedi (1983)	$30,490,000
The Phantom Menace (1999)	$64,810,000
Attack of the Clones (2002)	$80,000,000
Revenge of the Sith (2005)	$108,000,000

Source: the-numbers.com

Pie Charts

5.7

OBJECTIVES

A Read a pie chart.

B Construct a pie chart.

TICKET TO SUCCESS

Keep these questions in mind as you read through the section. Then respond in your own words and in complete sentences.

1. Why are pie charts important?
2. If a circle is divided into 20 equal slices, each of the slices represents what percent of the total area enclosed by the circle?
3. If a 10 GB portable hard drive contains 5.25 GB of data, then how much of the drive is free space? Construct a pie chart to support your answer.
4. Explain how you would construct a pie chart of monthly expenses for a person who spends $700 on rent, $200 on food, and $100 on entertainment.

Currently, the majority of energy consumption in the United States stems from fossil fuels, while less than 7% is from renewable energy resources. With a growing concern over environmental impact and surplus depletion, people are lobbying to increase the use of renewables. Suppose the sources of renewable energy can be presented in the form of a pie chart. Pie charts are another way in which to visualize numerical information. We have already worked with pie charts in previous chapters and seen how they work well to represent data in the form of fractions. Now we'll see how they lend themselves well to information that adds up to 100% and are common in the world around us. In fact, it is hard to pick up a newspaper or magazine without seeing a pie chart.

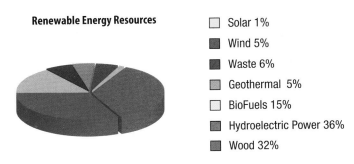

Renewable Energy Resources

- Solar 1%
- Wind 5%
- Waste 6%
- Geothermal 5%
- BioFuels 15%
- Hydroelectric Power 36%
- Wood 32%

5.7 Pie Charts

In this section, we will review how to read a pie chart and learn how to construct a chart using given data.

A Reading a Pie Chart

Some of this introductory material will be review. We want to begin our study of pie charts by reading information from pie charts.

EXAMPLE 1 The pie chart shows the class rank of the members of a drama club. Use the pie chart to answer the following questions.

a. Find the total membership of the club.
b. Find the ratio of freshmen to total number of members.
c. Find the ratio of seniors to juniors.

SOLUTION a. To find the total membership in the club, we add the numbers in all sections of the pie chart.

$$9 + 11 + 15 + 10 = 45 \text{ members}$$

b. The ratio of freshmen to total members is

$$\frac{\text{Number of freshmen}}{\text{Total number of members}} = \frac{11}{45}$$

c. The ratio of seniors to juniors is

$$\frac{\text{Number of seniors}}{\text{Number of juniors}} = \frac{9}{15} = \frac{3}{5}$$

EXAMPLE 2 The pie chart shows the results of a survey on how often people check their e-mail. Use the pie chart to answer the following questions. Suppose 500 people participated in the survey.

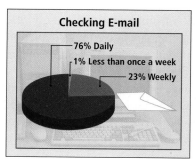

Source: UCLA Center for Communication Policy

a. How many people in the survey check their e-mail daily?
b. How many people check their e-mail once a week or less often?

SOLUTION a. To find out how many people in the survey check their e-mail daily, we need to find 76% of 500.

$$0.76(500) = 380 \text{ of the people surveyed check their e-mail daily}$$

PRACTICE PROBLEMS

1. Work Example 1 again if one more junior joins the club.

2. Work Example 2 again if 600 people responded to the survey.

Answers
1. a. 46   b. $\frac{11}{46}$ c. $\frac{9}{16}$
2. a. 456 people   b. 144 people

b. The people checking their e-mail weekly or less often account for 23% + 1% = 24%. To find out how many of the 500 people are in this category, we must find 24% of 500.

0.24(500) = 120 of the people surveyed check their e-mail weekly or less often ∎

B Constructing Pie Charts

EXAMPLE 3 Construct a pie chart that shows the free space and used space for a 1 terabyte (roughly 1,000 GB) hard drive that contains 446 GB of data.

SOLUTION 1 Using a Template As mentioned previously, pie charts are constructed with percents. Therefore we must first convert data to percents. To find the percent of used space, we divide the amount of used space by the amount of total space. We have

$$\frac{446}{1000} = 0.446 \text{ which is 45\% to the nearest percent}$$

The area of each section of the template on the left is 5% of the area of the whole circle. If we shade 9 sections of the template, we will have shaded 45% of the area of the whole circle.

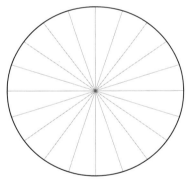

PIE CHART TEMPLATE Each slice is 5% of the area of the circle.

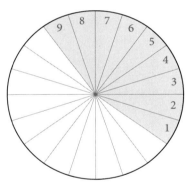

CREATING A PIE CHART To shade 45% of the circle, we shade 9 sections of the template.

The shaded area represents 45%, which is the amount of used disk space. The rest of the circle must represent the 55% free space on the disk. Shading each area with a different color and labeling each, we have our pie chart.

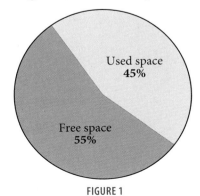

FIGURE 1

3. Construct a pie chart that shows the used space and free space on a 1 terabyte hard drive that contains 750 GB of data.

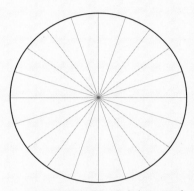

PIE CHART TEMPLATE Each slice is 5% of the area of the circle.

Answer
3. See Solutions Section.

SOLUTION 2 Using a Protractor Since a pie chart is a circle, and a circle contains 360°, we must now convert our data to degrees. We do this by multiplying our percents in decimal form by 360. We have

$$(0.45)360° = 162°$$

Now we place a protractor on top of a circle. First we draw a line from the center of the circle to 0° as shown in Figure 2. Now we measure and mark 162° from our starting point, as shown in Figure 3.

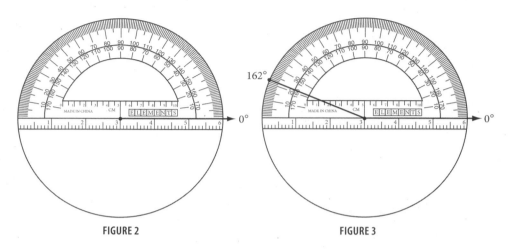

FIGURE 2   FIGURE 3

Finally we draw a line from the center of the circle to this mark, as shown in Figure 4. Then we shade and label the two regions as shown in Figure 5.

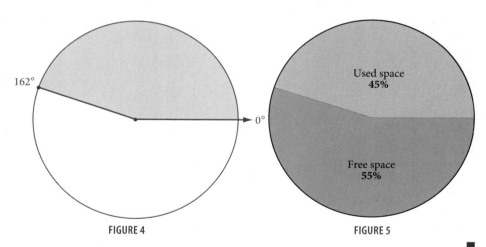

FIGURE 4   FIGURE 5

Problem Set 5.7

Moving Toward Success

"Neither comprehension nor learning can take place in an atmosphere of anxiety."

—Rose F. Kennedy, 1890–1995, American author and mother to President John F. Kennedy

1. Do you feel anxiety when you are taking a test? Why or why not?
2. What might you do to remain calm and boost your confidence if you feel anxious during a test?

A [Examples 1, 2]

1. **College Sophomores with Jobs** The pie chart shows the results of surveying 200 college sophomores to find out how many hours they worked per week at a job.

    a. Find the ratio of sophomores who work more than 15 hours a week to total sophomores.

    b. Find the ratio of sophomores who don't have a job to sophomores who work more than 15 hours a week.

    c. Find the ratio of sophomores with jobs to total sophomores.

    d. Find the ratio of sophomores with jobs to sophomores without jobs.

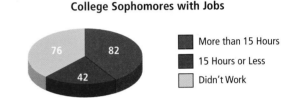

2. **Favorite Dip Flavor** The pie chart shows the results of a survey on favorite dip flavor.

    a. What is the most preferred dip flavor?

    b. Which dip flavor is preferred second most?

    c. Which dip flavor is least preferred?

    d. What percentage of people preferred ranch?

    e. What percentage of people preferred onion or dill?

    f. If 50 people responded to the survey, how many people preferred ranch?

    g. If 50 people responded to the survey, how many people preferred dill? (Round your answer to the nearest whole number.)

3. **Hoover Dam** The pie chart from this chapter's introduction shows the allocation percentages of electricity produced by the Hoover Dam.

   a. What percentage of electricity is allocated to Southern California Edison and Misc. Southern California Cities?

   b. What percentage does the state of Arizona recieve?

   c. Which location receives a greater percentage of electricity: the state of Nevada or Los Angeles, CA?

POWER ALLOCATION	
Location	Percentage Allocated
Metropolitan Water District of Southern California	28.54%
State of Nevada	23.37%
State of Arizona	18.95%
Los Angeles, CA	15.42%
Southern California Edison	5.54%
Boulder City, NV	1.77%
Misc Southern California Cities	6.41%

   Source: U.S. Department of the Interior, Bureaau of Reclamation

   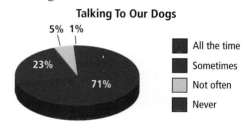

   d. What is the combined total percentage for the state of California?

4. **Talking to Our Dogs** A survey showed that most dog owners talk to their dogs.

   a. What percentage of dog owners say they never talk to their dogs?

   b. What percentage of dog owners say they talk to their dogs all the time?

   c. What percentage of dog owners say they talk to their dogs sometimes or not often?

5. **Monthly Car Payments** Suppose 3,000 people responded to a survey on car loan payments, the results of which are shown in the pie chart. Find the number of people whose monthly payments would be the following:

   a. $700 or more

   b. Less than $300

   c. $500 or more

   d. $300 to $699

   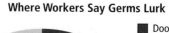

6. **Where Workers Say Germs Lurk** A survey asked workers where they thought the most germ-contaminated spot in the workplace was. Suppose the survey took place at a large company with 4,200 employees. Use the pie chart to determine the number of employees who would vote for each of the following as the most germ-contaminated areas.

   a. Keyboards

   b. Doorknobs

   c. Restrooms or other

   d. Telephones or doorknobs

B [Example 3]

7. **Grade Distribution** Student scores, for a class of 20, on a recent math test are shown in the table below. Complete the table and construct a pie chart that shows the number of As, Bs, and Cs earned on the test. Use the template provided here or use a protractor.

GRADE DISTRIBUTION		
Grade	Number	Percent
A	5	
B	8	
C	7	
Total	20	

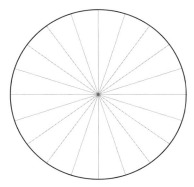

PIE CHART TEMPLATE   Each slice is 5% of the area of the circle.

8. **Building Sizes** The Lean and Mean Gym Company recently ran a promotion for their four locations in the county. The table shows the locations along with the amount of square feet at each location. Use the information to complete the table and construct a pie chart, using the template provided here or using a protractor.

GYM LOCATION AND SIZE		
Location	Square Feet	Percent
Downtown	35,000	
Uptown	85,000	
Lakeside	25,000	
Mall	75,000	
Total	220,000	

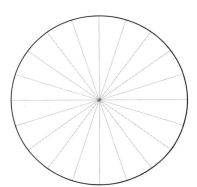

PIE CHART TEMPLATE   Each slice is 5% of the area of the circle.

9. **Room Sizes** Scott and Amy are building their dream house. The size of the house will be 2,400 square feet. The table below shows the size of each room. Use the information to complete the table and construct a pie chart, using the template provided here or using a protractor.

ROOM SIZES		
Room	Square Feet	Percent
Kitchen	400	
Dining room	310	
Bedrooms	890	
Living room	600	
Bathrooms	200	
Total	2,400	

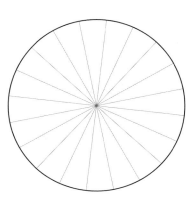

PIE CHART TEMPLATE   Each slice is 5% of the area of the circle.

10. **Airline Seating** Suppose the table below gives the number of seats in each of the three classes of seating on an American Airlines Boeing 777 airliner. Complete the table and create a pie chart from the information in the table.

AIRLINE SEATING		
Seating Class	Number of Seats	Percent
First	18	
Business	42	
Coach	163	

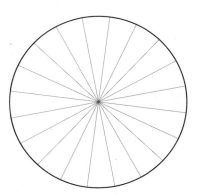

PIE CHART TEMPLATE Each slice is 5% of the area of the circle.

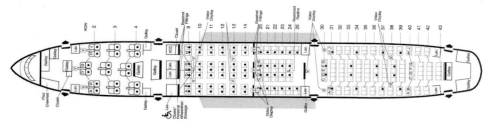

Maintaining Your Skills

Multiply.

11. $8 \cdot \dfrac{1}{3}$

12. $9 \cdot \dfrac{1}{3}$

13. $25 \cdot \dfrac{1}{1,000}$

14. $25 \cdot \dfrac{1}{100}$

15. $36.5 \cdot \dfrac{1}{100} \cdot 10$

16. $36.5 \cdot \dfrac{1}{1,000} \cdot 100$

17. $248 \cdot \dfrac{1}{10} \cdot \dfrac{1}{10}$

18. $969 \cdot \dfrac{1}{10} \cdot \dfrac{1}{10}$

19. $48 \cdot \dfrac{1}{12} \cdot \dfrac{1}{3}$

20. $56 \cdot \dfrac{1}{12} \cdot \dfrac{1}{2}$

Chapter 5 Summary

EXAMPLES

The Meaning of Percent [5.1]

1. 42% means 42 per hundred or $\frac{42}{100}$.

Percent means "per hundred." It is a way of comparing numbers to the number 100.

Changing Percents to Decimals [5.1]

2. 75% = 0.75

To change a percent to a decimal, drop the percent symbol (%), and move the decimal point two places to the *left*.

Changing Decimals to Percents [5.1]

3. 0.25 = 25%

To change a decimal to a percent, move the decimal point two places to the *right,* and use the % symbol.

Changing Percents to Fractions [5.1]

4. $6\% = \frac{6}{100} = \frac{3}{50}$

To change a percent to a fraction, drop the % symbol, and use a denominator of 100. Reduce the resulting fraction to lowest terms if necessary.

Changing Fractions to Percents [5.1]

5. $\frac{3}{4} = 0.75 = 75\%$

or

$\frac{9}{10} = \frac{90}{100} = 90\%$

To change a fraction to a percent, either write the fraction as a decimal and then change the decimal to a percent, or write the fraction as an equivalent fraction with denominator 100, drop the 100, and use the % symbol.

Basic Word Problems Involving Percents [5.2]

6. Translating to equations, we have
  Type A: $n = 0.14(68)$
  Type B: $75n = 25$
  Type C: $25 = 0.40n$

There are three basic types of word problems:

Type A: What number is 14% of 68?

Type B: What percent of 75 is 25?

Type C: 25 is 40% of what number?

To solve them, we write *is* as =, *of* as · (multiply), and *what number* or *what percent* as n. We then solve the resulting equation to find the answer to the original question.

Applications of Percent [5.3, 5.4, 5.5, 5.6]

There are many different kinds of application problems involving percent. They include problems on income tax, sales tax, commission, discount, percent increase and decrease, and interest. Generally, to solve these problems, we restate them as an equivalent problem of Type A, B, or C from the previous page. Problems involving simple interest can be solved using the formula

$$I = P \cdot R \cdot T$$

where I = interest, P = principal, R = interest rate, and T = time (in years). It is standard procedure with simple interest problems to use 360 days = 1 year.

Pie Charts [5.7]

A pie chart is another way to give a visual representation of the information in a table.

Seating Class	Number Of Seats
First	18
Business	42
Coach	163

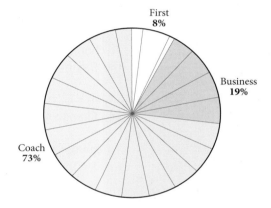

⊘ COMMON MISTAKES

1. A common mistake is forgetting to change a percent to a decimal when working problems that involve percents in the calculations. We always change percents to decimals before doing any calculations.
2. Moving the decimal point in the wrong direction when converting percents to decimals or decimals to percents is another common mistake. Remember, *percent* means "per hundred." Rewriting a number expressed as a percent as a decimal will make the numerical part smaller.

    25% = 0.25

Chapter 5 Review

Write each percent as a decimal. [5.1]

1. 35%
2. 17.8%
3. 5%
4. 0.2%

Write each decimal as a percent. [5.1]

5. 0.95
6. 0.8
7. 0.495
8. 1.65

Write each percent as a fraction or mixed number in lowest terms. [5.1]

9. 75%
10. 4%
11. 145%
12. 2.5%

Write each fraction or mixed number as a percent. [5.1]

13. $\frac{3}{10}$
14. $\frac{5}{8}$
15. $\frac{2}{3}$
16. $4\frac{3}{4}$

Solve the following problems. [5.2]

17. What number is 60% of 28?
18. What number is 122% of 55?
19. What percent of 38 is 19?
20. What percent of 19 is 38?
21. 24 is 30% of what number?
22. 16 is 8% of what number?

23. **Survey** Suppose 45 out of 60 people surveyed believe a college education will increase a person's earning potential. What percent believe this? [5.3]

24. **Commission** A real estate agent gets a commission of 6% on all houses he sells. If his total sales for December are $420,000, how much money does he make? [5.4]

25. **Total Price** A sewing machine that normally sells for $600 is on sale for 25% off. If the sales tax rate is 6%, what is the total price of the sewing machine if it is purchased during the sale? [5.4, 5.5]

26. **Total Price** A tennis racket that normally sells for $240 is on sale for 25% off. If the sales tax rate is 5%, what is the total price of the tennis racket if it is purchased during the sale? [5.5]

27. **Discount** A lawnmower that usually sells for $175 is marked down to $140. What is the discount? What is the discount rate? [5.5]

28. **Percent Increase** At the beginning of the summer, the price for a gallon of regular gasoline is $4.25. By the end of summer, the price has increased 16%. What is the new price of a gallon of regular gasoline? Round to the nearest cent. [5.5]

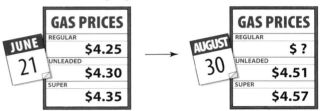

29. **Percent Decrease** A gallon of regular gasoline is selling for $4.45 in September. If the price decreases 14% in October, what is the new price for a gallon of regular gasoline? Round to the nearest cent. [5.5]

30. **Medical Costs** The table shows the average yearly cost of visits to the doctor, as reported in *USA Today*. What is the percent increase in cost from 1990 to 2000? Round to the nearest tenth of a percent. [5.5]

MEDICAL COSTS	
Year	Average Annual Cost
1990	$583
1995	$739
2000	$906
2005	$1,172

Source: USA Today

31. **Discount** A washing machine that usually sells for $300 is marked down to $240. What is the discount? What is the discount rate? [5.4, 5.5]

Chapter 5 Cumulative Review

Simplify.

1. 6,801
   539
   + 374

2. 5,038
   − 2,769

3. 52(867)

4. 1,023 ÷ 15

5. $4.73\overline{)156.09}$

6. $\dfrac{7}{8} - \dfrac{5}{8}$

7. $\left(\dfrac{5}{6}\right)^3$

8. 4.551 + 3.08

9. 5 − 3.678

10. 1.2(0.21)

11. $\dfrac{7}{15} \cdot \dfrac{5}{14}$

12. $\dfrac{8}{27} \div \dfrac{20}{63}$

13. $\dfrac{3}{8} + \dfrac{7}{12}$

14. $8\dfrac{1}{5} - 5\dfrac{7}{10}$

15. $9 \cdot 4\dfrac{2}{3}$

16. Subtract $5\dfrac{3}{8}$ from 10.375.

17. Find the quotient of $1\dfrac{1}{2}$ and $\dfrac{1}{4}$.

18. Translate into symbols, and then simplify: Twice the sum of 2 and 9.

19. Write the ratio of 3 to 12 as a fraction in lowest terms.

20. If 1 mile is 5,280 feet, how many feet are there in 2.5 miles?

21. If 1 square yard is 1,296 square inches, how many square inches are in $\dfrac{1}{2}$ square yard?

22. Write $\dfrac{1}{8}$ as a percent.

23. Convert 46% to a fraction.

24. Solve the equation $\dfrac{2}{x} = \dfrac{5}{8}$.

25. 3 · 52 + 2 · 42 − 5 · 23

26. What number is 5% of 32?

27. 55 is what percent of 275?

28. 8.8 is 15% of what number?

29. **Unit Pricing** If a six-pack of Coke costs $2.79, what is the price per can to the nearest cent?

30. **Unit Pricing** A quart of 2% reduced-fat milk contains four 1-cup servings. If the quart costs $1.61, find the price per serving to the nearest cent.

31. **Temperature** Use the formula $C = \dfrac{5(F - 32)}{9}$ to find the temperature in degrees Celsius when the Fahrenheit temperature is 212°F.

32. **Percent Increase** Kendra is earning $1,600 a month when she receives a raise to $1,800 a month. What is the percent increase in her monthly salary?

33. **Driving Distance** If Ethan drives his car 230 miles in 4 hours, how far will he drive in 6 hours if he drives at the same rate?

34. **Movie Tickets** A movie theater has a total of 250 seats. If they have a sellout crowd for a matinee and each ticket costs $7.25, how much money will ticket sales bring in that afternoon?

35. **Geometry** Find the perimeter and area of a square with side 8.5 inches.

36. **Average** If a basketball team has scores of 64, 76, 98, 55, and 102 in their first five games, find the mean score.

37. **Hourly Pay** Jean tutors in the math lab and earns $56 in one week. If she works 8 hours that week, what is her hourly pay?

Use the illustration to answer the following questions.

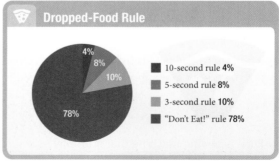

Dropped-Food Rule
- 10-second rule 4%
- 5-second rule 8%
- 3-second rule 10%
- "Don't Eat!" rule 78%

38. What is the ratio of people who believe in the 5-second rule to the people who believe in the 3-second rule?

39. What is the ratio of people who believe in the "Don't Eat" rule to the people who believe in the 10-second rule?

Chapter 5 Test

Write each percent as a decimal. [5.1]

1. 18%   2. 4%   3. 0.5%

Write each decimal as a percent. [5.1]

4. 0.45   5. 0.7   6. 1.35

Write each percent as a fraction or a mixed number in lowest terms. [5.1]

7. 65%   8. 146%   9. 3.5%

Write each number as a percent. [5.1]

10. $\frac{7}{20}$ 11. $\frac{3}{8}$ 12. $1\frac{3}{4}$

13. What number is 75% of 60? [5.2]

14. What percent of 40 is 18? [5.2]

15. 16 is 20% of what number? [5.2]

16. **Driver's Test** On a 25-question driver's test, a student answered 23 questions correctly. What percent of the questions did the student answer correctly? [5.3]

17. **Commission** A salesperson gets an 8% commission rate on all computers she sells. If she sells $12,000 in computers in 1 day, what is her commission? [5.4]

18. **Discount** A washing machine that usually sells for $250 is marked down to $210. What is the discount? What is the discount rate? [5.5]

19. **Percent Increase** A driver gets into a car accident and his insurance increases by 12%. If he paid $950 before the accident, how much is he paying now? [5.5]

20. **Total Price** A tennis racket that normally sells for $280 is on sale for 25% off. If the sales tax rate is 5%, what is the total price of the tennis racket if it is purchased during the sale? [5.4, 5.5]

21. **Simple Interest** If $5,000 is invested at 8% simple interest for 3 months, how much interest is earned? [5.6]

22. **Compound Interest** How much interest will be earned on a savings account that pays 10% compounded annually, if $12,000 is invested for 2 years? [5.6]

Use the illustration seen earlier in the chapter to answer problem 23.

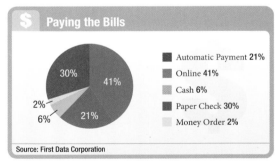

23. What is the ratio of people who use an automatic payment to those who write a check? [5.7]

# Chapter 5 Projects

PERCENT

Group Project

Sales Tax

Number of People 2

Time Needed 5 minutes

Equipment Pencil, paper, and calculator.

Background All of us spend time buying clothes and eating meals at restaurants. In all of these situations, it is good practice to check receipts. This project is intended to give you practice creating receipts of your own.

Procedure Fill in the missing parts of each receipt.

SALES RECEIPT	
Jeans	29.99
Sales Tax (7.75%)	
Total	

SALES RECEIPT	
2 Buffet Dinners @ 9.99	19.98
Discount (10%)	
Total	

SALES RECEIPT	
Computer	400.00
Discount: 30% off	
Discounted Price	
Sales Tax (6%)	
Total	

SALES RECEIPT	
Couch	
Sales Tax (7%)	
Total	588.50

Research Project

Credit Card Debt

Credit card companies are offering credit cards to college students who would not be able to get a card under normal credit card criteria (due to lack of credit history and low income). The credit card industry sees young people as a valuable market because research shows that they remain loyal to their first cards as they grow older. Nellie Mae found that 76% of undergraduate students had credit cards in 2008. For many of these students, lack of financial experience or education leads to serious debt. According to Nellie Mae, undergraduates with credit cards carried an average balance of $2,200 in 2010. About 45% of card-carrying college students don't pay their balances in full every month. Choose a credit card and find out the minimum monthly payment and the APR (annual percentage rate). Compute the minimum monthly payment and interest charges for a balance of $2,200.

Stockbytes/SuperStock

A Glimpse of Algebra

Being able to evaluate expressions is a very important skill in algebra. This glimpse of algebra will focus on this concept and give you some extra practice at evaluating expressions.

To evaluate an expression, such as $5x + 4$, when we know that x is 7, we simply substitute 7 for x in the expression $5x + 4$ and then simplify the result.

$$\begin{aligned} \text{When} \rightarrow\ & x = 7 \\ \text{the expression} \rightarrow\ & 5x + 4 \\ \text{becomes} \rightarrow\ & 5(7) + 4 \\ & 35 + 4 \\ & = 39 \end{aligned}$$

Here are some examples.

EXAMPLE 1 Find the value of the expression $4x + 3x - 8$ when x is 2.

SOLUTION Substituting 2 for x in the expression, we have

$$\begin{aligned} 4(2) + 3(2) - 8 &= 8 + 6 - 8 \\ &= 14 - 8 \\ &= 6 \end{aligned}$$

We say that $4x + 3x - 8$ becomes 6 when x is 2. ∎

EXAMPLE 2 Find the value of the following expression when a is 5.

$$\frac{4a + 8}{3a - 8}$$

SOLUTION Replacing a with 5 in the expression, we have

$$\frac{4(5) + 8}{3(5) - 8} = \frac{20 + 8}{15 - 8} = \frac{28}{7} = 4$$

∎

EXAMPLE 3 Find the value of $x^2 + 5x + 6$ when x is 4.

SOLUTION When x is 4, the expression $x^2 + 5x + 6$ becomes

$$\begin{aligned} (4)^2 + 5(4) + 6 &= 16 + 20 + 6 \\ &= 42 \end{aligned}$$

∎

EXAMPLE 4 Find the value of $\dfrac{4x + y}{4x - y}$ when x is 5 and y is 2.

SOLUTION This time we have two different variables. We replace x with 5 and y with 2 to get

$$\frac{4(5) + 2}{4(5) - 2} = \frac{20 + 2}{20 - 2} = \frac{22}{18} = \frac{11}{9}$$

∎

PRACTICE PROBLEMS

1. Find the value of the expression $6x + 3x - 10$ when x is 3.

2. Find the value of the following expression when a is 10.
$$\frac{4a + 20}{5a - 20}$$

3. Find the value of $x^2 + 6x + 8$ when x is 3.

4. Find the value of the following expression when x is 10 and y is 4.
$$\frac{3x + y}{3x - y}$$

Answers
1. 17
2. 2
3. 35
4. $\frac{17}{13}$

5. Find the value of
$(4x + 1)(4x - 1)$ when x is 2.

6. Find the value of the following expression when x is 5.
$$\frac{x^3 + 8}{x + 2}$$

Chapter 5 Percent

EXAMPLE 5 Find the value of $(2x + 3)(2x - 3)$ when x is 4.

SOLUTION Replacing x with 4 in the expression, we have

$$(2 \cdot \mathbf{4} + 3)(2 \cdot \mathbf{4} - 3) = (8 + 3)(8 - 3)$$
$$= (11)(5)$$
$$= 55$$

EXAMPLE 6 Find the value of $\frac{x^3 - 8}{x - 2}$ when x is 5.

SOLUTION We substitute 5 for x and then simplify.

$$\frac{\mathbf{5}^3 - 8}{\mathbf{5} - 2} = \frac{125 - 8}{3}$$
$$= \frac{117}{3}$$
$$= 39$$

Answers
5. 63
6. 19

A Glimpse of Algebra Problems

Find the value of each of the following expressions for the given values of the variables.

1. $6x + 2x - 7$ when x is 2

2. $8x + 10x - 5$ when x is 3

3. $4x + 6x + 8x$ when x is 10

4. $9x + 2x + 20x$ when x is 5

5. $\dfrac{4a + 20}{5a - 20}$ when a is 5

6. $\dfrac{4a + 8}{3a - 8}$ when a is 8

7. $\dfrac{2a + 3a + 1}{4a + 5a + 3}$ when a is 3

8. $\dfrac{7a + a + 4}{6a + 2a + 3}$ when a is 10

9. $x^2 + 5x + 6$ when x is 2

10. $x^2 + 6x + 8$ when x is 6

11. $x^2 + 10x + 25$ when x is 1

12. $x^2 + 10x + 25$ when x is 0

13. $\dfrac{3x + y}{3x - y}$ when x is 5 and y is 2

14. $\dfrac{5x - y}{5x + y}$ when x is 10 and y is 5

15. $\dfrac{4x + 6y}{6x + 4y}$ when x is 5 and y is 4

16. $\dfrac{8x - 3y}{3x + 8y}$ when x is 5 and y is 10

17. $(3x + 2)(3x - 2)$ when x is 4

18. $(5x + 4)(5x - 4)$ when x is 2

19. $(2x + 3)^2$ when x is 1

20. $(2x + 3)^3$ when x is 2

21. $\dfrac{x^3 + 1}{x + 1}$ when x is 2

22. $\dfrac{x^3 - 1}{x - 1}$ when x is 4

23. $\dfrac{x^3 - 8}{x^2 + 2x + 4}$ when x is 3

24. $\dfrac{x^3 + 8}{x^2 - 2x + 4}$ when x is 3

25. $\dfrac{x^4 - 16}{x^2 + 4}$ when x is 5

26. $\dfrac{x^4 - 16}{x + 2}$ when x is 3

Measurement

The Google Earth image above shows the Nile River in Africa. The Nile is the longest river in the world, measuring 4,160 miles and stretching across ten different countries. It serves as a lifeline to the inhabitants of those countries, such as a vital means of transportation and an irrigation source for nearby farmland. Look at the table below. Information about the Nile is listed in two different systems of measurements.

THE NILE RIVER	English Units	Metric Units
Length	4,160 mi	6,698 km
Nile Delta Area	1,004 mi^2	36,000 km^2
Flow Rate (monsoon season)	285,829 ft^3/s	8,100 m^3/s
Average Summer Temperature	86°F	30°C

Source: worldwildlife.org

In this chapter, we look at the process we use to convert from one set of units, such as miles per hour, to another set of units, such as kilometers per hour. Regardless of the units in question, we will be using the same method in all cases. This method is called *unit analysis*.

Chapter Outline

6.1 Unit Analysis I: Length

6.2 Unit Analysis II: Area and Volume

6.3 Unit Analysis III: Weight

6.4 Converting Between the Two Systems, and Temperature

6.5 Operations with Time and Mixed Units

Preview

Key Words	Definition
Conversion Factor	A quantity used to convert a measurement to a different unit of measure.
Unit Analysis	A method that uses conversion factors to convert between units of measure.

Chapter Outline

6.1 Unit Analysis I: Length
- A Convert between lengths in the U.S. system.
- B Convert between lengths in the metric system.
- C Solve application problems involving unit analysis.

6.2 Unit Analysis II: Area and Volume
- A Convert between areas using the U.S. system.
- B Convert between areas using the metric system.
- C Convert between volumes using the U.S. system.
- D Convert between volumes using the metric system.

6.3 Unit Analysis III: Weight
- A Convert between weights using the U.S. system.
- B Convert between weights using the metric system.

6.4 Converting Between the Two Systems, and Temperature
- A Convert between the U.S. and metric systems.
- B Convert temperatures between the Fahrenheit and Celsius scales.

6.5 Operations with Time and Mixed Units
- A Convert mixed units to a single unit.
- B Add and subtract mixed units.
- C Use multiplication with mixed units.

Unit Analysis I: Length

6.1

OBJECTIVES

A Convert between lengths in the U.S. system.

B Convert between lengths in the metric system.

C Solve application problems involving unit analysis.

TICKET TO SUCCESS

Keep these questions in mind as you read through the section. Then respond in your own words and in complete sentences.

1. What are the units of length for the U.S. system?
2. Give examples of metric units of length.
3. Explain the four steps of unit analysis.
4. Write an application problem that involves a conversion factor of inches to feet.

FIFA, which is the international governing body for the game of soccer, requires that a standard soccer field be a length between 90 to 120 meters (100 to 130 yards) and a width between 45 to 90 meters (50 to 100 yards). In this section, we will become more familiar with the units used to measure length. We will look at the U.S. system of measurement and the metric system of measurement, which will also help us make sense of the above soccer field dimensions.

A U.S. Units of Length

Measuring the *length* of an object is done by assigning a number to its length. To let other people know what that number represents, we include with it a unit of measure. The most common units used to represent length in the U.S. system are inches, feet, yards, and miles. The basic unit of length is the foot. The other units are defined in terms of feet, as Table 1 shows.

TABLE 1

12 inches (in.) =	1 foot (ft)
1 yard (yd) =	3 feet
1 mile (mi) =	5,280 feet

As you can see from the table, the abbreviations for inches, feet, yards, and miles are in., ft, yd, and mi, respectively. What we haven't indicated, even though you may not have realized it, is what 1 foot represents. We have defined all our units associated with length in terms of feet, but we haven't said what a foot is.

There is a long history of the evolution of what is now called a foot. At different times in the past, a foot has represented different arbitrary lengths. Currently, a foot is defined to be exactly 0.3048 meter (the basic measure of length in the metric system), where a meter is 1,650,763.73 wavelengths of the orange-red line in the spectrum of krypton-86 in a vacuum (this doesn't mean much to me either). The reason a foot and a meter are defined this way is that we always want them to measure the same length. Because the wavelength of the orange-red line in the spectrum of krypton-86 will always remain the same, so will the length that a foot represents.

Now that we have said what we mean by 1 foot (even though we may not understand the technical definition), we can go on and look at some examples that involve converting from one kind of unit to another.

PRACTICE PROBLEMS

1. Convert 8 feet to inches.

EXAMPLE 1 Convert 5 feet to inches.

SOLUTION Because 1 foot = 12 inches, we can multiply 5 by 12 inches to get

$$5 \text{ feet} = 5 \times 12 \text{ inches}$$
$$= 60 \text{ inches}$$

This method of converting from feet to inches probably seems fairly simple. But as we go further in this chapter, the conversions from one kind of unit to another will become more complicated. For these more complicated problems, we need another way to show conversions so that we can be certain to end them with the correct unit of measure. For example, since 1 ft = 12 in., we can say that there are 12 in. per 1 ft or 1 ft per 12 in. That is,

$$\frac{12 \text{ in.}}{1 \text{ ft}} \longleftarrow \text{Per} \quad \text{or} \quad \frac{1 \text{ ft}}{12 \text{ in.}} \longleftarrow \text{Per}$$

We call the expressions $\frac{12 \text{ in.}}{1 \text{ ft}}$ and $\frac{1 \text{ ft}}{12 \text{ in.}}$ *conversion factors*. The fraction bar is read as "per." Both conversion factors are really just the number 1. That is,

$$\frac{12 \text{ in.}}{1 \text{ ft}} = \frac{12 \text{ in.}}{12 \text{ in.}} = 1$$

We already know that multiplying a number by 1 leaves the number unchanged. So, to convert from one unit to the other, we can multiply by one of the conversion factors without changing value. Both the conversion factors above say the same thing about the units feet and inches. They both indicate that there are 12 inches in every foot. The one we choose to multiply by depends on what units we are starting with and what units we want to end up with. If we start with feet and we want to end up with inches, we multiply by the conversion factor

$$\frac{12 \text{ in.}}{1 \text{ ft}}$$

The units of feet will divide out and leave us with inches.

$$5 \text{ feet} = 5 \text{ ft} \times \frac{12 \text{ in.}}{1 \text{ ft}}$$
$$= 5 \times 12 \text{ in.}$$
$$= 60 \text{ in.}$$

Answer
1. 96 in.

The key to this method of conversion lies in setting the problem up so that the correct units divide out to simplify the expression. We are treating units such as feet in the same way we treated factors when reducing fractions. If a factor is common to the numerator and the denominator, we can divide it out and simplify the fraction. The same idea holds for units such as feet.

We can rewrite Table 1 so that it shows the conversion factors associated with units of length, as shown in Table 2.

TABLE 2
UNITS OF LENGTH IN THE U.S. SYSTEM

The relationship between	is	To convert one to the other, multiply by
feet and inches	12 in. = 1 ft	$\frac{12 \text{ in.}}{1 \text{ ft}}$ or $\frac{1 \text{ ft}}{12 \text{ in.}}$
feet and yards	1 yd = 3 ft	$\frac{3 \text{ ft}}{1 \text{ yd}}$ or $\frac{1 \text{ yd}}{3 \text{ ft}}$
feet and miles	1 mi = 5,280 ft	$\frac{5,280 \text{ ft}}{1 \text{ mi}}$ or $\frac{1 \text{ mi}}{5,280 \text{ ft}}$

NOTE
We will use this method of converting from one kind of unit to another throughout the rest of this chapter. You should practice using it until you are comfortable with it and can use it correctly. However, it is not the only method of converting units. You may see shortcuts that will allow you to get results more quickly. Use shortcuts if you wish, so long as you can consistently get correct answers and are not using your shortcuts because you don't understand our method of conversion. Use the method of conversion as given here until you are good at it; then use shortcuts if you want to.

Use the conversion factors in Table 2 to work through the next two examples.

EXAMPLE 2 The most common ceiling height in houses is 8 feet. How many yards is this?

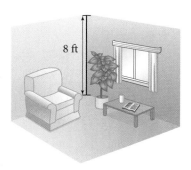

SOLUTION To convert 8 feet to yards, we multiply by the conversion factor $\frac{1 \text{ yd}}{3 \text{ ft}}$ so that feet will divide out and we will be left with yards.

$8 \text{ ft} = 8 \text{ ft} \times \frac{1 \text{ yd}}{3 \text{ ft}}$ Multiply by the correct conversion factor.

$= \frac{8}{3} \text{ yd}$ $8 \times \frac{1}{3} = \frac{8}{3}$

$= 2\frac{2}{3} \text{ yd}$ Or 2.67 yd rounded to the nearest hundredth

EXAMPLE 3 A football field is 100 yards long. How many inches long is a football field?

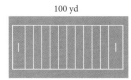

2. The roof of a two-story house is 26 feet above the ground. How many yards is this?

3. How many inches are in 220 yards?

Answers
2. $8\frac{2}{3}$ yd, or 8.67 yd rounded to the nearest hundredth
3. 7,920 in.

SOLUTION In this example, we must convert yards to feet and then feet to inches. (To make this example more interesting, we are pretending we don't know that there are 36 inches in a yard.) We choose the conversion factors that will allow all the units except inches to divide out.

$$100 \text{ yd} = 100 \text{ yd} \times \frac{3 \text{ ft}}{1 \text{ yd}} \times \frac{12 \text{ in.}}{1 \text{ ft}}$$
$$= 100 \times 3 \times 12 \text{ in.}$$
$$= 3{,}600 \text{ in.}$$

B Metric Units of Length

In the metric system, the standard unit of length is a meter. A meter is a little longer than a yard (about 3.4 inches longer). The other units of length in the metric system are written in terms of a meter. The metric system uses prefixes to indicate what part of the basic unit of measure is being used. For example, in *milli*meter the prefix *milli* means "one thousandth" of a meter. Table 3 gives the meanings of the most common metric prefixes.

TABLE 3
THE MEANING OF METRIC PREFIXES

Prefix	Meaning
milli	0.001
centi	0.01
deci	0.1
deka	10
hecto	100
kilo	1,000

We can use these prefixes to write the other units of length and conversion factors for the metric system, as given in Table 4.

TABLE 4
METRIC UNITS OF LENGTH

The relationship between	is	To convert one to the other, multiply by	
millimeters (mm) and meters (m)	1,000 mm = 1 m	$\frac{1{,}000 \text{ mm}}{1 \text{ m}}$ or	$\frac{1 \text{ m}}{1{,}000 \text{ mm}}$
centimeters (cm) and meters	100 cm = 1 m	$\frac{100 \text{ cm}}{1 \text{ m}}$ or	$\frac{1 \text{ m}}{100 \text{ cm}}$
decimeters (dm) and meters	10 dm = 1 m	$\frac{10 \text{ dm}}{1 \text{ m}}$ or	$\frac{1 \text{ m}}{10 \text{ dm}}$
dekameters (dam) and meters	1 dam = 10 m	$\frac{10 \text{ m}}{1 \text{ dam}}$ or	$\frac{1 \text{ dam}}{10 \text{ m}}$
hectometers (hm) and meters	1 hm = 100 m	$\frac{100 \text{ m}}{1 \text{ hm}}$ or	$\frac{1 \text{ hm}}{100 \text{ m}}$
kilometers (km) and meters	1 km = 1,000 m	$\frac{1{,}000 \text{ m}}{1 \text{ km}}$ or	$\frac{1 \text{ km}}{1{,}000 \text{ m}}$

We use the same method to convert between units in the metric system as we did with the U.S. system. We choose the conversion factor that will allow the units we start with to divide out, leaving the units we want to end up with.

EXAMPLE 4 Convert 25 millimeters to meters.

SOLUTION To convert from millimeters to meters, we multiply by the conversion factor $\frac{1 \text{ m}}{1{,}000 \text{ mm}}$.

$$25 \text{ mm} = 25 \text{ mm} \times \frac{1 \text{ m}}{1{,}000 \text{ mm}}$$

$$= \frac{25 \text{ m}}{1{,}000}$$

$$= 0.025 \text{ m}$$

4. Convert 67 centimeters to meters.

EXAMPLE 5 Convert 36.5 centimeters to decimeters.

SOLUTION We convert centimeters to meters and then meters to decimeters.

$$36.5 \text{ cm} = 36.5 \text{ cm} \times \frac{1 \text{ m}}{100 \text{ cm}} \times \frac{10 \text{ dm}}{1 \text{ m}}$$

$$= \frac{36.5 \times 10}{100} \text{ dm}$$

$$= 3.65 \text{ dm}$$

5. Convert 78.4 mm to decimeters.

The most common units of length in the metric system are millimeters, centimeters, meters, and kilometers. The other units of length we have listed in our table of metric lengths are not as widely used. The method we have used to convert from one unit of length to another in Examples 2–5 is called *unit analysis*. If you take a chemistry class, you will see it used many times. The same is true of many other science classes as well.

We can summarize the procedure used in unit analysis with the following steps:

Strategy **Unit Analysis**

Step 1 Identify the units you are starting with.

Step 2 Identify the units you want to end with.

Step 3 Find conversion factors that will bridge the starting units and the ending units.

Step 4 Set up the multiplication problem so that all units except the units you want to end with will divide out.

C Applications

Let's apply the unit analysis strategy to a couple more examples.

EXAMPLE 6 A sheep rancher is making new lambing pens for the upcoming lambing season. Each pen is a rectangle 6 feet wide and 8 feet long. The fencing material he wants to use sells for $1.36 per foot. If he is planning to build five separate lambing pens (they are separate because he wants a walkway between them), how much will he have to spend for fencing material?

6. The rancher in Example 6 decides to build six pens instead of five and upgrades his fencing material so that it costs $1.72 per foot. How much does it cost him to build the six pens?

Answers
4. 0.67 m
5. 0.784 dm
6. $288.96

SOLUTION To find the amount of fencing material he needs for one pen, we find the perimeter of a pen.

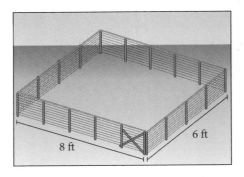

Perimeter = 6 + 6 + 8 + 8 = 28 feet

We set up the solution to the problem using unit analysis. Our starting unit is *pens* and our ending unit is *dollars*. Here are the conversion factors that will form a bridge between pens and dollars.

$$1 \text{ pen} = 28 \text{ feet of fencing}$$
$$1 \text{ foot of fencing} = 1.36 \text{ dollars}$$

Next we write the multiplication problem, using the conversion factors, that will allow all the units except dollars to divide out.

$$5 \text{ pens} = 5 \text{ pens} \times \frac{28 \text{ feet of fencing}}{1 \text{ pen}} \times \frac{1.36 \text{ dollars}}{1 \text{ foot of fencing}}$$
$$= 5 \times 28 \times 1.36 \text{ dollars}$$
$$= \$190.40$$

■

EXAMPLE 7 A number of years ago, a ski resort in Vermont advertised their new high-speed chair lift as "the world's fastest chair lift, with a speed of 1,100 feet per second." Show why the speed cannot be correct.

SOLUTION To solve this problem, we can convert feet per second into miles per hour, a unit of measure we are more familiar with on an intuitive level. Here are the conversion factors we will use:

$$1 \text{ mile} = 5,280 \text{ feet}$$

$$1 \text{ hour} = 60 \text{ minutes}$$

$$1 \text{ minute} = 60 \text{ seconds}$$

$$1,100 \text{ ft/second} = \frac{1,100 \text{ feet}}{1 \text{ second}} \times \frac{1 \text{ mile}}{5,280 \text{ feet}} \times \frac{60 \text{ seconds}}{1 \text{ minute}} \times \frac{60 \text{ minutes}}{1 \text{ hour}}$$

$$= \frac{1,100 \times 60 \times 60 \text{ miles}}{5,280 \text{ hours}}$$

$$= 750 \text{ miles/hour}$$

■

7. Assume that the mistake in the advertisement is that feet per second should read feet per minute. Is 1,100 feet per minute a reasonable speed for a chair lift?

Answer
7. 12.5 mi/hr is a reasonable speed for a chair lift.

Problem Set 6.1

Moving Toward Success

"When you write down your ideas you automatically focus your full attention on them. Few if any of us can write one thought and think another at the same time. Thus a pencil and paper make excellent concentration tools."

—Michael Leboeuf, 1942–present, American business author

1. Do you think it is important to take notes as you read this book? Explain.
2. Research note-taking techniques online. What are some things you could do to make your notes more helpful?

A Make the following conversions in the U.S. system by multiplying by the appropriate conversion factor. Write your answers as whole numbers or mixed numbers. [Examples 1–3]

1. 5 ft to inches
2. 9 ft to inches
3. 10 ft to inches
4. 20 ft to inches

5. 2 yd to feet
6. 8 yd to feet
7. 4.5 yd to inches
8. 9.5 yd to inches

9. 27 in. to feet
10. 36 in. to feet
11. 2.5 mi to feet
12. 6.75 mi to feet

13. 48 in. to yards
14. 56 in. to yards

B Make the following conversions in the metric system by multiplying by the appropriate conversion factor. Write your answers as whole numbers or decimals. [Examples 4, 5]

15. 18 m to centimeters
16. 18 m to millimeters
17. 4.8 km to meters
18. 8.9 km to meters

19. 5 dm to centimeters
20. 12 dm to millimeters
21. 248 m to kilometers
22. 969 m to kilometers

23. 67 cm to millimeters
24. 67 mm to centimeters
25. 3,498 cm to meters
26. 4,388 dm to meters

27. 63.4 cm to decimeters
28. 89.5 cm to decimeters

Applying the Concepts

29. Lighthouses The chart shows some of the tallest lighthouses in the United States. Convert the height of Cape Hatteras Lighthouse to yards.

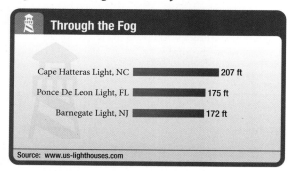

30. Gaming Consoles The chart shows how much energy is consumed by the top gaming consoles when in use.

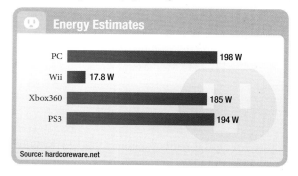

Convert the wattage of the following console to kilowatts.
a. PC
b. Xbox 360
c. Wii

31. Softball If the distance between first and second base in softball is 60 feet, how many yards is it from first base to second base?

32. Soccer According to international soccer regulations, a standard soccer field must be between 45 and 90 meters wide. What is the regulation width in centimeters?

33. High Jump If a person high jumps 6 feet 8 inches, how many inches is the jump?

34. Desk Width A desk is 48 inches wide. What is the width in yards?

35. Ceiling Height Suppose the ceiling of a home is 2.44 meters above the floor. Express the height of the ceiling in centimeters.

36. Tower Height A transmitting tower is 100 feet tall. How many inches is that?

37. Surveying A unit of measure sometimes used in surveying is the *chain*. There are 80 chains in 1 mile. How many chains are in 37 miles?

38. Surveying Another unit of measure used in surveying is a *link*; 1 link is about 8 inches. About how many links are there in 5 feet?

39. Metric System A very small unit of measure in the metric system is the *micron* (abbreviated µm). There are 1,000 µm in 1 millimeter. How many microns are in 12 centimeters?

40. Metric System Another very small unit of measure in the metric system is the *angstrom* (abbreviated Å). There are 10,000,000 Å in 1 millimeter. How many angstroms are in 15 decimeters?

41. Horse Racing In horse racing, 1 *furlong* is 220 yards. How many feet are in 12 furlongs?

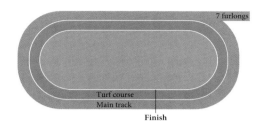

42. Speed of a Bullet A bullet from a machine gun on a B-17 Flying Fortress in World War II had a muzzle speed of 1,750 feet/second. Convert 1,750 feet/second to miles/hour. (Round to the nearest whole number.)

43. Speed Limit The maximum speed limit on part of Highway 101 in California is 55 miles/hour. Convert 55 miles/hour to feet/second. (Round to the nearest tenth.)

44. Speed Limit The maximum speed limit on part of Highway 5 in California is 65 miles/hour. Convert 65 miles/hour to feet/second. (Round to the nearest tenth.)

45. Track and Field A person who runs the 100-yard dash in 10.5 seconds has an average speed of 9.52 yards/second. Convert 9.52 yards/second to miles/hour. (Round to the nearest tenth.)

46. Track and Field A person who runs a mile in 8 minutes has an average speed of 0.125 miles/minute. Convert 0.125 miles/minute to miles/hour.

47. Speed of a Bullet The bullet from a rifle leaves the barrel traveling 1,500 feet/second. Convert 1,500 feet/second to miles/hour. (Round to the nearest whole number.)

48. Sailing A *fathom* is 6 feet. How many yards are in 19 fathoms?

Calculator Problems

Set up the following conversions as you have been doing. Then perform the calculations on a calculator.

49. Change 751 miles to feet.

50. Change 639.87 centimeters to meters.

51. Change 4,982 yards to inches.

52. Change 379 millimeters to kilometers.

53. Mount Whitney is the highest point in California. It is 14,494 feet above sea level. Give its height in miles to the nearest tenth.

54. The tallest mountain in the United States is Mount McKinley in Alaska. It is 20,320 feet tall. Give its height in miles to the nearest tenth.

55. California has 3,427 miles of shoreline. How many feet is this?

56. The tip of the TV tower at the top of the Empire State Building in New York City is 1,472 feet above the ground. Express this height in miles to the nearest hundredth.

Getting Ready for the Next Section

Perform the indicated operations.

57. 12×12 **58.** 36×24 **59.** $1 \times 4 \times 2$ **60.** $5 \times 4 \times 2$

61. $10 \times 10 \times 10$ **62.** $100 \times 100 \times 100$ **63.** $75 \times 43{,}560$ **64.** $55 \times 43{,}560$

65. $864 \div 144$ **66.** $1{,}728 \div 144$ **67.** $256 \div 640$ **68.** $960 \div 240$

69. $45 \times \dfrac{9}{1}$ **70.** $36 \times \dfrac{9}{1}$ **71.** $1{,}800 \times \dfrac{1}{4}$ **72.** $2{,}000 \times \dfrac{1}{4} \times \dfrac{1}{10}$

73. 1.5×30 **74.** 1.5×45 **75.** $2.2 \times 1{,}000$ **76.** $3.5 \times 1{,}000$

77. 67.5×9 **78.** 43.5×9

Maintaining Your Skills

Write your answers as whole numbers, proper fractions, or mixed numbers.

Find each product. (Multiply.)

79. $\dfrac{2}{3} \cdot \dfrac{1}{2}$ **80.** $\dfrac{7}{9} \cdot \dfrac{3}{14}$ **81.** $8 \cdot \dfrac{3}{4}$ **82.** $12 \cdot \dfrac{1}{3}$ **83.** $1\dfrac{1}{2} \cdot 2\dfrac{1}{3}$ **84.** $\dfrac{1}{6} \cdot 4\dfrac{2}{3}$

Find each quotient. (Divide.)

85. $\dfrac{3}{4} \div \dfrac{1}{8}$ **86.** $\dfrac{3}{5} \div \dfrac{6}{25}$ **87.** $4 \div \dfrac{2}{3}$ **88.** $1 \div \dfrac{1}{3}$ **89.** $1\dfrac{3}{4} \div 2\dfrac{1}{2}$ **90.** $\dfrac{9}{8} \div 1\dfrac{7}{8}$

Extending the Concepts

91. Fitness Walking The guidelines for fitness indicate that a person who walks 10,000 steps daily is physically fit. According to *The Walking Site* on the Internet, "The average person's stride length is approximately 2.5 feet long. That means it takes just over 2,000 steps to walk one mile, and 10,000 steps is close to 5 miles." Use your knowledge of unit analysis to determine if these facts are correct.

Unit Analysis II: Area and Volume

6.2

OBJECTIVES

A Convert between areas using the U.S. system.

B Convert between areas using the metric system.

C Convert between volumes using the U.S. system.

D Convert between volumes using the metric system.

TICKET TO SUCCESS

Keep these questions in mind as you read through the section. Then respond in your own words and in complete sentences.

1. Write the formula for the area of each of the following:
   a. a square of side s
   b. a rectangle with length l and width w
2. Explain the difference between square inches and cubic inches.
3. Fill in the numerators below so that each conversion factor is equal to 1.
   a. ___qt / 1 gal
   b. ___mL / 1 liter
   c. ___acres / 1 mi^2
4. What is the basic unit of measure for volume in the metric system?

Suppose you are remodeling your bathroom and plan to lay a tile floor with alternating blue and white square tiles. The tiles you have chosen measure 1 inch long by 1 inch wide. You know the square footage of your bathroom is 50 ft². How many tiles do you need to purchase to cover your floor? We will begin this chapter with converting between areas using the U.S. system, which will give us the tools to answer this tile question.

Figure 1 below gives a summary of the geometric objects we have worked with in previous chapters, along with the formulas for finding the area of each object.

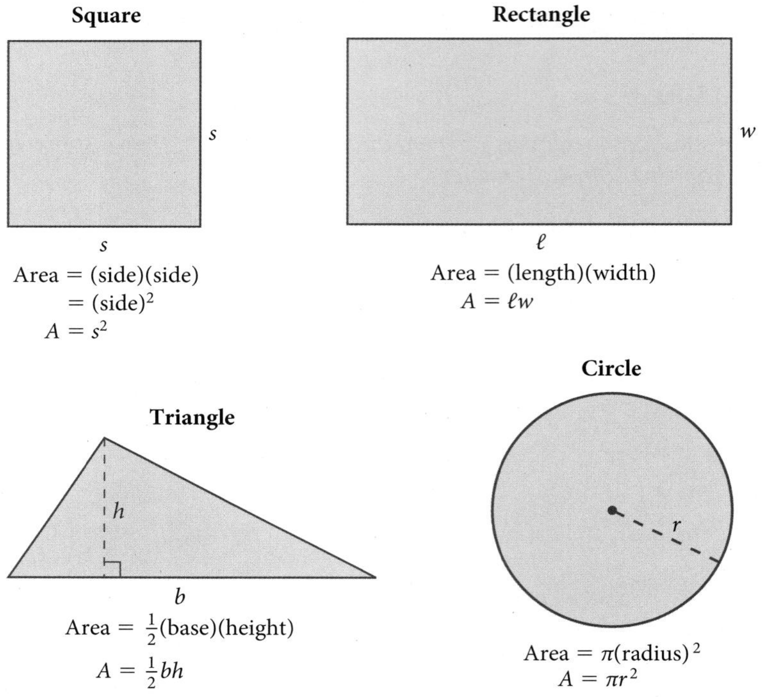

Square

s

Area = (side)(side)
     = (side)^2
$A = s^2$

Rectangle

ℓ, w

Area = (length)(width)
$A = \ell w$

Triangle

h, b

Area = $\frac{1}{2}$(base)(height)
$A = \frac{1}{2}bh$

Circle

r

Area = π(radius)^2
$A = \pi r^2$

FIGURE 1

A Area: The U.S. System

EXAMPLE 1   Find the number of square inches in 1 square foot.

SOLUTION   We can think of 1 square foot as 1 ft² = 1 ft × 1 ft. To convert from feet to inches, we use the conversion factor 1 foot = 12 inches. Because the unit foot appears twice in 1 ft², we multiply by our conversion factor twice.

$$1 \text{ ft}^2 = 1 \text{ ft} \times 1 \text{ ft} \times \frac{12 \text{ in.}}{1 \text{ ft}} \times \frac{12 \text{ in.}}{1 \text{ ft}} = 12 \text{ in.} \times 12 \text{ in.} = 144 \text{ in}^2$$

Now that we know that 1 ft² is the same as 144 in², we can use this fact as a conversion factor to convert between square feet and square inches. Depending on which units we are converting from, we would use either

$$\frac{144 \text{ in}^2}{1 \text{ ft}^2} \quad \text{or} \quad \frac{1 \text{ ft}^2}{144 \text{ in}^2}$$

EXAMPLE 2   A rectangular poster measures 36 inches by 24 inches. How many square feet of wall space will the poster cover?

SOLUTION   One way to work this problem is to find the number of square inches of wall space the poster covers, and then convert square inches to square feet.

PRACTICE PROBLEMS

1. Find the number of square feet in 1 square yard.

2. If the poster in Example 2 is surrounded by a frame 6 inches wide, find the number of square feet of wall space covered by the framed poster.

Answers
1. 1 yd² = 9 ft²
2. 12 ft²

Area of poster = length × width = 36 in. × 24 in. = 864 in^2

To finish the problem, we convert square inches to square feet.

$$864 \text{ in}^2 = 864 \text{ in}^2 \times \frac{1 \text{ ft}^2}{144 \text{ in}^2}$$

$$= \frac{864}{144} \text{ ft}^2$$

$$= 6 \text{ ft}^2$$

Table 1 gives the most common units of area in the U.S. system of measurement, along with the corresponding conversion factors.

TABLE 1

U.S. UNITS OF AREA

The relationship between	is	To convert one to the other, multiply by
square inches and square feet	144 in^2 = 1 ft^2	$\frac{144 \text{ in}^2}{1 \text{ ft}^2}$ or $\frac{1 \text{ ft}^2}{144 \text{ in}^2}$
square yards and square feet	9 ft^2 = 1 yd^2	$\frac{9 \text{ ft}^2}{1 \text{ yd}^2}$ or $\frac{1 \text{ yd}^2}{9 \text{ ft}^2}$
acres and square feet	1 acre = 43,560 ft^2	$\frac{43,560 \text{ ft}^2}{1 \text{ acre}}$ or $\frac{1 \text{ acre}}{43,560 \text{ ft}^2}$
acres and square miles	640 acres = 1 mi^2	$\frac{640 \text{ acres}}{1 \text{ mi}^2}$ or $\frac{1 \text{ mi}^2}{640 \text{ acres}}$

EXAMPLE 3 A dressmaker orders a bolt of material that is 1.5 yards wide and 30 yards long. How many square feet of material were ordered?

SOLUTION The area of the material in square yards is

$$\text{Area} = 1.5 \text{ yd} \times 30 \text{ yd}$$
$$= 45 \text{ yd}^2$$

Converting this to square feet, we have

$$45 \text{ yd}^2 = 45 \text{ yd}^2 \times \frac{9 \text{ ft}^2}{1 \text{ yd}^2}$$
$$= 405 \text{ ft}^2$$

EXAMPLE 4 A farmer has 75 acres of land. How many square feet of land does the farmer have?

3. The same dressmaker orders a bolt of material that is 1.5 yards wide and 45 yards long. How many square feet of material were ordered?

4. A farmer has 55 acres of land. How many square feet of land does the farmer have?

Answers
3. 607.5 ft^2
4. 2,395,800 ft^2

SOLUTION Changing acres to square feet, we have

$$75 \text{ acres} = 75 \text{ acres} \times \frac{43{,}560 \text{ ft}^2}{1 \text{ acre}}$$

$$= 75 \times 43{,}560 \text{ ft}^2$$

$$= 3{,}267{,}000 \text{ ft}^2 \quad \blacksquare$$

EXAMPLE 5 A new shopping center is to be constructed on 256 acres of land. How many square miles is this?

SOLUTION Multiplying by the conversion factor that will allow acres to divide out, we have

$$256 \text{ acres} = 256 \text{ acres} \times \frac{1 \text{ mi}^2}{640 \text{ acres}}$$

$$= \frac{256}{640} \text{ mi}^2$$

$$= 0.4 \text{ mi}^2 \quad \blacksquare$$

B Area: The Metric System

Units of area in the metric system are considerably simpler than those in the U.S. system because metric units are given in terms of powers of 10. Table 2 lists the conversion factors that are most commonly used.

TABLE 2

METRIC UNITS OF AREA

The relationship between	is	To convert one to the other, multiply by
square millimeters and square centimeters	$1 \text{ cm}^2 = 100 \text{ mm}^2$	$\frac{100 \text{ mm}^2}{1 \text{ cm}^2}$ or $\frac{1 \text{ cm}^2}{100 \text{ mm}^2}$
square centimeters and square decimeters	$1 \text{ dm}^2 = 100 \text{ cm}^2$	$\frac{100 \text{ cm}^2}{1 \text{ dm}^2}$ or $\frac{1 \text{ dm}^2}{100 \text{ cm}^2}$
square decimeters and square meters	$1 \text{ m}^2 = 100 \text{ dm}^2$	$\frac{100 \text{ dm}^2}{1 \text{ m}^2}$ or $\frac{1 \text{ m}^2}{100 \text{ dm}^2}$
square meters and ares (a)	$1 \text{ a} = 100 \text{ m}^2$	$\frac{100 \text{ m}^2}{1 \text{ a}}$ or $\frac{1 \text{ a}}{100 \text{ m}^2}$
ares and hectares (ha)	$1 \text{ ha} = 100 \text{ a}$	$\frac{100 \text{ a}}{1 \text{ ha}}$ or $\frac{1 \text{ ha}}{100 \text{ a}}$

EXAMPLE 6 How many square millimeters are in 1 square meter?

SOLUTION We start with 1 m² and end up with square millimeters.

$$1 \text{ m}^2 = 1 \text{ m}^2 \times \frac{100 \text{ dm}^2}{1 \text{ m}^2} \times \frac{100 \text{ cm}^2}{1 \text{ dm}^2} \times \frac{100 \text{ mm}^2}{1 \text{ cm}^2}$$

$$= 100 \times 100 \times 100 \text{ mm}^2$$

$$= 1{,}000{,}000 \text{ mm}^2 \quad \blacksquare$$

---

5. A school is to be constructed on 960 acres of land. How many square miles is this?

6. How many square centimeters are in 1 square meter?

Answers
5. 1.5 mi²
6. 10,000 cm²

C Volume: The U.S. System

Table 3 lists the units of volume in the U.S. system and their conversion factors.

TABLE 3

UNITS OF VOLUME IN THE U.S. SYSTEM

The relationship between	is	To convert one to the other, multiply by
cubic inches (in³) and cubic feet (ft³)	1 ft³ = 1,728 in³	$\frac{1,728 \text{ in}^3}{1 \text{ ft}^3}$ or $\frac{1 \text{ ft}^3}{1,728 \text{ in}^3}$
cubic feet and cubic yards (yd³)	1 yd³ = 27 ft³	$\frac{27 \text{ ft}^3}{1 \text{ yd}^3}$ or $\frac{1 \text{ yd}^3}{27 \text{ ft}^3}$
fluid ounces (fl oz) and pints (pt)	1 pt = 16 fl oz	$\frac{16 \text{ fl oz}}{1 \text{ pt}}$ or $\frac{1 \text{ pt}}{16 \text{ fl oz}}$
pints and quarts (qt)	1 qt = 2 pt	$\frac{2 \text{ pt}}{1 \text{ qt}}$ or $\frac{1 \text{ qt}}{2 \text{ pt}}$
quarts and gallons (gal)	1 gal = 4 qt	$\frac{4 \text{ qt}}{1 \text{ gal}}$ or $\frac{1 \text{ gal}}{4 \text{ qt}}$

EXAMPLE 7 What is the capacity (volume) in pints of a 1-gallon container of milk?

SOLUTION We change from gallons to quarts and then quarts to pints by multiplying by the appropriate conversion factors as given in Table 3.

$$1 \text{ gal} = 1 \text{ gal} \times \frac{4 \text{ qt}}{1 \text{ gal}} \times \frac{2 \text{ pt}}{1 \text{ qt}}$$

$$= 1 \times 4 \times 2 \text{ pt}$$

$$= 8 \text{ pt}$$

A 1-gallon container has the same capacity as 8 one-pint containers.

EXAMPLE 8 A dairy herd produces 1,800 quarts of milk each day. How many gallons is this equivalent to?

SOLUTION Converting 1,800 quarts to gallons, we have

$$1,800 \text{ qt} = 1,800 \text{ qt} \times \frac{1 \text{ gal}}{4 \text{ qt}}$$

$$= \frac{1,800}{4} \text{ gal}$$

$$= 450 \text{ gal}$$

We see that 1,800 quarts is equivalent to 450 gallons.

7. How many pints are in a 5-gallon pail?

8. A dairy herd produces 2,000 quarts of milk each day. How many 10-gallon containers will this milk fill?

Answers
7. 40 pt
8. 50 containers

D Volume: The Metric System

In the metric system, the basic unit of measure for volume is the liter. A liter is the volume enclosed by a cube that is 10 cm on each edge, as shown in Figure 2. We can see that a liter is equivalent to 1,000 cm³.

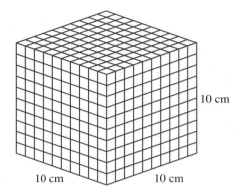

1 liter = 10 cm × 10 cm × 10 cm
= 1,000 cm³

FIGURE 2

The other units of volume in the metric system use the same prefixes we encountered previously. The units with prefixes centi, deci, and deka are not as common as the others, so in Table 4 we include only liters, milliliters, hectoliters, and kiloliters.

TABLE 4

METRIC UNITS OF VOLUME

The relationship between	is	To convert one to the other, multiply by
milliliters (mL) and liters	1 liter (L) = 1,000 mL	$\dfrac{1{,}000 \text{ mL}}{1 \text{ liter}}$ or $\dfrac{1 \text{ liter}}{1{,}000 \text{ mL}}$
hectoliters (hL) and liters	100 liters = 1 hL	$\dfrac{100 \text{ liters}}{1 \text{ hL}}$ or $\dfrac{1 \text{ hL}}{100 \text{ liters}}$
kiloliters (kL) and liters	1,000 liters (L) = 1 kL	$\dfrac{1{,}000 \text{ liters}}{1 \text{ kL}}$ or $\dfrac{1 \text{ kL}}{1{,}000 \text{ liters}}$

Here is an example of conversion from one unit of volume to another in the metric system.

EXAMPLE 9 A sports car has a 2.2-liter engine. What is the displacement (volume) of the engine in milliliters?

SOLUTION Using the appropriate conversion factor from Table 4, we have

$$2.2 \text{ liters} = 2.2 \text{ liters} \times \frac{1{,}000 \text{ mL}}{1 \text{ liter}}$$

$$= 2.2 \times 1{,}000 \text{ mL}$$

$$= 2{,}200 \text{ mL}$$

NOTE

As you can see from the table and the discussion above, a cubic centimeter (cm³) and a milliliter (mL) are equal. Both are one thousandth of a liter. It is also common in some fields (like medicine) to abbreviate the term cubic centimeter as cc. Although we will use the notation mL when discussing volume in the metric system, you should be aware that 1 mL = 1 cm³ = 1 cc.

9. A 3.5-liter engine will have a volume of how many milliliters?

Answer
9. 3,500 mL

Problem Set 6.2

Moving Toward Success

"Go confidently in the direction of your dreams. Live the life you have imagined."
—Henry David Thoreau, 1817–1862, American essayist and poet

1. What are the three main ways you waste your time?
2. How can you prevent these three things from interrupting your study time?

A Use the tables given in this section to make the following conversions. Be sure to show the conversion factor used in each case. [Examples 1–5]

1. 3 ft² to square inches
2. 5 ft² to square inches

3. 288 in² to square feet
4. 720 in² to square feet

5. 30 acres to square feet
6. 92 acres to square feet

7. 2 mi² to acres
8. 7 mi² to acres

9. 1,920 acres to square miles
10. 3,200 acres to square miles

11. 12 yd² to square feet
12. 20 yd² to square feet

B [Example 6]

13. 17 cm² to square millimeters
14. 150 mm² to square centimeters

15. 2.8 m² to square centimeters
16. 10 dm² to square millimeters

17. 1,200 mm² to square meters
18. 19.79 cm² to square meters

19. 5 ares to square meters
20. 12 ares to square centimeters

21. 7 hectares to ares
22. 3.6 hectares to ares

23. 342 ares to hectares

24. 986 ares to hectares

Make the following conversions using the conversion factors given in Tables 3 and 4.

C [Examples 7, 8]

25. 5 yd³ to cubic feet

26. 3.8 yd³ to cubic feet

27. 3 pt to fluid ounces

28. 8 pt to fluid ounces

29. 2 gal to quarts

30. 12 gal to quarts

31. 2.5 gal to pints

32. 7 gal to pints

33. 15 qt to fluid ounces

34. 5.9 qt to fluid ounces

35. 64 pt to gallons

36. 256 pt to gallons

37. 12 pt to quarts

38. 18 pt to quarts

39. 243 ft³ to cubic yards

40. 864 ft³ to cubic yards

D [Example 9]

41. 5 L to milliliters

42. 9.6 L to milliliters

43. 127 mL to liters

44. 93.8 mL to liters

45. 4 kL to milliliters

46. 3 kL to milliliters

47. 14.92 kL to liters

48. 4.71 kL to liters

Applying the Concepts

49. Nile River This chart is from the introduction to the chapter and shows some facts about the Nile River. What is the area of the Nile River delta in acres?

THE NILE RIVER		
	English Units	**Metric Units**
Length	4,160 mi	6,695 km
Nile Delta Area	1,004 mi²	36,000 km²
Flow Rate (monsoon season)	285,829 ft³/s	8,100 m³/s
Average Summer Temperature	86°F	30°C

Source: worldwildlife.org

50. Google Earth The Google Earth image shows an aerial view of a crop circle found near Wroughton, England. If the crop circle has a radius of about 59 meters, how many ares does it cover? Round to the nearest are. Use 3.14 for π.

51. Swimming Pool A public swimming pool measures 100 meters by 30 meters and is rectangular. What is the area of the pool in ares?

52. Construction A family decides to put tiles in the entryway of their home. The entryway has an area of 6 square meters. If each tile is 5 centimeters by 5 centimeters, how many tiles will it take to cover the entryway?

53. Landscaping A landscaper is putting in a brick patio. The area of the patio is 110 square meters. If the bricks measure 10 centimeters by 20 centimeters, how many bricks will it take to make the patio? Assume no space between bricks.

54. Sewing A dressmaker is using a pattern that requires 2 square yards of material. If the material is on a bolt that is 54 inches wide, how long a piece of material must be cut from the bolt to be sure there is enough material for the pattern?

55. Filling Coffee Cups If a regular-size coffee cup holds about $\frac{1}{2}$ pint, about how many cups can be filled from a 1-gallon coffee maker?

56. Filling Glasses If a regular-size drinking glass holds about 0.25 liter of liquid, how many glasses can be filled from a 750-milliliter container?

57. Tiling a room If you are tiling a room with tiles that are one inch by one inch and the room has an area of 50 square feet, how many tiles do you have to buy?

58. Volume of a Tank The gasoline tank on a car holds 18 gallons of gas. What is the volume of the tank in quarts?

59. Filling Glasses How many 8-fluid-ounce glasses of water will it take to fill a 3-gallon aquarium?

60. Filling a Container How many 5-milliliter test tubes filled with water will it take to fill a 1-liter container?

Calculator Problems

Set up the following problems as you have been doing. Then use a calculator to perform the actual calculations. Round answers to two decimal places where appropriate.

61. Geography Lake Superior is the largest of the Great Lakes. It covers 31,700 square miles of area. What is the area of Lake Superior in acres?

62. Geography The state of California consists of 156,360 square miles of land and 2,330 square miles of water. Write the total area (both land and water) in acres.

63. Geography Death Valley National Monument contains 2,067,795 acres of land. How many square miles is this?

64. Geography The Badlands National Monument in South Dakota was established in 1929. It covers 243,302 acres of land. What is the area in square miles?

65. Convert 93.4 qt to gallons.

66. Convert 7,362 fl oz to gallons.

67. How many cubic feet are contained in 796 cubic yards?

68. The engine of a car has a displacement of 440 cubic inches. What is the displacement in cubic feet?

Getting Ready for the Next Section

Perform the indicated operations.

69. 12×16

70. 15×16

71. $3 \times 2{,}000$

72. $5 \times 2{,}000$

73. $3 \times 1{,}000 \times 100$

74. $5 \times 1{,}000 \times 100$

75. $12{,}500 \times \dfrac{1}{1{,}000}$

76. $15{,}000 \times \dfrac{1}{1{,}000}$

Maintaining Your Skills

The following problems review addition and subtraction with fractions and mixed numbers.

77. $\dfrac{3}{8} + \dfrac{1}{4}$

78. $\dfrac{1}{2} + \dfrac{1}{4}$

79. $3\dfrac{1}{2} + 5\dfrac{1}{2}$

80. $6\dfrac{7}{8} + 1\dfrac{5}{8}$

81. $\dfrac{7}{15} - \dfrac{2}{15}$

82. $\dfrac{5}{8} - \dfrac{1}{4}$

83. $\dfrac{5}{36} - \dfrac{1}{48}$

84. $\dfrac{7}{39} - \dfrac{2}{65}$

6.3 Unit Analysis III: Weight

OBJECTIVES

A Convert between weights using the U.S. system.

B Convert between weights using the metric system.

TICKET TO SUCCESS

Keep these questions in mind as you read through the section. Then respond in your own words and in complete sentences.

1. What are the most common units of weight in the U.S. system?
2. Write the conversion factor used to convert from pounds to ounces.
3. What is the basic unit of weight in the metric system?
4. Write the conversion factor used to convert from milligrams to grams.

Go into any grocery store in the United States today and much of the produce is priced by weight, therefore, by the pound. Head over to the deli counter to request some sliced lunch meat and the scale may determine the meat's weight in ounces. Pounds and ounces are measurements of weight in the U.S. system. However, in other parts of the world, these same products are measured in grams and kilograms. In this section, we'll learn and work with the standard units of weight for the U.S. system and the metric system.

A Weights: The U.S. System

The most common units of weight in the U.S. system are ounces, pounds, and tons. The relationships among these units are given in Table 1.

TABLE 1
UNITS OF WEIGHT IN THE U.S. SYSTEM

The relationship between	is	To convert one to the other, multiply by
ounces (oz) and pounds (lb)	1 lb = 16 oz	$\dfrac{16 \text{ oz}}{1 \text{ lb}}$ or $\dfrac{1 \text{ lb}}{16 \text{ oz}}$
pounds and tons (T)	1 T = 2,000 lb	$\dfrac{2{,}000 \text{ lb}}{1 \text{ T}}$ or $\dfrac{1 \text{ T}}{2{,}000 \text{ lb}}$

PRACTICE PROBLEMS

1. Convert 15 pounds to ounces.

EXAMPLE 1 Convert 12 pounds to ounces.

SOLUTION Using the conversion factor from the table, and applying the method we have been using, we have

$$12 \text{ lb} = 12 \text{ lb} \times \frac{16 \text{ oz}}{1 \text{ lb}}$$
$$= 12 \times 16 \text{ oz}$$
$$= 192 \text{ oz}$$

12 pounds is equivalent to 192 ounces. ∎

2. Convert 5 tons to pounds.

EXAMPLE 2 Convert 3 tons to pounds.

SOLUTION We use the conversion factor from the table. We have

$$3 \text{ T} = 3 \text{ T} \times \frac{2{,}000 \text{ lb}}{1 \text{ T}}$$
$$= 6{,}000 \text{ lb}$$

6,000 pounds is the equivalent of 3 tons. ∎

B Weights: The Metric System

In the metric system, the basic unit of weight is a gram. We use the same prefixes we have already used to write the other metric units in terms of grams. Table 2 lists the most common metric units of weight and their conversion factors.

TABLE 2
METRIC UNITS OF WEIGHT

The relationship between	is	To convert one to the other, multiply by
milligrams (mg) and grams (g)	1 g = 1,000 mg	$\frac{1{,}000 \text{ mg}}{1 \text{ g}}$ or $\frac{1 \text{ g}}{1{,}000 \text{ mg}}$
centigrams (cg) and grams	1 g = 100 cg	$\frac{100 \text{ cg}}{1 \text{ g}}$ or $\frac{1 \text{ g}}{100 \text{ cg}}$
kilograms (kg) and grams	1,000 g = 1 kg	$\frac{1{,}000 \text{ g}}{1 \text{ kg}}$ or $\frac{1 \text{ kg}}{1{,}000 \text{ g}}$
metric tons (t) and kilograms	1,000 kg = 1 t	$\frac{1{,}000 \text{ kg}}{1 \text{ t}}$ or $\frac{1 \text{ t}}{1{,}000 \text{ kg}}$

3. Convert 5 kilograms to milligrams.

EXAMPLE 3 Convert 3 kilograms to centigrams.

SOLUTION We convert kilograms to grams and then grams to centigrams.

$$3 \text{ kg} = 3 \text{ kg} \times \frac{1{,}000 \text{ g}}{1 \text{ kg}} \times \frac{100 \text{ cg}}{1 \text{ g}}$$
$$= 3 \times 1{,}000 \times 100 \text{ cg}$$
$$= 300{,}000 \text{ cg}$$

∎

Answers
1. 240 oz
2. 10,000 lb
3. 5,000,000 mg

Problem Set 6.3

Moving Toward Success

"Being busy does not always mean real work. The object of all work is production or accomplishment and to either of these ends there must be forethought, system, planning, intelligence, and honest purpose, as well as perspiration. Seeming to do is not doing."

—Thomas Alva Edison, 1847–1931, American inventor

1. What time during the day do you feel productive? Why?
2. How can you incorporate studying for this class during those productive hours?

Use the conversion factors in Tables 1 and 2 to make the following conversions.

A [Examples 1, 2]

1. 8 lb to ounces
2. 5 lb to ounces
3. 2 T to pounds

4. 5 T to pounds
5. 192 oz to pounds
6. 176 oz to pounds

7. 1,800 lb to tons
8. 10,200 lb to tons
9. 1 T to ounces

10. 3 T to ounces
11. $3\frac{1}{2}$ lb to ounces
12. $5\frac{1}{4}$ lb to ounces

13. $6\frac{1}{2}$ T to pounds
14. $4\frac{1}{5}$ T to pounds

B [Example 3]

15. 2 kg to grams
16. 5 kg to grams
17. 4 cg to milligrams

18. 3 cg to milligrams
19. 2 kg to centigrams
20. 5 kg to centigrams

21. 5.08 g to centigrams
22. 7.14 g to centigrams
23. 450 cg to grams

24. 979 cg to grams

25. 478.95 mg to centigrams

26. 659.43 mg to centigrams

27. 1,578 mg to grams

28. 1,979 mg to grams

29. 42,000 cg to kilograms

30. 97,000 cg to kilograms

Applying the Concepts

31. Fish Oil A bottle of fish oil contains 60 soft gels, each containing 800 mg of the omega-3 fatty acid. How many total grams of the omega-3 fatty acid are in this bottle?

32. Fish Oil A bottle of fish oil contains 50 soft gels, each containing 300 mg of the omega-6 fatty acid. How many total grams of the omega-6 fatty acid are in this bottle?

33. B-Complex A certain B-complex vitamin supplement contains 50 mg of riboflavin, or vitamin B_2. A bottle contains 80 vitamins. How many total grams of riboflavin are in this bottle?

34. B-Complex A certain B-complex vitamin supplement contains 30 mg of thiamine, or vitamin B_1. A bottle contains 80 vitamins. How many total grams of thiamine are in this bottle?

35. Aspirin A bottle of low-strength aspirin contains 120 tablets. Each tablet contains 81 mg of aspirin. How many total grams of aspirin are in this bottle?

36. Aspirin A bottle of maximum-strength aspirin contains 90 tablets. Each tablet contains 500 mg of aspirin. How many total grams of aspirin are in this bottle?

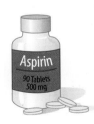

37. Vitamin C A certain brand of vitamin C contains 500 mg per tablet. A bottle contains 240 tablets. How many total grams of vitamin C are in this bottle?

38. Vitamin C A certain brand of vitamin C contains 600 mg per tablet. A bottle contains 150 vitamins. How many total grams of vitamin C are in this bottle?

39. One pound of apples cost $2.50. If you buy 3 pounds and 12 ounces of apples, how much do the apples cost? Round to the nearest cent.

40. One pound of grapefruit costs $2.80. If you buy 2 pounds and 2 ounces of grapefruit, how much does the grapefruit cost?

Coca-Cola Bottles The soft drink Coke is sold throughout the world. Although the size of the bottle varies between different countries, a "six-pack" is sold everywhere. For each of the problems below, find the number of liters in a "6-pack" from the given bottle size.

Country	Bottle size	Liters in a 6-pack
41. Estonia	500 mL	
42. Israel	350 mL	
43. Jordan	250 mL	
44. Kenya	300 mL	

Source: thecoca-colacompany.com

45. Nursing A patient is prescribed a dosage of Ceclor® of 561 mg. How many grams is the dosage?

46. Nursing A patient is prescribed a dosage of 425 mg. How many grams is the dosage?

47. Nursing Dilatrate®-SR comes in 40 milligram capsules. Use this information to determine how many capsules should be given for the prescribed dosages.
 a. 120 mg
 b. 40 mg
 c. 80 mg

48. Nursing A brand of methyldopa comes in 250 milligram tablets. Use this information to determine how many capsules should be given for the prescribed dosages.
 a. 0.125 gram
 b. 750 milligrams
 c. 0.5 gram

Getting Ready for the Next Section

Perform the indicated operations.

49. 8 × 2.54

50. 9 × 3.28

51. 3 × 1.06 × 2

52. 3 × 5 × 3.79

53. 80.5 ÷ 1.61

54. 96.6 ÷ 1.61

55. 125 ÷ 2.50

56. 165 ÷ 2.2

57. 2,000 ÷ 16.39
(Round your answer to the nearest whole number.)

58. 2,200 ÷ 16.39
(Round your answer to the nearest whole number.)

59. $\frac{9}{5}(120) + 32$

60. $\frac{9}{5}(40) + 32$

61. $\frac{5(102 - 30)}{9}$

62. $\frac{5(105 - 42)}{9}$

Maintaining Your Skills

Write each decimal as an equivalent proper fraction or mixed number.

63. 0.18

64. 0.04

65. 0.09

66. 0.045

67. 0.8

68. 0.08

69. 1.75

70. 3.125

Write each fraction or mixed number as a decimal.

71. $\frac{3}{4}$

72. $\frac{9}{10}$

73. $\frac{17}{20}$

74. $\frac{1}{8}$

75. $\frac{3}{5}$

76. $\frac{7}{8}$

77. $3\frac{5}{8}$

78. $1\frac{1}{16}$

Use the definition of exponents to simplify each expression.

80. $\left(\frac{5}{9}\right)^2$

81. $\left(2\frac{1}{2}\right)^2$

82. $\left(\frac{1}{3}\right)^4$

85. $(2.5)^2$

86. $(0.5)^4$

Converting Between the Two Systems, and Temperature

6.4

OBJECTIVES

A Convert between the U.S. and metric systems.

B Convert temperatures between the Fahrenheit and Celsius scales.

TICKET TO SUCCESS

Keep these questions in mind as you read through the section. Then respond in your own words and in complete sentences.

1. Write the equality that gives the relationship between centimeters and inches.
2. Write the equality that gives the relationship between grams and ounces.
3. How does the Fahrenheit scale relate to the Celsius scale?
4. Is it a hot day if the temperature outside is 37°C?

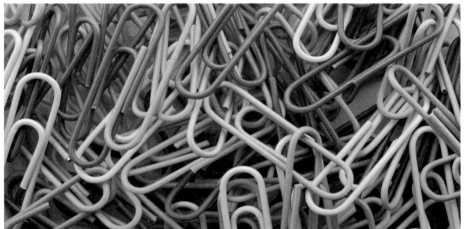

A Converting Between the U.S. and Metric Systems

Because most of us have always used the U.S. system of measurement in our everyday lives, we are much more familiar with it on an intuitive level than we are with the metric system. We have an intuitive idea of how long feet and inches are, how much a pound weighs, and what a square yard of material looks like. The metric system is actually much easier to use than the U.S. system. The reason some of us have such a hard time with the metric system is that we don't have the feel for it that we do for the U.S. system. We have trouble visualizing how long a meter is or how much a gram weighs. The following list is intended to give you something to associate with each basic unit of measurement in the metric system.

1. A meter is just a little longer than a yard.
2. The length of the edge of a sugar cube is about 1 centimeter.
3. A liter is just a little larger than a quart.
4. A sugar cube has a volume of approximately 1 milliliter.
5. A paper clip weighs about 1 gram.
6. A 2-pound can of coffee weighs about 1 kilogram.

The following table includes the most common conversion factors between the metric and the U.S. system of measurement.

TABLE 1

ACTUAL CONVERSION FACTORS BETWEEN THE METRIC AND U.S. SYSTEMS OF MEASUREMENT

The relationship between	is	To convert one to the other, multiply by
Length		
inches and centimeters	2.54 cm = 1 in.	$\dfrac{2.54 \text{ cm}}{1 \text{ in.}}$ or $\dfrac{1 \text{ in.}}{2.54 \text{ cm}}$
feet and meters	1 m = 3.28 ft	$\dfrac{3.28 \text{ ft}}{1 \text{ m}}$ or $\dfrac{1 \text{ m}}{3.28 \text{ ft}}$
miles and kilometers	1.61 km = 1 mi	$\dfrac{1.61 \text{ km}}{1 \text{ mi}}$ or $\dfrac{1 \text{ mi}}{1.61 \text{ km}}$
Area		
square inches and square centimeters	6.45 cm² = 1 in²	$\dfrac{6.45 \text{ cm}^2}{1 \text{ in}^2}$ or $\dfrac{1 \text{ in}^2}{6.45 \text{ cm}^2}$
square meters and square yards	1.196 yd² = 1 m²	$\dfrac{1.196 \text{ yd}^2}{1 \text{ m}^2}$ or $\dfrac{1 \text{ m}^2}{1.196 \text{ yd}^2}$
acres and hectares	1 ha = 2.47 acres	$\dfrac{2.47 \text{ acres}}{1 \text{ ha}}$ or $\dfrac{1 \text{ ha}}{2.47 \text{ acres}}$
Volume		
cubic inches and milliliters	16.39 mL = 1 in³	$\dfrac{16.39 \text{ mL}}{1 \text{ in}^3}$ or $\dfrac{1 \text{ in}^3}{16.39 \text{ mL}}$
liters and quarts	1.06 qt = 1 liter	$\dfrac{1.06 \text{ qt}}{1 \text{ liter}}$ or $\dfrac{1 \text{ liter}}{1.06 \text{ qt}}$
gallons and liters	3.79 liters = 1 gal	$\dfrac{3.79 \text{ liters}}{1 \text{ gal}}$ or $\dfrac{1 \text{ gal}}{3.79 \text{ liters}}$
Weight		
ounces and grams	28.3 g = 1 oz	$\dfrac{28.3 \text{ g}}{1 \text{ oz}}$ or $\dfrac{1 \text{ oz}}{28.3 \text{ g}}$
kilograms and pounds	2.20 lb = 1 kg	$\dfrac{2.20 \text{ lb}}{1 \text{ kg}}$ or $\dfrac{1 \text{ kg}}{2.20 \text{ lb}}$

There are many other conversion factors that we could have included in Table 1. We have listed only the most common ones. Almost all of them are approximations. That is, most of the conversion factors are decimals that have been rounded to the nearest hundredth. If we want more accuracy, we obtain a table that has more digits in the conversion factors. For now, we use the conversion factors from Table 1 to work through the next examples.

EXAMPLE 1   Convert 8 inches to centimeters.

SOLUTION   Choosing the appropriate conversion factor from Table 1, we have

$$8 \text{ in.} = 8 \text{ in.} \times \frac{2.54 \text{ cm}}{1 \text{ in.}}$$
$$= 8 \times 2.54 \text{ cm}$$
$$= 20.32 \text{ cm}$$

EXAMPLE 2   Convert 80.5 kilometers to miles.

SOLUTION   Using the conversion factor that takes us from kilometers to miles, we have

$$80.5 \text{ km} = 80.5 \text{ km} \times \frac{1 \text{ mi}}{1.61 \text{ km}}$$
$$= \frac{80.5}{1.61} \text{ mi}$$
$$= 50 \text{ mi}$$

PRACTICE PROBLEMS

1. Convert 10 inches to centimeters.

2. Convert 16.4 feet to meters.

Answers
1. 25.4 cm
2. 5 m

So 50 miles is equivalent to 80.5 kilometers. If we travel at 50 miles per hour in a car, we are moving at the rate of 80.5 kilometers per hour. ∎

EXAMPLE 3 Convert 3 liters to pints.

SOLUTION Because Table 1 doesn't list a conversion factor that will take us directly from liters to pints, we first convert liters to quarts, and then convert quarts to pints.

$$3 \text{ liters} = 3 \text{ liters} \times \frac{1.06 \text{ qt}}{1 \text{ liter}} \times \frac{2 \text{ pt}}{1 \text{ qt}}$$

$$= 3 \times 1.06 \times 2 \text{ pt}$$
$$= 6.36 \text{ pt}$$

∎

3. Convert 10 liters to gallons. Round to the nearest hundredth.

EXAMPLE 4 The engine in a car has a 2-liter displacement. What is the displacement in cubic inches?

SOLUTION We convert liters to milliliters and then milliliters to cubic inches.

$$2 \text{ liters} = 2 \text{ liters} \times \frac{1,000 \text{ mL}}{1 \text{ liter}} \times \frac{1 \text{ in}^3}{16.39 \text{ mL}}$$

$$= \frac{2 \times 1,000}{16.39} \text{ in}^3 \quad \text{This calculation should be done on a calculator.}$$

$$= 122 \text{ in}^3 \quad \text{Round to the nearest cubic inch.}$$

∎

4. The engine in a car has a 2.2-liter displacement. What is the displacement in cubic inches (to the nearest cubic inch)?

EXAMPLE 5 If a person weighs 125 pounds, what is her weight in kilograms?

SOLUTION Converting from pounds to kilograms, we have

$$125 \text{ lb} = 125 \text{ lb} \times \frac{1 \text{ kg}}{2.20 \text{ lb}}$$

$$= \frac{125}{2.20} \text{ kg}$$

$$= 56.8 \text{ kg} \quad \text{Round to the nearest tenth.}$$

∎

5. A person who weighs 165 pounds weighs how many kilograms?

B Temperature

We end this section with a discussion of temperature in both systems of measurement.

In the U.S. system, we measure temperature on the Fahrenheit scale. On this scale, water boils at 212 degrees and freezes at 32 degrees. When we write 32 degrees measured on the Fahrenheit scale, we use the notation

32°F   Read, "32 degrees Fahrenheit."

In the metric system, the scale we use to measure temperature is the Celsius scale (formerly called the centigrade scale). On this scale, water boils at 100 degrees and freezes at 0 degrees. When we write 100 degrees measured on the Celsius scale, we use the notation

100°C   Read, "100 degrees Celsius."

Answers
3. 2.64 gal
4. 134 in³
5. 75 kg

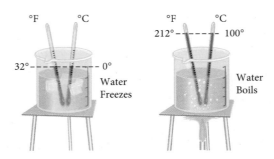

Table 2 gives the formulas, in both symbols and words, that are used to convert between the two scales.

TABLE 2		
To Convert From	Formula In Symbols	Formula In Words
Fahrenheit to Celsius	$C = \frac{5}{9}(F - 32)$	Subtract 32, multiply by 5, and then divide by 9.
Celsius to Fahrenheit	$F = \frac{9}{5}C + 32$	Multiply by $\frac{9}{5}$, and then add 32.

The following examples show how we use the formulas given in Table 3.

EXAMPLE 6 Convert 120°C to degrees Fahrenheit.

SOLUTION We use the formula

$$F = \frac{9}{5}C + 32$$

and replace C with 120.

When → $C = 120$

the formula → $F = \frac{9}{5}C + 32$

becomes → $F = \frac{9}{5}(120) + 32$

$$F = 216 + 32$$
$$F = 248$$

We see that 120°C is equivalent to 248°F; they both mean the same temperature. ∎

EXAMPLE 7 A man with the flu has a temperature of 102°F. What is his temperature on the Celsius scale?

SOLUTION When → $F = 102$

the formula → $C = \frac{5(F - 32)}{9}$

becomes → $C = \frac{5(102 - 32)}{9}$

$$C = \frac{5(70)}{9}$$

$C = 38.9$ Round to the nearest tenth.

The man's temperature, rounded to the nearest tenth, is 38.9°C on the Celsius scale. ∎

---

6. Convert 40°C to degrees Fahrenheit.

7. A child is running a temperature of 101.6°F. What is her temperature, to the nearest tenth of a degree, on the Celsius scale?

Answers
6. 104°F
7. 38.7°C

Problem Set 6.4

Moving Toward Success

"Respect your fellow human being, treat them fairly, disagree with them honestly, enjoy their friendship, explore your thoughts about one another candidly, work together for a common goal and help one another achieve it."

—Bill Bradley, 1943–present, American retired NBA basketball player and senator

1. What does it mean to be an active listener? How does being an active listener help you succeed in this course?
2. How can you improve your listening skills during this class?

A B Use Tables 1 and 3 to make the following conversions. [Examples 1–7]

1. 6 in. to centimeters
2. 1 ft to centimeters
3. 4 m to feet
4. 2 km to feet
5. 6 m to yards
6. 15 mi to kilometers
7. 20 mi to meters (round to the nearest hundred meters)
8. 600 m to yards
9. 5 m² to square yards (round to the nearest hundredth)
10. 2 in² to square centimeters (round to the nearest tenth)
11. 10 ha to acres
12. 50 a to acres
13. 500 in³ to milliliters
14. 400 in³ to liters
15. 2 L to quarts
16. 15 L to quarts
17. 20 gal to liters
18. 15 gal to liters
19. 12 oz to grams
20. 1 lb to grams (round to the nearest 10 grams)
21. 15 kg to pounds
22. 10 kg to ounces
23. 185°C to degrees Fahrenheit
24. 20°C to degrees Fahrenheit
25. 86°F to degrees Celsius
26. 122°F to degrees Celsius

Applying the Concepts

27. Temperature The chart shows the temperatures for some of the world's hottest places. Convert the temperature in Al'Aziziyah to Celsius.

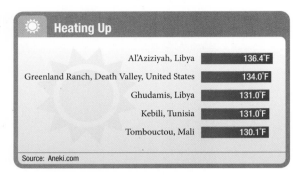

28. Google Earth The Google Earth image is of Lake Clark National Park in Alaska. Lake Clark has an average temperature of 40 degrees Fahrenheit. What is its average temperature in Celsius to the nearest degree?

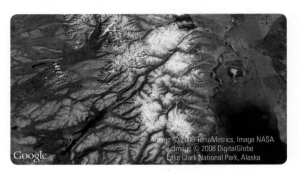

Nile River Recall the following table from the introduction to this chapter.

29. Using the conversion factor of 1.61 km = 1 mi, show the work needed to calculate the Nile's length in miles to length in meters.

30. How many inches long is the Nile?

THE NILE RIVER		
	English Units	Metric Units
Length	4,160 mi	6,698 km
Nile Delta Area	1,004 mi²	36,000 km²
Flow Rate (monsoon season)	285,829 ft³/s	8,100 m³/s
Average Summer Temperature	86°F	30°C
Source: worldwildlife.org		

Nursing Liquid medication is usually given in milligrams per milliliter. Use the information to find the amount a patient should take for a prescribed dosage.

31. Vantin© has a dosage strength of 100 mg/5 mL. If a patient is prescribed a dosage of 150 mg, how many milliliters should she take?

32. A brand of amoxicillin has a dosage strength of 125 mg/5 mL. If a patient is prescribed a dosage of 25 mg, how many milliliters should she take?

Calculator Problems

Set up the following problems as we have set up the examples in this section. Then use a calculator for the calculations and round your answers to the nearest hundredth.

33. 10 cm to inches

34. 100 mi to kilometers

35. 25 ft to meters

36. 400 mL to cubic inches

37. 49 qt to liters

38. 65 L to gallons

39. 500 g to ounces

40. 100 lb to kilograms

41. **Weight** Give your weight in kilograms.

42. **Height** Give your height in meters and centimeters.

43. **Sports** The 100-yard dash is a popular race in track. How far is 100 yards in meters?

44. **Engine Displacement** A 351-cubic-inch engine has a displacement of how many liters?

45. **Sewing** 25 square yards of material is how many square meters?

46. **Weight** How many grams does a 5 lb 4 oz roast weigh?

47. **Speed** 55 miles per hour is equivalent to how many kilometers per hour?

48. **Capacity** A 1-quart container holds how many liters?

49. **Sports** A high jumper jumps 6 ft 8 in. How many meters is this?

50. **Farming** A farmer owns 57 acres of land. How many hectares is that?

51. **Body Temperature** A person has a temperature of 101°F. What is the person's temperature, to the nearest tenth, on the Celsius scale?

52. **Air Temperature** If the temperature outside is 30°C, is it a better day for water skiing or for snow skiing?

Getting Ready for the Next Section

Perform the indicated operations.

53. 15 + 60

54. 25 + 60

55. 37
+ 45

56. 27
+ 46

57. 3 + 0.25

58. 2 + 0.75

59. 82 − 60

60. 73 − 60

61. 75
− 34

62. 85
− 42

63. 12 × 4

64. 8 × 4

65. 3 × 60 + 15

66. 2 × 65 + 45

67. $3 + 17 \times \frac{1}{65}$

68. $2 + 45 \times \frac{1}{60}$

69. If fish costs $6.00 per pound, find the cost of 15 pounds.

70. If fish costs $5.00 per pound, find the cost of 14 pounds.

Maintaining Your Skills

Find the mean and the range for each set of numbers.

71. 5, 7, 9, 11

72. 6, 8, 10, 12

73. 1, 4, 5, 10, 10

74. 2, 4, 4, 6, 9

Find the median and the range for each set of numbers.

75. 15, 18, 21, 24, 29

76. 20, 30, 35, 45, 50

77. 32, 38, 42, 48

78. 53, 61, 67, 75

Find the mode and the range for each set of numbers.

79. 20, 15, 14, 13, 14, 18

80. 17, 31, 31, 26, 31, 29

81. A student has quiz scores of 65, 72, 70, 88, 70, and 73. Find each of the following:
 a. Mean score
 b. Median score
 c. Mode of the scores
 d. Range of scores

82. A person has bowling scores of 207, 224, 195, 207, 185, and 182. Find each of the following:
 a. Mean score
 b. Median score
 c. Mode of the scores
 d. Range of scores

6.5 Operations with Time and Mixed Units

OBJECTIVES

A Convert mixed units to a single unit.

B Add and subtract mixed units.

C Use multiplication with mixed units.

TICKET TO SUCCESS

Keep these questions in mind as you read through the section. Then respond in your own words and in complete sentences.

1. Explain the difference between saying *2 and a half hours* and saying *2 hours and 30 minutes*.
2. How are operations with mixed units of measure similar to operations with mixed numbers?
3. Why do we borrow a 60 from the minutes column for the seconds column when subtracting in Example 3?
4. Give an example of when you may have to use multiplication with mixed units of measure.

Many occupations require the use of a time card. A time card records the number of hours and minutes at work. At the end of a work week, the hours and minutes are totaled separately, and then the minutes are converted to hours.

In this section, we will perform operations with mixed units of measure. Mixed units are used when we use 2 hours 30 minutes, rather than two and a half hours, or 5 feet 9 inches, rather than five and three-quarter feet. As you will see, many of these types of problems arise in everyday life.

A Converting Time to Single Units

The relationship between	is	To convert from one to the other, multiply by
minutes and seconds	1 min = 60 sec	$\dfrac{1 \text{ min}}{60 \text{ sec}}$ or $\dfrac{60 \text{ sec}}{1 \text{ min}}$
hours and minutes	1 hr = 60 min	$\dfrac{1 \text{ hr}}{60 \text{ min}}$ or $\dfrac{60 \text{ min}}{1 \text{ hr}}$

Chapter 6 Measurement

We use the information in the table to solve the following problems.

EXAMPLE 1 Convert 3 hours 15 minutes to
a. minutes.   b. hours.

SOLUTION a. To convert to minutes, we multiply the hours by the conversion factor and then add minutes.

$$3 \text{ hr } 15 \text{ min} = 3 \text{ hr} \times \frac{60 \text{ min}}{1 \text{ hr}} + 15 \text{ min}$$
$$= 180 \text{ min} + 15 \text{ min}$$
$$= 195 \text{ min}$$

b. To convert to hours, we multiply the minutes by the conversion factor and then add hours.

$$3 \text{ hr } 15 \text{ min} = 3 \text{ hr} + 15 \text{ min} \times \frac{1 \text{ hr}}{60 \text{ min}}$$
$$= 3 \text{ hr} + 0.25 \text{ hr}$$
$$= 3.25 \text{ hr}$$

B Addition and Subtraction with Mixed Units

Adding mixed units is similar to adding mixed fractions in that we add each type of unit separately. That is, we align the whole numbers with the whole numbers and the fractions with the fractions.

EXAMPLE 2 Add 5 minutes 37 seconds and 7 minutes 45 seconds.

SOLUTION First, we align the units properly.

```
   5 min   37 sec
 + 7 min   45 sec
  12 min   82 sec
```

Since there are 60 seconds in every minute, we write 82 seconds as 1 minute 22 seconds. We have

$$12 \text{ min } 82 \text{ sec} = 12 \text{ min} + 1 \text{ min } 22 \text{ sec}$$
$$= 13 \text{ min } 22 \text{ sec}$$

Again, the idea of adding units separately is similar to adding mixed fractions. When we subtract units of time, we "borrow" 60 seconds from the minutes column, or 60 minutes from the hours column. The next example shows how we do this.

EXAMPLE 3 Subtract 34 minutes from 8 hours 15 minutes.

SOLUTION Again, we first line up the numbers in the hours column, and then the numbers in the minutes column.

```
  8 hr   15 min   ⇒   7 hr   75 min
-        34 min   -          34 min
                      7 hr   41 min
```

PRACTICE PROBLEMS

1. Convert 2 hours 45 minutes to
   a. minutes
   b. hours

2. Add 4 min 27 sec and 8 min 46 sec.

3. Subtract 42 min from 6 hr 25 min.

Answers
1. a. 165 minutes b. 2.75 hours
2. 13 min 13 sec
3. 5 hr 43 min

C Multiplication with Mixed Units

Next, we see how to multiply using mixed units of measure.

EXAMPLE 4   Jake purchases 4 halibut. The fish cost $6.00 per pound, and each weighs 3 lb 12 oz. What is the cost of the fish?

SOLUTION   First, we multiply each unit by 4, the number of halibut purchased.

$$\begin{array}{r} 3\text{ lb} \quad 12\text{ oz} \\ \times \qquad\qquad 4 \\ \hline 12\text{ lb} \quad 48\text{ oz} \end{array}$$

To convert the 48 ounces to pounds, we multiply the ounces by the conversion factor.

$$\begin{aligned} 12\text{ lb }48\text{ oz} &= 12\text{ lb} + 48\text{ oz} \times \frac{1\text{ lb}}{16\text{ oz}} \\ &= 12\text{ lb} + 3\text{ lb} \\ &= 15\text{ lb} \end{aligned}$$

Finally, we multiply the 15 lb and $6.00/lb for a total price of $90.00. ■

4. Rob is purchasing 4 halibut. The fish cost $5.00 per pound, and each weighs 3 lb 8 oz. What is the cost of the fish?

EXAMPLE 5   A grizzly bear cub was 1 foot 8 inches long. Fully grown, it is 5 times as long. How long is the fully grown grizzly bear?

5. A baby giraffe was 5 feet 8 inches tall. Fully grown, it is 3 times as tall. How tall is the fully grown giraffe?

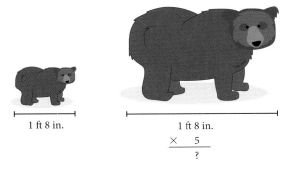

$$\begin{array}{r} 1\text{ ft }8\text{ in.} \\ \times \quad 5 \\ \hline ? \end{array}$$

SOLUTION   First, we multiply each unit by 5.

$$\begin{array}{r} 1\text{ ft} \quad 8\text{ in.} \\ \times \qquad\quad 5 \\ \hline 5\text{ ft} \quad 40\text{ in.} \end{array}$$

To convert the 40 inches to feet, we multiply the inches by the conversion factor.

$$\begin{aligned} 5\text{ ft }40\text{ in} &= 5\text{ ft} + 40\text{ in.} \times \frac{1\text{ ft}}{12\text{ in.}} \\ &= 5\text{ ft} + 3\text{ ft} + 4\text{ in.} \\ &= 8\text{ ft }4\text{ in.} \end{aligned}$$ ■

Answers
4. $70
5. 17 feet

Problem Set 6.5

Moving Toward Success

"To insure good health: eat lightly, breathe deeply, live moderately, cultivate cheerfulness, and maintain an interest in life."

—William Londen, Author

1. Why do you think eating healthy balanced meals leads to success in math?
2. Why do you think scheduling physical activity during your day can help you be successful in math?

A Use the tables of conversion factors given in this section and other sections in this chapter to make the following conversions. (Round your answers to the nearest hundredth.) [Example 1]

1. 4 hours 30 minutes to
   a. Minutes
   b. Hours

2. 2 hours 45 minutes to
   a. Minutes
   b. Hours

3. 5 hours 20 minutes to
   a. Minutes
   b. Hours

4. 4 hours 40 minutes to
   a. Minutes
   b. Hours

5. 6 minutes 30 seconds to
   a. Seconds
   b. Minutes

6. 8 minutes 45 seconds to
   a. Seconds
   b. Minutes

7. 5 minutes 20 seconds to
   a. Seconds
   b. Minutes

8. 4 minutes 40 seconds to
   a. Seconds
   b. Minutes

9. 2 pounds 8 ounces to
   a. Ounces
   b. Pounds

10. 3 pounds 4 ounces to
    a. Ounces
    b. Pounds

11. 4 pounds 12 ounces to
    a. Ounces
    b. Pounds

12. 5 pounds 16 ounces to
    a. Ounces
    b. Pounds

13. 4 feet 6 inches to
    a. Inches
    b. Feet

14. 3 feet 3 inches to
    a. Inches
    b. Feet

15. 5 feet 9 inches to
    a. Inches
    b. Feet

16. 3 feet 4 inches to
    a. Inches
    b. Feet

17. 2 gallons 1 quart to
    a. Quarts
    b. Gallons

18. 3 gallons 2 quarts to
    a. Quarts
    b. Gallons

B Perform the indicated operation. Again, remember to use the appropriate conversion factor. [Examples 2, 3]

19. Add 4 hours 47 minutes and 6 hours 13 minutes.

20. Add 5 hours 39 minutes and 2 hours 21 minutes.

21. Add 8 feet 10 inches and 13 feet 6 inches.

22. Add 16 feet 7 inches and 7 feet 9 inches.

23. Add 4 pounds 12 ounces and 6 pounds 4 ounces.

24. Add 11 pounds 9 ounces and 3 pounds 7 ounces.

25. Subtract 2 hours 35 minutes from 8 hours 15 minutes.

26. Subtract 3 hours 47 minutes from 5 hours 33 minutes.

27. Subtract 3 hours 43 minutes from 7 hours 30 minutes.

28. Subtract 1 hour 44 minutes from 6 hours 22 minutes.

29. Subtract 4 hours 17 minutes from 5 hours 9 minutes.

30. Subtract 2 hours 54 minutes from 3 hours 7 minutes.

31. Multiply 2 hours 36 minutes by 4.

32. Multiply 3 hours 48 minutes by 3.

33. Multiply 6 pounds 12 ounces by 5.

34. Multiply 4 pounds 6 ounces by 6.

35. Multiply 2 feet 9 inches by 4.

36. Multiply 3 feet 4 inches by 6.

Applying the Concepts

37. Fifth Avenue Mile The chart shows the times of the five fastest runners for 2009's Continental Airlines Fifth Avenue Mile. How much faster was Andy Baddeley than Bernard Lagat?

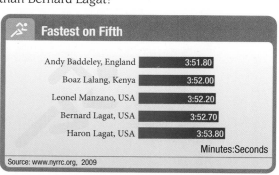

38. Cars The chart shows the fastest cars in the world. Convert the speed of the McLaren F1 to feet per second.

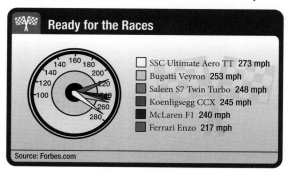

Triathlon The Ironman Triathlon World Championship, held each October in Kona on the island of Hawaii, consists of three parts: a 2.4-mile ocean swim, a 112-mile bike race, and a 26.2-mile marathon. The table shows some results from the 2010 race event.

Triathlete	Swim Time (Hr:Min:Sec)	Bike Time (Hr:Min:Sec)	Run Time (Hr:Min:Sec)	Total Time (Hr:Min:Sec)
Chris McCormack	0:51:36	4:31:51	2:43:31	
Andreas Raelert	0:51:27	4:32:27	2:44:25	

Source: ironman.com

39. Fill in the total time column.

40. How much faster was Chris's total time than Andreas's?

41. How much faster was Andreas's swim time than Chris's?

42. How much faster was Chris's run time than Andreas's?

43. Cost of Fish Fredrick is purchasing four whole salmon. The fish cost $4.00 per pound, and each weighs 6 lb 8 oz. What is the cost of the fish?

44. Cost of Steak Mike is purchasing eight top sirloin steaks. The meat costs $4.00 per pound, and each steak weighs 1 lb 4 oz. What is the total cost of the steaks?

45. Stationary Bike Maggie rides a stationary bike for 1 hour and 15 minutes, 4 days a week. After 2 weeks, how many hours has she spent riding the stationary bike?

46. Gardening Scott works in his garden for 1 hour and 5 minutes, 3 days a week. After 4 weeks, how many hours has Scott spent gardening?

47. Cost of Fabric Allison is making a quilt. She buys 3 yards and 1 foot each of six different fabrics. The fabrics cost $7.50 a yard. How much will Allison spend?

48. Cost of Lumber Trish is building a fence. She buys six fence posts at the lumberyard, each measuring 5 ft 4 in. The lumber costs $3 per foot. How much will Trish spend?

49. Cost of Avocados Jacqueline is buying six avocados. Each avocado weighs 8 oz. How much will they cost her if avocados cost $2.00 a pound?

50. Cost of Apples Mary is purchasing 12 apples. Each apple weighs 4 oz. If the cost of the apples is $1.50 a pound, how much will Mary pay?

Maintaining Your Skills

51. Caffeine Content Suppose the following bar chart shows the amount of caffeine in five different soft drinks. Use the information in the bar chart to fill in the table.

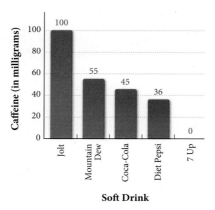

CAFFEINE CONTENT IN SOFT DRINKS	
Drink	Caffeine (in milligrams)
Jolt	
Mountain Dew	
Coca-Cola	
Diet Pepsi	
7 Up	

52. Exercise Suppose the following bar chart shows the number of calories burned in 1 hour of exercise by a person who weighs 150 pounds. Use the information in the bar chart to fill in the table.

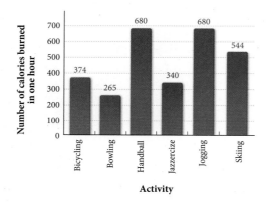

CALORIES BURNED BY A 150-POUND PERSON IN ONE HOUR	
Activity	Calories
Bicycling	
Bowling	
Handball	
Jazzercise	
Jogging	
Skiing	

Extending the Concepts

53. In 2011, the horse Animal Kingdom won the Kentucky Derby with a time of 2:02.04, or 2 minutes and 2.04 seconds. The record time for the Kentucky Derby is still held by Secretariat, who won the race with a time of 1:59.40 in 1973. How much faster did Secretariat run than Animal Kingdom?

54. In 2010, the horse Drosselmeyer won the Belmont Stakes with a time of 2:31.57, or two minutes and 31.57 seconds. The record time for the Belmont Stakes is still held by Secretariat, who won the race with a time of 2:24.00 in 1973. How much faster did Secretariat run than Drosselmeyer?

Chapter 6 Summary

EXAMPLES

1. Convert 5 feet to inches.

$$5 \text{ ft} = 5 \text{ ft} \times \frac{12 \text{ in.}}{1 \text{ ft}}$$
$$= 5 \times 12 \text{ in.}$$
$$= 60 \text{ in.}$$

■ Conversion Factors [6.1, 6.2, 6.3, 6.4, 6.5]

To convert from one kind of unit to another, we choose an appropriate conversion factor from one of the tables given in this chapter. For example, if we want to convert 5 feet to inches, we look for conversion factors that give the relationship between feet and inches. There are two conversion factors for feet and inches:

$$\frac{12 \text{ in.}}{1 \text{ ft}} \quad \text{and} \quad \frac{1 \text{ ft}}{12 \text{ in.}}$$

■ Length [6.1]

2. Convert 8 feet to yards.

$$8 \text{ ft} = 8 \text{ ft} \times \frac{1 \text{ yd}}{3 \text{ ft}}$$
$$= \frac{8}{3} \text{ yd}$$
$$= 2\frac{2}{3} \text{ yd}$$

U.S. SYSTEM		
The relationship between	is	To convert from one to the other, multiply by
feet and inches	12 in. = 1 ft	$\frac{12 \text{ in.}}{1 \text{ ft}}$ or $\frac{1 \text{ ft}}{12 \text{ in.}}$
feet and yards	1 yd = 3 ft	$\frac{3 \text{ ft}}{1 \text{ yd}}$ or $\frac{1 \text{ yd}}{3 \text{ ft}}$
feet and miles	1 mi = 5,280 ft	$\frac{5,280 \text{ ft}}{1 \text{ mi}}$ or $\frac{1 \text{ mi}}{5,280 \text{ ft}}$

3. Convert 25 millimeters to meters.

$$25 \text{ mm} = 25 \text{ mm} \times \frac{1 \text{ m}}{1,000 \text{ mm}}$$
$$= \frac{25 \text{ m}}{1,000}$$
$$= 0.025 \text{ m}$$

METRIC SYSTEM		
The relationship between	is	To convert from one to the other, multiply by
millimeters (mm) and meters (m)	1,000 mm = 1 m	$\frac{1,000 \text{ mm}}{1 \text{ m}}$ or $\frac{1 \text{ m}}{1,000 \text{ mm}}$
centimeters (cm) and meters	100 cm = 1 m	$\frac{100 \text{ cm}}{1 \text{ m}}$ or $\frac{1 \text{ m}}{100 \text{ cm}}$
decimeters (dm) and meters	10 dm = 1 m	$\frac{10 \text{ dm}}{1 \text{ m}}$ or $\frac{1 \text{ m}}{10 \text{ dm}}$
dekameters (dam) and meters	1 dam = 10 m	$\frac{10 \text{ m}}{1 \text{ dam}}$ or $\frac{1 \text{ dam}}{10 \text{ m}}$
hectometers (hm) and meters	1 hm = 100 m	$\frac{100 \text{ m}}{1 \text{ hm}}$ or $\frac{1 \text{ hm}}{100 \text{ m}}$
kilometers (km) and meters	1 km = 1,000 m	$\frac{1,000 \text{ m}}{1 \text{ km}}$ or $\frac{1 \text{ km}}{1,000 \text{ m}}$

Area [6.2]

4. Convert 256 acres to square miles.

$$256 \text{ acres} = 256 \text{ acres} \times \frac{1 \text{ mi}^2}{640 \text{ acres}}$$
$$= \frac{256}{640} \text{ mi}^2$$
$$= 0.4 \text{ mi}^2$$

U.S. SYSTEM		
The relationship between	is	To convert from one to the other, multiply by
square inches and square feet	144 in² = 1 ft²	$\frac{144 \text{ in}^2}{1 \text{ ft}^2}$ or $\frac{1 \text{ ft}^2}{144 \text{ in}^2}$
square yards and square feet	9 ft² = 1 yd²	$\frac{9 \text{ ft}^2}{1 \text{ yd}^2}$ or $\frac{1 \text{ yd}^2}{9 \text{ ft}^2}$
acres and square feet	1 acre = 43,560 ft²	$\frac{43,560 \text{ ft}^2}{1 \text{ acre}}$ or $\frac{1 \text{ acre}}{43,560 \text{ ft}^2}$
acres and square miles	640 acres = 1 mi²	$\frac{640 \text{ acres}}{1 \text{ mi}^2}$ or $\frac{1 \text{ mi}^2}{640 \text{ acres}}$

METRIC SYSTEM		
The relationship between	is	To convert from one to the other, multiply by
square millimeters and square centimeters	1 cm² = 100 mm²	$\frac{100 \text{ mm}^2}{1 \text{ cm}^2}$ or $\frac{1 \text{ cm}^2}{100 \text{ mm}^2}$
square centimeters and square decimeters	1 dm² = 100 cm²	$\frac{100 \text{ cm}^2}{1 \text{ dm}^2}$ or $\frac{1 \text{ dm}^2}{100 \text{ cm}^2}$
square decimeters and square meters	1 m² = 100 dm²	$\frac{100 \text{ dm}^2}{1 \text{ m}^2}$ or $\frac{1 \text{ m}^2}{100 \text{ dm}^2}$
square meters and ares (a)	1 a = 100 m²	$\frac{100 \text{ m}^2}{1 \text{ a}}$ or $\frac{1 \text{ a}}{100 \text{ m}^2}$
ares and hectares (ha)	1 ha = 100 a	$\frac{100 \text{ a}}{1 \text{ ha}}$ or $\frac{1 \text{ ha}}{100 \text{ a}}$

Volume [6.2]

U.S. SYSTEM		
The relationship between	is	To convert from one to the other, multiply by
cubic inches (in³) and cubic feet (ft³)	1 ft³ = 1,728 in³	$\frac{1,728 \text{ in}^3}{1 \text{ ft}^3}$ or $\frac{1 \text{ ft}^3}{1,728 \text{ in}^3}$
cubic feet and cubic yards (yd³)	1 yd³ = 27 ft³	$\frac{27 \text{ ft}^3}{1 \text{ yd}^3}$ or $\frac{1 \text{ yd}^3}{27 \text{ ft}^3}$
fluid ounces (fl oz) and pints (pt)	1 pt = 16 fl oz	$\frac{16 \text{ fl oz}}{1 \text{ pt}}$ or $\frac{1 \text{ pt}}{16 \text{ fl oz}}$
pints and quarts (qt)	1 qt = 2 pt	$\frac{2 \text{ pt}}{1 \text{ qt}}$ or $\frac{1 \text{ qt}}{2 \text{ pt}}$
quarts and gallons (gal)	1 gal = 4 qt	$\frac{4 \text{ qt}}{1 \text{ gal}}$ or $\frac{1 \text{ gal}}{4 \text{ qt}}$

5. Convert 2.2 liters to milliliters.

$$2.2 \text{ liters} = 2.2 \cancel{\text{ liters}} \times \frac{1{,}000 \text{ mL}}{1 \cancel{\text{ liters}}}$$
$$= 2.2 \times 1{,}000 \text{ mL}$$
$$= 2{,}200 \text{ mL}$$

METRIC SYSTEM		
The relationship between	is	To convert from one to the other, multiply by
milliliters (mL) and liters	1 liter (L) = 1,000 mL	$\frac{1{,}000 \text{ mL}}{1 \text{ liter}}$ or $\frac{1 \text{ liter}}{1{,}000 \text{ mL}}$
hectoliters (hL) and liters	100 liters = 1 hL	$\frac{100 \text{ liters}}{1 \text{ hL}}$ or $\frac{1 \text{ hL}}{100 \text{ liters}}$
kiloliters (kL) and liters	1,000 liters (L) = 1 kL	$\frac{1{,}000 \text{ liters}}{1 \text{ kL}}$ or $\frac{1 \text{ kL}}{1{,}000 \text{ liters}}$

Weight [6.3]

6. Convert 12 pounds to ounces.

$$12 \text{ lb} = 12 \cancel{\text{ lb}} \times \frac{16 \text{ oz}}{1 \cancel{\text{ lb}}}$$
$$= 12 \times 16 \text{ oz}$$
$$= 192 \text{ oz}$$

U.S. SYSTEM		
The relationship between	is	To convert from one to the other, multiply by
ounces (oz) and pounds (lb)	1 lb = 16 oz	$\frac{16 \text{ oz}}{1 \text{ lb}}$ or $\frac{1 \text{ lb}}{16 \text{ oz}}$
pounds and tons (T)	1 T = 2,000 lb	$\frac{2{,}000 \text{ lb}}{1 \text{ T}}$ or $\frac{1 \text{ T}}{2{,}000 \text{ lb}}$

7. Convert 3 kilograms to centigrams.

$$3 \text{ kg} = 3 \cancel{\text{ kg}} \times \frac{1{,}000 \cancel{\text{ g}}}{1 \cancel{\text{ kg}}} \times \frac{100 \text{ cg}}{1 \cancel{\text{ g}}}$$
$$= 3 \times 1{,}000 \times 100 \text{ cg}$$
$$= 300{,}000 \text{ cg}$$

METRIC SYSTEM		
The relationship between	is	To convert from one to the other, multiply by
milligrams (mg) and grams (g)	1 g = 1,000 mg	$\frac{1{,}000 \text{ mg}}{1 \text{ g}}$ or $\frac{1 \text{ g}}{1{,}000 \text{ mg}}$
centigrams (cg) and grams	1 g = 100 cg	$\frac{100 \text{ cg}}{1 \text{ g}}$ or $\frac{1 \text{ g}}{100 \text{ cg}}$
kilograms (kg) and grams	1,000 g = 1 kg	$\frac{1{,}000 \text{ g}}{1 \text{ kg}}$ or $\frac{1 \text{ kg}}{1{,}000 \text{ g}}$
metric tons (t) and kilograms	1,000 kg = 1 t	$\frac{1{,}000 \text{ kg}}{1 \text{ t}}$ or $\frac{1 \text{ t}}{1{,}000 \text{ kg}}$

Converting Between the U.S. and Metric Systems [6.4]

8. Convert 8 inches to centimeters.

$$8 \text{ in.} = 8 \text{ in.} \times \frac{2.54 \text{ cm}}{1 \text{ in.}}$$
$$= 8 \times 2.54 \text{ cm}$$
$$= 20.32 \text{ cm}$$

CONVERSION FACTORS

The relationship between	is	To convert from one to the other, multiply by
Length		
inches and centimeters	2.54 cm = 1 in.	$\frac{2.54 \text{ cm}}{1 \text{ in.}}$ or $\frac{1 \text{ in.}}{2.54 \text{ cm}}$
feet and meters	1 m = 3.28 ft	$\frac{3.28 \text{ ft}}{1 \text{ m}}$ or $\frac{1 \text{ m}}{3.28 \text{ ft}}$
miles and kilometers	1.61 km = 1 mi	$\frac{1.61 \text{ km}}{1 \text{ mi}}$ or $\frac{1 \text{ mi}}{1.61 \text{ km}}$
Area		
square inches and square centimeters	6.45 cm² = 1 in²	$\frac{6.45 \text{ cm}^2}{1 \text{ in}^2}$ or $\frac{1 \text{ in}^2}{6.45 \text{ cm}^2}$
square meters and square yards	1.196 yd² = 1 m²	$\frac{1.196 \text{ yd}^2}{1 \text{ m}^2}$ or $\frac{1 \text{ m}^2}{1.196 \text{ yd}^2}$
acres and hectares	1 ha = 2.47 acres	$\frac{2.47 \text{ acres}}{1 \text{ ha}}$ or $\frac{1 \text{ ha}}{2.47 \text{ acres}}$
Volume		
cubic inches and milliliters	16.39 mL = 1 in³	$\frac{16.39 \text{ mL}}{1 \text{ in}^3}$ or $\frac{1 \text{ in}^3}{16.39 \text{ mL}}$
liters and quarts	1.06 qt = 1 liter	$\frac{1.06 \text{ qt}}{1 \text{ liter}}$ or $\frac{1 \text{ liter}}{1.06 \text{ qt}}$
gallons and liters	3.79 liters = 1 gal	$\frac{3.79 \text{ liters}}{1 \text{ gal}}$ or $\frac{1 \text{ gal}}{3.79 \text{ liters}}$
Weight		
ounces and grams	28.3 g = 1 oz	$\frac{28.3 \text{ g}}{1 \text{ oz}}$ or $\frac{1 \text{ oz}}{28.3 \text{ g}}$
kilograms and pounds	2.20 lb = 1 kg	$\frac{2.20 \text{ lb}}{1 \text{ kg}}$ or $\frac{1 \text{ kg}}{2.20 \text{ lb}}$

Temperature [6.4]

9. Convert 120°C to degrees Fahrenheit.

$$F = \frac{9}{5}C + 32$$
$$= \frac{9}{5}(120) + 32$$
$$= 216 + 32$$
$$= 248 \text{ °F}$$

To Convert From	Formula In Symbols	Formula In Words
Fahrenheit to Celsius	$C = \frac{5(F - 32)}{9}$	Subtract 32, multiply by 5, and then divide by 9.
Celsius to Fahrenheit	$F = \frac{9}{5}C + 32$	Multiply by $\frac{9}{5}$, and then add 32.

Time [6.5]

10. Convert 3 hours 45 minutes to minutes.

$$= 3 \text{ hr} \times \frac{60 \text{ min}}{1 \text{ hr}} + 45 \text{ min}$$
$$= 180 \text{ min} + 45 \text{ min}$$
$$= 225 \text{ min}$$

The relationship between	is	To convert from one to the other, multiply by
minutes and seconds	1 min = 60 sec	$\frac{1 \text{ min}}{60 \text{ sec}}$ or $\frac{60 \text{ sec}}{1 \text{ min}}$
hours and minutes	1 hr = 60 min	$\frac{1 \text{ hr}}{60 \text{ min}}$ or $\frac{60 \text{ min}}{1 \text{ hr}}$

Chapter 6 Review

Use the tables given in this chapter to make the following conversions. [6.1-6.4]

1. 12 ft to inches
2. 18 ft to yards
3. 49 cm to meters
4. 2 km to decimeters
5. 10 acres to square feet
6. 7,800 m² to ares
7. 4 ft² to square inches
8. 7 qt to pints
9. 24 qt to gallons
10. 5 L to milliliters
11. 8 lb to ounces
12. 2 lb 4 oz to ounces
13. 5 kg to grams
14. 5 t to kilograms
15. 4 in. to centimeters
16. 7 mi to kilometers
17. 7 L to quarts
18. 5 gal to liters
19. 5 oz to grams
20. 9 kg to pounds
21. 120°C to degrees Fahrenheit
22. 122°F to degrees Celsius

Work the following problems. Round answers to the nearest hundredth where necessary. [6.1 – 6.4]

23. A case of soft drinks holds 24 cans. If each can holds 355 ml, how many liters are there in the whole case?
24. Change 862 mi to feet.
25. Glacier Bay National Monument covers 2,805,269 acres. What is the area in square miles?
26. How many ounces does a 134-lb person weigh?
27. Change 250 mi to kilometers.
28. How many grams is 7 lb 8 oz?
29. **Construction** A 12-square-meter patio is to be built using bricks that measure 10 centimeters by 20 centimeters. How many bricks will be needed to cover the patio? [6.2]

30. **Capacity** If a regular drinking glass holds 0.25 liter of liquid, how many glasses can be filled from a 6.5-liter container? [6.2]
31. **Filling an Aquarium** How many 8-fluid-ounce glasses of water will it take to fill a 5-gallon aquarium? [6.2]
32. **Comparing Area** On April 3, 2000, *USA Today* changed the size of its paper. Previous to this date, each page of the paper was $13\frac{1}{2}$ inches wide and $22\frac{1}{4}$ inches long, giving each page an area of $300\frac{3}{8}$ in². Convert this area to square feet. [6.2]
33. **Speed** A car is traveling at a speed of 188 kilometers per hour. What is the speed in miles per hour? Round to the nearest whole number. [6.4]
34. **Volcanoes** Pyroclastic flows are high speed avalanches of volcanic gases and ash that accompany some volcano eruptions. Pyroclastic flows have been known to travel at more than 80 kilometers per hour. [6.4]

    a. Convert 80 km/hr to miles per hour. Round to the nearest whole number.
    b. Could you outrun a pyroclastic flow on foot, on a bicycle, or in a car?
35. **Speed** A race car is traveling at 200 miles per hour. What is the speed in kilometers per hour? [6.4]
36. Convert 4 hours 45 minutes to [6.5]
    a. minutes
    b. hours
37. Add 4 pounds 4 ounces and 8 pounds 12 ounces. [6.5]
38. **Cost of Fish** Mark is purchasing two whole salmon. The fish cost $5.00 per pound, and each weighs 12 lb 8 oz. What is the cost of the fish? [6.5]

Chapter 6 Cumulative Review

Simplify:

1. 7,520
    599
    + 8,640

2. 6,000
    − 3,999

3. 156 ÷ 13

4. 9(7 · 2)

5. 64)31,362

6. 2^8

7. $12 + 81 ÷ 3^2$

8. $\dfrac{329}{47}$

9. 25 + 13

10. (10 + 4) + (212 − 100)

11. $\dfrac{39}{3}$

12. 10.5(2.7)

13. 5.4 + 2.58 + 3.09

14. 45.7 − 2.86

15. 2.5)40.5

16. $\left(\dfrac{1}{4}\right)^3 \left(\dfrac{1}{2}\right)^2$

17. $17 ÷ \left(\dfrac{1}{3}\right)^2$

18. $\dfrac{8}{25} + \dfrac{7}{50}$

19. $\left(16 ÷ 1\dfrac{1}{4}\right) ÷ 2$

20. $15 - 3\dfrac{1}{2}$

21. $\dfrac{3}{8}(2.4) - \dfrac{3}{5}(0.25)$

22. $\dfrac{5}{6}(3.6) - \dfrac{3}{4}(3.2)$

23. $13 + \dfrac{3}{14} ÷ \dfrac{5}{42}$

Solve.

24. 2 · x = 15

25. 46 = 4 · y

26. $\dfrac{2}{3} = \dfrac{12}{x}$

Solve.

27. Find the perimeter and area of the figure below.

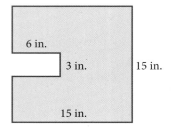

28. Find the perimeter of the figure below.

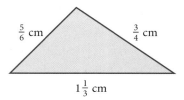

29. Find the difference between 62 and 15.

30. If a car travels 142 miles in $2\dfrac{1}{2}$ hours, what is its rate in miles per hour?

31. What number is 24% of 7,450?

32. Factor 126 into a product of prime factors.

33. Find $\dfrac{2}{3}$ of the product of 7 and 9.

34. If 5,280 feet = 1 mile, convert 3,432 feet to miles.

Use the chart to answer problem 35.

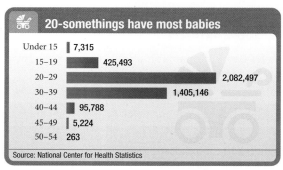

35. What percentage of babies were born to mothers who were between the ages of 20 and 39. Round to the nearest percent.

Chapter 6 Test

Use the tables in the chapter to make the following conversions. [6.1 – 6.4]

1. 7 yd to feet

2. 750 m to kilometers

3. 3 acres to square feet

4. 432 in² to square feet

5. 10 L to milliliters

6. 5 mi to kilometers

7. 10 L to quarts

8. 80°F to degrees Celsius (round to the nearest tenth)

9. 1.2 T to pounds

10. 144 oz to pounds

11. 504 in² to square feet

12. 4.5 a to square meters

13. 175.5 ft³ to cubic yards

14. 40°C to degrees Fahrenheit

Work the following problems. Round answers to the nearest hundredth. [6.1 – 6.4]

15. How many gallons are there in a 1-liter bottle of cola?

16. Change 579 yd to inches.

17. A car engine has a displacement of 409 in³. What is the displacement in cubic feet?

18. Change 75 qt to liters.

19. Change 245 ft to meters.

20. How many liters are contained in an 8-quart container?

21. 16 cm to inches

22. 5 mi to kilometers

23. 17 in² to square centimeters

24. 7 ha to acres

25. 5 qt to liters

26. 17 lb to kilograms

27. **Construction** A 40-square-foot pantry floor is to be tiled using tiles that measure 8 inches by 8 inches. How many tiles will be needed to cover the pantry floor? [6.2]

28. **Filling an Aquarium** How many 12-fluid-ounce glasses of water will it take to fill a 6-gallon aquarium? [6.2]

29. Convert 5 hours 30 minutes to [6.5]
    a. minutes
    b. hours

30. Add 3 pounds 4 ounces and 7 pounds 12 ounces. [6.5]

The chart shows the gas mileage in miles per gallon for several popular automobiles and trucks. Use the information to answer problem 31.

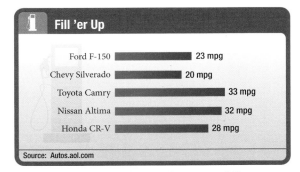

Fill 'er Up

Ford F-150 — 23 mpg
Chevy Silverado — 20 mpg
Toyota Camry — 33 mpg
Nissan Altima — 32 mpg
Honda CR-V — 28 mpg

Source: Autos.aol.com

31. Convert the mileage for Ford F-150 to kilometers per liter. Round to the nearest tenth.

Chapter 6 Projects

MEASUREMENT

Group Project

Body Mass Index

Number of People 2

Time Needed 25 minutes

Equipment Pencil, paper, and calculator

Background Body mass index (BMI) is computed by using a mathematical formula in which one's weight in kilograms is divided by the square of one's height in meters. According to the Centers for Disease Control and Prevention, a healthy BMI for adults is between 18.5 and 24.9. Children aged 2–20 have a healthy BMI if they are in the 5th to 84th percentile for their age and sex. A high BMI is predictive of cardiovascular disease.

Procedure Fill the empty boxes in the given chart with the BMIs of the corresponding weights and heights using the following conversion factors.

1 inch = 2.54 cm

1 meter = 100 cm

1 kg = 2.2 lb

Height Weight (lbs)	4'10"	5'4"	5'9"	6'1"
100				
120		21		
140				
200				

Example 5'4", 120 lbs

1. Convert height to inches.

$$5 \text{ feet} \times \frac{12 \text{ in.}}{1 \text{ ft}} = 60 \text{ in.}$$

$$5'4" = 64 \text{ in.}$$

Then convert height to meters.

$$64 \text{ in.} \times \frac{2.54 \text{ cm}}{1 \text{ in.}} = 162.56 \text{ cm}$$

$$162.56 \text{ cm} \times \frac{1 \text{ m}}{100 \text{ cm}} = 1.6256 \text{ m}$$

2. Convert weight to kilograms.

$$120 \text{ lbs} \times \frac{1 \text{ kg}}{2.2 \text{ lbs}} \approx 54.5 \text{ kg}$$

3. Compute $\frac{\text{weight in kg}}{(\text{height in m})^2}$.

$$\frac{54.5}{(1.6256)^2} \approx 21$$

Research Project

Richard Alfred Tapia

Richard A. Tapia is a mathematician and professor at Rice University in Houston, Texas, where he is Noah Harding Professor of Computational and Applied Mathematics. His parents immigrated from Mexico, separately, as teenagers to provide better educational opportunities for themselves and future generations. Born in Los Angeles, Tapia was the first in his family to attend college. In addition to being internationally known for his research, Tapia has helped his department at Rice become a national leader in awarding Ph.D. degrees to women and minority recipients. Research the life and work of Dr. Tapia. Summarize your results in an essay.

Courtesy of Rice University

Challenge Project

PARKING ON CAMPUS

If you have ever tried to park on a college campus, you know that it can be difficult and often expensive. Many colleges have reorganized lots, built structures, and added biking and walking paths to their campuses to help aid student commuters. Because every college campus varies in size and shape, so do the parking lots. Furthermore, because parking lots vary, the number of cars each lot can hold varies, too. Let's examine a parking lot on a local college campus and calculate its area.

STEP 1 Use Google Earth to locate your college campus. As an example, we'll use Cuesta College, in San Luis Obispo, California.

STEP 2 Use the **Ruler** tool set to **Miles** to measure the perimeter of one parking lot on your campus. Make sure to record the measurement for each side of the parking lot. As you can see from our example parking lot in Figure 1, your lot may not be a perfect rectangle or square.

FIGURE 1

STEP 3 Now find the area of the parking lot. If your lot is not a perfect square, rectangle, or triangle, you will need to divide the lot into smaller, more manageable geometric shapes. See Figure 2 for an example. You may need to divide your lot into more (or fewer) shapes than what is shown. Calculate the areas of the smaller shapes, then add them all together to find the total area of the lot.

FIGURE 2

STEP 4 Lastly, change your perimeter measurement from miles to feet (rounding to the nearest foot), and recalculate the total area.

Introduction to Algebra

Chapter Outline

7.1 Positive and Negative Numbers

7.2 Addition with Negative Numbers

7.3 Subtraction with Negative Numbers

7.4 Multiplication with Negative Numbers

7.5 Division with Negative Numbers

7.6 Simplifying Algebraic Expressions

The Grand Canyon in Arizona is a large gorge created by the Colorado River over millions of years. Much of the Grand Canyon is located in Grand Canyon National Park, which receives more than four million visitors per year. Visitors come from all over the world to hike the canyon's trails and view the magnificent rock formations. Many of the hiking trails have significant changes in altitude. Sometimes we represent changes in altitude with negative numbers.

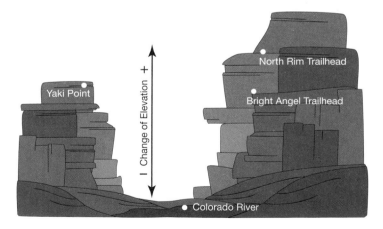

Suppose we were to assign a zero value to Grand Canyon's Bright Angel trailhead. The North Rim trailhead is 1,470 feet higher in elevation than the Bright Angel trailhead, therefore, we could assign a positive value of 1,470 to the North Rim trailhead. However, the Colorado River is approximately 3,985 feet below the Bright Angel trailhead. We could then assign the river a negative value of -3,985. In this chapter, we will apply our knowledge of basic mathematical operations to problems that involve negative numbers, which will help us make more sense of this example.

Preview

Key Words	Definition
Origin	The location on a number line that represents 0
Absolute Value	The distance from 0 on the number line
Opposites	Two numbers that are the same distance from 0 on the number line but in opposite directions
Subtraction	Addition of the opposite of a number
Algebraic Expression	A combination of constants and variables joined by arithmetic symbols
Similar Terms	Expressions that contain the same variable parts

Chapter Outline

7.1 Positive and Negative Numbers
A Use the number line and inequality symbols to compare numbers.
B Find the absolute value of a number.
C Find the opposite of a number.

7.2 Addition with Negative Numbers
A Use the number line to add positive and negative numbers.
B Add positive and negative numbers using a rule.

7.3 Subtraction with Negative Numbers
A Subtract numbers by thinking of subtraction as addition of the opposite.

7.4 Multiplication with Negative Numbers
A Multiply positive and negative numbers.
B Apply the rule for order of operations to expressions containing positive and negative numbers.

7.5 Division with Negative Numbers
A Divide positive and negative numbers.
B Apply the rule for order of operations to expressions that contain positive and negative numbers.

7.6 Simplifying Algebraic Expressions
A Simplify expressions by using the associative property.
B Apply the distributive property to expressions containing numbers and variables.
C Use the distributive property to combine similar terms.
D Use the formulas for area and perimeter of squares and rectangles.

7.1 Positive and Negative Numbers

OBJECTIVES

A Use the number line and inequality symbols to compare numbers.

B Find the absolute value of a number.

C Find the opposite of a number.

TICKET TO SUCCESS

Keep these questions in mind as you read through the section. Then respond in your own words and in complete sentences.

1. Write the statement "3 is less than 5" in symbols.
2. What is the notation for inequality symbols?
3. What is absolute value?
4. Using the number line, how can you tell if two numbers are opposites?

Before the late nineteenth century, time zones did not exist. Each town would set its clocks according to the motions of the Sun. It was not until the late 1800s that a system of worldwide time zones was developed. This system divides the earth into 24 time zones with Greenwich, England, designated as the center of the time zones (Greenwich mean time, or GMT). The Royal Observatory in Greenwich (pictured above) is the location of the prime meridian and is assigned a value of zero. Each of the World Time Zones is assigned a positive or negative number ranging from −12 to +12 depending on its position east or west of Greenwich, England.

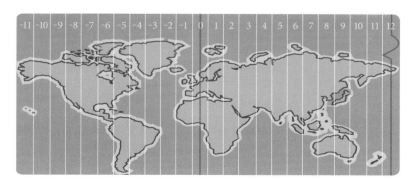

If New York is 5 time zones to the left of GMT, this would be noted as −5:00 GMT.

A Comparing Numbers

To see the relationship between negative and positive numbers, we can extend the number line as shown in Figure 1. We first draw a straight line and label a convenient point with 0. This is called the *origin*, and it is usually in the middle of the line. Then we label positive numbers to the right (as we have done previously), and negative numbers to the left.

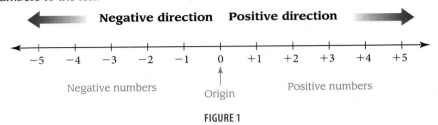

FIGURE 1

The numbers increase going from left to right. If we move to the right, we are moving in the positive direction. If we move to the left, we are moving in the negative direction. *Any number to the left of another number is considered to be less than the number to its right.*

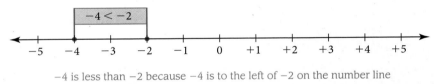

−4 is less than −2 because −4 is to the left of −2 on the number line

FIGURE 2

We see from the line that every negative number is less than every positive number.

In algebra, we can use inequality symbols when comparing numbers.

> **Notation Inequalities**
>
> For any whole numbers a and b
>
> 1. $a < b$ is read "a is less than b" and is true when a is to the **left** of b on a number line
> 2. $a > b$ is read "a is greater than b" and is true when a is to the **right** of b on a number line

As you can see, the inequality symbols always point to the smaller of the two numbers being compared. Here are some examples that illustrate how we use the inequality symbols.

EXAMPLE 1 $3 < 5$ is read "3 is less than 5." Note that it would also be correct to write $5 > 3$. Both statements, "3 is less than 5" and "5 is greater than 3," have the same meaning. The inequality symbols always point to the smaller number. ∎

EXAMPLE 2 $0 > 100$ is a false statement, because 0 is less than 100, not greater than 100. To write a true inequality statement using the numbers 0 and 100, we would have to write either $0 < 100$ or $100 > 0$. ∎

NOTE
A number, other than 0, with no sign (+ or −) in front of it is assumed to be positive. That is, $5 = +5$.

PRACTICE PROBLEMS
Write each of the following in words.
1. $2 < 8$

2. $5 > 10$

Answers
1. 2 is less than 8.
2. 5 is greater than 10, a false statement.

EXAMPLE 3 $-3 < 5$ is a true statement, because -3 is to the left of 5 on the number line, and, therefore, it must be less than 5. Another statement that means the same thing is $5 > -3$. ∎

EXAMPLE 4 $-5 < -2$ is a true statement, because -5 is to the left of -2 on the number line, meaning that -5 is less than -2. Both statements $-5 < -2$ and $-2 > -5$ have the same meaning; they both say that -5 is a smaller number than -2. ∎

B Absolute Value

It is sometimes convenient to talk about only the numerical part of a number and disregard the sign (+ or −) in front of it. The following definition gives us a way of doing this.

> **Definition**
> The **absolute value** of a number is its distance from 0 on the number line. We denote the absolute value of a number with vertical lines. For example, the absolute value of -3 is written $|-3|$.

The absolute value of a number is never negative because it is a distance, and a distance is always measured in positive units (unless it happens to be 0).
Here are some examples of absolute value problems.

EXAMPLE 5 $|5| = 5$ The number 5 is 5 units from 0. ∎

EXAMPLE 6 $|-3| = 3$ The number -3 is 3 units from 0. ∎

EXAMPLE 7 $|-7| = 7$ The number -7 is 7 units from 0. ∎

C Opposites

> **Definition**
> Two numbers that are the same distance from 0 but in opposite directions from 0 are called **opposites.** The notation for the opposite of a is $-a$.

EXAMPLE 8 Give the opposite of each of the following numbers:

$5, 7, 1, -5, -8$

SOLUTION   The opposite of 5 is -5.
The opposite of 7 is -7.
The opposite of 1 is -1.
The opposite of -5 is $-(-5)$, or 5.
The opposite of -8 is $-(-8)$, or 8. ∎

Write each of the following in words.

3. $-4 < 4$

4. $-7 < -2$

Simplify.

5. $|6|$

6. $|-6|$

7. $|-8|$

8. Give the opposite of each of the following numbers: 8, 10, 0, -4.

Answers
3. -4 is less than 4.
4. -7 is less than -2.
5. 6
6. 6
7. 8
8. $-8, -10, 0, 4$

We see from the previous example that the opposite of every positive number is a negative number; likewise, the opposite of every negative number is a positive number. The last two parts of Example 8 illustrate the following property:

> **Property** Opposite of a Negative
> If a represents any positive number, then it is *always* true that
> $$-(-a) = a$$

In other words, this property states that the opposite of a negative number is a positive number.

It should be evident now that the symbols + and − can be used to indicate several different ideas in mathematics. In the past, we have used them to indicate addition and subtraction. They can also be used to indicate the direction a number is from 0 on the number line. For instance, the number +3 (read "positive 3") is the number that is 3 units from zero in the positive direction. On the other hand, the number −3 (read "negative 3") is the number that is 3 units from 0 in the negative direction. The symbol − can also be used to indicate the opposite of a number, as in −(−2) = 2. The interpretation of the symbols + and − depends on the situation in which they are used. For example,

3 + 5	The + sign indicates addition.
7 − 2	The − sign indicates subtraction.
−7	The − sign is read "negative" 7.
−(−5)	The first − sign is read "the opposite of." The second − sign is read "negative" 5.

This may seem confusing at first, but as you work through the problems in this chapter you will get used to the different interpretations of the symbols + and −.

We should mention here that the set of whole numbers along with their opposites forms the set of *integers*. That is,

$$\text{Integers} = \{\ldots, -3, -2, -1, 0, 1, 2, 3, \ldots\}$$

NOTE In some books, opposites are called additive inverses.

Problem Set 7.1

"There are no failures—just experiences and your reactions to them."
—Tom Krause, 1934–present, motivational speaker and teacher

1. How might a positive attitude toward this class benefit you?
2. How do you plan on staying positive before, during, and after a test?

A Write each of the following in words. [Example 1–4]

1. $4 < 7$
2. $0 < 10$
3. $5 > -2$
4. $8 > -8$
5. $-10 < -3$
6. $-20 < -5$
7. $0 > -4$
8. $0 > -100$

Write each of the following in symbols.

9. 30 is greater than -30.
10. -30 is less than 30.
11. -10 is less than 0.
12. 0 is greater than -10.
13. -3 is greater than -15.
14. -15 is less than -3.

A Place either $<$ or $>$ between each of the following pairs of numbers so that the resulting statement is true. [Examples 1–4]

15. 3   7
16. 17   0
17. 7 -5
18. 2 -13
19. -6 0
20. -14 0
21. -12 -2
22. -20 -1
23. $-\dfrac{1}{2}$ $-\dfrac{3}{4}$
24. $-\dfrac{6}{7}$ $\dfrac{5}{6}$
25. -0.75 -0.25
26. -1 -3.5
27. -0.1 -0.01
28. -0.04 -0.4
29. -3 $|6|$
30. $|8|$ -2
31. 15 $|-4|$
32. 20 $|-6|$
33. $|-2|$ $|-7|$
34. $|-3|$ $|-1|$

B Find each of the following absolute values. [Examples 5–7]

35. $|2|$ **36.** $|7|$ **37.** $|100|$ **38.** $|10{,}000|$ **39.** $|-8|$ **40.** $|-9|$

41. $|-231|$ **42.** $|-457|$ **43.** $\left|-\dfrac{3}{4}\right|$ **44.** $\left|-\dfrac{1}{10}\right|$ **45.** $|-200|$ **46.** $|-350|$

47. $|8|$ **48.** $|9|$ **49.** $|231|$ **50.** $|457|$

C Give the opposite of each of the following numbers. [Example 8]

51. 3   **52.** −5   **53.** −2   **54.** 15   **55.** 75   **56.** −32

57. 0   **58.** 1   **59.** −0.123   **60.** −3.45   **61.** $\dfrac{7}{8}$ **62.** $\dfrac{1}{100}$

Simplify each of the following.

63. $-(-2)$ **64.** $-(-5)$ **65.** $-(-8)$ **66.** $-(-3)$

67. $-|-2|$ **68.** $-|-5|$ **69.** $-|-8|$ **70.** $-|-3|$

71. What number is its own opposite?

72. Is $|a| = a$ always a true statement?

73. If n is a negative number, is $-n$ positive or negative?

74. If n is a positive number, is $-n$ positive or negative?

Estimating

Work Problems 75–80 mentally, without pencil and paper or a calculator.

75. Is −60 closer to 0 or −100?

76. Is −20 closer to 0 or −30?

77. Is −10 closer to −20 or 20?

78. Is −20 closer to −40 or 10?

79. Is −362 closer to −360 or −370?

80. Is −368 closer to −360 or −370?

Applying the Concepts

81. The London Eye has a height of 450 feet. Describe the location of someone standing on the ground in relation to someone at the top of the London Eye.

82. The Eiffel Tower has several levels visitors can walk around on. The first is 57 meters above the ground, the second is 115 meters high, and the third level is 276 meters high. What is the location of someone standing on the first level in relation to someone standing on the third level?

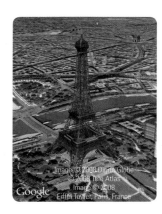

83. Recall our discussion of the Grand Canyon at the beginning of this chapter. The Bright Angel trail at Grand Canyon National Park ends at Indian Garden, 3,060 feet below the trailhead. Write this as a negative number with respect to the trailhead.

84. The South Kaibab Trail at Grand Canyon National Park ends at Cedar Ridge, 1,140 feet below the trailhead. Write this as a negative number with respect to the trailhead.

85. Car Depreciation Depreciation refers to the decline in a car's market value during the time you own the car. According to sources such as Kelley Blue Book and Edmunds.com, not all cars depreciate at the same rate. Suppose you pay $25,000 for a new car which has a high rate of depreciation. Your car loses about $5,000 in value per year. Represent this loss in value as a negative number. A car with a low rate of depreciation loses about $2,750 in value each year. Represent this loss as a negative number.

86. Census Figures In 2011, the U.S. Census Bureau released population change estimates for the ten most populous cities from 2000 to 2010. Chicago, Illinois had the largest population loss. The city's population fell by 200,148 people. Represent the loss of population for Chicago as a negative number.

87. Temperature and Altitude Yamina is flying from Phoenix to San Francisco on a Boeing 737 jet. When the plane reaches an altitude of 33,000 feet, the temperature outside the plane is 61 degrees below zero Fahrenheit. Represent this temperature with a negative number. If the temperature outside the plane gets warmer by 10 degrees, what will the new temperature be?

88. Temperature Change At 11:00 in the morning in Superior, Wisconsin, Jim notices the temperature is 15 degrees below zero Fahrenheit. Write this temperature as a negative number. At noon it has warmed up by 8 degrees. What is the temperature at noon?

89. Temperature Change At 10:00 in the morning in White Bear Lake, Minnesota, Zach notices the temperature is 5 degrees below zero Fahrenheit. Write this temperature as a negative number. By noon the temperature has dropped another 10 degrees. What is the temperature at noon?

90. Snorkeling Steve is snorkeling in the ocean near his home in Maui. At one point he is 6 feet below the surface. Represent this situation with a negative number. If he descends another 6 feet, what negative number will represent his new position?

91. Time Zones New Orleans, Louisiana, is 1 time zone west of New York City. Represent this time zone as a negative number, as discussed in the introduction to this section.

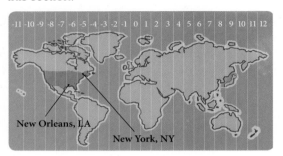

92. Time Zones Seattle, Washington, is 2 time zones west of New Orleans, Louisiana. Represent this time zone as a negative number, as discussed in the introduction to this section.

Table 2 lists various wind chill temperatures. The top row gives air temperature, while the first column gives wind speed in miles per hour. Suppose the numbers within the table indicate how cold the weather will feel. For example, if the thermometer reads 30°F and the wind is blowing at 15 miles per hour, the wind chill temperature is 9°F.

TABLE 2
WIND CHILL TEMPERATURES

Wind Speed	Air temperatures (°F)							
	30°	25°	20°	15°	10°	5°	0°	−5°
10 mph	16°	10°	3°	−3°	−9°	−15°	−22°	−27°
15 mph	9°	2°	−5°	−11°	−18°	−25°	−31°	−38°
20 mph	4°	−3°	−10°	−17°	−24°	−31°	−39°	−46°
25 mph	1°	−7°	−15°	−22°	−29°	−36°	−44°	−51°
30 mph	−2°	−10°	−18°	−25°	−33°	−41°	−49°	−56°

93. Wind Chill Find the wind chill temperature if the thermometer reads 25°F and the wind is blowing at 25 miles per hour.

94. Wind Chill Find the wind chill temperature if the thermometer reads 10°F and the wind is blowing at 25 miles per hour.

95. Wind Chill Which will feel colder: a day with an air temperature of 10°F and a 25-mph wind, or a day with an air temperature of −5°F and a 10-mph wind?

96. Wind Chill Which will feel colder: a day with an air temperature of 15°F and a 20-mph wind, or a day with an air temperature of 5°F and a 10-mph wind?

Suppose Table 3 lists the record low temperatures for each month of the year for a city in New York. Table 4 lists the record high temperatures for the same city.

TABLE 3 RECORD LOW TEMPERATURES	
Month	Temperature
January	−36°F
February	−30°F
March	−14°F
April	−2°F
May	19°F
June	22°F
July	35°F
August	30°F
September	19°F
October	15°F
November	−11°F
December	−26°F

TABLE 4 RECORD HIGH TEMPERATURES	
Month	Temperature
January	54°F
February	59°F
March	69°F
April	82°F
May	90°F
June	93°F
July	97°F
August	93°F
September	90°F
October	87°F
November	67°F
December	60°F

97. Temperature Figure 5 below is a bar chart of the information in Table 3 above. Use the template in Figure 6 to construct a scatter diagram of the same information. Then connect the dots in the scatter diagram to obtain a line graph. (Notice that we have used the numbers 1 through 12 to represent the months January through December.)

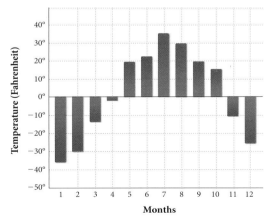

FIGURE 5 A bar chart of Table 3

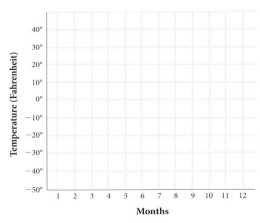

FIGURE 6 A scatter diagram, then a line graph of Table 3

98. Temperature Figure 7 below is a bar chart of the information in Table 4 from the previous page. Use the template in Figure 8 to construct a scatter diagram of the same information. Then connect the dots in the scatter diagram to obtain a line graph. (Again, we have used the numbers 1 through 12 to represent the months January through December.)

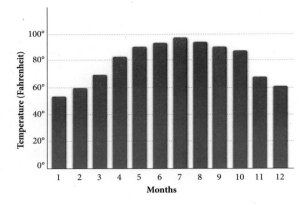

FIGURE 7 A bar chart of Table 4

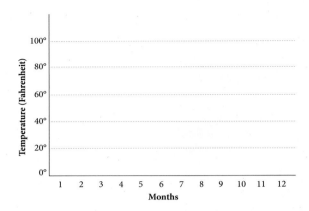

FIGURE 8 A scatter diagram, then a line graph of Table 4

Getting Ready for the Next Section

Add or subtract.

99. $10 + 15$

100. $12 + 15$

101. $15 - 10$

102. $15 - 12$

103. $10 - 5 - 3 + 4$

104. $12 - 3 - 7 + 5$

105. $[3 + 10] + [8 - 2]$

106. $[2 + 12] + [7 - 5]$

107. $276 + 32 + 4{,}005$

108. $17 + 3 + 152 + 1{,}200$

109. $635 - 579$

110. $2{,}987 - 1{,}130$

Maintaining Your Skills

Complete each statement using the commutative property of addition.

111. $3 + 5 =$

112. $9 + x =$

Complete each statement using the associative property of addition.

113. $7 + (2 + 6) =$

114. $(x + 3) + 5 =$

Write each of the following in symbols.

115. The sum of x and 4

116. The sum of x and 4 is 9.

117. 5 more than y

118. x increased by 8

7.2 Addition with Negative Numbers

OBJECTIVES

A Use the number line to add positive and negative numbers.

B Add positive and negative numbers using a rule.

TICKET TO SUCCESS

Keep these questions in mind as you read through the section. Then respond in your own words and in complete sentences.

1. How does the number line help us visualize addition problems?
2. Explain how you would use the number line to add 3 and 5.
3. How do you add two numbers with the same sign?
4. How do you add two numbers with different signs?

Suppose you are in Las Vegas playing blackjack and you lose $3 on the first hand and then you lose $5 on the next hand. If you represent winning with positive numbers and losing with negative numbers, how will you represent the results from your first two hands? Since you lost $3 and $5 for a total of $8, one way to represent the situation is with addition of negative numbers.

$$(-\$3) + (-\$5) = -\$8$$

From this example, we see that the sum of two negative numbers is a negative number. To generalize addition of positive and negative numbers, we can use the number line.

A Adding with a Number Line

We can think of each number on the number line as having two characteristics: (1) a *distance* from 0 (absolute value) and (2) a *direction* from 0 (positive or negative). The distance from 0 is represented by the numerical part of the number (like the 5 in the number -5), and its direction is represented by the $+$ or $-$ sign in front of the number.

We can visualize addition of numbers on the number line by thinking in terms of distance and direction from 0. Let's begin with a simple problem we know the answer to. We interpret the sum $3 + 5$ on the number line as follows:

1. The first number is 3, which tells us to "start at the origin, and move 3 units in the positive direction."
2. The $+$ sign is read "and then move."

NOTE
This method of adding numbers may seem a little complicated at first, but it will allow us to add numbers we couldn't otherwise add.

3. The 5 means "5 units in the positive direction."

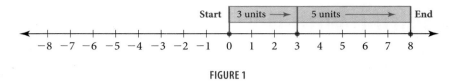

FIGURE 1

Figure 1 shows these steps. To summarize, 3 + 5 means to start at the origin (0), move 3 units in the *positive* direction, and then move 5 units in the *positive* direction. We end up at 8, which is the sum we are looking for: 3 + 5 = 8.

EXAMPLE 1 Add 3 + (−5) using the number line.

SOLUTION We start at the origin, move 3 units in the positive direction, and then move 5 units in the negative direction, as shown in Figure 2. The last arrow ends at −2, which must be the sum of 3 and −5. That is,

$$3 + (-5) = -2$$

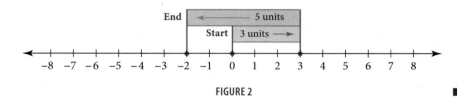

FIGURE 2

EXAMPLE 2 Add −3 + 5 using the number line.

SOLUTION We start at the origin, move 3 units in the negative direction, and then move 5 units in the positive direction, as shown in Figure 3. We end up at 2, which is the sum of −3 and 5. That is,

$$-3 + 5 = 2$$

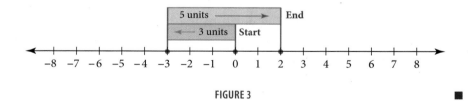

FIGURE 3

EXAMPLE 3 Add −3 + (−5) using the number line.

SOLUTION We start at the origin, move 3 units in the negative direction, and then move 5 more units in the negative direction. This is shown on the number line in Figure 4. As you can see, the last arrow ends at −8. We must conclude that the sum of −3 and −5 is −8. That is,

$$-3 + (-5) = -8$$

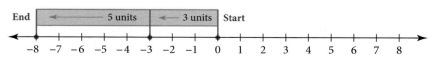

FIGURE 4

PRACTICE PROBLEMS

1. Add: 2 + (−5).

2. Add: −2 + 5.

3. Add: −2 + (−5).

Answers
1. −3
2. 3
3. −7

Adding numbers on the number line as we have done in these first three examples gives us a way of visualizing addition of positive and negative numbers. We eventually want to be able to write a rule for addition of positive and negative numbers that doesn't involve the number line. The number line is a way of justifying the rule we will eventually write. Here is a summary of the results we have so far:

$$3 + 5 = 8 \qquad -3 + 5 = 2$$
$$3 + (-5) = -2 \qquad -3 + (-5) = -8$$

Examine these results to see if you notice any pattern in the answers.

Here are some more addition problems and their number lines.

EXAMPLE 4 $4 + 7 = 11$

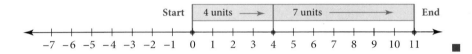

4. Add: $2 + 6$.

EXAMPLE 5 $4 + (-7) = -3$

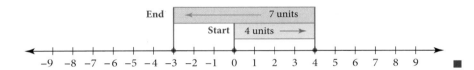

5. Add: $2 + (-6)$.

EXAMPLE 6 $-4 + 7 = 3$

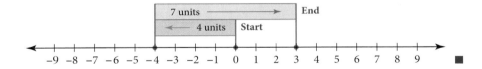

6. Add: $-2 + 6$.

EXAMPLE 7 $-4 + (-7) = -11$

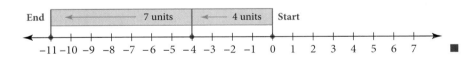

7. Add: $-2 + (-6)$.

B A Rule for Addition

A summary of the results of these last four examples looks like this:

$$4 + 7 = 11$$
$$4 + (-7) = -3$$
$$-4 + 7 = 3$$
$$-4 + (-7) = -11$$

Answers
4. 8
5. −4
6. 4
7. −8

Looking over all the examples in this section, and noticing how the results in the problems are related, we can write the following rule for adding any two numbers:

> **Rule** Addition of Any Two Numbers
> 1. To add two numbers with the *same* sign: Simply add their absolute values, and use the common sign. If both numbers are positive, the answer is positive. If both numbers are negative, the answer is negative.
> 2. To add two numbers with *different* signs: Subtract the smaller absolute value from the larger absolute value. The answer will have the sign of the number with the larger absolute value.

NOTE
This rule covers all possible addition problems involving positive and negative numbers. You *must* memorize it. After you have worked some problems, the rule will seem almost automatic.

The following examples show how the rule is used. You will find that the rule for addition is consistent with all the results obtained using the number line.

EXAMPLE 8 Add all combinations of positive and negative 10 and 15.

SOLUTION
$$10 + 15 = 25$$
$$10 + (-15) = -5$$
$$-10 + 15 = 5$$
$$-10 + (-15) = -25$$

Notice that when we add two numbers with the same sign, the answer also has that sign. When the signs are not the same, the answer has the sign of the number with the larger absolute value.

Once you have become familiar with the rule for adding positive and negative numbers, you can apply it to more complicated sums.

8. Add all combinations of positive and negative 12 and 15.

EXAMPLE 9 Simplify: $10 + (-5) + (-3) + 4$.

SOLUTION Adding left to right, we have

$$10 + (-5) + (-3) + 4 = 5 + (-3) + 4 \quad 10 + (-5) = 5$$
$$= 2 + 4 \quad 5 + (-3) = 2$$
$$= 6$$

9. Simplify: $12 + (-3) + (-7) + 5$.

EXAMPLE 10 Simplify: $[-3 + (-10)] + [8 + (-2)]$.

SOLUTION We begin by adding the numbers inside the brackets.

$$[-3 + (-10)] + [8 + (-2)] = [-13] + [6]$$
$$= -7$$

10. Simplify:
$[-2 + (-12)] + [7 + (-5)]$.

EXAMPLE 11 Add: $-4.75 + (-2.25)$.

SOLUTION Because both signs are negative, we add absolute values. The answer will be negative.

$$-4.75 + (-2.25) = -7$$

11. Add: $-5.76 + (-3.24)$.

Answers
8. See Solutions Section.
9. 7
10. -12
11. -9.00

7.2 Addition with Negative Numbers

EXAMPLE 12 Add: $3.42 + (-6.89)$.

SOLUTION The signs are different, so we subtract the smaller absolute value from the larger absolute value. The answer will be negative, because 6.89 is larger than 3.42 and the sign in front of 6.89 is $-$.

$$3.42 + (-6.89) = -3.47$$ ■

EXAMPLE 13 Add: $\dfrac{3}{8} + \left(-\dfrac{1}{8}\right)$.

SOLUTION We subtract absolute values. The answer will be positive, because $\dfrac{3}{8}$ is positive.

$$\dfrac{3}{8} + \left(-\dfrac{1}{8}\right) = \dfrac{2}{8}$$

$$= \dfrac{1}{4} \qquad \text{Reduce to lowest terms.}$$ ■

EXAMPLE 14 Add: $\dfrac{1}{10} + \left(-\dfrac{4}{5}\right) + \left(-\dfrac{3}{20}\right)$.

SOLUTION To begin, change each fraction to an equivalent fraction with an LCD of 20.

$$\dfrac{1}{10} + \left(-\dfrac{4}{5}\right) + \left(-\dfrac{3}{20}\right) = \dfrac{1 \cdot 2}{10 \cdot 2} + \left(-\dfrac{4 \cdot 4}{5 \cdot 4}\right) + \left(-\dfrac{3}{20}\right)$$

$$= \dfrac{2}{20} + \left(-\dfrac{16}{20}\right) + \left(-\dfrac{3}{20}\right)$$

$$= -\dfrac{14}{20} + \left(-\dfrac{3}{20}\right)$$

$$= -\dfrac{17}{20}$$ ■

USING TECHNOLOGY

Calculators

There are a number of different ways in which calculators display negative numbers. Some calculators use a key labeled $\boxed{+/-}$, whereas others use a key labeled $\boxed{(-)}$. You will need to consult with the manual that came with your calculator to see how your calculator does the job.

Here are a couple of ways to find the sum $-10 + (-15)$ on a calculator:

Scientific Calculator: 10 $\boxed{+/-}$ $\boxed{+}$ 15 $\boxed{+/-}$ $\boxed{=}$

Graphing Calculator: $\boxed{(-)}$ 10 $\boxed{+}$ $\boxed{(-)}$ 15 $\boxed{ENT}$

---

12. Add: $6.88 + (-8.55)$.

13. Add: $\dfrac{5}{6} + \left(-\dfrac{2}{6}\right)$.

14. Add: $\dfrac{1}{2} + \left(-\dfrac{3}{4}\right) + \left(-\dfrac{5}{8}\right)$.

Answers
12. -1.67
13. $\dfrac{1}{2}$
14. $-\dfrac{7}{8}$

Problem Set 7.2

Moving Toward Success

"Take rest; a field that has rested gives a bountiful crop."
—Ovid, 43 BC–17 AD, Ancient Roman classical poet and author of Metamorphoses

1. Why is it important to still get good restful sleep between studying and going to class?
2. We recommend that you treat this class as a full-time job. Why?

A Draw a number line from -10 to $+10$ and use it to add the following numbers. [Examples 1–7]

1. $2 + 3$
2. $2 + (-3)$
3. $-2 + 3$
4. $-2 + (-3)$
5. $5 + (-7)$
6. $-5 + 7$
7. $-4 + (-2)$
8. $-8 + (-2)$
9. $10 + (-6)$
10. $-9 + 3$
11. $7 + (-3)$
12. $-7 + 3$
13. $-4 + (-5)$
14. $-2 + (-7)$

B Combine the following by using the rule for addition of positive and negative numbers. (Your goal is to be fast and accurate at addition, with the latter being more important.) [Example 8]

15. $7 + 8$
16. $9 + 12$
17. $5 + (-8)$
18. $4 + (-11)$
19. $-6 + (-5)$
20. $-7 + (-2)$
21. $-10 + 3$
22. $-14 + 7$
23. $-1 + (-2)$
24. $-5 + (-4)$
25. $-11 + (-5)$
26. $-16 + (-10)$
27. $4 + (-12)$
28. $9 + (-1)$
29. $-85 + (-42)$
30. $-96 + (-31)$
31. $-121 + 170$
32. $-130 + 158$
33. $-375 + 409$
34. $-765 + 213$

Complete the following tables.

35.

First Number a	Second Number b	Their Sum $a+b$
5	−3	
5	−4	
5	−5	
5	−6	
5	−7	

36.

First Number a	Second Number b	Their Sum $a+b$
−5	3	
−5	4	
−5	5	
−5	6	
−5	7	

37.

First Number x	Second Number y	Their Sum $x+y$
−5	−3	
−5	−4	
−5	−5	
−5	−6	
−5	−7	

38.

First Number x	Second Number y	Their Sum $x+y$
30	−20	
−30	20	
−30	−20	
30	20	
−30	0	

B Add the following numbers left to right. [Example 9]

39. 24 + (−6) + (−8)

40. 35 + (−5) + (−30)

41. −201 + (−143) + (−101)

42. −27 + (−56) + (−89)

43. −321 + 752 + (−324)

44. −571 + 437 + (−502)

45. −2 + (−5) + (−6) + (−7)

46. −8 + (−3) + (−4) + (−7)

47. 15 + (−30) + 18 + (−20)

48. 20 + (−15) + 30 + (−18)

49. −78 + (−42) + 57 + 13

50. −89 + (−51) + 65 + 17

B Use the rule for order of operations to simplify each of the following. [Example 10]

51. (−8 + 5) + (−6 + 2)

52. (−3 + 1) + (−9 + 4)

53. (−10 + 4) + (−3 + 12)

54. (−11 + 5) + (−3 + 2)

55. 20 + (−30 + 50) + 10

56. 30 + (−40 + 20) + 50

57. 108 + (−456 + 275)

58. 106 + (−512 + 318)

59. [5 + (−8)] + [3 + (−11)]

60. [8 + (−2)] + [5 + (−7)]

61. [57 + (−35)] + [19 + (−24)]

62. [63 + (−27)] + [18 + (−24)]

B Use the rule for addition of numbers to add the following fractions and decimals. [Examples 11–14]

63. −1.3 + (−2.5)

64. −9.1 + (−4.5)

65. 24.8 + (−10.4)

66. 29.5 + (−21.3)

67. −5.35 + 2.35 + (−6.89)

68. −9.48 + 5.48 + (−4.28)

69. $-\frac{5}{6} + \left(-\frac{1}{6}\right)$

70. $-\frac{7}{9} + \left(-\frac{2}{9}\right)$

71. $\frac{3}{7} + \left(-\frac{5}{7}\right)$

72. $\frac{11}{13} + \left(-\frac{12}{13}\right)$

73. $-\frac{2}{5} + \frac{3}{5} + \left(-\frac{4}{5}\right)$

74. $-\frac{6}{7} + \frac{4}{7} + \left(-\frac{1}{7}\right)$

75. −3.8 + 2.54 + 0.4

76. −9.6 + 5.15 + 0.8

77. −2.89 + (−1.4) + 0.09

78. −3.99 + (−1.42) + 0.06

79. $\frac{1}{2} + \left(-\frac{3}{4}\right)$

80. $\frac{3}{5} + \left(-\frac{7}{10}\right)$

81. Find the sum of −8, −10, and −3.

82. Find the sum of −4, 17, and −6.

83. What number do you add to 8 to get 3?

84. What number do you add to 10 to get 4?

85. What number do you add to −3 to get −7?

86. What number do you add to −5 to get −8?

87. What number do you add to −4 to get 3?

88. What number do you add to −7 to get 2?

89. If the sum of −3 and 5 is increased by 8, what number results?

90. If the sum of −9 and −2 is increased by 10, what number results?

Applying the Concepts

91. One of the trails at the Grand Canyon starts at Bright Angel Trailhead and then drops 4,060 feet to the Colorado River and then climbs 4,440 feet to Yaki Point. What is the trail's ending position in relation to the Bright Angel Trailhead? If the trail ends below the starting position, write the answer as a negative number.

92. One of the trails in the Grand Canyon starts at the North Rim trailhead and drops 5,490 feet to the Colorado River. Then the trail climbs 4,060 feet to the Bright Angel Trailhead. What is the Bright Angel Trailhead's position in relation to the North Rim Trailhead? If the trail ends below the starting position, write the answer as a negative number.

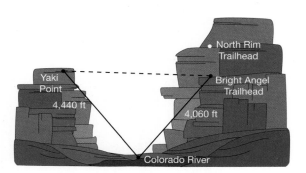
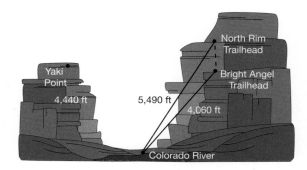

93. Checkbook Balance Ethan has a balance of −$40 in his checkbook. If he deposits $100 and then writes a check for $50, what is the new balance in his checkbook?

94. Blackjack You win $15 during a game of blackjack. The next two games you lose $35 and $12 respectively. The last game you play, you win $17. What is your net amount of money lost or won? Write a loss as a negative number.

Getting Ready for the Next Section

Give the opposite of each number.

95. 2 **96.** 3 **97.** −4 **98.** −5 **99.** $\frac{2}{5}$

100. $\frac{3}{8}$ **101.** −30 **102.** −15 **103.** 60.3 **104.** 70.4

105. Subtract 3 from 5.

106. Subtract 2 from 8.

107. Find the difference of 7 and 4.

108. Find the difference of 8 and 6.

Maintaining Your Skills

The problems below review subtraction with whole numbers.

Subtract.

109. 763 − 159 **110.** 1,007 − 136 **111.** 465 − 462 − 3 **112.** 481 − 479 − 2

Write each of the following statements in symbols.

113. The difference of 10 and x.

114. The difference of x and 10.

115. 17 subtracted from y.

116. y subtracted from 17.

Subtraction with Negative Numbers

7.3

OBJECTIVES

A Subtract numbers by thinking of subtraction as addition of the opposite.

TICKET TO SUCCESS

Keep these questions in mind as you read through the section. Then respond in your own words and in complete sentences.

1. Explain the definition for subtraction using words and symbols.
2. Explain the process you would use to subtract 2 from -7.
3. Why can we think of subtraction as addition of the opposite?
4. Write an addition problem that is equivalent to the subtraction problem $-20 - (-30)$.

How would we represent the final balance in a checkbook if the original balance was $20 and we wrote a check for $30? The final balance would be $-$$10. We can summarize the whole situation with subtraction:

$$\$20 - \$30 = -\$10$$

NUMBER	DATE	DESCRIPTION OF TRANSACTION	PAYMENT/DEBIT (-)	DEPOSIT/CREDIT (+)	BALANCE
					$20 00
1501	9/15	Campus Bookstore	$30 00		-$10 00

RECORD ALL CHARGES OR CREDITS THAT AFFECT YOUR ACCOUNT

From this we see that subtracting 30 from 20 gives us -10. Another example that gives the same answer but involves addition is this:

$$20 + (-30) = -10$$

7.3 Subtraction with Negative Numbers

A Subtraction

From the two examples above, we find that subtracting 30 gives the same result as adding −30. We use this kind of reasoning to give a definition for subtraction that will allow us to use the rules we developed for addition to do our subtraction problems. Here is that definition:

> **Definition**
>
> **Subtraction** If a and b represent any two numbers, then it is always true that
> $$a - b = a + (-b)$$
> To subtract b, add its opposite, $-b$.
>
> *In words:* Subtracting a number is equivalent to adding its opposite.

Let's see if this definition conflicts with what we already know to be true about subtraction.

EXAMPLE 1   Subtract: $5 - 2$.

SOLUTION   From previous experience, we know that
$$5 - 2 = 3$$

We can get the same answer by using the definition we just gave for subtraction. Instead of subtracting 2, we can add its opposite, −2. Here is how it looks:

$5 - 2 = 5 + (-2)$ Change subtraction to addition of the opposite.

$ = 3$ Apply the rule for addition of positive and negative numbers.

The result in Example 1 is the same whether we use our previous knowledge of subtraction or the new definition. The new definition is essential when the problems begin to get more complicated.

EXAMPLE 2   Subtract: $-7 - 2$.

SOLUTION   We have never subtracted a positive number from a negative number before. We must apply our definition of subtraction.

$-7 - 2 = -7 + (-2)$ Instead of subtracting 2, we add its opposite, −2.

$ = -9$ Apply the rule for addition.

EXAMPLE 3   Subtract: $-10 - 5$.

SOLUTION   We apply the definition of subtraction (if you don't know the definition of subtraction yet, go back and read it) and add as usual.

$-10 - 5 = -10 + (-5)$ Definition of subtraction

$ = -15$ Add.

PRACTICE PROBLEMS

1. Subtract: $7 - 3$.

> **NOTE**
>
> This definition of subtraction may seem a little strange at first. In Example 1, you will notice that using the definition gives us the same results we are used to getting with subtraction. As we progress further into the section, we will use the definition to subtract numbers we haven't been able to subtract before.

2. Subtract: $-7 - 3$.

> **NOTE**
>
> A real-life analogy to Example 2 would be: "If the temperature were 7° below 0 and then it dropped another 2°, what would the temperature be then?"

3. Subtract: $-8 - 6$.

Answers
1. 4
2. −10
3. −14

EXAMPLE 4   Subtract: $12 - (-6)$.

SOLUTION   The first $-$ sign is read "subtract," and the second one is read "negative." The problem in words is "12 subtract negative 6." We can use the definition of subtraction to change this to the addition of positive 6.

$$12 - (-6) = 12 + 6 \qquad \text{Subtracting } -6 \text{ is equivalent to adding } +6.$$
$$= 18 \qquad \text{Add.}$$

EXAMPLE 5   Subtract: $-20 - (-30)$.

SOLUTION   Instead of subtracting -30, we can use the definition of subtraction to write the problem again as the addition of 30.

$$-20 - (-30) = -20 + 30 \qquad \text{Definition of subtraction}$$
$$= 10 \qquad \text{Add.}$$

Examples 1–5 illustrate all the possible combinations of subtraction with positive and negative numbers. There are no new rules for subtraction. We apply the definition to change each subtraction problem into an equivalent addition problem. The rule for addition can then be used to obtain the correct answer.

EXAMPLE 6   The following table shows the relationship between subtraction and addition:

Subtraction	Addition of the Opposite	Answer
$7 - 9$	$7 + (-9)$	-2
$-7 - 9$	$-7 + (-9)$	-16
$7 - (-9)$	$7 + 9$	16
$-7 - (-9)$	$-7 + 9$	2
$15 - 10$	$15 + (-10)$	5
$-15 - 10$	$-15 + (-10)$	-25
$15 - (-10)$	$15 + 10$	25
$-15 - (-10)$	$-15 + 10$	-5

EXAMPLE 7   Combine: $-3 + 6 - 2$.

SOLUTION   The first step is to change subtraction to addition of the opposite. After that has been done, we add from left to right.

$$-3 + 6 - 2 = -3 + 6 + (-2) \qquad \text{Subtracting 2 is equivalent to adding } -2.$$
$$= 3 + (-2) \qquad \text{Add left to right.}$$
$$= 1$$

---

4. Subtract: $10 - (-6)$.

5. Subtract: $-10 - (-15)$.

NOTE
Examples 4 and 5 may give results you are not used to getting. But you must realize that the results are correct. That is, $12 - (-6)$ is 18, and $-20 - (-30)$ is 10. If you think these results should be different, then you are not thinking of subtraction correctly.

6. Subtract each of the following.
 a. $8 - 5$
 b. $-8 - 5$
 c. $8 - (-5)$
 d. $-8 - (-5)$
 e. $12 - 10$
 f. $-12 - 10$
 g. $12 - (-10)$
 h. $-12 - (-10)$

7. Combine: $-4 + 6 - 7$.

Answers
4. 16
5. 5
6. a. 3 **b.** -13 **c.** 13 **d.** -3
   e. 2 **f.** -22 **g.** 22 **h.** -2
7. -5

Chapter 7 Introduction to Algebra

8. Combine: $15 - (-5) - 8$.

EXAMPLE 8 Combine: $10 - (-4) - 8$.

SOLUTION Changing subtraction to addition of the opposite, we have
$$10 - (-4) - 8 = 10 + 4 + (-8)$$
$$= 14 + (-8)$$
$$= 6$$ ∎

9. Subtract 2 from -8.

EXAMPLE 9 Subtract 3 from -5.

SOLUTION Subtracting 3 is equivalent to adding -3.
$$-5 - 3 = -5 + (-3) = -8$$

Subtracting 3 from -5 gives us -8. ∎

10. Subtract -5 from 7.

EXAMPLE 10 Subtract -4 from 9.

SOLUTION Subtracting -4 is the same as adding $+4$.
$$9 - (-4) = 9 + 4 = 13$$

Subtracting -4 from 9 gives us 13. ∎

11. Find the difference of -8 and -6.

EXAMPLE 11 Find the difference of -7 and -4.

SOLUTION Subtracting -4 from -7 looks like this:
$$-7 - (-4) = -7 + 4 = -3$$

The difference of -7 and -4 is -3. ∎

Subtract.
12. $-57.8 - 70.4$

EXAMPLE 12 Subtract 60.3 from -49.8.

SOLUTION $-49.8 - 60.3 = -49.8 + (-60.3)$
$$= -110.1$$ ∎

13. $-\dfrac{5}{8} - \dfrac{3}{8}$

EXAMPLE 13 Find the difference of $-\dfrac{3}{5}$ and $\dfrac{2}{5}$.

SOLUTION $-\dfrac{3}{5} - \dfrac{2}{5} = -\dfrac{3}{5} + \left(-\dfrac{2}{5}\right)$
$$= -\dfrac{5}{5}$$
$$= -1$$ ∎

Subtraction and Taking Away

Some people may believe that the answer to $-5 - 9$ should be -4 or 4, not -14. If this is happening to you, you are probably thinking of subtraction in terms of taking one number away from another. Thinking of subtraction in this way works well with positive numbers if you always subtract the smaller number from the larger. In algebra, however, we encounter many situations other than this. The definition of subtraction, that $a - b = a + (-b)$ clearly indicates the correct way to use subtraction. That is, when working subtraction problems, you should think "addition of the opposite," not "taking one number away from another."

Answers
8. 12
9. -10
10. 12
11. -2
12. -128.2
13. -1

Problem Set 7.3

Moving Toward Success

"Confidence is the most important single factor in this game, and no matter how great your natural talent, there is only one way to obtain and sustain it: work."

— Eleanor Roosevelt, 1884–1962, American humanitarian and First Lady

1. Why should you begin each study session with the most difficult topics?
2. When studying, should you take breaks? Why or why not?

A Subtract. [Examples 1–5, 12, 13]

1. $7 - 5$
2. $5 - 7$
3. $8 - 6$
4. $6 - 8$

5. $-3 - 5$
6. $-5 - 3$
7. $-4 - 1$
8. $-1 - 4$

9. $5 - (-2)$
10. $2 - (-5)$
11. $3 - (-9)$
12. $9 - (-3)$

13. $-4 - (-7)$
14. $-7 - (-4)$
15. $-10 - (-3)$
16. $-3 - (-10)$

17. $15 - 18$
18. $20 - 32$
19. $100 - 113$
20. $121 - 21$

21. $-30 - 20$
22. $-50 - 60$
23. $-79 - 21$
24. $-86 - 31$

25. $156 - (-243)$
26. $292 - (-841)$
27. $-35 - (-14)$
28. $-29 - (-4)$

29. $-9.01 - 2.4$
30. $-8.23 - 5.4$
31. $-0.89 - 1.01$
32. $-0.42 - 2.04$

33. $-\dfrac{1}{6} - \dfrac{5}{6}$ **34.** $-\dfrac{4}{7} - \dfrac{3}{7}$ **35.** $\dfrac{5}{12} - \dfrac{5}{6}$ **36.** $\dfrac{7}{15} - \dfrac{4}{5}$

37. $-\dfrac{13}{70} - \dfrac{23}{42}$ **38.** $-\dfrac{17}{60} - \dfrac{17}{90}$

A Simplify as much as possible by first changing all subtractions to addition of the opposite and then adding left to right. [Examples 7, 8]

39. $4 - 5 - 6$ **40.** $7 - 3 - 2$ **41.** $-8 + 3 - 4$ **42.** $-10 - 1 + 16$

43. $-8 - 4 - 2$ **44.** $-7 - 3 - 6$ **45.** $12 - 30 - 47$ **46.** $-29 - 53 - 37$

47. $33 - (-22) - 66$ **48.** $44 - (-11) + 55$ **49.** $101 - (-95) + 6$ **50.** $-211 - (-207) + 3$

51. $-900 + 400 - (-100)$ **52.** $-300 + 600 - (-200)$ **53.** $-3.4 - 5.6 - 8.5$ **54.** $-2.1 - 3.1 - 4.1$

55. $\dfrac{1}{2} - \dfrac{1}{3} - \dfrac{1}{4}$ **56.** $\dfrac{1}{5} - \dfrac{1}{6} - \dfrac{1}{7}$

A Translate each of the following and simplify the result. [Examples 9–11]

57. Subtract -6 from 5.

58. Subtract 8 from -2.

59. Find the difference of -5 and -1.

60. Find the difference of -7 and -3.

61. Subtract -4 from the sum of -8 and 12.

62. Subtract -7 from the sum of 7 and -12.

63. What number do you subtract from -3 to get -9?

64. What number do you subtract from 5 to get 8?

Estimating

Work Problems 65–70 mentally, without pencil and paper or a calculator.

65. The answer to the problem $52 - 49$ is closest to which of the following numbers?
  a. 100   b. 0   c. -100

66. The answer to the problem $-52 - 49$ is closest to which of the following numbers?
  a. 100   b. 0   c. -100

67. The answer to the problem $52 - (-49)$ is closest to which of the following numbers?
  a. 100   b. 0   c. -100

68. The answer to the problem $-52 - (-49)$ is closest to which of the following numbers?
  a. 100   b. 0   c. -100

69. Is $-161 - (-62)$ closer to -200 or -100?

70. Is $-553 - 50$ closer to -600 or -500?

Applying the Concepts

71. The chart shows the record low temperatures for the Grand Canyon near North Rim, Arizona. What is the temperature difference between January and July?

How Low Can You Go?

Temperatures (°F) by month:
- Jan: -27
- Feb: -29
- Mar: -9
- Apr: 3
- May: 10
- Jun: 20
- Jul: 28
- Aug: 30
- Sep: 11
- Oct: 3
- Nov: -16
- Dec: -22

Source: weather.sg.msn.com

72. The table shows the lowest and highest points in the Grand Canyon and Death Valley. What is the difference between the lowest point in the Grand Canyon and the lowest point in Death Valley?

LOWEST AND HIGHEST POINTS	
Location	Elevation (ft.)
GRAND CANYON	
Point Imperial	8,803
Lake Mead	1,200
DEATH VALLEY	
Telescope Point	11,049
Badwater Basin	-282

Source: U.S. National Park Services

73. The highest point in Grand Canyon National Park is at Point Imperial with an elevation of 8,803 feet. The lowest point in the park is at Lake Mead at 1,200 feet. What is the difference between the highest and the lowest points?

74. Suppose you had a balance of $153 on your credit card. When you pay your bills at the end of the month, you pay the credit card company $200. What is the balance on your credit card?

75. Tracking Inventory By definition, inventory is the total amount of goods contained in a store or warehouse at any given time. It is helpful for store owners to know the number of items they have available for sale in order to accommodate customer demand. This table shows the beginning inventory on May 1st and tracks the number of items bought and sold for one month. Determine the number of items in inventory at the end of the month.

Date	Transaction	Number of Units Available	Number of Units Sold
May 1	Beginning Inventory	400	
May 3	Purchase	100	
May 8	Sale		700
May 15	Purchase	600	
May 19	Purchase	200	
May 25	Sale		400
May 27	Sale		300
May 31	Ending Inventory		

76. Profit and Loss You own a small business which provides computer support to homeowners who wish to create their own in-house computer network. In addition to setting up the network you also maintain and troubleshoot home PCs. Business gets off to a slow start. You record a profit of $2,298 during the first quarter of the year, a loss of $2,854 during the second quarter, a profit of $3,057 during the third quarter, and a profit of $1,250 for the last quarter of the year. Do you end the year with a net profit or a net loss? Represent that profit or loss as a positive or negative value.

Energy Estimates The chart shows energy consumed in watts of several gaming systems during use and while in idle mode. Use the information to answer the following problems.

77. Find the difference between the energy consumed by a PC in use and an Xbox 360 in use.

78. Find the difference between the energy consumed by a PS3 in idle mode and a Wii in idle mode.

79. Find the difference between a Wii in use and the PC in use.

80. Find the difference between a PS3 in use and an Xbox 360 in idle mode.

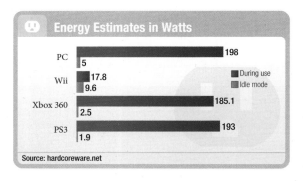

Repeated below is the table of wind chill temperatures that we used previously in this chapter. Use it for Problems 81–84.

Wind speed	AIR TEMPERATURE (°F)							
	30°	25°	20°	15°	10°	5°	0°	−5°
10 mph	16°	10°	3°	−3°	−9°	−15°	−22°	−27°
15 mph	9°	2°	−5°	−11°	−18°	−25°	−31°	−38°
20 mph	4°	−3°	−10°	−17°	−24°	−31°	−39°	−46°
25 mph	1°	−7°	−15°	−22°	−29°	−36°	−44°	−51°
30 mph	−2°	−10°	−18°	−25°	−33°	−41°	−49°	−56°

81. Wind Chill If the temperature outside is 15°F, what is the difference in wind chill temperature between a 15-mile-per-hour wind and a 25-mile-per-hour wind?

82. Wind Chill If the temperature outside is 0°F, what is the difference in wind chill temperature between a 15-mile-per-hour wind and a 25-mile-per-hour wind?

83. Wind Chill Find the difference in temperature between a day in which the air temperature is 20°F and the wind is blowing at 10 miles per hour and a day in which the air temperature is 10°F and the wind is blowing at 20 miles per hour.

84. Wind Chill Find the difference in temperature between a day in which the air temperature is 0°F and the wind is blowing at 10 miles per hour and a day in which the air temperature is −5°F and the wind is blowing at 20 miles per hour.

Use Tables 1 and 2 to work Problems 85–88.

TABLE 1 RECORD LOW TEMPERATURES	
Month	Temperature
January	−36°F
February	−30°F
March	−14°F
April	−2°F
May	19°F
June	22°F
July	35°F
August	30°F
September	19°F
October	15°F
November	−11°F
December	−26°F

TABLE 2 RECORD HIGH TEMPERATURES	
Month	Temperature
January	54°F
February	59°F
March	69°F
April	82°F
May	90°F
June	93°F
July	97°F
August	93°F
September	90°F
October	87°F
November	67°F
December	60°F

85. Temperature Difference Find the difference between the record high temperature and the record low temperature for the month of December.

86. Temperature Difference Find the difference between the record high temperature and the record low temperature for the month of March.

87. Temperature Difference Find the difference between the record low temperatures of March and December.

88. Temperature Difference Find the difference between the record high temperatures of March and December.

Getting Ready for the Next Section

Perform the indicated operations.

89. 3(2)(5) **90.** 5(2)(4) **91.** 6^2 **92.** 8^2

93. 4^3 **94.** 3^3 **95.** 6(3 + 5) **96.** 2(5 + 8)

97. 3(9 − 2) + 4(7 − 2) **98.** 2(5 − 3) − 7(4 − 2) **99.** (3 + 7)(6 − 2) **100.** (6 + 1)(9 − 4)

Simplify each of the following.

101. 2 + 3(4 + 1) **102.** 6 + 5(2 + 3) **103.** (6 + 2)(6 − 2) **104.** (7 + 1)(7 − 1)

105. 5^2 **106.** 2^3 **107.** $2^3 \cdot 3^2$ **108.** $2^3 + 3^2$

Maintaining Your Skills

Write each of the following in symbols.

109. The product of 3 and 5.

110. The product of 5 and 3.

111. The product of 7 and *x*.

112. The product of 2 and *y*.

Rewrite the following using the commutative property of multiplication.

113. 3(5) =

114. 7(*x*) =

Rewrite the following using the associative property of multiplication.

115. 5(7 · 8) =

116. 4(6 · *y*) =

Apply the distributive property to each expression and then simplify the result.

117. 2(3 + 4)

118. 5(6 + 7)

7.4 Multiplication with Negative Numbers

OBJECTIVES

A Multiply positive and negative numbers.

B Apply the rule for order of operations to expressions containing positive and negative numbers.

TICKET TO SUCCESS

Keep these questions in mind as you read through the section. Then respond in your own words and in complete sentences.

1. Write the multiplication problem 3(−5) as an addition problem.
2. If you multiplied two numbers with the same signs, what sign would the answer have?
3. Explain how you would multiply any two numbers.
4. Write a multiplication problem in which you would need to use the order of operations to find an answer.

Suppose you buy three shares of a certain stock on Monday, and by Friday the price per share has dropped $5. How much money have you lost? The answer is $15. Because it is a loss, we can express it as −$15. The multiplication problem below can be used to describe the relationship among the numbers.

$$\underbrace{3}_{\text{3 shares}} \underbrace{(-5)}_{\text{each loses \$5}} = \underbrace{-15}_{\text{for a total of }-\$15}$$

From this we conclude that it is reasonable to say that the product of a positive number and a negative number is a negative number.

A Multiplication with Negatives

In order to generalize multiplication with negative numbers, recall that we first defined multiplication by whole numbers to be repeated addition. That is,

$$\underbrace{3}_{\text{Multiplication}} \cdot 5 = \underbrace{5 + 5 + 5}_{\text{Repeated addition}}$$

This concept is very helpful when it comes to developing the rule for multiplication problems that involve negative numbers. For the first example we look at what happens when we multiply a negative number by a positive number.

PRACTICE PROBLEMS

1. Multiply: $2(-6)$.

2. Multiply: $-2(6)$.

3. Multiply: $-2(-6)$.

NOTE
We want to be able to justify everything we do in mathematics. The discussion here tells us *why* $-3(-5) = 15$.

Answers
1. -12
2. -12
3. 12

EXAMPLE 1 Multiply: $3(-5)$.

SOLUTION Writing this product as repeated addition, we have
$$3(-5) = (-5) + (-5) + (-5)$$
$$= -10 + (-5)$$
$$= -15$$

The result, -15, is obtained by adding the three negative 5s. ∎

EXAMPLE 2 Multiply: $-3(5)$.

SOLUTION In order to write this multiplication problem in terms of repeated addition, we will have to reverse the order of the two numbers. This is easily done, because multiplication is a commutative operation.

$$-3(5) = 5(-3) \quad \text{Commutative property}$$
$$= (-3) + (-3) + (-3) + (-3) + (-3) \quad \text{Repeated addition}$$
$$= -15 \quad \text{Add.}$$

The product of -3 and 5 is -15. ∎

EXAMPLE 3 Multiply: $-3(-5)$.

SOLUTION It is impossible to write this product in terms of repeated addition. We will find the answer to $-3(-5)$ by solving a different problem. Look at the following problem:

$$-3[5 + (-5)] = -3[0] = 0$$

The result is 0, because multiplying by 0 always produces 0. Now we can work the same problem another way, and in the process find the answer to $-3(-5)$. Applying the distributive property to the same expression, we have

$$-3[5 + (-5)] = -3(5) + (-3)(-5) \quad \text{Distributive property}$$
$$= -15 + (?) \quad -3(5) = -15$$

The question mark must be 15, because we already know that the answer to the problem is 0, and 15 is the only number we can add to -15 to get 0. So our problem is solved.

$$-3(-5) = 15$$
∎

Table 1 gives a summary of what we have done so far in this section.

TABLE 1

Original Numbers Have	For Example	The Answer Is
Same signs	$3(5) = 15$	Positive
Different signs	$-3(5) = -15$	Negative
Different signs	$3(-5) = -15$	Negative
Same signs	$-3(-5) = 15$	Positive

From the examples we have done so far in this section and their summaries in Table 1, we can write a rule for multiplication of positive and negative numbers. This rule should be memorized. By the time you have finished reading this section and working the problems at the end of the section, you should be fast and accurate when multiplying positive and negative numbers.

7.4 Multiplication with Negative Numbers

Rule **Multiplication of Any Two Numbers**

To multiply any two numbers, we multiply their absolute values.
1. The answer is *positive* if both the original numbers have the same sign. That is, the product of two numbers with the same sign is positive.
2. The answer is *negative* if the original two numbers have different signs. The product of two numbers with different signs is negative.

EXAMPLE 4 $2(4) = 8$ Like signs; positive answer

EXAMPLE 5 $-2(-4) = 8$ Like signs; positive answer

EXAMPLE 6 $2(-4) = -8$ Unlike signs; negative answer

EXAMPLE 7 $-2(4) = -8$ Unlike signs; negative answer

EXAMPLE 8 $7(-6) = -42$ Unlike signs; negative answer

EXAMPLE 9 $-5(-8) = 40$ Like signs; positive answer

EXAMPLE 10 $-3(2)(-5) = -6(-5)$ Multiply -3 and 2 to get -6.
$= 30$

Here are a few more multiplication problems involving negative fractions and decimals.

EXAMPLE 11 $\left(\frac{2}{3}\right)\left(-\frac{3}{5}\right) = -\frac{6}{15} = -\frac{2}{5}$ The rule for multiplication also holds for fractions.

EXAMPLE 12 $\left(-\frac{7}{8}\right)\left(-\frac{5}{14}\right) = \frac{35}{112} = \frac{5}{16}$

EXAMPLE 13 $(-5)(3.4) = -17.0$ The rule for multiplication also holds for decimals.

EXAMPLE 14 $(-0.4)(-0.8) = 0.32$

EXAMPLE 15 Use the definition of exponents to expand each expression. Then simplify by multiplying.

   a. $(-6)^2 = (-6)(-6)$ Definition of exponents
          $= 36$ Multiply.

   b. $-6^2 = -6 \cdot 6$ Definition of exponents
          $= -36$ Multiply.

   c. $(-4)^3 = (-4)(-4)(-4)$ Definition of exponents
          $= -64$ Multiply.

   d. $-4^3 = -4 \cdot 4 \cdot 4$ Definition of exponents
          $= -64$ Multiply.

Multiply.
4. $3(2)$
5. $-3(-2)$
6. $3(-2)$
7. $-3(2)$
8. $8(-9)$
9. $-6(-4)$
10. $-5(2)(-4)$
11. $\frac{3}{4}\left(-\frac{4}{7}\right)$
12. $\left(-\frac{5}{6}\right)\left(-\frac{9}{20}\right)$
13. $(-3)(6.7)$
14. $(-0.6)(-0.5)$
15. Use the definition of exponents to expand each expression. Then simplify by multiplying.
   a. $(-8)^2$
   b. -8^2
   c. $(-3)^3$
   d. -3^3

Answers
4. 6
5. 6
6. -6
7. -6
8. -72
9. 24
10. 40
11. $-\frac{3}{7}$
12. $\frac{3}{8}$
13. -20.1
14. 0.30
15. **a.** 64 **b.** -64 **c.** -27 **d.** -27

In Example 15, the base is a negative number in parts a and c, but not in Parts b and d. We know this is true because of the use of parentheses.

B Order of Operations

Now let's apply the rule for order of operations to the following examples that involve positive and negative numbers. Remember, the rule for order of operations specifies that we are to work inside the parentheses first, and then simplify numbers containing exponents. After this, we multiply and divide, left to right. The last step is to add and subtract, left to right.

EXAMPLE 16 Simplify: $-6[3 + (-5)]$.

SOLUTION We begin inside the brackets and work our way out.

$$-6[3 + (-5)] = -6[-2]$$
$$= 12$$

EXAMPLE 17 Simplify: $-4 + 5(-6 + 2)$.

SOLUTION Simplifying inside the parentheses first, we have

$$-4 + 5(-6 + 2) = -4 + 5(-4) \quad \text{Simplify inside parentheses.}$$
$$= -4 + (-20) \quad \text{Multiply.}$$
$$= -24 \quad \text{Add.}$$

EXAMPLE 18 Simplify: $-2(7) + 3(-6)$.

SOLUTION Multiplying left to right before we add gives us

$$-2(7) + 3(-6) = -14 + (-18)$$
$$= -32$$

EXAMPLE 19 Simplify: $-3(2 - 9) + 4(-7 - 2)$.

SOLUTION We begin by subtracting inside the parentheses.

$$-3(2 - 9) + 4(-7 - 2) = -3(-7) + 4(-9)$$
$$= 21 + (-36)$$
$$= -15$$

EXAMPLE 20 Simplify: $(-3 - 7)(2 - 6)$.

SOLUTION Again, we begin by simplifying inside the parentheses.

$$(-3 - 7)(2 - 6) = (-10)(-4)$$
$$= 40$$

USING TECHNOLOGY

Calculator Note

Here is how we work the problem shown in Example 16 on a calculator. (The $\boxed{\times}$ key on the first line may, or may not, be necessary. Try your calculator without it and see.)

Scientific Calculator: $\boxed{(}\ 3\ \boxed{+/-}\ \boxed{-}\ 7\ \boxed{)}\ \boxed{\times}\ \boxed{(}\ 2\ \boxed{-}\ 6\ \boxed{)}\ \boxed{=}$

Graphing Calculator: $\boxed{(}\ \boxed{(-)}\ 3\ \boxed{-}\ 7\ \boxed{)}\ \boxed{(}\ 2\ \boxed{-}\ 6\ \boxed{)}\ \boxed{ENT}$

---

16. Simplify: $-2[5 + (-8)]$.

17. Simplify: $-3 + 4(-7 + 3)$.

18. Simplify: $-3(5) + 4(-4)$.

19. Simplify: $-2(3 - 5) - 7(-2 - 4)$.

20. Simplify: $(-6 - 1)(4 - 9)$.

Answers
16. 6
17. −19
18. −31
19. 46
20. 35

Problem Set 7.4

Moving Toward Success

"The five essential entrepreneurial skills for success are concentration, discrimination, organization, innovation and communication."

—Michael Faraday, 1791–1867, English physicist

1. Which of the following study methods should you use in this class?
   a. Review class notes.
   b. Work problems.
   c. Make and review flashcards.
   d. Outline chapters.
   e. Study with a group.
   f. All the above
2. One study technique is to review topics in a different order than presented in the chapter. How might this be helpful?

A Find each of the following products. (Multiply.) [Examples 1–14]

1. $7(-8)$
2. $-3(5)$
3. $-6(10)$
4. $4(-8)$

5. $-7(-8)$
6. $-4(-7)$
7. $-9(-9)$
8. $-6(-3)$

9. $-2.1(4.3)$
10. $-6.8(5.7)$
11. $-\dfrac{4}{5}\left(-\dfrac{15}{28}\right)$
12. $-\dfrac{8}{9}\left(-\dfrac{27}{32}\right)$

13. $-12\left(\dfrac{2}{3}\right)$
14. $-18\left(\dfrac{5}{6}\right)$
15. $3(-2)(4)$
16. $5(-1)(3)$

17. $-4(3)(-2)$
18. $-4(5)(-6)$
19. $-1(-2)(-3)$
20. $-2(-3)(-4)$

A Use the definition of exponents to expand each of the following expressions. Then multiply according to the rule for multiplication. [Example 15]

21. a. $(-4)^2$ b. -4^2
22. a. $(-5)^2$ b. -5^2
23. a. $(-5)^3$ b. -5^3
24. a. $(-4)^3$ b. -4^3
25. a. $(-2)^4$ b. -2^4
26. a. $(-1)^4$ b. -1^4

Complete the following tables. Remember, if $x = -5$, then $x^2 = (-5)^2 = 25$. [Example 15]

27.

Number x	Square x^2
−3	
−2	
−1	
0	
1	
2	
3	

28.

Number x	Cube x^3
−3	
−2	
−1	
0	
1	
2	
3	

29.

First Number x	Second Number y	Their Product xy
6	2	
6	1	
6	0	
6	−1	
6	−2	

30.

First Number a	Second Number b	Their Product ab
−5	3	
−5	2	
−5	1	
−5	0	
−5	−1	
−5	−2	
−5	−3	

B Use the rule for order of operations along with the rules for addition, subtraction, and multiplication to simplify each of the following expressions. [Examples 16–20]

31. $4(-3 + 2)$ **32.** $7(-6 + 3)$ **33.** $-10(-2 - 3)$ **34.** $-5(-6 - 2)$

35. $-3 + 2(5 - 3)$ **36.** $-7 + 3(6 - 2)$ **37.** $-7 + 2[-5 - 9]$ **38.** $-8 + 3[-4 - 1]$

39. $2(-5) + 3(-4)$ **40.** $6(-1) + 2(-7)$ **41.** $3(-2)4 + 3(-2)$ **42.** $2(-1)(-3) + 4(-6)$

43. $(8 - 3)(2 - 7)$ **44.** $(9 - 3)(2 - 6)$ **45.** $(2 - 5)(3 - 6)$ **46.** $(3 - 7)(2 - 8)$

47. $3(5 - 8) + 4(6 - 7)$ **48.** $2(3 - 7) + 3(5 - 6)$ **49.** $-2(8 - 10) + 3(4 - 9)$ **50.** $-3(6 - 9) + 2(3 - 8)$

51. $-3(4 - 7) - 2(-3 - 2)$ **52.** $-5(-2 - 8) - 4(6 - 10)$ **53.** $3(-2)(6 - 7)$ **54.** $4(-3)(2 - 5)$

55. Find the product of -3, -2, and -1.

56. Find the product of -7, -1, and 0.

57. What number do you multiply by -3 to get 12?

58. What number do you multiply by -7 to get -21?

59. Subtract -3 from the product of -5 and 4.

60. Subtract 5 from the product of -8 and 1.

Work Problems 61–68 mentally, without pencil and paper or a calculator.

61. The product $-32(-522)$ is closest to which of the following numbers?
   a. 15,000   **b.** -500 **c.** $-1,500$ **d.** $-15,000$

62. The product $32(-522)$ is closest to which of the following numbers?
   a. 15,000   **b.** -500 **c.** $-1,500$ **d.** $-15,000$

63. The product $-47(470)$ is closest to which of the following numbers?
   a. 25,000   **b.** 420   **c.** $-2,500$ **d.** $-25,000$

64. The product $-47(-470)$ is closest to which of the following numbers?
   a. 25,000   **b.** 420   **c.** $-2,500$ **d.** $-25,000$

65. The product $-222(-987)$ is closest to which of the following numbers?
   a. 200,000   **b.** 800   **c.** -800 **d.** $-1,200$

66. The sum $-222 + (-987)$ is closest to which of the following numbers?
   a. 200,000   **b.** 800   **c.** -800 **d.** $-1,200$

67. The difference $-222 - (-987)$ is closest to which of the following numbers?
   a. 200,000   **b.** 800   **c.** -800 **d.** $-1,200$

68. The difference $-222 - 987$ is closest to which of the following numbers?
   a. 200,000   **b.** 800   **c.** -800 **d.** $-1,200$

Applying the Concepts

69. The chart shows the record low temperatures for the Grand Canyon near North Rim, Arizona. Write the record low temperature for March.

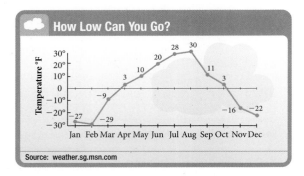

70. The chart shows the most visited countries each year.

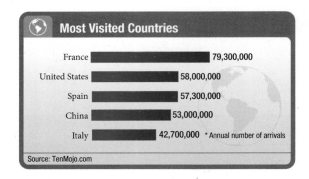

a. What is the average number of monthly visitors to Spain?

b. What is the difference between the average monthly visitors to Italy compared to France? What does your answer mean in words?.

c. Use negative numbers to write an expression for the average number of monthly visitors over three months to Italy compared to France.

71. Temperature Change A hot-air balloon is rising to its cruising altitude. Suppose the air temperature around the balloon drops 4 degrees each time the balloon rises 1,000 feet. What is the net change in air temperature around the balloon as it rises from 2,000 feet to 6,000 feet?

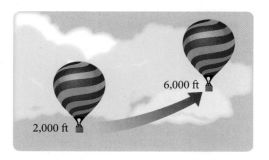

72. Temperature Change A small airplane is rising to its cruising altitude. Suppose the air temperature around the plane drops 4 degrees each time the plane increases its altitude by 1,000 feet. What is the net change in air temperature around the plane as it rises from 5,000 feet to 12,000 feet?

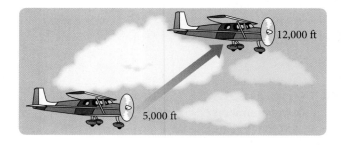

73. Expense Account A business woman has a travel expense account of $1,000. If she spends $75 a week for 8 weeks, what will the balance of her expense account be?

74. Six months ago, you bought five shares of stock for $27. However, the value of each stock has dropped to $16. Express your loss as a negative number.

Getting Ready for the Next Section

Perform the indicated operations.

75. $35 \div 5$

76. $32 \div 4$

77. $\dfrac{20}{4}$

78. $\dfrac{30}{5}$

79. $12 - 17$

80. $7 - 11$

81. $(6 \cdot 3) \div 2$

82. $(8 \cdot 5) \div 4$

83. $80 \div 10 \div 2$

84. $80 \div 2 \div 10$

85. $15 + 5(4) \div 10$

86. $[20 + 6(2)] \div (11 - 7)$

87. $4(10^2) + 20 \div 4$

88. $3(4^2) + 10 \div 5$

Maintaining Your Skills

Write each of the following statements in symbols.

89. The quotient of 12 and 6

90. The quotient of x and 5

Rewrite each of the following multiplication problems as an equivalent division problem.

91. $2(3) = 6$

92. $5 \cdot 4 = 20$

Rewrite each of the following division problems as an equivalent multiplication problem.

93. $10 \div 5 = 2$

94. $\dfrac{63}{9} = 7$

Divide.

95. $4{,}984 \div 56$

96. $4{,}994 \div 56$

Extending the Concepts

In Chapter 1, we defined a geometric sequence to be a sequence of numbers in which each number, after the first number, is obtained from the previous number by multiplying by the same amount each time.

Find the next two terms in each of the following geometric sequences.

97. 2, −6, 18, . . . **98.** 1, −4, 16, . . . **99.** −2, 6, −18, . . . **100.** −1, 4, −16, . . .

Simplify each of the following according to the rule for order of operations.

101. $5(-2)^2 - 3(-2)^3$ **102.** $8(-1)^3 - 6(-3)^2$ **103.** $7 - 3(4 - 8)$

104. $6 - 2(9 - 11)$ **105.** $5 - 2[3 - 4(6 - 8)]$ **106.** $7 - 4[6 - 3(2 - 9)]$

7.5 Division with Negative Numbers

OBJECTIVES

A Divide positive and negative numbers.

B Apply the rule for order of operations to expressions that contain positive and negative numbers.

TICKET TO SUCCESS

Keep these questions in mind as you read through the section. Then respond in your own words and in complete sentences.

1. Write a multiplication problem that is equivalent to the division problem $\frac{-12}{4} = -3$.
2. Write a multiplication problem that is equivalent to the division problem $\frac{-12}{-4} = 3$.
3. What generalization can we make about the signs of quotients that contain negative signs?
4. Write a division problem in which you would need to use the order of operations to find an answer.

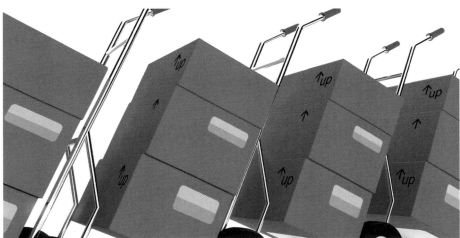

Suppose four friends invest equal amounts of money in a moving truck to start a small business. After 2 years the truck has dropped $10,000 in value. If we represent this change with the number −$10,000, then the loss to each of the four partners can be found with division:

$$(-\$10{,}000) \div 4 = -\$2{,}500$$

From this example, it seems reasonable to assume that a negative number divided by a positive number will give a negative answer.

A Division with Negatives

To cover all the possible situations we can encounter with division of negative numbers, we use the relationship between multiplication and division. If we let n be the answer to the problem $12 \div (-2)$, then we know that

$$12 \div (-2) = n \quad \text{and} \quad -2(n) = 12$$

From our work with multiplication, we know that n must be -6 in the multiplication problem above, because -6 is the only number we can multiply -2 by to get 12. Because of the relationship between the two problems above, it must be true that 12 divided by -2 is -6.

The following pairs of problems show more quotients of positive and negative numbers. In each case, the multiplication problem on the right justifies the answer to the division problem on the left.

$$6 \div 3 = 2 \quad \text{because} \quad 3(2) = 6$$
$$6 \div (-3) = -2 \quad \text{because} \quad -3(-2) = 6$$
$$-6 \div 3 = -2 \quad \text{because} \quad 3(-2) = -6$$
$$-6 \div (-3) = 2 \quad \text{because} \quad -3(2) = -6$$

The results given above can be used to write the rule for division with negative numbers.

> **Rule Division of Any Two Numbers**
>
> To divide any two numbers, we divide their absolute values.
> 1. The answer is *positive* if both the original numbers have the same sign. That is, the quotient of two numbers with the same signs is positive.
> 2. The answer is *negative* if the original two numbers have different signs. That is, the quotient of two numbers with different signs is negative.

EXAMPLE 1 $-12 \div 4 = -3$ Unlike signs, negative answer

EXAMPLE 2 $12 \div (-4) = -3$ Unlike signs; negative answer

EXAMPLE 3 $-12 \div (-4) = 3$ Like signs; positive answer

EXAMPLE 4 $\dfrac{12}{-4} = -3$ Unlike signs; negative answer

EXAMPLE 5 $\dfrac{-20}{-4} = 5$ Like signs; positive answer

PRACTICE PROBLEMS

Divide.
1. $-8 \div 2$

2. $8 \div (-2)$

3. $-8 \div (-2)$

4. $\dfrac{20}{-5}$

5. $\dfrac{-30}{-5}$

Answers
1. -4
2. -4
3. 4
4. -4
5. 6

From the examples we have done so far, we can make the following generalization about quotients that contain negative signs:

> If a and b are numbers and b is not equal to 0, then
> $$-\frac{a}{b} = \frac{a}{-b} = \frac{-a}{b} \quad \text{and} \quad \frac{-a}{-b} = \frac{a}{b}$$

B Order of Operations

The last examples in this section involve more than one operation. We use the rules developed previously in this chapter and the rule for order of operations to simplify each.

EXAMPLE 6 Simplify: $\dfrac{6(-3)}{-2}$.

SOLUTION We begin by multiplying 6 and -3.

$$\frac{6(-3)}{-2} = \frac{-18}{-2} \quad \text{Multiply: } 6(-3) = -18.$$
$$= 9 \quad \text{Like signs; positive answer}$$

6. Simplify: $\dfrac{8(-5)}{-4}$.

EXAMPLE 7 Simplify: $\dfrac{-15 + 5(-4)}{12 - 17}$.

SOLUTION Simplifying above and below the fraction bar, we have

$$\frac{-15 + 5(-4)}{12 - 17} = \frac{-15 + (-20)}{-5} = \frac{-35}{-5} = 7$$

7. Simplify: $\dfrac{-20 + 6(-2)}{7 - 11}$.

EXAMPLE 8 Simplify: $-4(10^2) + 20 \div (-4)$.

SOLUTION Applying the rule for order of operations, we have

$$-4(10^2) + 20 \div (-4) = -4(100) + 20 \div (-4) \quad \text{Work exponents first.}$$
$$= -400 + (-5) \quad \text{Multiply and divide.}$$
$$= -405 \quad \text{Add.}$$

8. Simplify: $-3(4^2) + 10 \div (-5)$.

EXAMPLE 9 Simplify: $-80 \div 10 \div 2$.

SOLUTION In a situation like this, the rule for order of operations states that we are to divide left to right.

$$-80 \div 10 \div 2 = -8 \div 2 \quad \text{Divide } -80 \text{ by } 10.$$
$$= -4$$

9. Simplify: $-80 \div 2 \div 10$.

Answers
6. 10
7. 8
8. -50
9. -4

Problem Set 7.5

Moving Toward Success

"Creativity is inventing, experimenting, growing, taking risks, breaking rules, making mistakes, and having fun."
—Mary Lou Cook, American community activist and author

1. What are some fun things you would rather be doing than studying?
2. How can you work some of those things into your day and still meet your study schedule?

A Find each of the following quotients. (Divide.) [Examples 1–5]

1. $-15 \div 5$
2. $15 \div (-3)$
3. $20 \div (-4)$
4. $-20 \div 4$
5. $-30 \div (-10)$
6. $-50 \div (-25)$
7. $\dfrac{-14}{-7}$
8. $\dfrac{-18}{-6}$
9. $\dfrac{12}{-3}$
10. $\dfrac{12}{-4}$
11. $-22 \div 11$
12. $-35 \div 7$
13. $\dfrac{0}{-3}$
14. $\dfrac{0}{-5}$
15. $125 \div (-25)$
16. $-144 \div (-9)$

Complete the following tables.

17.

First Number a	Second Number b	Their Quotient $\dfrac{a}{b}$
100	−5	
100	−10	
100	−25	
100	−50	

18.

First Number a	Second Number b	Their Quotient $\dfrac{a}{b}$
24	−4	
24	−3	
24	−2	
24	−1	

19.

First Number a	Second Number b	Their Quotient $\dfrac{a}{b}$
−100	−5	
−100	5	
100	−5	
100	5	

20.

First Number a	Second Number b	Their Quotient $\dfrac{a}{b}$
−24	−2	
−24	−4	
−24	−6	
−24	−8	

21. Find the quotient of −25 and 5.

22. Find the quotient of −38 and −19.

23. What number do you divide by −5 to get −7?

24. What number do you divide by 6 to get −7?

25. Subtract −3 from the quotient of 27 and 9.

26. Subtract −7 from the quotient of −72 and −9.

B Use any of the rules developed in this chapter and the rule for order of operations to simplify each of the following expressions as much as possible. [Examples 6–9]

27. $\dfrac{4(-7)}{-28}$

28. $\dfrac{6(-3)}{-18}$

29. $\dfrac{-3(-10)}{-5}$

30. $\dfrac{-4(-12)}{-6}$

31. $\dfrac{2(-3)}{6-3}$

32. $\dfrac{2(-3)}{3-6}$

33. $\dfrac{4-8}{8-4}$

34. $\dfrac{9-5}{5-9}$

35. $\dfrac{2(-3)+10}{-4}$

36. $\dfrac{7(-2)-6}{-10}$

37. $\dfrac{2+3(-6)}{4-12}$

38. $\dfrac{3+9(-1)}{5-7}$

39. $\dfrac{6(-7)+3(-2)}{20-4}$

40. $\dfrac{9(-8)+5(-1)}{12-1}$

41. $\dfrac{3(-7)(-4)}{6(-2)}$

42. $\dfrac{-2(4)(-8)}{(-2)(-2)}$

43. $(-5)^2 + 20 \div 4$

44. $6^2 + 36 \div 9$

45. $100 \div (-5)^2$

46. $400 \div (-4)^2$

47. $-100 \div 10 \div 2$

48. $-500 \div 50 \div 10$

49. $-100 \div (10 \div 2)$

50. $-500 \div (50 \div 10)$

51. $(-100 \div 10) \div 2$

52. $(-500 \div 50) \div 10$

Estimating

Work Problems 53–60 mentally, without pencil and paper or a calculator.

53. Is $397 \div (-401)$ closer to 1 or −1?

54. Is $-751 \div (-749)$ closer to 1 or −1?

55. The quotient $-121 \div 27$ is closest to which of the following numbers?
  a. −150   **b.** −100   **c.** −4   **d.** 6

56. The quotient $1,000 \div (-337)$ is closest to which of the following numbers?
  a. 663   **b.** −3   **c.** −30   **d.** −663

57. Which number is closest to the sum $-151 + (-49)$?
  a. −200   **b.** −100   **c.** 3   **d.** 7,500

58. Which number is closest to $-151 - (-49)$?
  a. −200   **b.** −100   **c.** 3   **d.** 7,500

59. Which number is closest to the product $-151(-49)$?
  a. −200   **b.** −100   **c.** 3   **d.** 7,500

60. Which number is closest to the quotient $-151 \div (-49)$?
  a. −200   **b.** −100   **c.** 3   **d.** 7,500

Applying the Concepts

The chart shows the annual cost of living, in dollars, for some cities throughout the world. Expenses can also be written as negative numbers. Use the information from the chart to answer the following questions.

61. Find the monthly cost of living in Moscow. Use negative numbers and round to the nearest cent.

62. Find the cost of living for three months in Osaka. Use negative numbers and round to the nearest cent.

63. Temperature Line Graph Suppose the table below gives the low temperature for each day of one week in White Bear Lake, Minnesota. Use the template to draw a line graph of the information in the table.

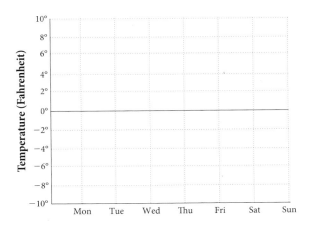

LOW TEMPERATURES IN WHITE BEAR LAKE, MINNESOTA	
Day	Temperature
Monday	10 °F
Tuesday	8 °F
Wednesday	−5 °F
Thursday	−3 °F
Friday	−8 °F
Saturday	5 °F
Sunday	7 °F

64. Temperature Line Graph Suppose the table below gives the low temperature for each day of one week in Fairbanks, Alaska. Use the template to draw a line graph of the information in the table.

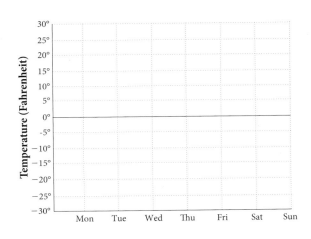

LOW TEMPERATURES IN FAIRBANKS, ALASKA	
Day	Temperature
Monday	−26 °F
Tuesday	−5 °F
Wednesday	9 °F
Thursday	12 °F
Friday	3 °F
Saturday	−15 °F
Sunday	−20 °F

Getting Ready for the Next Section

The problems below review some of the properties of addition and multiplication we covered in Chapter 1.

Rewrite each expression using the commutative property of addition or multiplication.

65. $3 + x$

66. $4y$

Rewrite each expression using the associative property of addition or multiplication.

67. $5 + (7 + a)$

68. $(x + 4) + 6$

69. $3(4y)$

70. $(3y)8$

Apply the distributive property to each expression.

71. $5(3 + 7)$

72. $8(4 + 2)$

Simplify.

73. 6^2

74. 12^2

75. 4^3

76. 5^2

77. $2(100) + 2(75)$

78. $2(100) + 2(53)$

79. $100(75)$

80. $100(53)$

Maintaining Your Skills

The problems below review addition, subtraction, multiplication, and division of positive and negative numbers, as covered in this chapter.

Perform the indicated operations.

81. $8 + (-4)$

82. $-8 + 4$

83. $-8 + (-4)$

84. $-8 - 4$

85. $8 - (-4)$

86. $-8 - (-4)$

87. $8(-4)$

88. $-8(4)$

89. $-8(-4)$

90. $8 \div (-4)$

91. $-8 \div 4$

92. $-8 \div (-4)$

Extending the Concepts

Find the next term in each sequence below.

93. $32, -16, 8, \ldots$

94. $243, -81, 27, \ldots$

95. $-32, 16, -8, \ldots$

96. $-243, 81, -27, \ldots$

Simplify each of the following expressions.

97. $\dfrac{6 - 3(2 - 11)}{6 - 3(2 + 11)}$

98. $\dfrac{8 + 4(3 - 5)}{8 - 4(3 + 5)}$

99. $\dfrac{6 - (3 - 4) - 3}{1 - 2 - 3}$

100. $\dfrac{7 - (3 - 6) - 4}{-1 - 2 - 3}$

7.6 Simplifying Algebraic Expressions

OBJECTIVES

A Simplify expressions by using the associative property.

B Apply the distributive property to expressions containing numbers and variables.

C Use the distributive property to combine similar terms.

D Use the formulas for area and perimeter of squares and rectangles.

TICKET TO SUCCESS

Keep these questions in mind as you read through the section. Then respond in your own words and in complete sentences.

1. Without actually multiplying, how do you apply the associative property to the expression $4(5x)$?
2. What are similar terms?
3. Explain why $2a - a$ is a, rather than 1.
4. Can two rectangles with the same perimeter have different areas? Explain your answer.

Suppose we have 24 yards of fencing that we are to use to build a rectangular dog run. If we want the dog run to have the largest area possible then we want the rectangle, with a perimeter of 24 yards, that encloses the largest area. The diagram below shows six dog runs, each of which has a perimeter of 24 yards. Notice how the length decreases as the width increases.

Dog runs with a perimeter = 24 yards

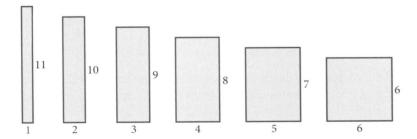

In this section, we want to simplify expressions containing variables—that is, algebraic expressions. An algebraic expression is a combination of constants and variables joined by arithmetic operations such as addition, subtraction, multiplication, and division.

NOTE
An algebraic expression does not contain an equals sign.

A Using the Associative Property

To begin, let's review how we use the associative properties for addition and multiplication to simplify expressions.

Consider the expression $4(5x)$. We can apply the associative property of multiplication to this expression to change the grouping so that the 4 and the 5 are grouped together, instead of the 5 and the x. Here's how it looks:

$$4(5x) = (4 \cdot 5)x \quad \text{Associative property}$$
$$= 20x \quad \text{Multiply: } 4 \cdot 5 = 20.$$

We have simplified the expression to $20x$, which in most cases in algebra will be easier to work with than the original expression.

Here are some more examples:

EXAMPLE 1
$$7(3a) = (7 \cdot 3)a \quad \text{Associative property}$$
$$= 21a \quad \text{7 times 3 is 21.} \quad ■$$

EXAMPLE 2
$$-2(5x) = (-2 \cdot 5)x \quad \text{Associative property}$$
$$= -10x \quad \text{The product of } -2 \text{ and } 5 \text{ is } -10. \quad ■$$

EXAMPLE 3
$$3(-4y) = [3(-4)]y \quad \text{Associative property}$$
$$= -12y \quad \text{3 times } -4 \text{ is } -12. \quad ■$$

We can use the associative property of addition to simplify expressions also.

EXAMPLE 4
$$3 + (8 + x) = (3 + 8) + x \quad \text{Associative property}$$
$$= 11 + x \quad \text{The sum of 3 and 8 is 11.} \quad ■$$

EXAMPLE 5
$$(2x + 5) + 10 = 2x + (5 + 10) \quad \text{Associative property}$$
$$= 2x + 15 \quad \text{Add.} \quad ■$$

B Using the Distributive Property

In Chapter 1, we introduced the distributive property. In symbols, it looks like this:
$$a(b + c) = ab + ac$$

Because subtraction is defined as addition of the opposite, the distributive property holds for subtraction as well as addition. That is,
$$a(b - c) = ab - ac$$

We say that multiplication distributes over addition and subtraction. Here are some examples that review how the distributive property is applied to expressions that contain variables.

EXAMPLE 6
$$4(x + 5) = 4(x) + 4(5) \quad \text{Distributive property}$$
$$= 4x + 20 \quad \text{Multiply.} \quad ■$$

PRACTICE PROBLEMS

Multiply.
1. $5(7a)$

2. $-3(9x)$

3. $5(-8y)$

Simplify.
4. $6 + (9 + x)$

Simplify.
5. $(3x + 7) + 4$

Apply the distributive property.
6. $6(x + 4)$

Answers
1. $35a$
2. $-27x$
3. $-40y$
4. $15 + x$
5. $3x + 11$
6. $6x + 24$

7.6 Simplifying Algebraic Expressions

EXAMPLE 7
$$2(a - 3) = 2(a) - 2(3) \quad \text{Distributive property}$$
$$= 2a - 6 \quad \text{Multiply.}$$

In Examples 1–3, we simplified expressions such as $4(5x)$ by using the associative property of multiplication. Here are some examples that use a combination of the associative property and the distributive property.

EXAMPLE 8
$$4(5x + 3) = 4(5x) + 4(3) \quad \text{Distributive property}$$
$$= (4 \cdot 5)x + 4(3) \quad \text{Associative property}$$
$$= 20x + 12 \quad \text{Multiply.}$$

EXAMPLE 9
$$7(3a - 6) = 7(3a) - 7(6) \quad \text{Distributive property}$$
$$= 21a - 42 \quad \text{Associative property and multiplication}$$

EXAMPLE 10
$$5(2x + 3y) = 5(2x) + 5(3y) \quad \text{Distributive property}$$
$$= 10x + 15y \quad \text{Associative property and multiplication}$$

C Similar Terms

We can also use the distributive property to simplify expressions like $4x + 3x$. Because multiplication is a commutative operation, we can also rewrite the distributive property like this:

$$b \cdot a + c \cdot a = (b + c)a$$

Applying the distributive property in this form to the expression $4x + 3x$, we have

$$4x + 3x = (4 + 3)x \quad \text{Distributive property}$$
$$= 7x \quad \text{Add.}$$

Expressions like $4x$ and $3x$ are called *similar terms* because the variable parts are the same. Some other examples of similar terms are $5y$ and $-6y$ and the terms $7a$, $-13a$, and $\frac{3}{4}a$. To simplify an algebraic expression (an expression that involves both numbers and variables), we combine similar terms by applying the distributive property. Table 1 shows several pairs of similar terms and how they can be combined using the distributive property.

TABLE 1

Original Expression		Apply Distributive Property		Simplified Expression
$4x + 3x$	=	$(4 + 3)x$	=	$7x$
$7a + a$	=	$(7 + 1)a$	=	$8a$
$-5x + 7x$	=	$(-5 + 7)x$	=	$2x$
$8y - y$	=	$(8 - 1)y$	=	$7y$
$-4a - 2a$	=	$(-4 - 2)a$	=	$-6a$
$3x - 7x$	=	$(3 - 7)x$	=	$-4x$

Apply the distributive property.

7. $7(a - 5)$

8. $6(4x + 5)$

9. $3(8a - 4)$

10. $8(3x + 4y)$

Answers
7. $7a - 35$
8. $24x + 30$
9. $24a - 12$
10. $24x + 32y$

Apply the distributive property.
11. $-3b + 5b$

12. $5a - 4 - 3a$

As you can see from the table, the distributive property can be applied to any combination of positive and negative terms so long as they are similar terms.

EXAMPLE 11
$$-4x + 3x = (-4 + 3)x \quad \text{Distributive property}$$
$$= (-1)x \quad \text{Add.}$$
$$= -x$$

EXAMPLE 12
$$14y - 3x - 10y = 14y - 10y - 3x \quad \text{Commutative property of addition}$$
$$= (14 - 10)y - 3x \quad \text{Distributive property}$$
$$= 4y - 3x \quad \text{Add.}$$

D Algebraic Expressions Representing Area and Perimeter

Below are a square with a side of length s and a rectangle with a length of l and a width of w. Table 2 gives the formulas for the area and perimeter of each.

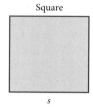
Square

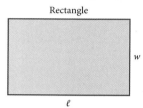
Rectangle

TABLE 2

	Square	Rectangle
Area A	s^2	ℓw
Perimeter P	$4s$	$2\ell + 2w$

EXAMPLE 13 Find the area and perimeter of a square with a side 6 inches long.

SOLUTION Substituting 6 for s in the formulas for area and perimeter of a square, we have

$$\text{Area} = A = s^2 = 6^2 = 36 \text{ square inches}$$
$$\text{Perimeter} = P = 4s = 4(6) = 24 \text{ inches}$$

13. Find the area and perimeter of a square if its side is 12 feet long.

EXAMPLE 14 Suppose a soccer field is 100 yards long and 75 yards wide. Find the area and perimeter.

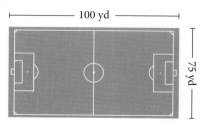

14. A football field is 100 yards long and approximately 53 yards wide. Find the area and perimeter.

SOLUTION Substituting 100 for l and 75 for w in the formulas for area and perimeter of a rectangle, we have

$$\text{Area} = A = lw = 100(75) = 7,500 \text{ square yards}$$
$$\text{Perimeter} = P = 2l + 2w = 2(100) + 2(75) = 200 + 150 = 350 \text{ yards}$$

Answers
11. $2b$
12. $2a - 4$
13. $A = 144$ sq ft, $P = 48$ ft
14. $A = 5,300$ sq yd, $P = 306$ yd

Problem Set 7.6

Moving Toward Success

"Most successful men have not achieved their distinction by having some new talent or opportunity presented to them. They have developed the opportunity that was at hand."
—Bruce Barton, 1886–1967, American author, advertising executive, and politician

1. Is it important to reward yourself for doing something well in this class? Why or why not?
2. How might you reward yourself for doing something well in this class?

A Apply the associative property to each expression, and then simplify the result. [Examples 1–5]

1. $5(4a)$
2. $8(9a)$
3. $6(8a)$
4. $3(2a)$

5. $-6(3x)$
6. $-2(7x)$
7. $-3(9x)$
8. $-4(6x)$

9. $5(-2y)$
10. $3(-8y)$
11. $6(-10y)$
12. $5(-5y)$

13. $2 + (3 + x)$
14. $9 + (6 + x)$
15. $5 + (8 + x)$
16. $3 + (9 + x)$

17. $4 + (6 + y)$
18. $2 + (8 + y)$
19. $7 + (1 + y)$
20. $4 + (1 + y)$

21. $(5x + 2) + 4$
22. $(8x + 3) + 10$
23. $(6y + 4) + 3$
24. $(3y + 7) + 8$

25. $(12a + 2) + 19$
26. $(6a + 3) + 14$
27. $(7x + 8) + 20$
28. $(14x + 3) + 15$

B Apply the distributive property to each expression, and then simplify. [Examples 6–10]

29. $7(x + 5)$
30. $8(x + 3)$
31. $6(a - 7)$
32. $4(a - 9)$

33. $2(x - y)$
34. $5(x - a)$
35. $4(5 + x)$
36. $8(3 + x)$

37. $3(2x + 5)$
38. $8(5x + 4)$
39. $6(3a + 1)$
40. $4(8a + 3)$

41. $2(6x - 3y)$
42. $7(5x - y)$
43. $5(7 - 4y)$
44. $8(6 - 3y)$

C Use the distributive property to combine similar terms. [Examples 11, 12]

45. $3x + 5x$
46. $7x + 8x$
47. $3a + a$
48. $8a + a$

49. $-2x + 6x$
50. $-3x + 9x$
51. $6y - y$
52. $3y - y$

53. $-8a - 2a$
54. $-7a - 5a$
55. $4x - 9x$
56. $5x - 11x$

Applying the Concepts

The chart shows costs for solar and wind energy equipment. Use the chart to work problems 57 and 58.

57. A farmer is replacing several turbines on his windmills. He plans to replace *x* turbines, and he is going to get a discount of $300 for each turbine he buys. Also, he'll get a $250 rebate on his entire purchase. Write an expression that describes this situation and then simplify it.

Solar Versus Wind Energy Costs

Solar Energy Equipment Cost:		Wind Energy Equipment Cost:	
Modules	$6200	Turbine	$3300
Fixed Rack	$1570	Tower	$3000
Charge Controller	$971	Cable	$715
Cable	$440		
TOTAL	$9181	TOTAL	$7015

Source: a Limited

58. A homeowner is replacing 4 solar modules. She is going to receive a discount of some amount *x* for each module and a $350 mail-in rebate. Write an expression that describes this situation and then simplify it.

D Area and Perimeter Find the area and perimeter of each square if the length of each side is as given below. [Example 13]

59. *s* = 6 feet   **60.** *s* = 14 yards   **61.** *s* = 9 inches   **62.** *s* = 15 meters

D Area and Perimeter Find the area and perimeter for a rectangle if the length and width are as given below. [Example 14]

63. *l* = 20 inches, *w* = 10 inches   **64.** *l* = 40 yards, *w* = 20 yards

65. *l* = 25 feet, *w* = 12 feet   **66.** *l* = 210 meters, *w* = 120 meters

Temperature Scales In the metric system, the scale we use to measure temperature is the Celsius scale. On this scale water boils at 100 degrees and freezes at 0 degrees. When we write 100 degrees measured on the Celsius scale, we use the notation 100°C, which is read "100 degrees Celsius." If we know the temperature in degrees Fahrenheit, we can convert to degrees Celsius by using the formula

$$C = \frac{5(F - 32)}{9}$$

where *F* is the temperature in degrees Fahrenheit. Use this formula to find the temperature in degrees Celsius for each of the following Fahrenheit temperatures.

67. 68°F   **68.** 59°F   **69.** 41°F   **70.** 23°F   **71.** 14°F   **72.** 32°F

Chapter 7 Summary

EXAMPLES

Absolute Value [7.1]

1. $|3| = 3$ and $|-3| = 3$

The *absolute value* of a number is its distance from 0 on the number line. It is the numerical part of a number. The absolute value of a number is never negative.

Opposites [7.1]

2. $-(5) = -5$ and $-(-5) = 5$

Two numbers are called *opposites* if they are the same distance from 0 on the number line but in opposite directions from 0. The opposite of a positive number is a negative number, and the opposite of a negative number is a positive number.

Addition of Positive and Negative Numbers [7.2]

3. a. $3 + 5 = 8$
$-3 + (-5) = -8$

b. $5 + (-3) = 2$
$-5 + 3 = -2$

1. To add two numbers with *the same sign:* Simply add absolute values and use the common sign. If both numbers are positive, the answer is positive. If both numbers are negative, the answer is negative.

2. To add two numbers with *different signs:* Subtract the smaller absolute value from the larger absolute value. The answer has the same sign as the number with the larger absolute value.

Subtraction [7.3]

4. $3 - 5 = 3 + (-5) = -2$
$-3 - 5 = -3 + (-5) = -8$
$3 - (-5) = 3 + 5 = 8$
$-3 - (-5) = -3 + 5 = 2$

Subtracting a number is equivalent to adding its opposite. If a and b represent numbers, then subtraction is defined in terms of addition as follows:

$$a - b = a + (-b)$$
$\uparrow \uparrow$
Subtraction   Addition of the opposite

Multiplication with Positive and Negative Numbers [7.4]

5. $3(5) = 15$
$3(-5) = -15$
$-3(5) = -15$
$-3(-5) = 15$

To multiply two numbers, multiply their absolute values.
1. The answer is *positive* if both numbers have the same sign.
2. The answer is *negative* if the numbers have different signs.

Division with Positive and Negative Numbers [7.5]

6. $\dfrac{12}{4} = 3$

$\dfrac{-12}{4} = -3$

$\dfrac{12}{-4} = -3$

$\dfrac{-12}{-4} = 3$

The rule for assigning the correct sign to the answer in a division problem is the same as the rule for multiplication. That is, like signs give a positive answer, and unlike signs give a negative answer.

Simplifying Expressions [7.6]

7. Simplify.
   a. $-2(5x) = (-2 \cdot 5)x = -10x$
   b. $4(2a - 8) = 4(2a) - 4(8)$
   $ = 8a - 32$

We simplify algebraic expressions by applying the commutative, associative, and distributive properties.

Combining Similar Terms [7.6]

8. Combine similar terms.
   a. $5x + 7x = (5 + 7)x = 12x$
   b. $2y - 8y = (2 - 8)y = -6y$

We combine similar terms by applying the distributive property.

Chapter 7 Review

Give the opposite of each number. [7.1]

1. 17
2. −32
3. −4.6
4. $\dfrac{3}{5}$

For each pair of numbers, name the smaller number. [7.1]

5. 6; −6
6. −8; −3
7. $|-3|$; 2
8. $|-4|$; $|6|$

Simplify each expression. [7.1]

9. $-(-4)$
10. $-|-4|$
11. $|-6|$
12. $|19|$

13. Name two numbers that are 7 units from −8 on the number line. [7.1]

Perform the indicated operations. [7.2, 7.3, 7.4, 7.5]

14. $5 + (-7)$
15. $-3 + 8$
16. $-345 + (-626)$
17. $-23 + 58$
18. $7 - 9 - 4 - 6$
19. $-7 - 5 - 2 - 3$
20. $4 - (-3)$
21. $30 - 42$
22. $5(-4)$
23. $-4(-3)$
24. $(56)(-31)$
25. $(20)(-4)$
26. $\dfrac{-14}{-7}$
27. $\dfrac{-25}{5}$

Simplify the following expressions as much as possible. [7.2, 7.3, 7.4, 7.5]

28. $(-6)^2$
29. $\left(-\dfrac{3}{4}\right)^2$
30. $(-2)^3$
31. $(-0.2)^4$
32. $7 + 4(6 - 9)$
33. $(-3)(-4) + 2(-5)$
34. $(7 - 3)(7 - 9)$
35. $3(-6) + 8(2 - 5)$
36. $\dfrac{8 - 4}{-8 + 4}$
37. $\dfrac{-4 + 2(-5)}{6 - 4}$
38. $\dfrac{8(-2) + 5(-4)}{12 - 3}$
39. $\dfrac{-2(5) + 4(-3)}{10 - 8}$

40. Give the sum of −19 and −23. [7.2]
41. Give the sum of −78 and −51. [7.2]

42. **Gambling** A gambler wins $58 Saturday night and then loses $86 on Sunday. Use positive and negative numbers to describe this situation. Then give the gambler's net loss or gain as a positive or negative number. [7.2]

43. Find the difference of −6 and 5. [7.3]
44. Subtract −8 from −10. [7.3]
45. **Temperature** On Wednesday, the temperature reaches a high of 17° above 0 and a low of 7° below 0. What is the difference between the high and low temperatures for Wednesday? [7.3]

46. What is the product of −9 and 3? [7.4]
47. What is −3 times the sum of −9 and −4? [7.2, 7.4]
48. Divide the product of 8 and −4 by −16. [7.4, 7.5]
49. Give the quotient of −38 and 2. [7.5]

Indicate whether each statement is *True* or *False*. [7.2, 7.3, 7.4, 7.5]

50. $\dfrac{-10}{-5} = -2$
51. $10 - (-5) = 15$
52. $2(-3) = -3 + (-3)$
53. $-6 - (-2) = -8$

Use the associative properties to simplify each expression. [7.6]

54. $(3x + 4) + 8$
55. $8(3x)$
56. $-3(7a)$
57. $6(-5y)$

Apply the distributive property and then simplify if possible. [7.6]

58. $4(x + 3)$
59. $2(x - 5)$
60. $7(3y - 8)$
61. $3(2a + 5b)$

Combine similar terms. [7.6]

62. $7x - 4x$
63. $-8a + 10a$
64. $5y - y$
65. $12x + 4x$

Chapter 7 Cumulative Review

Simplify.

1. $\dfrac{3}{5} + \dfrac{2}{7}$

2. $\dfrac{7}{6} - \dfrac{5}{6}$

3. $613 - 297$

4. $\left(3\dfrac{1}{2} + \dfrac{1}{3}\right)\left(4\dfrac{1}{6} - \dfrac{2}{3}\right)$

5. $\left(\dfrac{2}{3}\right)^4$

6. $53(807)$

7. $\dfrac{5}{8} \div (-10)$

8. $6 - 3^2 + (6 - 9)$

9. Round 37.6451 to the nearest hundredth.

10. Change $4\dfrac{7}{8}$ to an improper fraction.

11. Write the number 38,609 in words.

12. Identify the property or properties used in the following: $5(x + 9) = 5(x) + 5(9)$

Simplify.

13. $6(3)^3 - 9(2)^2$

14. $\sqrt{\dfrac{36}{49}}$

15. $\dfrac{3}{8} - \dfrac{1}{4} + \dfrac{5}{6}$

16. $(0.2)^3 + (0.3)^2$

17. $\dfrac{9 - 5}{-9 + 5}$

18. $-(-6)$

Write each ratio as a fraction in lowest terms.

19. 24 seconds to 1 minute

20. $\dfrac{2}{3}$ to $\dfrac{3}{4}$

21. Change $\dfrac{49}{6}$ to a mixed number.

22. Write $4\dfrac{7}{8}$ as a decimal.

23. Change $\dfrac{3}{8}$ to a percent.

24. Change 76% to a fraction.

25. What is 2.5% of 40?

26. 17 is what percent of 42.5?

Make the following conversions.

27. 350 m to kilometers

28. 14 gal to liters

29. Write $\dfrac{14}{25}$ as a decimal.

30. 10 is 50% of what number?

31. Reduce $\dfrac{99}{36}$.

32. **Sale Price** A dress that normally sells for $129 is on sale for 20% off the normal price. What is the sale price of the dress?

33. **Ratio** If the ratio of men to women in a self-defense class is 3 to 4, and there are 15 men in the class, how many women are in the class?

34. **Surfboard Length** A surfing company decides that a surfboard would be more efficient if its length were reduced by $3\dfrac{5}{8}$ inches. If the original length was 7 feet $\dfrac{3}{16}$ inches, what will be the new length of the board (in inches)?

35. **Average Distance** A bicyclist on a cross-country trip travels 72 miles the first day, 113 miles the second day, 108 miles the third day, and 95 miles the fourth day. What is her average distance traveled during the four days?

36. **Area and Perimeter** Find the area and perimeter of the triangle below.

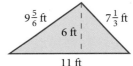

37. **Cost of Chocolate** If white chocolate sells for $4.32 per pound, how much will 2.5 pounds cost?

38. **Number Line** The distance between two numbers on the number line is 9. If one of the numbers is -4, what are the two possibilities for the other number?

The chart shows how many number one songs each of the artists have had. Use the information to answer problems 39-40.

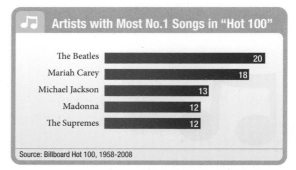

39. What is the difference between the number of number one songs from Madonna compared to the Beatles?

40. What is the ratio of number one songs for The Supremes to Mariah Carey?

Chapter 7 Test

Give the opposite of each number. [7.1]

1. 14
2. $-\dfrac{2}{3}$

Place an inequality symbol (< or >) between each pair of numbers so that the resulting statement is true. [7.1]

3. $-1 \quad -4$
4. $|-4| \quad |2|$

Simplify each expression. [7.1]

5. $-(-7)$
6. $-|-2|$

Perform the indicated operations. [7.2, 7.3, 7.4, 7.5]

7. $8 + (-17)$
8. $-4.2 - 1.7$
9. $-\dfrac{2}{3} + \left(-\dfrac{4}{5}\right)$
10. $-65 - (-29)$
11. $(-6)(-7)$
12. $-\dfrac{1}{3}(-18)$
13. $\dfrac{-80}{16}$
14. $\dfrac{-35}{-7}$
15. $-4.3 + 0.7$
16. $-\dfrac{4}{9} \div \dfrac{16}{21}$
17. $-\dfrac{7}{3}(6)$
18. $-\dfrac{42}{6}$
19. $\dfrac{3}{5} - \dfrac{9}{10}$
20. $\left(-3\dfrac{1}{16}\right)\left(-\dfrac{8}{21}\right)$

Simplify the following expressions as much as possible. [7.2, 7.3, 7.4, 7.5]

21. $(-3)^2$
22. $(-2)^3$
23. $(-7)(3) + (-2)(-5)$
24. $(8 - 5)(6 - 11)$
25. $\dfrac{-5 + 3(-3)}{5 - 7}$
26. $\dfrac{-3(2) + 5(-2)}{7 - 3}$
27. $(-4)^3 - (6 + 4)$
28. $(-3)(8) - (-7)(-3)$
29. $3 - 2(5 - 3^2)$
30. $\dfrac{5}{3}(-6) - 7(-2)$
31. $\dfrac{-7 + 3(4)}{16 + 2(-3)}$
32. $3(5 - 9) + 4^2$
33. $(-5)^2 - 3(5 - 3)$
34. $\dfrac{8(-2) - 6(7 - 3)}{2(5 + 3)}$

35. Give the sum of -15 and -46. [7.2]

36. Subtract -5 from -12. [7.3]

37. What is the product of -8 and -3? [7.4]

38. Give the quotient of 45 and -9. [7.5]

39. Two times the sum of 6 and -13. [7.2, 7.4]

40. Three times the difference of 5 and 9 is increased by 7. [7.2, 7.3, 7.4]

41. **Gambling** A gambler loses $100 Saturday night and wins $65 on Sunday. Give the gambler's net loss or gain as a negative or positive number. [7.2]

42. **Temperature** On Friday, the temperature reaches a high of 21° above 0 and a low of 4° below 0. What is the difference between the high and low temperatures for Friday? [7.3]

43. **Broadway shows** The chart shows the number of shows performed by these Broadway plays. Use the information to describe the difference in the number of shows performed by A Chorus Line and The Phantom of the Opera. Use words to explain what your answer means.

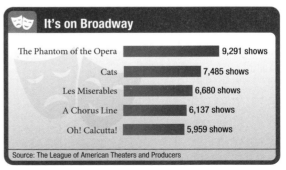

It's on Broadway

The Phantom of the Opera	9,291 shows
Cats	7,485 shows
Les Miserables	6,680 shows
A Chorus Line	6,137 shows
Oh! Calcutta!	5,959 shows

Source: The League of American Theaters and Producers

Chapter 7 Projects

INTRODUCTION TO ALGEBRA

Group Project

Random Motion

Number of People 3

Time Needed 15 minutes

Equipment Coins, dice, pencil, and paper

Background Microscopic atoms and molecules move randomly. We use random movement models to help us understand their motion. Random motion also helps us understand things like the stock market and computer science.

Mathematicians have studied questions such as where the ant is likely to end up after taking a certain number of steps. Lets say, in a random walk, an ant starts at a lamppost and takes steps of equal length along the street. We can think of the lamppost as the origin. The ant either takes a step in the negative or positive direction.

Stage	Coin	Die	Position of Ant
0	—	—	0
1			
2			
3			
4			
5			
6			
7			
8			
9			
10			

Procedure You will use a coin and die to simulate random motion. The ant will start at 0 on the number line. Roll the die and flip the coin. The ant will move the number of steps shown on the die. If the coin comes up heads, the ant moves in the positive direction. If the coin comes up tails, the ant moves in the negative direction. Repeat this process 10 times. Start each stage from the ending position of the previous stage. For example, if the ant ends up at −3 after Stage 1, then in Stage 2 the ant starts at −3. Record your results in the table. When you have completed your table, compare your results with other groups.

Research Project

David Harold Blackwell

In 1941, at age 22, David Blackwell earned his doctorate, becoming the seventh African American to earn a Ph.D. in mathematics. In high school, Blackwell did not care for algebra and trigonometry. When he took a course in analysis, he really became interested in math. Although Blackwell faced a good deal of racism during his career, he became a successful teacher, author, and mathematician. Research the life and work of Dr. Blackwell, and then present your results in an essay.

David Harold Blackwell

Solving Equations

Chapter Outline

8.1 The Distributive Property and Algebraic Expressions

8.2 The Addition Property of Equality

8.3 The Multiplication Property of Equality

8.4 Linear Equations in One Variable

8.5 Applications

8.6 Evaluating Formulas

Central Park in New York City was the first landscaped public park in the United States. In order to begin construction of the park in 1858, the city had to evict 1,600 residents off the commissioned land. By the park's completion in 1873, 10 million cartloads of material had been removed from the land to make way for 18,500 cubic yards of topsoil and 4 million trees, shrubs, and plants brought in as landscaping. Today, more than 25 million people visit the park each year.

Central Park, New York City	Area
Total Area	843 acres
Lawns	250 acres
Lakes and streams	150 acres
Woodlands	130 acres

Source: centralparknyc.org

Central Park is $\frac{1}{2}$ a mile wide and covers an area of 1.4 square miles. A person who jogs around the perimeter of the park will cover approximately 6.6 miles. Because the park can be modeled with a rectangle, we can use these numbers to find the length x of the park. In fact, solving either of the two equations below will give us the length.

$$\frac{1}{2}x = 1.4 \qquad 2x + 2 \cdot \frac{1}{2} = 6.6$$

In this chapter, we will learn how to take the numbers and relationships given in the paragraph above and translate them into equations like the ones above. Before we do that, we will learn how to solve these equations, and many others as well.

Preview

Key Words	Definition
Angle	Two rays with the same endpoint
Equation	Two expressions of the same value separated by an equals sign
Solution	A number that when used in place of the variable makes the equation a true statement
Variable Term	Any term that contains a variable
Constant Term	Any term that contains only numbers
Formula	An equation with more than one variable

Chapter Outline

8.1 The Distributive Property and Algebraic Expressions
- A Apply the distributive property to an expression.
- B Combine similar terms.
- C Find the value of an algebraic expression.
- D Solve applications involving complementary and supplementary angles.

8.2 The Addition Property of Equality
- A Identify a solution to an equation.
- B Use the addition property of equality to solve linear equations.

8.3 The Multiplication Property of Equality
- A Use the multiplication property of equality to solve equations.

8.4 Linear Equations in One Variable
- A Solve linear equations with one variable.
- B Solve linear equations involving fractions and decimals.

8.5 Applications
- A Use the Blueprint for Problem Solving to solve a variety of application problems.

8.6 Evaluating Formulas
- A Solve a formula for a given variable.
- B Solve problems using the rate equation.

The Distributive Property and Algebraic Expressions

8.1

OBJECTIVES

A Apply the distributive property to an expression.

B Combine similar terms.

C Find the value of an algebraic expression.

D Solve applications involving complementary and supplementary angles.

TICKET TO SUCCESS

Keep these questions in mind as you read through the section. Then respond in your own words and in complete sentences.

1. Show how the distributive property can be applied to an expression that involves negative numbers.
2. What property allows $x(5 + 3)$ to be rewritten as $5x + 3x$? How would you combine these similar terms?
3. What is the difference between an acute angle and an obtuse angle?
4. Briefly explain the difference between complementary angles and supplementary angles.

Suppose you spent $24.50 on espresso drinks for your coworkers. You returned to the office with one beverage tray that held four drinks, and another tray that held 3 drinks. Each drink cost the same amount and your coworkers need to repay you. How much money do you get from each coworker? To solve this problem, we must set up the following equation:

$$4x + 3x = 24.50$$

To solve for x, we must recall how to apply the distributive property and how to combine similar terms. We will do both in this section.

First, we recall that the distributive property from Section 1.5 can be used to find the area of a rectangle using two different methods.

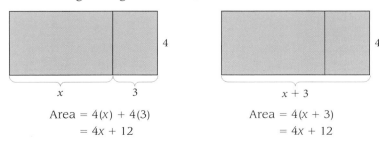

Since the areas are equal, the equation $4(x + 3) = 4(x) + 4(3)$ is a true statement.

A The Distributive Property

Let's practice applying the distributive property to more expressions.

EXAMPLE 1 Apply the distributive property to the expression:

$$5(x + 3)$$

SOLUTION Distributing the 5 over x and 3, we have

$$5(x + 3) = 5(x) + 5(3) \quad \text{Distributive property}$$
$$= 5x + 15 \quad \text{Multiply.}$$

Remember, $5x$ means "5 times x."

The distributive property can be applied to more complicated expressions involving negative numbers.

EXAMPLE 2 Multiply: $-4(3x + 5)$.

SOLUTION Multiplying both the $3x$ and the 5 by -4, we have

$$-4(3x + 5) = -4(3x) + (-4)5 \quad \text{Distributive property}$$
$$= -12x + (-20) \quad \text{Multiply.}$$
$$= -12x - 20 \quad \text{Definition of subtraction}$$

Notice, first of all, that when we apply the distributive property here, we multiply through by -4. It is important to include the sign with the number when we use the distributive property. Second, when we multiply -4 and $3x$, the result is $-12x$ because

$$-4(3x) = (-4 \cdot 3)x \quad \text{Associative property}$$
$$= -12x \quad \text{Multiply.}$$

EXAMPLE 3 Multiply: $\frac{1}{3}(3x - 12)$.

SOLUTION

$$\frac{1}{3}(3x - 12) = \frac{1}{3}(3x) - \frac{1}{3}(12) \quad \text{Distributive property}$$
$$= 1x - \frac{12}{3} \quad \text{Simplify.}$$
$$= x - 4 \quad \text{Divide.}$$

We can also use the distributive property to simplify expressions like $4x + 3x$. Because multiplication is a commutative operation, we can rewrite the distributive property like this:

$$b \cdot a + c \cdot a = (b + c)a$$

Applying the distributive property in this form to the expression $4x + 3x$, we have

$$4x + 3x = (4 + 3)x \quad \text{Distributive property}$$
$$= 7x \quad \text{Add.}$$

PRACTICE PROBLEMS

1. Apply the distributive property to the expression $6(x + 4)$.

2. Multiply: $-3(2x + 4)$.

3. Multiply: $\frac{1}{2}(2x - 4)$.

Answers
1. $6x + 24$
2. $-6x - 12$
3. $x - 2$

B Similar Terms

Recall that expressions like $4x$ and $3x$ are called *similar terms* or like terms, because the variable parts are the same. Some other examples of similar terms are $5y$ and $-6y$, and the terms $7a$, $-13a$, and $\frac{3}{4}a$. To simplify an algebraic expression (an expression that involves both numbers and variables), we combine similar terms by applying the distributive property. Table 1 reviews how we combine similar terms using the distributive property.

NOTE
We are using the word *term* in a different sense here than we did with fractions. (The terms of a fraction are the numerator and the denominator.)

TABLE 1

Original Expression		Apply Distributive Property		Simplified Expression
$4x + 3x$	=	$(4 + 3)x$	=	$7x$
$7a + a$	=	$(7 + 1)a$	=	$8a$
$-5x + 7x$	=	$(-5 + 7)x$	=	$2x$
$8y - y$	=	$(8 - 1)y$	=	$7y$
$-4a - 2a$	=	$(-4 - 2)a$	=	$-6a$
$3x - 7x$	=	$(3 - 7)x$	=	$-4x$

As you can see from the table, the distributive property can be applied to any combination of positive and negative terms so long as they are similar terms.

EXAMPLE 4 Simplify: $5x - 2 + 3x + 7$.

SOLUTION We begin by changing subtraction to addition of the opposite and applying the commutative property to rearrange the order of the terms. We want similar terms to be written next to each other.

$$\begin{aligned} 5x - 2 + 3x + 7 &= 5x + 3x + (-2) + 7 && \text{Commutative property} \\ &= (5 + 3)x + (-2) + 7 && \text{Distributive property} \\ &= 8x + 5 && \text{Add.} \end{aligned}$$

Notice that we take the negative sign in front of the 2 with the 2 when we rearrange terms. How do we justify doing this? ■

4. Simplify: $6x - 2 + 3x + 8$.

EXAMPLE 5 Simplify: $3(4x + 5) + 6$.

SOLUTION We begin by distributing the 3 across the sum of $4x$ and 5. Then we combine similar terms.

$$\begin{aligned} 3(4x + 5) + 6 &= 12x + 15 + 6 && \text{Distributive property} \\ &= 12x + 21 && \text{Add 15 and 6.} \end{aligned}$$ ■

5. Simplify: $2(4x + 3) + 7$.

EXAMPLE 6 Simplify: $2(3x + 1) + 4(2x - 5)$.

SOLUTION Again, we apply the distributive property first; then we combine similar terms. Here is the solution showing only the essential steps:

$$\begin{aligned} 2(3x + 1) + 4(2x - 5) &= 6x + 2 + 8x - 20 && \text{Distributive property} \\ &= 14x - 18 && \text{Combine similar terms.} \end{aligned}$$

6. Simplify: $3(2x + 1) + 5(4x - 3)$.

■

Answers
4. $9x + 6$
5. $8x + 13$
6. $26x - 12$

C The Value of an Algebraic Expression

An expression such as $3x + 5$ will take on different values depending on what x is. If we were to let x equal 2, the expression $3x + 5$ would become 11. On the other hand, if x is 10, the same expression has a value of 35.

When →	$x = 2$	When →	$x = 10$
the expression →	$3x + 5$	the expression →	$3x + 5$
becomes →	$3(2) + 5$	becomes →	$3(10) + 5$
	$= 6 + 5$		$= 30 + 5$
	$= 11$		$= 35$

EXAMPLE 7 Find the value of each of the following expressions by replacing the variable with the given number.

a. $3x - 1$, when $x = 2$
$$3(2) - 1 = 6 - 1 = 5$$

b. $2x - 3 + 4x$, when $x = -1$
$$2(-1) - 3 + 4(-1) = -2 - 3 + (-4) = -9$$

c. $y^2 - 6y + 9$, when $y = 4$
$$4^2 - 6(4) + 9 = 16 - 24 + 9 = 5$$

EXAMPLE 8 Find the area of a 30-W solar panel shown here with a length of 15 inches and a width of $10 + 3x$ inches.

SOLUTION Previously we worked with area, so we know that Area = (length)(width). Using the values for length and width, we have

$$A = lw$$
$$A = 15(10 + 3x) \quad \text{Length} = 15;\ \text{width} = 10 + 3x$$
$$= 150 + 45x \text{ in}^2 \quad \text{Distributive property}$$

The area of this solar panel is $150 + 45x$ square inches. ∎

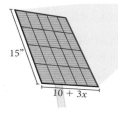

D Angles

> **FACTS FROM GEOMETRY**   **Angles**
>
> An angle is formed by two straight lines, called rays, with the same endpoint. The common endpoint is called the *vertex* of the angle, and the rays are called the *sides* of the angle.
>
> In Figure 1, angle θ (theta) is formed by the two rays OA and OB. The vertex of θ is O. Angle θ is also denoted as angle AOB, where the letter associated with the vertex is always the middle letter in the three letters used to denote the angle.
>
> **Degree Measure** The angle formed by rotating a ray through one complete revolution about its endpoint (Figure 2) has a measure of 360 degrees, which we write as 360°.

---

7. a. Find the value of
  $4x - 7$ when $x = 3$.

b. Find the value of
  $2x - 5 + 6x$ when $x = -2$.

c. Find the value of
  $y^2 - 10y + 25$ when
  $y = -2$.

8. Find the area of a 30-W solar panel with a length of 25 cm and a width of $8 + 2x$ cm.

Answers
7. a. 5 **b.** −21 **c.** 49
8. $200 + 50x$ cm²

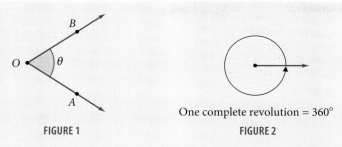

FIGURE 1

One complete revolution = 360°

FIGURE 2

One degree of angle measure, written 1°, is $\frac{1}{360}$ of a complete rotation of a ray about its endpoint; there are 360° in one full rotation. (The number 360 was decided upon by early civilizations because it was believed that the Earth was at the center of the universe and the Sun would rotate once around the Earth every 360 days.) Similarly, 180° is half of a complete rotation, and 90° is a quarter of a full rotation. Angles that measure 90° are called *right angles*, and angles that measure 180° are called *straight angles*. If an angle measures between 0° and 90° it is called an *acute angle*, and an angle that measures between 90° and 180° is an *obtuse angle*. Figure 3 illustrates further.

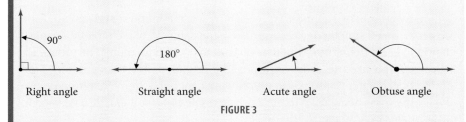

FIGURE 3

Complementary Angles and Supplementary Angles If two angles add up to 90°, we call them *complementary angles*, and each is called the *complement* of the other. If two angles have a sum of 180°, we call them *supplementary angles*, and each is called the supplement of the other. Figure 4 illustrates the relationship between angles that are complementary and angles that are supplementary.

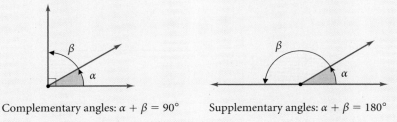

FIGURE 4

EXAMPLE 9 Find x in each of the following diagrams.

a.

x, 30°

Complementary angles

b.

x, 45°

Supplementary angles

9. Find x in each of the following diagrams.

a.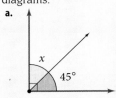

x, 45°

Complementary angles

b.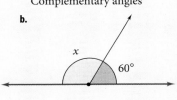

x, 60°

Supplementary angles

Answers
9. a. 45° **b.** 120°

SOLUTION We use subtraction to find each angle.

a. Because the two angles are complementary, we can find x by subtracting 30° from 90°.

$$x = 90° - 30° = 60°$$

We say 30° and 60° are complementary angles. The complement of 30° is 60°.

b. The two angles in the diagram are supplementary. To find x, we subtract 45° from 180°.

$$x = 180° - 45° = 135°$$

We say 45° and 135° are supplementary angles. The supplement of 45° is 135°. ∎

USING TECHNOLOGY

Protractors

When we think of technology, we think of computers and calculators. However, some simpler devices are also in the category of technology, because they help us do things that would be difficult to do without them. The protractor below can be used to draw and measure angles as we briefly touched on during our Chapter 5 discussion on drawing pie charts. In the diagram below, the protractor is being used to measure an angle of 120°. It can also be used to draw angles of any size.

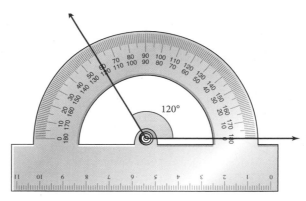

If you have a protractor, use it to draw the following angles: 30°, 45°, 60°, 120°, 135°, and 150°. Then imagine how you would draw these angles without a protractor.

Problem Set 8.1

Moving Toward Success

"Don't wait until everything is just right. It will never be perfect. There will always be challenges, obstacles and less than perfect conditions. So what. Get started now. With each step you take, you will grow stronger and stronger, more and more skilled, more and more self-confident and more and more successful."

—Mark Victor Hansen, 1948–present, American motivational speaker and author

1. Explain what it means to feel burned out.
2. Why is it important to recognize if you are feeling burned out? How might you help yourself feel less burned out?

A For review, use the distributive property to combine each of the following pairs of similar terms. [Examples 1–3]

1. $2x + 8x$
2. $3x + 7x$
3. $-4y + 5y$
4. $-3y + 10y$

5. $4a - a$
6. $9a - a$
7. $8(x + 2)$
8. $8(x - 2)$

9. $2(3a + 7)$
10. $5(3a + 2)$
11. $\frac{1}{3}(3x + 6)$
12. $\frac{1}{2}(2x + 4)$

B Simplify the following expressions by combining similar terms. In some cases the order of the terms must be rearranged first by using the commutative property. [Examples 4–6]

13. $4x + 2x + 3 + 8$
14. $7x + 5x + 2 + 9$
15. $7x - 5x + 6 - 4$
16. $10x - 7x + 9 - 6$

17. $-2a + a + 7 + 5$
18. $-8a + 3a + 12 + 1$
19. $6y - 2y - 5 + 1$
20. $4y - 3y - 7 + 2$

21. $4x + 2x - 8x + 4$
22. $6x + 5x - 12x + 6$
23. $9x - x - 5 - 1$
24. $2x - x - 3 - 8$

25. $2a + 4 + 3a + 5$ **26.** $9a + 1 + 2a + 6$ **27.** $3x + 2 - 4x + 1$ **28.** $7x + 5 - 2x + 6$

29. $12y + 3 + 5y$ **30.** $8y + 1 + 6y$ **31.** $4a - 3 - 5a + 2a$ **32.** $6a - 4 - 2a + 6a$

Apply the distributive property to each expression and then simplify.

33. $2(3x + 4) + 8$ **34.** $2(5x + 1) + 10$ **35.** $5(2x - 3) + 4$ **36.** $6(4x - 2) + 7$

37. $8(2y + 4) + 3y$ **38.** $2(5y + 1) + 2y$ **39.** $6(4y - 3) + 6y$ **40.** $5(2y - 6) + 4y$

41. $2(x + 3) + 4(x + 2)$ **42.** $3(x + 1) + 2(x + 5)$ **43.** $3(2a + 4) + 7(3a - 1)$ **44.** $7(2a + 2) + 4(5a - 1)$

C Find the value of each of the following expressions when $x = 5$. [Example 7]

45. $2x + 4$ **46.** $3x + 2$ **47.** $7x - 8$ **48.** $8x - 9$

49. $-4x + 1$ **50.** $-3x + 7$ **51.** $-8 + 3x$ **52.** $-7 + 2x$

Find the value of each of the following expressions when $a = -2$.

53. $2a + 5$ **54.** $3a + 4$ **55.** $-7a + 4$ **56.** $-9a + 3$

57. $-a + 10$ **58.** $-a + 8$ **59.** $-4 + 3a$ **60.** $-6 + 5a$

Find the value of each of the following expressions when $x = 3$. You may substitute 3 for x in each expression the way it is written, or you may simplify each expression first and then substitute 3 for x.

61. $3x + 5x + 4$ **62.** $6x + 8x + 7$ **63.** $9x + x + 3 + 7$ **64.** $5x + 3x + 2 + 4$

65. $4x + 3 + 2x + 5$ **66.** $7x + 6 + 2x + 9$ **67.** $3x - 8 + 2x - 3$ **68.** $7x - 2 + 4x - 1$

Find the value of $12x - 3$ for each of the following values of x.

69. $\dfrac{1}{2}$ **70.** $\dfrac{1}{3}$ **71.** $\dfrac{1}{4}$ **72.** $\dfrac{1}{6}$

73. $\dfrac{3}{2}$ **74.** $\dfrac{2}{3}$ **75.** $\dfrac{3}{4}$ **76.** $\dfrac{5}{6}$

Use the distributive property to write two equivalent expressions for the area of each figure.

77. 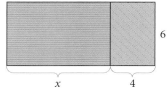   **78.**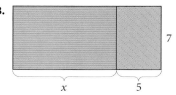

Write an expression for the perimeter of each figure.

79. 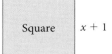   **80.**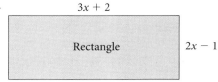

81. 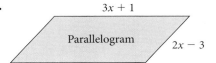   **82.**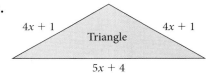

Applying the Concepts

83. Buildings This Google Earth image shows the Leaning Tower of Pisa. Most buildings stand at a right angle, but this tower is sinking on one side. The angle of inclination is the angle between the ground and the tower. If the angle between the tower and the vertical is 5° what is the angle of inclination?

84. Geometry This Google Earth image shows the Pentagon. The interior angles of a regular pentagon are all the same and sum to 540°. Find the size of each angle.

Find x in each figure and decide if the two angles are complementary or supplementary. [Example 9]

85.

86.

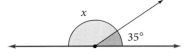

87.

88.

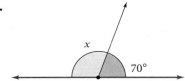

89. Luke earns $12 per hour working as a math tutor. We can express the amount he earns each week for working x hours with the expression 12x. Indicate with a yes or no, which of the following could be one of Luke's paychecks before deductions. If you answer no, explain your answer.

   a. $60 for working five hours

   b. $100 for working nine hours

   c. $80 for working seven hours

   d. $168 for working 14 hours

90. Kelly earns $15 per hour working as a graphic designer. We can express the amount she earns each week for working x hours with the expression 15x. Indicate with a yes or no which of the following could be one of Kelly's paychecks before deductions. If you answer no, explain your answer.

   a. $75 for working five hours

   b. $125 for working nine hours

   c. $90 for working six hours

   d. $500 for working 35 hours

91. Temperature and Altitude On a certain day, the temperature on the ground is 72 degrees Fahrenheit, and the temperature at an altitude of A feet above the ground is found from the expression $72 - \frac{A}{300}$. Find the temperature at the following altitudes.
  a. 12,000 feet   b. 15,000 feet   c. 27,000 feet

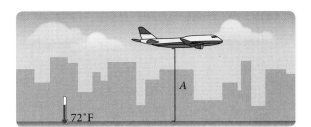

92. Perimeter of a Rectangle As you know, the expression $2l + 2w$ gives the perimeter of a rectangle with length l and width w. The garden below has a width of 3.5 feet and a length of 8 feet. What is the length of the fence that surrounds the garden?

93. Cost of Bottled Water A water bottling company charges $7 per month for their water dispenser and $2 for each gallon of water delivered. If you have g gallons of water delivered in a month, then the expression $7 + 2g$ gives the amount of your bill for that month. Find the monthly bill for each of the following deliveries.
  a. 10 gallons   b. 20 gallons

94. Cellular Phone Rates A cellular phone company charges $35 per month plus 25 cents for each minute, or fraction of a minute, that you use one of their cellular phones. The expression $\frac{3500 + 25t}{100}$ gives the amount of money, in dollars, you will pay for using one of their phones for t minutes a month. Find the monthly bill for using one of their phones
  a. 20 minutes in a month.   b. 40 minutes in a month.

95. Espresso Suppose you spent $24.50 on espresso drinks for your coworkers. You returned to the office with one beverage tray that held four drinks, and another tray that held 3 drinks. Each drink cost the same amount and your coworkers need to repay you. How much money do you get from each coworker?

96. Pizza You and your friends order an 18-inch pizza and a 14-inch pizza for a total of $35. If the larger pizza was divided into 12 slices and the smaller pizza was divided into 8 slices, how much does each slice cost?

Getting Ready for the Next Section

Add.

97. $4 + (-4)$ **98.** $2 + (-2)$ **99.** $-2 + (-4)$ **100.** $-2 + (-5)$ **101.** $-5 + 2$

102. $-3 + 12$ **103.** $\dfrac{5}{8} + \dfrac{3}{4}$ **104.** $\dfrac{5}{6} + \dfrac{2}{3}$ **105.** $-\dfrac{3}{4} + \dfrac{3}{4}$ **106.** $-\dfrac{2}{3} + \dfrac{2}{3}$

Simplify.

107. $x + 0$ **108.** $y + 0$ **109.** $y + 4 - 6$ **110.** $y + 6 - 2$

Maintaining Your Skills

Give the opposite of each number.

111. 9 **112.** 12 **113.** -6 **114.** -5

Problems 115–120 review material we covered in Chapter 1. Match each statement on the left with the property that justifies it on the right.

115. $2(6 + 5) = 2(6) + 2(5)$ **a.** Distributive property

116. $3 + (4 + 1) = (3 + 4) + 1$ **b.** Associative property

117. $x + 5 = 5 + x$ **c.** Commutative property

118. $(a + 3) + 2 = a + (3 + 2)$ **d.** Commutative and associative properties

119. $(x + 5) + 1 = 1 + (x + 5)$

120. $(a + 4) + 2 = (4 + 2) + a$

Perform the indicated operation.

121. $-\dfrac{5}{4}\left(\dfrac{8}{15}\right)$ **122.** $-\dfrac{4}{3}\left(\dfrac{6}{5}\right)$ **123.** $12 \div \dfrac{2}{3}$

124. $6 \div \dfrac{3}{5}$ **125.** $\dfrac{2}{3} - \dfrac{3}{4}$ **126.** $\dfrac{3}{5} - \dfrac{5}{8}$

8.2 The Addition Property of Equality

OBJECTIVES

A Identify a solution to an equation.

B Use the addition property of equality to solve linear equations.

TICKET TO SUCCESS

Keep these questions in mind as you read through the section. Then respond in your own words and in complete sentences.

1. Define *solution*.
2. How would you check to see if a value for x is a solution to a given equation?
3. Briefly explain the addition property of equality.
4. Can the addition property of equality be applied to subtraction? Explain.

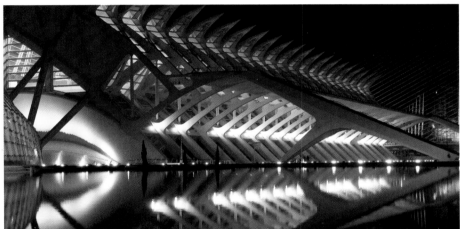

Previously we defined complementary angles as two angles whose sum is 90°. If A and B are complementary angles, then

$$A + B = 90°$$

If we know that $A = 30°$, then we can substitute 30° for A in the formula above to obtain the equation

$$30° + B = 90°$$

Complementary angles

In this section, we will learn how to solve equations like this one that involve addition and subtraction with one variable.

A Solutions to Equations

When one expression is set equal to another, the result is called an equation. Here is a formal definition:

> **Definition**
> Two expressions of the same value separated by an equals sign is an **equation**.

NOTE
Although an equation may have many solutions, the equations we work with in the first part of this chapter will always have a single solution.

Here are some examples of equations:
$$2x + 3 = 7$$
$$4^2 = 16$$
$$\frac{1}{2}(3x + 6) = 9$$
$$(x + 3)(x + 2) = x^2 + 5x + 6$$

We can solve an equation by finding the value for a variable that makes the equation a true statement. This value is called a *solution*.

> **Definition**
>
> A **solution** for an equation is a number that when used in place of the variable makes the equation a true statement.

For example, the equation $x + 3 = 7$ has as its solution the number 4, because replacing x with 4 in the equation gives a true statement.

$$\begin{array}{ll} \text{When} \rightarrow & x = 4 \\ \text{the equation} \rightarrow & x + 3 = 7 \\ \text{becomes} \rightarrow & 4 + 3 = 7 \\ & 7 = 7 \quad \text{A true statement} \end{array}$$

PRACTICE PROBLEMS

1. Show that $x = 3$ is the solution to the equation $5x - 4 = 11$.

EXAMPLE 1 Is $x = 5$ the solution to the equation $3x + 2 = 17$?

SOLUTION To see if it is, we replace x with 5 in the equation and find out if the result is a true statement.

$$\begin{array}{ll} \text{When} \rightarrow & x = 5 \\ \text{the equation} \rightarrow & 3x + 2 = 17 \\ \text{becomes} \rightarrow & 3(5) + 2 = 17 \\ & 15 + 2 = 17 \\ & 17 = 17 \quad \text{A true statement} \end{array}$$

Because the result is a true statement, we can conclude that $x = 5$ is the solution to $3x + 2 = 17$. ∎

2. Is $a = -3$ the solution to the equation $6a - 3 = 2a + 4$?

EXAMPLE 2 Is $a = -2$ the solution to the equation $7a + 4 = 3a - 2$?

SOLUTION
$$\begin{array}{ll} \text{When} \rightarrow & a = -2 \\ \text{the equation} \rightarrow & 7a + 4 = 3a - 2 \\ \text{becomes} \rightarrow & 7(-2) + 4 = 3(-2) - 2 \\ & -14 + 4 = -6 - 2 \\ & -10 = -8 \quad \text{A false statement} \end{array}$$

Because the result is a false statement, we must conclude that $a = -2$ is *not* the solution to the equation $7a + 4 = 3a - 2$. ∎

B Addition Property of Equality

We want to develop a process for solving equations with one variable. The most important property needed for solving the equations in this section is called the *addition property of equality*. The formal definition looks like this:

Answer
1. See Solutions Section.
2. No

8.2 The Addition Property of Equality

> **Addition Property of Equality**
> Let A, B, and C represent algebraic expressions.
>
> If $\quad A = B$
> then $\quad A + C = B + C$
>
> *In words:* Adding the same quantity to both sides of an equation never changes the solution to the equation.

This property is extremely useful in solving equations. Our goal in solving equations is to isolate the variable on one side of the equation. We want to end up with an equation of the form

$$x = \text{a number}$$

To do so, we use the addition property of equality. Remember to follow this basic rule of algebra: *Whatever is done to one side of an equation must be done to the other side in order to preserve the equality.*

EXAMPLE 3 Solve for x: $x + 4 = -2$.

SOLUTION We want to isolate x on one side of the equation. If we add -4 to both sides, the left side will be $x + 4 + (-4)$, which is $x + 0$ or just x.

$$
\begin{aligned}
x + 4 &= -2 \\
x + 4 + (-4) &= -2 + (-4) \quad &\text{Add } -4 \text{ to both sides.} \\
x + 0 &= -6 \quad &\text{Add.} \\
x &= -6 \quad &x + 0 = x
\end{aligned}
$$

The solution is -6. We can check it if we want to by replacing x with -6 in the original equation:

$$
\begin{aligned}
\text{When} &\rightarrow & x &= -6 \\
\text{the equation} &\rightarrow & x + 4 &= -2 \\
\text{becomes} &\rightarrow & -6 + 4 &= -2 \\
& & -2 &= -2 \quad \text{A true statement}
\end{aligned}
$$

EXAMPLE 4 Solve for a: $a - 3 = 5$.

SOLUTION
$$
\begin{aligned}
a - 3 &= 5 \\
a - 3 + \mathbf{3} &= 5 + \mathbf{3} \quad &\text{Add 3 to both sides.} \\
a + 0 &= 8 \\
a &= 8 \quad &a + 0 = a
\end{aligned}
$$

The solution to $a - 3 = 5$ is $a = 8$.

EXAMPLE 5 Solve for y: $y + 4 - 6 = 7 - 1$.

SOLUTION Before we apply the addition property of equality, we must simplify each side of the equation as much as possible.

$$
\begin{aligned}
y + 4 - 6 &= 7 - 1 \\
y - 2 &= 6 \quad &\text{Simplify each side.} \\
y - 2 + \mathbf{2} &= 6 + \mathbf{2} \quad &\text{Add 2 to both sides.} \\
y + 0 &= 8 \\
y &= 8 \quad &y + 0 = y
\end{aligned}
$$

3. Solve for x: $x + 5 = -2$.

> **NOTE**
> With some of the equations in this section, you will be able to see the solution just by looking at the equation. But it is important that you show all the steps used to solve the equations anyway. The equations you come across in the future will not be as easy to solve, so you should learn the steps involved very well.

4. Solve for a: $a - 2 = 7$.

5. Solve for y: $y + 6 - 2 = 8 - 9$.

Answers
3. -7
4. 9
5. -5

6. Solve for x: $5x - 3 - 4x = 4 - 7$.

EXAMPLE 6 Solve for x: $3x - 2 - 2x = 4 - 9$.

SOLUTION Simplifying each side as much as possible, we have

$$3x - 2 - 2x = 4 - 9$$
$$x - 2 = -5 \qquad 3x - 2x = x$$
$$x - 2 + \mathbf{2} = -5 + \mathbf{2} \qquad \text{Add 2 to both sides.}$$
$$x + 0 = -3$$
$$x = -3 \qquad x + 0 = x$$

7. Solve for x: $-5 - 7 = x + 2$.

EXAMPLE 7 Solve for x: $-3 - 6 = x + 4$.

SOLUTION The variable appears on the right side of the equation in this problem. This makes no difference; we can isolate x on either side of the equation. We can leave it on the right side if we like.

$$-3 - 6 = x + 4$$
$$-9 = x + 4 \qquad \text{Simplify the left side.}$$
$$-9 + \mathbf{(-4)} = x + 4 + \mathbf{(-4)} \qquad \text{Add } -4 \text{ to both sides.}$$
$$-13 = x + 0$$
$$-13 = x \qquad x + 0 = x$$

The statement $-13 = x$ is equivalent to the statement $x = -13$. In either case, the solution to our equation is -13.

8. Solve: $a - \dfrac{2}{3} = \dfrac{5}{6}$.

EXAMPLE 8 Solve for a: $a - \dfrac{3}{4} = \dfrac{5}{8}$.

SOLUTION To isolate a, we add $\dfrac{3}{4}$ to each side.

$$a - \frac{3}{4} = \frac{5}{8}$$
$$a - \frac{3}{4} + \mathbf{\frac{3}{4}} = \frac{5}{8} + \mathbf{\frac{3}{4}}$$
$$a = \frac{11}{8}$$

When solving equations we will leave answers like $\dfrac{11}{8}$ as improper fractions, rather than change them to mixed numbers.

9. Solve: $5(3a - 4) - 14a = 25$.

EXAMPLE 9 Solve for a: $4(2a - 3) - 7a = 2 - 5$.

SOLUTION We must begin by applying the distributive property to separate terms on the left side of the equation. Following that, we combine similar terms and then apply the addition property of equality.

$$4(2a - 3) - 7a = 2 - 5 \qquad \text{Original equation}$$
$$8a - 12 - 7a = 2 - 5 \qquad \text{Distributive property}$$
$$a - 12 = -3 \qquad \text{Simplify each side.}$$
$$a - 12 + \mathbf{12} = -3 + \mathbf{12} \qquad \text{Add 12 to each side.}$$
$$a = 9$$

Answers
6. 0
7. -14
8. $\dfrac{3}{2}$
9. 45

Problem Set 8.2

Moving Toward Success

"Talent is cheaper than table salt. What separates the talented individual from the successful one is a lot of hard work."
— Stephen King, 1947–present, American author

1. What do you think makes a good math student?
2. Can you be a good math student just by reading the textbook? Why or why not?

A Check to see if the number to the right of each of the following equations is the solution to the equation. [Examples 1, 2]

1. $2x + 1 = 5; 2$
2. $4x + 3 = 7; 1$
3. $3x + 4 = 19; 5$
4. $3x + 8 = 14; 2$
5. $2x - 4 = 2; 4$
6. $5x - 6 = 9; 3$
7. $2x + 1 = 3x + 3; -2$
8. $4x + 5 = 2x - 1; -6$
9. $x - 4 = 2x + 1; -4$
10. $x - 8 = 3x + 2; -5$

B Solve each equation. [Examples 3, 4, 8]

11. $x + 2 = 8$
12. $x + 3 = 5$
13. $x - 4 = 7$
14. $x - 6 = 2$
15. $a + 9 = -6$
16. $a + 3 = -1$
17. $x - 5 = -4$
18. $x - 8 = -3$
19. $y - 3 = -6$
20. $y - 5 = -1$
21. $a + \dfrac{1}{3} = -\dfrac{2}{3}$
22. $a + \dfrac{1}{4} = -\dfrac{3}{4}$
23. $x - \dfrac{3}{5} = \dfrac{4}{5}$
24. $x - \dfrac{7}{8} = \dfrac{3}{8}$
25. $y + 7.3 = -2.7$
26. $y + 8.2 = -2.8$

B Simplify each side of the following equations before applying the addition property. [Examples 5–7]

27. $x + 4 - 7 = 3 - 10$ **28.** $x + 6 - 2 = 5 - 12$ **29.** $x - 6 + 4 = -3 - 2$ **30.** $x - 8 + 2 = -7 - 1$

31. $3 - 5 = a - 4$ **32.** $2 - 6 = a - 1$ **33.** $3a + 7 - 2a = 1$ **34.** $5a + 6 - 4a = 4$

35. $6a - 2 - 5a = -9 + 1$ **36.** $7a - 6 - 6a = -3 + 1$ **37.** $8 - 5 = 3x - 2x + 4$ **38.** $10 - 6 = 8x - 7x + 6$

B The following equations contain parentheses. Apply the distributive property to remove the parentheses, then simplify each side before using the addition property of equality. [Example 9]

39. $2(x + 3) - x = 4$ **40.** $5(x + 1) - 4x = 2$ **41.** $-3(x - 4) + 4x = 3 - 7$ **42.** $-2(x - 5) + 3x = 4 - 9$

43. $5(2a + 1) - 9a = 8 - 6$ **44.** $4(2a - 1) - 7a = 9 - 5$ **45.** $-(x + 3) + 2x - 1 = 6$ **46.** $-(x - 7) + 2x - 8 = 4$

Find the value of x for each of the figures, given the perimeter.

47. $P = 36$

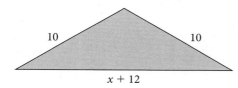

48. $P = 30$

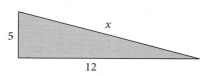

49. $P = 16$

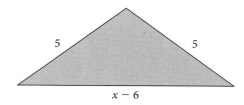

50. $P = 60$

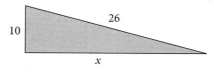

Applying the Concepts

Temperature The chart shows the temperatures for some of the world's hottest places. To convert from Celsius to Kelvin we use the formula $y = x + 273$, where y is the temperature in Kelvin and x is the temperature in Celsius. Use the formula to answer Questions 51 and 52.

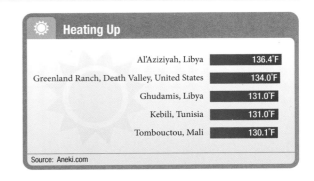

51. The hottest temperature in Al'Aziziyah was 331 Kelvin. Convert this to Celsius.

52. The hottest temperature in Kebili, Tunisia, was 328 Kelvin. Convert this to Celsius.

53. Geometry Two angles are complementary angles. If one of the angles is 23°, then solving the equation $x + 23° = 90°$ will give you the other angle. Solve the equation.

54. Geometry Two angles are supplementary angles. If one of the angles is 23°, then solving the equation $x + 23° = 180°$ will give you the other angle. Solve the equation.

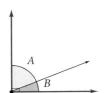

Complementary angles

55. Theater Tickets The El Portal Center for the Arts in North Hollywood, California, holds a maximum of 400 people. The two balconies hold 86 and 89 people each; the rest of the seats are at the stage level. Solving the equation $x + 86 + 89 = 400$ will give you the number of seats on the stage level.

  a. Solve the equation for x.

  b. If tickets on the stage level are $30 each, and tickets in either balcony are $25 each, what is the maximum amount of money the theater can bring in for a show?

56. Geometry The sum of the angles in the triangle on the swing set is 180°. Use this fact to write an equation containing x. Then solve the equation.

Getting Ready for the Next Section

Find the reciprocal of each number.

57. 4 **58.** 3 **59.** $\frac{1}{2}$ **60.** $\frac{1}{3}$ **61.** $\frac{2}{3}$ **62.** $\frac{3}{5}$

Multiply.

63. $2 \cdot \frac{1}{2}$ **64.** $\frac{1}{4} \cdot 4$ **65.** $-\frac{1}{3}(-3)$ **66.** $-\frac{1}{4}(-4)$

67. $\frac{3}{2}\left(\frac{2}{3}\right)$ **68.** $\frac{5}{3}\left(\frac{3}{5}\right)$ **69.** $\left(-\frac{5}{4}\right)\left(-\frac{4}{5}\right)$ **70.** $\left(-\frac{4}{3}\right)\left(-\frac{3}{4}\right)$

Simplify.

71. $1 \cdot x$ **72.** $1 \cdot a$ **73.** $4x - 11 + 3x$ **74.** $2x - 11 + 3x$

Maintaining Your Skills

75. $\frac{3}{2} + \frac{5}{10}$ **76.** $\frac{1}{3} + \frac{4}{12}$ **77.** $\frac{2}{7} + \frac{1}{14}$ **78.** $\frac{3}{8} + \frac{1}{16}$

79. $\frac{1}{3} - \frac{2}{5}$ **80.** $\frac{3}{4} - \frac{3}{7}$ **81.** $\frac{1}{6} - \frac{4}{3}$ **82.** $\frac{2}{5} - \frac{5}{10}$

Translating Translate each of the following into an equation, and then solve the equation.

83. The sum of x and 12 is 30.

84. The difference of x and 12 is 30.

85. The difference of 8 and 5 is equal to the sum of x and 7.

86. The sum of 8 and 5 is equal to the difference of x and 7.

The Multiplication Property of Equality

8.3

OBJECTIVES

A Use the multiplication property of equality to solve equations.

TICKET TO SUCCESS

Keep these questions in mind as you read through the section. Then respond in your own words and in complete sentences.

1. In words, briefly explain the multiplication property of equality.
2. Using symbols, explain the multiplication property of equality.
3. True or false? Multiplying both sides of an equation by the same nonzero quantity will never change the solution to the equation.
4. Dividing both sides of the equation $4x = -20$ by 4 is the same as multiplying both sides by what number?

In movie theaters, a film plays at 24 frames per second. Over the internet, that number is sometimes cut in half, to 12 frames per second, to make the file size smaller.

We can use a new property called the *multiplication property of equality* on the equation $240 = \frac{x}{12}$ to find the number of total frames, x, in a 240-second movie clip that plays at 12 frames per second.

In this section, we will continue to solve equations in one variable. We will again use the addition property of equality, but we will also use the multiplication property of equality to solve the equations in this section. We will state the multiplication property of equality and then see how it is used by looking at some examples.

A Multiplication Property of Equality

Here is the formal explanation of the multiplication property of equality:

> **Multiplication Property of Equality**
>
> Let A, B, and C represent algebraic expressions, with C not equal to 0.
>
> If $\quad A = B$
> then $\quad AC = BC$
>
> *In words:* Multiplying both sides of an equation by the same nonzero quantity will never change the solution to the equation.

PRACTICE PROBLEMS

1. Solve for x: $\frac{1}{3}x = 5$.

Now, because division is defined as multiplication by the reciprocal, we are also free to divide both sides of an equation by the same nonzero quantity and always be sure we have not changed the solution to the equation.

EXAMPLE 1 Solve for x: $\frac{1}{2}x = 3$.

SOLUTION Our goal here is the same as it was in Section 8.2. We want to isolate x (that is, $1x$) on one side of the equation. We have $\frac{1}{2}x$ on the left side. If we multiply both sides by 2, we will have $1x$ on the left side. Here is how it looks:

$$\frac{1}{2}x = 3$$

$$\mathbf{2}\left(\frac{1}{2}x\right) = \mathbf{2}(3) \qquad \text{Multiply both sides by 2.}$$

$$x = 6$$

To see why $2\left(\frac{1}{2}x\right)$ is equivalent to x, we use the associative property.

$$2\left(\frac{1}{2}x\right) = \left(2 \cdot \frac{1}{2}\right)x \qquad \text{Associative property}$$

$$= 1 \cdot x \qquad 2 \cdot \frac{1}{2} = 1$$

$$= x \qquad 1 \cdot x = x$$

Although we will not show this step when solving problems, it is implied. ∎

2. Solve for a: $\frac{1}{5}a + 3 = 7$.

EXAMPLE 2 Solve for a: $\frac{1}{3}a + 2 = 7$.

SOLUTION We begin by adding -2 to both sides to get $\frac{1}{3}a$ by itself. We then multiply by 3 to solve for a.

$$\frac{1}{3}a + 2 = 7.$$

$$\frac{1}{3}a + 2 + (\mathbf{-2}) = 7 + (\mathbf{-2}) \qquad \text{Add } -2 \text{ to both sides.}$$

$$\frac{1}{3}a = 5$$

$$\mathbf{3} \cdot \frac{1}{3}a = \mathbf{3} \cdot 5 \qquad \text{Multiply both sides by 3.}$$

$$a = 15$$

We can check our solution to see that it is correct:

$$\text{When} \rightarrow \qquad a = 15$$
$$\text{the equation} \rightarrow \qquad \frac{1}{3}a + 2 = 7$$
$$\text{becomes} \rightarrow \qquad \frac{1}{3}(15) + 2 = 7$$
$$5 + 2 = 7$$
$$7 = 7 \qquad \text{A true statement} \quad \blacksquare$$

Answers
1. 15
2. 20

EXAMPLE 3 Solve for y: $\frac{2}{3}y = 12$.

SOLUTION In this case, we multiply each side of the equation by the reciprocal of $\frac{2}{3}$, which is $\frac{3}{2}$.

$$\frac{2}{3}y = 12$$

$$\mathbf{\frac{3}{2}}\left(\frac{2}{3}y\right) = \mathbf{\frac{3}{2}}(12)$$

$$y = 18$$

The solution checks because $\frac{2}{3}$ of 18 is 12. ∎

Note The reciprocal of a negative number is also a negative number. Remember, reciprocals are two numbers that have a product of 1. Since 1 is a positive number, any two numbers we multiply to get 1 must both have the same sign. Here are some negative numbers and their reciprocals:

The reciprocal of -2 is $-\frac{1}{2}$.

The reciprocal of -7 is $-\frac{1}{7}$.

The reciprocal of $-\frac{1}{3}$ is -3.

The reciprocal of $-\frac{3}{4}$ is $-\frac{4}{3}$.

The reciprocal of $-\frac{9}{5}$ is $-\frac{5}{9}$.

EXAMPLE 4 Solve for x: $-\frac{4}{5}x = \frac{8}{15}$.

SOLUTION The reciprocal of $-\frac{4}{5}$ is $-\frac{5}{4}$.

$$-\frac{4}{5}x = \frac{8}{15}$$

$$-\mathbf{\frac{5}{4}}\left(-\frac{4}{5}x\right) = -\mathbf{\frac{5}{4}}\left(\frac{8}{15}\right) \quad \text{Multiply both sides by } -\frac{5}{4}.$$

$$x = -\frac{2}{3}$$

∎

Many times, it is convenient to divide both sides by a nonzero number to solve an equation, as the next example shows.

EXAMPLE 5 Solve for x: $4x = -20$.

SOLUTION If we divide both sides by 4, the left side will be just x, which is what we want. It is okay to divide both sides by 4 because division by 4 is equivalent to multiplication by $\frac{1}{4}$, and the multiplication property of equality states that we can multiply both sides by any number so long as it isn't 0.

$$4x = -20$$

$$\frac{4x}{\mathbf{4}} = \frac{-20}{\mathbf{4}} \quad \text{Divide both sides by 4.}$$

$$x = -5$$

3. Solve for y: $\frac{3}{5}y = 6$.

4. Solve for x: $-\frac{3}{4}x = \frac{6}{5}$.

5. Solve for x: $6x = -42$.

NOTE
If we multiply each side by $\frac{1}{4}$, the solution looks like this:

$$\frac{1}{4}(4x) = \frac{1}{4}(-20)$$

$$\left(\frac{1}{4} \cdot 4\right)x = -5$$

$$1x = -5$$

$$x = -5$$

Answers
3. 10
4. $-\frac{8}{5}$
5. -7

6. Solve for x: $-5x + 6 = -14$.

Because $4x$ means "4 times x," the factors in the numerator of $\frac{4x}{4}$ are 4 and x. Because the factor 4 is common to the numerator and the denominator, we divide it out to get just x. ∎

EXAMPLE 6 Solve for x: $-3x + 7 = -5$.

SOLUTION We begin by adding -7 to both sides to reduce the left side to $-3x$.

$$-3x + 7 = -5$$
$$-3x + 7 + (-7) = -5 + (-7) \quad \text{Add } -7 \text{ to both sides.}$$
$$-3x = -12$$
$$\frac{-3x}{-3} = \frac{-12}{-3} \quad \text{Divide both sides by } -3.$$
$$x = 4$$
∎

With more complicated equations, we simplify each side separately before applying the addition or multiplication properties of equality. The following examples illustrate this.

7. Solve for x:
$3x - 7x + 5 = 3 - 18$.

EXAMPLE 7 Solve for x: $5x - 8x + 3 = 4 - 10$.

SOLUTION We combine similar terms to simplify each side and then solve as usual.

$$5x - 8x + 3 = 4 - 10$$
$$-3x + 3 = -6 \quad \text{Simplify each side.}$$
$$-3x + 3 + (-3) = -6 + (-3) \quad \text{Add } -3 \text{ to both sides.}$$
$$-3x = -9$$
$$\frac{-3x}{-3} = \frac{-9}{-3} \quad \text{Divide both sides by } -3.$$
$$x = 3$$
∎

8. Solve for x:
$-5 + 4 = 2x - 11 + 3x$.

EXAMPLE 8 Solve for x: $-8 + 11 = 4x - 11 + 3x$.

SOLUTION We begin by simplifying each side separately.

$$-8 + 11 = 4x - 11 + 3x$$
$$3 = 7x - 11 \quad \text{Simplify both sides.}$$
$$3 + \mathbf{11} = 7x - 11 + \mathbf{11} \quad \text{Add 11 to both sides.}$$
$$14 = 7x$$
$$\frac{14}{\mathbf{7}} = \frac{7x}{\mathbf{7}} \quad \text{Divide both sides by 7.}$$
$$2 = x$$

Again, it makes no difference which side of the equation x ends up on, so long as it is just one x. ∎

Answers
6. 4
7. 5
8. 2

Problem Set 8.3

Moving Toward Success

"Dedication involves making the space to let young ideas take hold; every tree was once a seed and every company was once an idea."

—Zephyr Bloch-Jorgensen, 1970–present, Australian author

1. Why is practice important in mathematics?
2. Name one math concept you have had difficulty with in this course thus far. Did practice help you master it? Explain.

A Use the multiplication property of equality to solve each of the following equations. In each case, show all the steps. [Examples 1, 3–5]

1. $\frac{1}{4}x = 2$
2. $\frac{1}{3}x = 7$
3. $\frac{1}{2}x = -3$
4. $\frac{1}{5}x = -6$

5. $-\frac{1}{3}x = 2$
6. $-\frac{1}{3}x = 5$
7. $-\frac{1}{6}x = -1$
8. $-\frac{1}{2}x = -4$

9. $\frac{3}{4}y = 12$
10. $\frac{2}{3}y = 18$
11. $3a = 48$
12. $2a = 28$

13. $-\frac{3}{5}x = \frac{9}{10}$
14. $-\frac{4}{5}x = -\frac{8}{15}$
15. $5x = -35$
16. $7x = -35$

17. $-8y = 64$
18. $-9y = 27$
19. $-7x = -42$
20. $-6x = -42$

A Using the addition property of equality first, solve each of the following equations. [Examples 2, 6]

21. $3x - 1 = 5$
22. $2x + 4 = 6$
23. $-4a + 3 = -9$
24. $-5a + 10 = 50$

25. $6x - 5 = 19$
26. $7x - 5 = 30$
27. $\frac{1}{3}a + 3 = -5$
28. $\frac{1}{2}a + 2 = -7$

29. $-\frac{1}{4}a + 5 = 2$
30. $-\frac{1}{5}a + 3 = 7$
31. $2x - 4 = -20$
32. $3x - 5 = -26$

33. $\frac{2}{3}x - 4 = 6$
34. $\frac{3}{4}x - 2 = 7$
35. $-11a + 4 = -29$
36. $-12a + 1 = -47$

37. $-3y - 2 = 1$
38. $-2y - 8 = 2$
39. $-2x - 5 = -7$
40. $-3x - 6 = -36$

A Simplify each side of the following equations first, then solve. [Examples 7, 8]

41. $2x + 3x - 5 = 7 + 3$
42. $4x + 5x - 8 = 6 + 4$

43. $4x - 7 + 2x = 9 - 10$
44. $5x - 6 + 3x = -6 - 8$

45. $3a + 2a + a = 7 - 13$
46. $8a - 6a + a = 8 - 14$

47. $5x + 4x + 3x = 4 - 8$
48. $4x + 8x - 2x = 15 - 10$

49. $5 - 18 = 3y - 2y + 1$
50. $7 - 16 = 4y - 3y + 2$

Find the value of x for each of the figures, given the perimeter. The figures in problems 51 and 52 are squares.

51. $P = 72$

$2x$

52. $P = 96$

$3x$

53. $P = 80$

$3x$
$2x$

54. $P = 64$

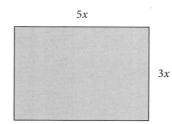
$5x$
$3x$

55. Central Park Recall our discussion of Central Park in New York City from the beginning of this chapter. The park's length in miles is given by the equation $\frac{1}{2}x = 1.4$. Use the multiplication property of equality to solve for x.

56. Movie Clips Recall our discussion on frames per second of a movie from the beginning of this section. Use the multiplication property of equality on the equation $240 = \frac{x}{12}$ to find the number of total frames, x, in a 240 second movie clip that plays at 12 frames per second

Applying the Concepts

57. Cars The chart shows the fastest cars in America. To convert miles per hour to feet per second, we use the formula $y = \frac{15}{22}x$, where x is the car's speed in feet per second and y is the speed in miles per hour. Find the speed of the McLaren F1 in feet per second. Round to the nearest tenth.

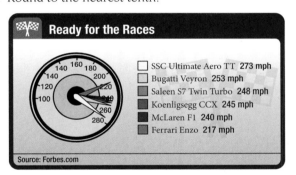

58. Mountains The chart shows the heights of the tallest mountains in the world. To convert the heights of the mountains in feet into miles, we use the formula $y = 5{,}280x$, where y is the height in feet and x is the height in miles. Find the height of K2 in miles. Round to the nearest tenth of a mile.

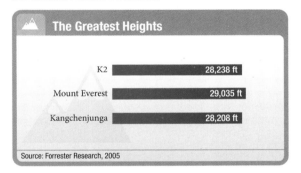

59. Streaming Video If you watch a movie trailer that is two an a half minutes long and the video is streamed at 24 frames per second, how many frames are in the movie clip?

60. Part-time Tuition Costs Many two-year colleges have a large number of students who take courses on a part-time basis. Students pay a charge for each credit hour taken plus an activity fee. Suppose the equation $\$1960 = \$175x + \$35$ can be used to determine the number of credit hours a student is taking during the upcoming semester. Solve this equation.

61. Super Bowl XLV According to Reuters, the Green Bay Packers' victory over the Pittsburgh Steelers in the 2011 Super Bowl XLV was the most watched Super Bowl ever, with 4.5 million more viewers than the previous year's record. The equation $217{,}500{,}000 = 2x - 4{,}500{,}000$ shows that the total number of viewers for both Super Bowl games was 217.5 million. Solve for x to determine how many viewers watched Super Bowl XLV.

62. Blending Gasoline In an attempt to save money at the gas pump, customers will combine two different octane gasolines to get a blend that is slightly higher in octane than regular gas but not as expensive as premium gas. The equation $14x + 120 - 6x = 200$ can be used to find out how many gallons of one octane are needed. Solve this equation.

$14x + 120 - 6x = 200$

Maintaining Your Skills

Translations Translate each sentence below into an equation, then solve the equation.

63. The sum of $2x$ and 5 is 19.

64. The sum of 8 and $3x$ is 2.

65. The difference of $5x$ and 6 is -9.

66. The difference of 9 and $6x$ is 21.

Getting Ready for the Next Section

Apply the distributive property to each of the following expressions.

67. $2(3a - 8)$ **68.** $4(2a - 5)$ **69.** $-3(5x - 1)$ **70.** $-2(7x - 3)$

Simplify each of the following expressions as much as possible.

71. $3(y - 5) + 6$ **72.** $5(y + 3) + 7$ **73.** $6(2x - 1) + 4x$ **74.** $8(3x - 2) + 4x$

8.4 Linear Equations in One Variable

OBJECTIVES

A Solve linear equations with one variable.

B Solve linear equations involving fractions and decimals.

TICKET TO SUCCESS

Keep these questions in mind as you read through the section. Then respond in your own words and in complete sentences.

1. What is the first step of solving a linear equation in one variable?
2. Define *constant term*.
3. Define *variable term*.
4. What is the last step of solving a linear equation in one variable? Why is this step important?

The Rhind Papyrus is an ancient Egyptian document, created around 1650 B.C., that contains some mathematical riddles. One problem on the Rhind Papyrus asked the reader to find a quantity such that when it is added to one-fourth of itself the sum is 15. The equation that describes this situation is

$$x + \frac{1}{4}x = 15$$

As you can see, this equation contains a fraction. One of the topics we will discuss in this section is how to solve equations that contain fractions.

In this chapter, we have been solving what are called *linear equations in one variable*. They are equations that contain only one variable, and that variable is always raised to the first power and never appears in a denominator. Here are some examples of linear equations in one variable:

$$3x + 2 = 17, \quad 7a + 4 = 3a - 2, \quad 2(3y - 5) = 6$$

Because of the work we have done in the first three sections of this chapter, we are now able to solve any linear equation in one variable. The steps outlined here can be used as a guide to solving these equations.

A Solving Linear Equations with One Variable

Strategy Solving a Linear Equation with One Variable

Step 1 Simplify each side of the equation as much as possible. This step is done using the commutative, associative, and distributive properties.

Step 2 Use the addition property of equality to get all variable terms on one side of the equation and all constant terms on the other, and then combine like terms. A *variable term* is any term that contains the variable. A *constant term* is any term that contains only a number.

Step 3 Use the multiplication property of equality to get the variable by itself on one side of the equation.

Step 4 Check your solution in the original equation if you think it is necessary.

> **NOTE**
> Once you have some practice at solving equations, these steps will seem almost automatic. Until that time, it is a good idea to pay close attention to these steps.

PRACTICE PROBLEMS

1. Solve: $4(x + 3) = -8$.

EXAMPLE 1   Solve: $3(x + 2) = -9$.

SOLUTION   We begin by applying the distributive property to the left side.

Step 1
$$3(x + 2) = -9$$
$$3x + 6 = -9 \quad \text{Distributive property}$$

Step 2
$$3x + 6 + (-6) = -9 + (-6) \quad \text{Add } -6 \text{ to both sides.}$$
$$3x = -15$$

Step 3
$$\frac{3x}{3} = \frac{-15}{3} \quad \text{Divide both sides by 3.}$$
$$x = -5$$

This general method of solving linear equations involves using the two properties developed in Sections 8.2 and 8.3. We can add any number to both sides of an equation or multiply (or divide) both sides by the same nonzero number and always be sure we have not changed the solution to the equation. The equations may change in form, but the solution to the equation stays the same. Looking back to Example 1, we can see that each equation looks a little different from the preceding one. What is interesting, and useful, is that each of the equations says the same thing about x. They all say that x is -5. The last equation, of course, is the easiest to read. That is why our goal is to end up with x isolated on one side of the equation.

2. Solve: $6a + 7 = 4a - 3$.

EXAMPLE 2   Solve: $4a + 5 = 2a - 7$.

SOLUTION   Neither side can be simplified any further. What we have to do is get the variable terms ($4a$ and $2a$) on the same side of the equation. We can eliminate the variable term from the right side by adding $-2a$ to both sides.

Step 2
$$4a + 5 = 2a - 7$$
$$4a + (-2a) + 5 = 2a + (-2a) - 7 \quad \text{Add } -2a \text{ to both sides.}$$
$$2a + 5 = -7$$
$$2a + 5 + (-5) = -7 + (-5) \quad \text{Add } -5 \text{ to both sides.}$$
$$2a = -12$$

Step 3
$$\frac{2a}{2} = \frac{-12}{2} \quad \text{Divide by 2.}$$
$$a = -6$$

Answers
1. -5
2. -5

8.4 Linear Equations in One Variable

EXAMPLE 3 Solve: $2(x - 4) + 5 = -11$.

SOLUTION We begin by applying the distributive property to multiply 2 and $x - 4$.

Step 1:
$$2(x - 4) + 5 = -11$$
$$2x - 8 + 5 = -11 \quad \text{Distributive property}$$
$$2x - 3 = -11$$

Step 2:
$$2x - 3 + \mathbf{3} = -11 + \mathbf{3} \quad \text{Add 3 to both sides.}$$
$$2x = -8$$

Step 3:
$$\frac{2x}{\mathbf{2}} = \frac{-8}{\mathbf{2}} \quad \text{Divide by 2.}$$
$$x = -4$$

3. Solve: $5(x - 2) + 3 = -12$.

EXAMPLE 4 Solve: $5(2x - 4) + 3 = 4x - 5$.

SOLUTION We apply the distributive property to multiply 5 and $2x - 4$. We then combine similar terms and solve as usual.

Step 1:
$$5(2x - 4) + 3 = 4x - 5$$
$$10x - 20 + 3 = 4x - 5 \quad \text{Distributive property}$$
$$10x - 17 = 4x - 5 \quad \text{Simplify the left side.}$$

Step 2:
$$10x + \mathbf{(-4x)} - 17 = 4x + \mathbf{(-4x)} - 5 \quad \text{Add } -4x \text{ to both sides.}$$
$$6x - 17 = -5$$
$$6x - 17 + \mathbf{17} = -5 + \mathbf{17} \quad \text{Add 17 to both sides.}$$
$$6x = 12$$

Step 3:
$$\frac{6x}{\mathbf{6}} = \frac{12}{\mathbf{6}} \quad \text{Divide by 6.}$$
$$x = 2$$

4. Solve: $3(4x - 5) + 6 = 3x + 9$.

B Equations Involving Fractions

We will now solve some equations that involve fractions. Because integers are usually easier to work with than fractions, we will begin each problem by clearing the equation we are trying to solve of all fractions. To do this, we will use the multiplication property of equality to multiply each side of the equation by the LCD for all fractions appearing in the equation. Here is an example:

EXAMPLE 5 Solve the equation $\frac{x}{2} + \frac{x}{6} = 8$.

SOLUTION The LCD for the fractions $\frac{x}{2}$ and $\frac{x}{6}$ is 6. It has the property that both 2 and 6 divide it evenly. Therefore, if we multiply both sides of the equation by 6, we will be left with an equation that does not involve fractions.

$$\mathbf{6}\left(\frac{x}{2} + \frac{x}{6}\right) = \mathbf{6}(8) \quad \text{Multiply each side by 6.}$$

$$6\left(\frac{x}{2}\right) + 6\left(\frac{x}{6}\right) = 6(8) \quad \text{Apply the distributive property.}$$

$$3x + x = 48$$
$$4x = 48 \quad \text{Combine similar terms.}$$
$$x = 12 \quad \text{Divide each side by 4.}$$

We could check our solution by substituting 12 for x in the original equation. If we do so, the result is a true statement. The solution is 12.

5. Solve: $\frac{x}{3} + \frac{x}{6} = 9$.

As you can see from Example 5, the most important step in solving an equation that involves fractions is the first step. In that first step, we multiply both sides of the equation by the LCD for all the fractions in the equation. After we have done

Answers
3. -1
4. 2
5. 18

6. Solve: $3x + \dfrac{1}{4} = \dfrac{5}{8}$.

7. Solve: $\dfrac{4}{x} + 3 = \dfrac{11}{5}$.

8. Solve: $\dfrac{1}{5}x - 2.4 = 8.3$.

9. Solve: $7a - 0.18 = 2a + 0.77$.

Answers
6. $\dfrac{1}{8}$
7. -5
8. 53.5
9. 0.19

---

so, the equation is clear of fractions because the LCD has the property that all the denominators divide it evenly.

EXAMPLE 6 Solve the equation $2x + \dfrac{1}{2} = \dfrac{3}{4}$.

SOLUTION This time the LCD is 4. We begin by multiplying both sides of the equation by 4 to clear the equation of fractions.

$$\mathbf{4}\left(2x + \dfrac{1}{2}\right) = \mathbf{4}\left(\dfrac{3}{4}\right) \quad \text{Multiply each side by the LCD, 4.}$$

$$4(2x) + 4\left(\dfrac{1}{2}\right) = 4\left(\dfrac{3}{4}\right) \quad \text{Apply the distributive property.}$$

$$8x + 2 = 3$$

$$8x = 1 \quad \text{Add } -2 \text{ to each side.}$$

$$x = \dfrac{1}{8} \quad \text{Divide each side by 8.} \quad \blacksquare$$

EXAMPLE 7 Solve for x: $\dfrac{3}{x} + 2 = \dfrac{1}{2}$. (Assume x is not 0.)

SOLUTION This time the LCD is $2x$. Following the steps we used in Examples 5 and 6, we have

$$\mathbf{2x}\left(\dfrac{3}{x} + 2\right) = \mathbf{2x}\left(\dfrac{1}{2}\right) \quad \text{Multiply through by the LCD, } 2x.$$

$$2x\left(\dfrac{3}{x}\right) + 2x(2) = 2x\left(\dfrac{1}{2}\right) \quad \text{Distributive property}$$

$$6 + 4x = x$$

$$6 = -3x \quad \text{Add } -4x \text{ to each side.}$$

$$-2 = x \quad \text{Divide each side by } -3. \quad \blacksquare$$

Equations Containing Decimals

EXAMPLE 8 Solve: $\dfrac{1}{2}x - 3.78 = 2.52$.

SOLUTION We begin by adding 3.78 to each side of the equation. Then we multiply each side by 2.

$$\dfrac{1}{2}x - 3.78 = 2.52$$

$$\dfrac{1}{2}x - 3.78 + \mathbf{3.78} = 2.52 + \mathbf{3.78} \quad \text{Add 3.78 to each side.}$$

$$\dfrac{1}{2}x = 6.30$$

$$\mathbf{2}\left(\dfrac{1}{2}x\right) = \mathbf{2}(6.30) \quad \text{Multiply each side by 2.}$$

$$x = 12.6 \quad \blacksquare$$

EXAMPLE 9 Solve: $5a - 0.42 = -3a + 0.98$.

SOLUTION We can isolate a on the left side of the equation by adding $3a$ to each side.

$$5a + \mathbf{3a} - 0.42 = -3a + \mathbf{3a} + 0.98 \quad \text{Add } 3a \text{ to each side.}$$

$$8a - 0.42 = 0.98$$

$$8a - 0.42 + \mathbf{0.42} = 0.98 + \mathbf{0.42} \quad \text{Add 0.42 to each side.}$$

$$8a = 1.40$$

$$\dfrac{8a}{\mathbf{8}} = \dfrac{1.40}{\mathbf{8}} \quad \text{Divide each side by 8.}$$

$$a = 0.175 \quad \blacksquare$$

Problem Set 8.4

Moving Toward Success

"We can always choose to perceive things differently. You can focus on what's wrong in your life, or you can focus on what's right."

—Marianne Williamson, 1952–present, Author and lecturer

1. Visualize yourself being successful on a test. Briefly describe your visualization.
2. How would visualizing yourself working through possible problems increase your potential for success during a test?

A Solve each equation using the methods shown in this section. [Examples 1–4]

1. $5(x + 1) = 20$
2. $4(x + 2) = 24$
3. $6(x - 3) = -6$
4. $7(x - 2) = -7$
5. $2x + 4 = 3x + 7$
6. $5x + 3 = 2x + (-3)$
7. $7y - 3 = 4y - 15$
8. $3y + 5 = 9y + 8$
9. $12x + 3 = -2x + 17$
10. $15x + 1 = -4x + 20$
11. $6x - 8 = -x - 8$
12. $7x - 5 = -x - 5$
13. $7(a - 1) + 4 = 11$
14. $3(a - 2) + 1 = 4$
15. $8(x + 5) - 6 = 18$
16. $7(x + 8) - 4 = 10$
17. $2(3x - 6) + 1 = 7$
18. $5(2x - 4) + 8 = 38$
19. $10(y + 1) + 4 = 3y + 7$
20. $12(y + 2) + 5 = 2y - 1$
21. $4(x - 6) + 1 = 2x - 9$
22. $7(x - 4) + 3 = 5x - 9$
23. $2(3x + 1) = 4(x - 1)$
24. $7(x - 8) = 2(x - 13)$
25. $3a + 4 = 2(a - 5) + 15$
26. $10a + 3 = 4(a - 1) + 1$
27. $9x - 6 = -3(x + 2) - 24$
28. $8x - 10 = -4(x + 3) + 2$
29. $3x - 5 = 11 + 2(x - 6)$
30. $5x - 7 = -7 + 2(x + 3)$

B Solve each equation by first finding the LCD for the fractions in the equation and then multiplying both sides of the equation by it. (Assume x is not 0 in Problems 39–46.) [Examples 5–7]

31. $\dfrac{x}{3} + \dfrac{x}{6} = 5$
32. $\dfrac{x}{2} - \dfrac{x}{4} = 3$
33. $\dfrac{x}{5} - x = 4$
34. $\dfrac{x}{3} + x = 8$

35. $3x + \dfrac{1}{2} = \dfrac{1}{4}$
36. $3x - \dfrac{1}{3} = \dfrac{1}{6}$
37. $\dfrac{x}{3} + \dfrac{1}{2} = -\dfrac{1}{2}$
38. $\dfrac{x}{2} + \dfrac{4}{3} = -\dfrac{2}{3}$

39. $\dfrac{4}{x} = \dfrac{1}{5}$
40. $\dfrac{2}{3} = \dfrac{6}{x}$
41. $\dfrac{3}{x} + 1 = \dfrac{2}{x}$
42. $\dfrac{4}{x} + 3 = \dfrac{1}{x}$

43. $\dfrac{3}{x} - \dfrac{2}{x} = \dfrac{1}{5}$
44. $\dfrac{7}{x} + \dfrac{1}{x} = 2$
45. $\dfrac{1}{x} - \dfrac{1}{2} = -\dfrac{1}{4}$
46. $\dfrac{3}{x} - \dfrac{4}{5} = -\dfrac{1}{5}$

Solve each equation. [Examples 8, 9]

47. $4x - 4.7 = 3.5$
48. $2x + 3.8 = -7.7$
49. $0.02 + 5y = -0.3$
50. $0.8 + 10y = -0.7$

51. $\dfrac{1}{3}x - 2.99 = 1.02$
52. $\dfrac{1}{7}x + 2.87 = -3.01$
53. $7n - 0.32 = 5n + 0.56$
54. $6n + 0.88 = 2n - 0.77$

55. $3a + 4.6 = 7a + 5.3$
56. $2a - 3.3 = 7a - 5.2$
57. $0.5x + 0.1(x + 20) = 3.2$
58. $0.1x + 0.5(x + 8) = 7$

Find the value of x for each of the figures, given the perimeter.

59. $P = 36$

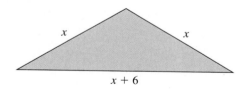

60. $P = 30$

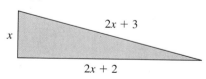

61. $P = 16$

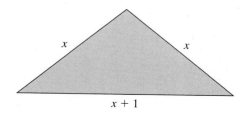

62. $P = 60$

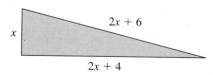

Applying the Concepts

63. Skyscrapers The chart shows the heights of the three tallest buildings in the world. The height of the Empire State Building, x, relative to the Burj Khalafia can be given by the equation $2716 = 1466 + x$. What is the height of the Empire State Building?

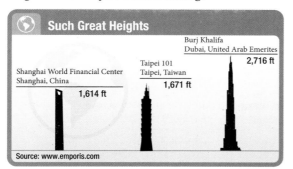

64. Sound The chart shows the decibel level of various sounds. The human threshold of pain, x, relative to the decibel level at a football stadium is given by the equation $117 = x - 3$. What is the human threshold of pain?

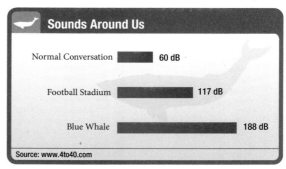

65. Geometry The figure shows part of a room. From a point on the floor, the angle of elevation to the top of the window is 45°, while the angle of elevation to the ceiling above the window is 58°. Solving either of the equations $58 - x = 45$ or $45 + x = 58$ will give us the number of degrees in the angle labeled $x°$. Solve both equations.

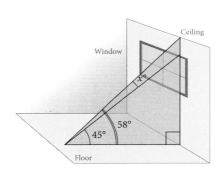

66. Rhind Papyrus As we mentioned in the introduction to this section, the Rhind Papyrus was created around 1650 B.C. and contains the riddle, "What quantity when added to one-fourth of itself becomes 15?" Solve this riddle by finding x in the equation below.

$$x + \frac{1}{4}x = 15$$

67. Math Tutoring Several students on campus decide to start a small business that offers tutoring services to students enrolled in mathematics courses. The following equation shows the amount of money they collected at the end of the month, assuming expenses of $400 and a charge of $30 an hour per student. Solve the equation $500 = 30x - 400$ to determine the number of hours students came in for tutoring in one month.

68. Shopping for a Calculator You find that you need to purchase a specific calculator for your college mathematics class. The equation $\$18.75 = p + \frac{1}{4}p$ shows the price charged by the college bookstore for a calculator after it has been marked up. How much did the bookstore pay the manufacturer for the calculator?

Getting Ready for the Next Section

Write the mathematical expressions that are equivalent to each of the following English phrases.

69. The sum of a number and 2

70. The sum of a number and 5

71. Twice a number

72. Three times a number

73. Twice the sum of a number and 6

74. Three times the sum of a number and 8

75. The difference of x and 4

76. The difference of 4 and x

77. The sum of twice a number and 5

78. The sum of three times a number and 4

Maintaining Your Skills

Place the correct symbol, $<$ or $>$, between the numbers.

79. 0.02   0.2   **80.** 0.3   0.03   **81.** 0.45   0.4   **82.** 0.5   0.56

Write the numbers in order from smallest to largest.

83. 0.01 0.013 0.03 0.003 0.031 0.001

84. 0.062 0.006 0.002 0.02 0.06 0.026

Extending the Concepts

85. Admission to the school basketball game is $4 for students and $6 for general admission. For the first game of the season, 100 more student tickets than general admission tickets were sold. The total amount of money collected was $2,400.

   a. Write an equation that will help us find the number of students in attendance.

   b. Solve this equation for x.

   c. What was the total attendance for the game?

8.5 Applications

OBJECTIVES

A Use the Blueprint for Problem Solving to solve a variety of application problems.

TICKET TO SUCCESS

Keep these questions in mind as you read through the section. Then respond in your own words and in complete sentences.

1. What is the first step of the Blueprint for Problem Solving?
2. What is the last step of the Blueprint for Problem Solving?
3. Why is the Blueprint for Problem Solving important?
4. Write a mathematical expression equivalent to the phrase "twice the sum of a number and ten."

Suppose you work part-time as a waiter at a local restaurant, where you are paid a rate of $10/hour. One week you work four shifts. Two of the shifts were each x hours long, and the other two were each 1 hour longer than the first two. If you received $220 at the end of the week, how many hours per shift did you work? The skills we have learned in this book allow us to set up the following equation for the above problem:

$$10(4x + 2) = 220$$

We can solve for x in this way:

$$\frac{\cancel{10}(4x + 2)}{\cancel{10}} = \frac{220}{10}$$

$$4x + 2 - \mathbf{2} = 22 - \mathbf{2}$$

$$4x = 20$$

$$x = 5$$

Therefore, you have worked two 5-hour shifts and two 6-hour shifts for a total of 22 hours. In this section, we will present a strategy that will help you set up and solve word problems such as this one.

However, as you begin reading through the examples in this section, you may find yourself asking why some of these problems seem so contrived. The title of the section is "Applications," but many of the problems here don't seem to have much to do with real life. You are right about that. Example 5 is what we refer to as an "age problem." Realistically, it is not the kind of problem you would expect to find if you choose a career in which you use algebra. However, solving age problems is good practice for someone with little experience with application problems, because the solution process has a form that can be applied to all similar age problems.

A The Blueprint for Problem Solving

To begin this section, we list the steps used in solving application problems. We call this strategy the *Blueprint for Problem Solving*. It is an outline that will overlay the solution process we use on all application problems.

> **Strategy Blueprint for Problem Solving**
>
> **Step 1** **Read** the problem, and then mentally **list** the items that are known and the items that are unknown.
>
> **Step 2** **Assign a variable** to one of the unknown items. (In most cases, this will amount to letting x equal the item that is asked for in the problem.) Then **translate** the other **information** in the problem to expressions involving the variable.
>
> **Step 3** **Reread** the problem, and then **write an equation**, using the items and variables listed in Steps 1 and 2, that describes the situation.
>
> **Step 4** **Solve the equation** found in Step 3.
>
> **Step 5** **Write** your **answer** using a complete sentence.
>
> **Step 6** **Reread** the problem, and **check** your solution with the original words in the problem.

NOTE
When you first start using this blueprint, you may find it helpful to write down the known and unknown items until you get the hang of mentally listing them.

There are a number of substeps within each of the steps in our blueprint. For instance, with Steps 1 and 2 it is always a good idea to draw a diagram or picture if it helps you to visualize the relationship between the items in the problem.

It is important for you to remember that solving application problems is more of an art than a science. Be flexible. No one strategy works all of the time. Try to stay away from looking for the "one way" to set up and solve a problem. Think of the blueprint for problem solving as guidelines that will help you organize your approach to these problems, rather than as a set of rules.

Number Problems

EXAMPLE 1 The sum of a number and 2 is 8. Find the number.

SOLUTION Using our blueprint for problem solving as an outline, we solve the problem as follows:

Step 1: *Read* the problem, and then mentally *list* the items that are known and the items that are unknown.

Known items: The numbers 2 and 8
Unknown item: The number in question

PRACTICE PROBLEMS

1. The sum of a number and 3 is 10. Find the number.

Answer
1. The number is 7.

Step 2: *Assign a variable* to one of the unknown items. Then *translate* the other *information* in the problem to expressions involving the variable.

> Let x = the number asked for in the problem.
> Then "The sum of a number and 2" translates to $x + 2$.

Step 3: *Reread* the problem, and then *write an equation,* using the items and variables listed in Steps 1 and 2, that describes the situation.

> With all word problems, the word "is" translates to =.
>
> The sum of x and 2 is 8.
> $x + 2 = 8$

Step 4: *Solve the equation* found in Step 3.

$$x + 2 = 8$$
$$x + 2 + (-2) = 8 + (-2) \quad \text{Add } -2 \text{ to each side.}$$
$$x = 6$$

Step 5: *Write* your *answer* using a complete sentence.

> The number is 6.

Step 6: *Reread* the problem, and *check* your solution with the original words in the problem.

> The sum of 6 and 2 is 8.   A true statement   ∎

To help with other problems of the type shown in Example 1, here are some common English words and phrases and their mathematical translations.

English	Algebra
The sum of a and b	$a + b$
The difference of a and b	$a - b$
The product of a and b	$a \cdot b$
The quotient of a and b	$\dfrac{a}{b}$
Of	$\cdot$ (multiply)
Is	= (equals)
A number	x
4 more than x	$x + 4$
4 times x	$4x$
4 less than x	$x - 4$

You may find some examples and problems in this section and the problem set that follows that you can solve without using algebra or our blueprint. It is very important that you solve those problems using the methods we are showing here. The purpose behind these problems is to give you experience using the blueprint as a guide to solving problems written in words. Your answers are much less important than the work that you show in obtaining your answer.

2. If 4 is added to the sum of twice a number and three times the number, the result is 34. Find the number.

Chapter 8 Solving Equations

EXAMPLE 2 If 5 is added to the sum of twice a number and three times the number, the result is 25. Find the number.

SOLUTION
Step 1: Read and list.

> *Known items:* The numbers 5 and 25, twice a number, and three times a number
> *Unknown item:* The number in question

Step 2: Assign a variable and translate the information.

> Let x = the number asked for in the problem.
> Then "The sum of twice a number and three times the number" translates to $2x + 3x$.

Step 3: Reread and write an equation.

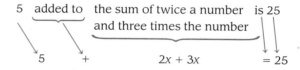

Step 4: Solve the equation.

$$5 + 2x + 3x = 25$$
$$5x + 5 = 25 \qquad \text{Simplify the left side.}$$
$$5x + 5 + (-5) = 25 + (-5) \qquad \text{Add } -5 \text{ to both sides.}$$
$$5x = 20$$
$$\frac{5x}{5} = \frac{20}{5} \qquad \text{Divide by 5.}$$
$$x = 4$$

Step 5: Write your answer.

> The number is 4.

Step 6: Reread and check.

> Twice **4** is 8, and three times **4** is 12. Their sum is $8 + 12 = 20$. Five added to this is 25. Therefore, 5 added to the sum of twice **4** and three times **4** is 25. ∎

Geometry Problems

EXAMPLE 3 The length of a rectangle is three times the width. The perimeter is 72 centimeters. Find the width and the length.

SOLUTION
Step 1: Read and list.

> *Known items:* The length is three times the width.
> The perimeter is 72 centimeters.
> *Unknown items:* The length and the width

3. The length of a rectangle is twice the width. The perimeter is 42 centimeters. Find the length and the width.

Step 2: Assign a variable, and translate the information.

Answers
2. The number is 6.
3. The width is 7 cm, and the length is 14 cm.

8.5 Applications

We let x = the width. Because the length is three times the width, the length must be $3x$. A picture will help.

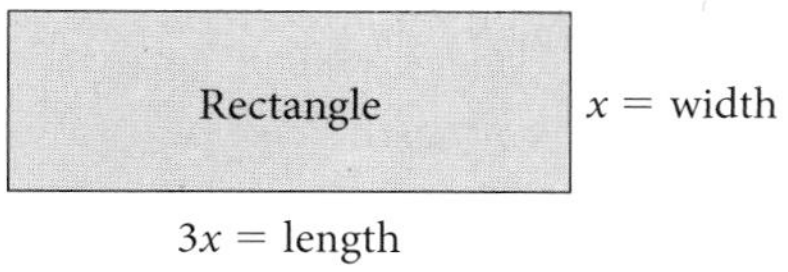

Rectangle x = width

$3x$ = length

FIGURE 1

Step 3: *Reread and write an equation.*

Because the perimeter is the sum of the sides, it must be $x + x + 3x + 3x$ (the sum of the four sides). But the perimeter is also given as 72 centimeters. Hence,

$$x + x + 3x + 3x = 72$$

Step 4: *Solve the equation.*

$$x + x + 3x + 3x = 72$$
$$8x = 72$$
$$x = 9$$

Step 5: *Write your answer.*

The width, x, is 9 centimeters. The length, $3x$, must be 27 centimeters.

Step 6: *Reread and check.*

From the diagram below, we see that these solutions check.

Length = 3 × width   Perimeter is 72.
27 = 3 · 9   9 + 9 + 27 + 27 = 72

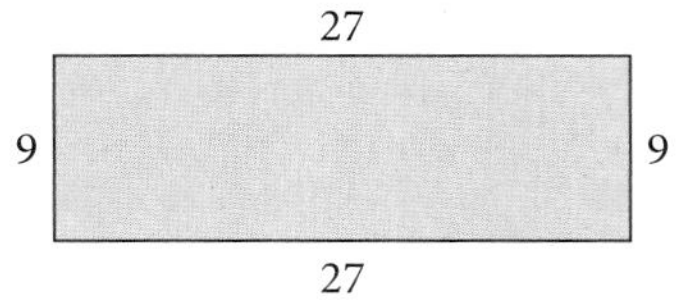

27
9   9
27

FIGURE 2

■

Next, we will review some facts about triangles that we introduced in a previous chapter.

FACTS FROM GEOMETRY Labeling Triangles and the Sum of the Angles in a Triangle

Recall that one way to label the important parts of a triangle is to label the vertices with capital letters and the sides with small letters, as shown in Figure 3.

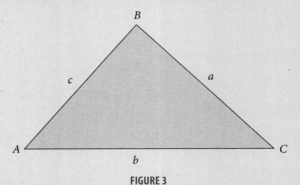

FIGURE 3

In Figure 3, notice that side a is opposite vertex A, side b is opposite vertex B, and side c is opposite vertex C. Also, because each vertex is the vertex of one of the angles of the triangle, we refer to the three interior angles as A, B, and C.

In any triangle, the sum of the interior angles is 180°. For the triangle shown in Figure 3, the relationship is written

$$A + B + C = 180°$$

We can apply the Blueprint for Problem Solving to the following triangle problem.

EXAMPLE 4 The angles in a triangle are such that one angle is twice the smallest angle, while the third angle is three times as large as the smallest angle. Find the measure of all three angles.

SOLUTION
Step 1: *Read and list.*

Known items: The sum of all three angles is 180°; one angle is twice the smallest angle; and the largest angle is three times the smallest angle.

Unknown items: The measure of each angle

Step 2: *Assign a variable and translate information.*

Let x be the smallest angle, then $2x$ will be the measure of another angle, and $3x$ will be the measure of the largest angle.

Step 3: *Reread and write an equation.*

When working with geometric objects, drawing a generic diagram will help us visualize what it is that we are asked to find. In Figure 4, we draw a triangle with angles A, B, and C.

---

4. The angles in a triangle are such that one angle is three times the smallest angle, while the largest angle is five times the smallest angle. Find the measure of all three angles.

Answer
4. The angles are 20°, 60°, and 100°.

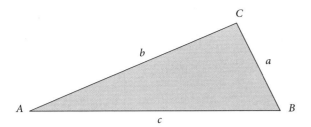

FIGURE 4

We can let the value of $A = x$, the value of $B = 2x$, and the value of $C = 3x$. We know that the sum of angles A, B, and C will be 180°, so our equation becomes

$$x + 2x + 3x = 180°$$

Step 4: Solve the equation.

$$x + 2x + 3x = 180°$$
$$6x = 180°$$
$$x = 30°$$

Step 5: Write the answer.

The smallest angle A measures 30°.
Angle B measures $2x$, or $2(30°) = 60°$.
Angle C measures $3x$, or $3(30°) = 90°$.

Step 6: Reread and check.

The angles must add to 180°:
$$A + B + C = 180°$$
$$30° + 60° + 90° = 180°$$
$$180° = 180°$$ Our answers check.   ■

Age Problem

EXAMPLE 5   Jo Ann is 22 years older than her daughter Stacey. In six years, the sum of their ages will be 42. How old are they now?

SOLUTION
Step 1: Read and list.

Known items: Jo Ann is 22 years older than Stacey. Six years from now their ages will add to 42.
Unknown items: Their ages now

Step 2: Assign a variable and translate the information.

Let x = Stacey's age now. Because Jo Ann is 22 years older than Stacey, her age is $x + 22$.

5. Joyce is 21 years older than her son Travis. In six years the sum of their ages will be 49. How old are they now?

Answer
5. Travis is 8; Joyce is 29.

Step 3: *Reread and write an equation.*

As an aid in writing the equation, we use the following table:

	Now	In Six years
Stacey	x	x + 6
Jo Ann	x + 22	x + 28

Their ages in six years will be their ages now plus 6.

Because the sum of their ages six years from now is 42, we write the equation as

$$(x + 6) + (x + 28) = 42$$

↑ Stacey's age in 6 years   ↑ Jo Ann's age in 6 years

Step 4: *Solve the equation.*

$$x + 6 + x + 28 = 42$$
$$2x + 34 = 42$$
$$2x = 8$$
$$x = 4$$

Step 5: *Write your answer.*

Stacey is now 4 years old, and Jo Ann is 4 + 22 = 26 years old.

Step 6: *Reread and check.*

To check, we see that in six years, Stacey will be 10, and Jo Ann will be 32. The sum of 10 and 32 is 42, which checks. ■

Car Rental Problem

EXAMPLE 6 A car rental company charges $11 per day and 16 cents per mile for their cars. If a car was rented for 1 day and the charge was $25.40, how many miles was the car driven?

SOLUTION
Step 1: *Read and list.*

Known items: Charges are $11 per day and 16 cents per mile. Car is rented for 1 day. Total charge is $25.40.
Unknown items: How many miles the car was driven

Step 2: *Assign a variable and translate information.*

If we let x = the number of miles driven, then the charge for the number of miles driven will be $0.16x$, the cost per mile times the number of miles.

Step 3: *Reread and write an equation.*

To find the total cost to rent the car, we add 11 to $0.16x$. Here is the equation that describes the situation:

$$\underbrace{\$11 \text{ per day}}_{11} + \underbrace{16 \text{ cents per mile}}_{0.16x} = \underbrace{\text{Total cost}}_{25.40}$$

---

6. If a car were rented from the company in Example 6 for 2 days and the total charge was $41, how many miles was the car driven?

Answers
6. The car was driven 118.75 miles.

Step 4: *Solve the equation.*

To solve the equation, we add −11 to each side and then divide each side by 0.16.

$$11 + (-11) + 0.16x = 25.40 + (-11) \quad \text{Add −11 to each side.}$$
$$0.16x = 14.40$$
$$\frac{0.16x}{0.16} = \frac{14.40}{0.16} \quad \text{Divide each side by 0.16.}$$
$$x = 90 \quad 14.40 \div 0.16 = 90$$

Step 5: *Write the answer.*

The car was driven 90 miles.

Step 6: *Reread and check.*

The charge for 1 day is $11. The 90 miles adds 90($0.16) = $14.40 to the 1-day charge. The total is $11 + $14.40 = $25.40, which checks with the total charge given in the problem. ∎

Coin Problem

EXAMPLE 7 Diane has $1.60 in dimes and nickels. If she has 7 more dimes than nickels, how many of each coin does she have?

SOLUTION

Step 1: *Read and list.*

Known items: We have dimes and nickels. There are 7 more dimes than nickels, and the total value of the coins is $1.60.
Unknown items: How many of each type of coin Diane has

Step 2: *Assign a variable and translate information.*

If we let x = the number of nickels, then the number of dimes must be $x + 7$, because Diane has 7 more dimes than nickels. Because each nickel is worth 5 cents, the amount of money she has in nickels is $0.05x$. Similarly, because each dime is worth 10 cents, the amount of money she has in dimes is $0.10(x + 7)$. Here is a table that summarizes what we have so far:

	Nickels	Dimes
Number of	x	$x + 7$
Value of	$0.05x$	$0.10(x+7)$

Step 3: *Reread and write an equation.*

Because the total value of all the coins is $1.60, the equation that describes this situation is

Amount of money in nickels + Amount of money in dimes = Total amount of money

$$0.05x + 0.10(x + 7) = 1.60$$

---

7. Amy has $1.75 in dimes and quarters. If she has 7 more dimes than quarters, how many of each coin does she have?

Answer
7. There are 3 quarters and 10 dimes.

Step 4: *Solve the equation.*

This time, let's show only the essential steps in the solution.

$0.05x + 0.10x + 0.70 = 1.60$	Distributive property
$0.15x + 0.70 = 1.60$	Add $0.05x$ and $0.10x$ to get $0.15x$.
$0.15x = 0.90$	Add -0.70 to each side.
$x = 6$	Divide each side by 0.15.

Step 5: *Write the answer.*

Because $x = 6$, Diane has 6 nickels. To find the number of dimes, we add 7 to the number of nickels (she has 7 more dimes than nickels). The number of dimes is $6 + 7 = 13$.

Step 6: *Reread and check.*

Here is a check of our results.

$$\begin{aligned}
\text{6 nickels are worth } 6(\$0.05) &= \$0.30 \\
\underline{\text{13 dimes are worth } 13(\$0.10) &= \$1.30} \\
\text{The total value is } &\$1.60
\end{aligned}$$
∎

Problem Set 8.5

Moving Toward Success

"You are the embodiment of the information you choose to accept and act upon. To change your circumstances you need to change your thinking and subsequent actions."
—Adlin Sinclair, British-born businessman, motivational speaker, and humanitarian

1. What are critical thinking skills?
2. How does mathematics influence your critical thinking skills?

Write each of the following English phrases in symbols using the variable x.

1. The sum of x and 3
2. The difference of x and 2
3. The sum of twice x and 1
4. The sum of three times x and 4
5. Five x decreased by 6
6. Twice the sum of x and 5
7. Three times the sum of x and 1
8. Four times the sum of twice x and 1
9. Five times the sum of three x and 4
10. Three x added to the sum of twice x and 1

Use the six steps in the Blueprint for Problem Solving to solve the following word problems. You may recognize the solution to some of them by just reading the problem. In all cases, be sure to assign a variable and write the equation used to describe the problem. Write your answer using a complete sentence.

Number Problems

11. The sum of a number and 3 is 5. Find the number.
12. If 2 is subtracted from a number, the result is 4. Find the number.
13. The sum of twice a number and 1 is -3. Find the number.
14. If three times a number is increased by 4, the result is -8. Find the number.
15. When 6 is subtracted from five times a number, the result is 9. Find the number.
16. Twice the sum of a number and 5 is 4. Find the number.
17. Three times the sum of a number and 1 is 18. Find the number.
18. Four times the sum of twice a number and 6 is -8. Find the number.
19. Five times the sum of three times a number and 4 is -10. Find the number.
20. If the sum of three times a number and two times the same number is increased by 1, the result is 16. Find the number.

Geometry Problems

21. The length of a rectangle is twice its width. The perimeter is 30 meters. Find the length and the width.

22. The width of a rectangle is 3 feet less than its length. If the perimeter is 22 feet, what is the width?

23. The perimeter of a square is 32 centimeters. What is the length of one side?

24. Two sides of a triangle are equal in length, and the third side is 10 inches. If the perimeter is 26 inches, how long are the two equal sides?

25. Two angles in a triangle are equal, and their sum is equal to the third angle in the triangle. What are the measures of each of the three interior angles?

26. One angle in a triangle measures twice the smallest angle, while the largest angle is six times the smallest angle. Find the measures of all three angles.

27. The smallest angle in a triangle is $\frac{1}{3}$ as large as the largest angle. The third angle is twice the smallest angle. Find the three angles.

28. One angle in a triangle is half the largest angle, but three times the smallest. Find all three angles.

Age Problems

29. Pat is 20 years older than his son Patrick. In 2 years, the sum of their ages will be 90. How old are they now? Fill in the table to help you find the equation.

	Now	In 2 Years
Patrick	x	
Pat		

30. Diane is 23 years older than her daughter Amy. In 5 years, the sum of their ages will be 91. How old are they now? Fill in the table to help you find the equation.

	Now	In 5 Years
Amy	x	
Diane		

31. Dale is 4 years older than Sue. Five years ago the sum of their ages was 64. How old are they now?

32. Pat is 2 years younger than his wife, Wynn. Ten years ago the sum of their ages was 48. How old are they now?

Rental Problems

33. A car rental company charges $10 a day and 16 cents per mile for their cars. If a car were rented for 1 day for a total charge of $23.92, how many miles was it driven?

34. A car rental company charges $12 a day and 18 cents per mile to rent their cars. If the total charge for a 1-day rental were $33.78, how many miles was the car driven?

35. A rental company charges $9 per day and 15 cents a mile for their cars. If a car were rented for 2 days for a total charge of $40.05, how many miles was it driven?

36. A car rental company charges $11 a day and 18 cents per mile to rent their cars. If the total charge for a 2-day rental were $61.60, how many miles was it driven?

Coin Problems

37. Mary has $2.20 in dimes and nickels. If she has 10 more dimes than nickels, how many of each coin does she have?

38. Bob has $1.65 in dimes and nickels. If he has 9 more nickels than dimes, how many of each coin does he have?

39. Suppose you have $9.60 in dimes and quarters. How many of each coin do you have if you have twice as many quarters as dimes?

40. A collection of dimes and quarters has a total value of $2.75. If there are 3 times as many dimes as quarters, how many of each coin is in the collection?

Miscellaneous Problems

41. Magic Square The sum of the numbers in each row, each column, and each diagonal of the square below is 15. Use this fact, along with the information in the first column of the square, to write an equation containing the variable x, then solve the equation to find x. Next, write and solve equations that will give you y and z.

x	1	y
3	5	7
4	z	2

42. Magic Square The sum of the numbers in each row, each column, and each diagonal of the square below is 3. Use this fact, along with the information in the second row of the square, to write an equation containing the variable a, then solve the equation to find a. Next, write and solve an equation that will allow you to find the value of b. Next, write and solve equations that will give you c and d.

4	d	b
a	1	3
0	c	-2

43. Wages A waitress works 5 shifts in one week and is paid $9 per hour. She works three weekday shifts that are x hours long and two weekend shifts that are two hours longer than her weekday shifts. If at the end of the work week she receives $216, how many hours are in each shift?

44. Ticket Sales Stacey is selling tickets to the school play. The tickets are $6 for adults and $4 for children. She sells twice as many adult tickets as children's tickets and brings in a total of $112. How many of each kind of ticket did she sell?

45. Central Park The chart from this chapter's introduction shows information about the area of Central Park in New York City. Using this information, write a linear equation that represents how much area in acres is left over for facilities, such as sports fields and concert stages. Then solve for the unknown in your equation.

Central Park, New York City	Area
Total Area	843 acres
Lawns	250 acres
Lakes and streams	150 acres
Woodlands	130 acres

Source: centralparknyc.org

46. Skyscrapers The chart shows the heights of some of the most famous US landmarks. The height of the proposed Freedom Tower at the World Trade Center in New York City, including the spire, is 114 feet less than 3 times the height of the Gateway Arch. What is the proposed height of Freedom Tower?

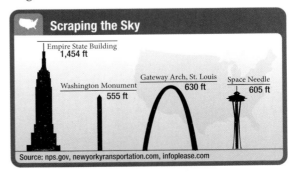

Getting Ready for the Next Section

Simplify.

47. $\frac{5}{9}(95 - 32)$

48. $\frac{5}{9}(77 - 32)$

49. Find the value of $90 - x$ when $x = 25$.

50. Find the value of $180 - x$ when $x = 25$.

51. Find the value of $2x + 6$ when $x = -2$

52. Find the value of $2x + 6$ when $x = 0$.

Solve.

53. $40 = 2l + 12$

54. $80 = 2l + 12$

55. $6 + 3y = 4$

56. $8 + 3y = 4$

Maintaining Your Skills

The problems below review some of the work you have done with percents.

Change each fraction to a decimal and then to a percent.

57. $\frac{3}{4}$

58. $\frac{5}{8}$

59. $1\frac{1}{5}$

60. $\frac{7}{10}$

Change each percent to a fraction and a decimal.

61. 37%

62. 18%

63. 3.4%

64. 125%

65. What number is 15% of 135?

66. 19 is what percent of 38?

67. 12 is 16% of what number?

Evaluating Formulas

8.6

OBJECTIVES

A Solve a formula for a given variable.

B Solve problems using the rate equation.

TICKET TO SUCCESS

Keep these questions in mind as you read through the section. Then respond in your own words and in complete sentences.

1. What is a formula?
2. How do you solve a formula for one of its variables?
3. What is the formula that gives the relationship between Fahrenheit and Celsius?
4. What is the rate equation?

In mathematics, a *formula* is an equation that contains more than one variable. The equation $P = 2w + 2l$ is an example of a formula. This formula tells us the relationship between the perimeter P of a rectangle, its length l, and its width w.

There are many formulas with which you may be familiar already. Perhaps you have used the formula $d = r \cdot t$ to find out how far you would go if you traveled at 50 miles an hour for 3 hours. If you take a chemistry class while you are in college, you will certainly use the formula that gives the relationship between the two temperature scales, Fahrenheit and Celsius.

$$F = \frac{9}{5}C + 32$$

A Formulas

> **Definition**
>
> In mathematics, a **formula** is an equation with more than one variable.

Although there are many kinds of problems we can work using formulas, we will limit ourselves to those that require only substitutions. The examples that follow illustrate this type of problem.

PRACTICE PROBLEMS

1. Suppose the livestock pen in Example 1 has a perimeter of 80 feet. If the width is still 6 feet, what is the new length?

EXAMPLE 1 The perimeter P of a rectangular livestock pen is 40 feet. If the width w is 6 feet, find the length.

SOLUTION First we substitute 40 for P and 6 for w in the formula $P = 2l + 2w$. Then we solve for l.

When → $P = 40$ and $w = 6$
the formula → $P = 2l + 2w$
becomes → $40 = 2l + 2(6)$
$40 = 2l + 12$ Multiply 2 and 6.
$28 = 2l$ Add -12 to each side.
$14 = l$ Multiply each side by $\frac{1}{2}$.

To summarize our results, if a rectangular pen has a perimeter of 40 feet and a width of 6 feet, then the length must be 14 feet. ∎

EXAMPLE 2 Use the formula $C = \frac{5}{9}(F - 32)$ to find C when F is 95 degrees.

SOLUTION Substituting 95 for F in the formula gives us the following:

When → $F = 95$
the formula → $C = \frac{5}{9}(F - 32)$
becomes → $C = \frac{5}{9}(95 - 32)$
$= \frac{5}{9}(63)$
$= \frac{5}{9} \cdot \frac{63}{1}$
$= \frac{315}{9}$
$= 35$

A temperature of 95 degrees Fahrenheit is the same as a temperature of 35 degrees Celsius. ∎

2. Use the formula in Example 2 to find C when F is 77 degrees.

NOTE
The formula we are using here,
$$C = \frac{5}{9}(F - 32),$$
is an alternative form of the formula we mentioned previously:
$$F = \frac{9}{5}C + 32$$
Both formulas describe the same relationship between the two temperature scales. If you go on to take an algebra class, you will learn how to convert one formula into the other.

EXAMPLE 3 Use the formula $y = 2x + 6$ to find y when x is -2.

SOLUTION Proceeding as we have in the previous examples, we have

When → $x = -2$
the formula → $y = 2x + 6$
becomes → $y = 2(-2) + 6$
$= -4 + 6$
$= 2$ ∎

3. Use the formula in Example 3 to find y when x is 0.

In some cases, evaluating a formula also involves solving an equation, as the next example illustrates.

Answers
1. 34 feet
2. 25 degrees Celsius
3. 6

EXAMPLE 4 Find y when x is 3 in the formula $2x + 3y = 4$.

SOLUTION First, we substitute 3 for x; then we solve the resulting equation for y.

When → $x = 3$
the equation → $2x + 3y = 4$
becomes → $2(3) + 3y = 4$
$6 + 3y = 4$
$3y = -2$ Add -6 to each side.
$y = -\dfrac{2}{3}$ Divide each side by 3.   ∎

4. Use the formula in Example 4 to find y when x is -3.

B Rate Equation

Now we will look at some problems that use what is called the *rate equation*. You use this equation on an intuitive level when you are estimating how long it will take you to drive long distances. For example, if you drive at 50 miles per hour for 2 hours, you will travel 100 miles. Here is the rate equation:

$$\text{Distance} = \text{rate} \cdot \text{time, or } d = r \cdot t$$

The rate equation has two equivalent forms, one of which is obtained by solving for r, while the other is obtained by solving for t. Here they are:

$$r = \dfrac{d}{t} \text{ and } t = \dfrac{d}{r}$$

The rate in this equation is also referred to as *average speed*.

EXAMPLE 5 At 1 p.m., Jordan leaves her house and drives at an average speed of 50 miles per hour to her sister's house. She arrives at 4 p.m.

  a. How many hours was the drive to her sister's house?
  b. How many miles from her sister does Jordan live?

SOLUTION

  a. If she left at 1:00 p.m. and arrived at 4:00 p.m., we simply subtract 1 from 4 for an answer of 3 hours.
  b. We are asked to find a distance in miles given a rate of 50 miles per hour and a time of 3 hours. We will use the rate equation, $d = r \cdot t$, to solve this. We have

  $d = 50$ miles per hour $\cdot$ 3 hours
  $d = 50(3)$
  $d = 150$ miles   ∎

Notice in Example 5, that we were asked to find a distance in miles, so our answer has a unit of miles. When we are asked to find a time, our answer will include a unit of time, like days, hours, minutes, or seconds.

When we are asked to find a rate, our answer will include units of rate, like miles per hour, feet per second, problems per minute, and so on.

5. At 9 a.m. Maggie leaves her house and drives at an average speed of 60 miles per hour to her sister's house. She arrives at 11 a.m.
  a. How many hours was the drive to her sister's house?
  b. How many miles from her sister does Maggie live?

Answers
4. $\dfrac{10}{3}$
5. a. 2 hours   **b.** 120 miles

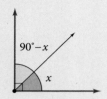

Complementary angles

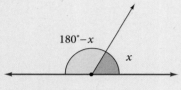

Supplementary angles

6. Find the complement and the supplement of 35°.

FACTS FROM GEOMETRY

Earlier we defined complementary angles as angles that add to 90°. That is, if x and y are complementary angles, then

$$x + y = 90°$$

If we solve this formula for y, we obtain a formula equivalent to our original formula:

$$y = 90° - x$$

Because y is the complement of x, we can generalize by saying that the complement of angle x is the angle $90° - x$. By a similar reasoning process, we can say that the supplement of angle x is the angle $180° - x$. To summarize, if x is an angle, then

the complement of x is $90° - x$, and
the supplement of x is $180° - x$.

If you go on to take a trigonometry class, you will see these formulas again.

EXAMPLE 6 Find the complement and the supplement of 25°.

SOLUTION We can use the formulas above with $x = 25°$.

The complement of 25° is $90° - 25° = 65°$.
The supplement of 25° is $180° - 25° = 155°$.

Answers
6. Complement = 55°;
Supplement = 145°

Problem Set 8.6

Moving Toward Success

"Success is not final, failure is not fatal: it is the courage to continue that counts."

—Winston Churchill, 1874–1965, British orator, author, and Prime Minister

1. What was your favorite thing about this class? What is one thing you wish you had done differently in this class?
2. How do you plan to apply the study skills learned during this class to your future classes?

A Suppose $y = 3x - 2$. In Problems 1–6, find y if:

1. $x = 3$
2. $x = -5$
3. $x = -\frac{1}{3}$
4. $x = \frac{2}{3}$
5. $x = 0$
6. $x = 5$

Suppose $x + y = 5$. In Problems 7–12, find x if:

7. $y = 2$
8. $y = -2$
9. $y = 0$
10. $y = 5$
11. $y = -3$
12. $y = 3$

Suppose $x + y = 3$. In Problems 13–18, find y if:

13. $x = 2$
14. $x = -2$
15. $x = 0$
16. $x = 3$
17. $x = \frac{1}{2}$
18. $x = -\frac{1}{2}$

Suppose $4x + 3y = 12$. In Problems 19–24, find y if:

19. $x = 3$
20. $x = -5$
21. $x = -\frac{1}{4}$
22. $x = \frac{3}{2}$
23. $x = 0$
24. $x = -3$

Suppose $4x + 3y = 12$. In Problems 25–30, find x if:

25. $y = 4$
26. $y = -4$
27. $y = -\frac{1}{3}$
28. $y = \frac{5}{3}$
29. $y = 0$
30. $y = -3$

A The formula for the area A of a rectangle with length l and width w is $A = l \cdot w$. Find A if: [Examples 1–4]

31. $l = 32$ feet and $w = 22$ feet
32. $l = 22$ feet and $w = 12$ feet

33. $l = \frac{3}{2}$ inch and $w = \frac{3}{4}$ inch
34. $l = \frac{3}{5}$ inch and $w = \frac{3}{10}$ inch

The formula $G = H \cdot R$ tells us how much gross pay G a person receives for working H hours at an hourly rate of pay R. In Problems 35-38, find G.

35. $H = 40$ hours and $R = \$6$

36. $H = 36$ hours and $R = \$8$

37. $H = 30$ hours and $R = \$9\frac{1}{2}$

38. $H = 20$ hours and $R = \$6\frac{3}{4}$

Because there are 3 feet in every yard, the formula $F = 3 \cdot Y$ will convert Y yards into F feet. In Problems 39-42, find F.

39. $Y = 4$ yards

40. $Y = 8$ yards

41. $Y = 2\frac{2}{3}$ yards

42. $Y = 6\frac{1}{3}$ yards

If you invest P dollars (P is for *principal*) at simple interest rate R for T years, the amount of interest you will earn is given by the formula $I = P \cdot R \cdot T$. In Problems 43 and 44, find I.

43. $P = \$1{,}000$, $R = \frac{7}{100}$, and $T = 2$ years

44. $P = \$2{,}000$, $R = \frac{6}{100}$, and $T = 2\frac{1}{2}$ years

In Problems 45-48, use the formula $P = 2w + 2l$ to find P.

45. $w = 10$ inches and $l = 19$ inches

46. $w = 12$ inches and $l = 22$ inches

47. $w = \frac{3}{4}$ foot and $l = \frac{7}{8}$ foot

48. $w = \frac{1}{2}$ foot and $l = \frac{3}{2}$ feet

We have mentioned the two temperature scales, Fahrenheit and Celsius. Table 1 is intended to give you a more intuitive idea of the relationship between the two temperatures scales.

TABLE 1
COMPARING TWO TEMPERATURE SCALES

Situation	Temperature (Fahrenheit)	Temperature (Celsius)
Water freezes	32°F	0°C
Room temperature	68°F	20°C
Normal body temperature	$98\frac{3}{5}$°F	37°C
Water boils	212°F	100°C
Bake cookies	365°F	185°C

Table 2 gives the formulas, in both symbols and words, that are used to convert between the two scales.

TABLE 2

FORMULAS FOR CONVERTING BETWEEN TEMPERATURE SCALES

To Convert From	Formula In Symbols	Formula In Words
Fahrenheit to Celsius	$C = \frac{5}{9}(F - 32)$	Subtract 32, then multiply by $\frac{5}{9}$.
Celsius to Fahrenheit	$F = \frac{9}{5}C + 32$	Multiply by $\frac{9}{5}$, then add 32.

49. Let $F = 212$ in the formula $C = \frac{5}{9}(F - 32)$, and solve for C. Does the value of C agree with the information in Table 1?

50. Let $C = 100$ in the formula $F = \frac{9}{5}C + 32$, and solve for F. Does the value of F agree with the information in Table 1?

51. Let $F = 68$ in the formula $C = \frac{5}{9}(F - 32)$, and solve for C. Does the value of C agree with the information in Table 1?

52. Let $C = 37$ in the formula $F = \frac{9}{5}C + 32$, and solve for F. Does the value of F agree with the information in Table 1?

53. Find C when F is $32°$.

54. Find C when F is $-4°$.

55. Find F when C is $-15°$.

56. Find F when C is $35°$.

As you know, the volume V enclosed by a rectangular solid with length l, width w, and height h is $V = l \cdot w \cdot h$. In Problems 57-60, find V if

57. $l = 6$ inches, $w = 12$ inches, and $h = 5$ inches

58. $l = 16$ inches, $w = 22$ inches, and $h = 15$ inches

59. $l = 6$ yards, $w = \frac{1}{2}$ yard, and $h = \frac{1}{3}$ yard

60. $l = 30$ yards, $w = \frac{5}{2}$ yards, and $h = \frac{5}{3}$ yards

B Maximum Heart Rate In exercise physiology, a person's maximum heart rate, in beats per minute, is found by subtracting his age in years from 220. So, if A represents your age in years, then your maximum heart rate is

$$M = 220 - A$$

Use this formula to complete the following tables.

61.

Age (years)	Maximum Heart Rate (beats per minute)
18	
19	
20	
21	
22	
23	

62.

Age (years)	Maximum Heart Rate (beats per minute)
15	
20	
25	
30	
35	
40	

Training Heart Rate A person's training heart rate, in beats per minute, is the person's resting heart rate plus 60% of the difference between maximum heart rate and his resting heart rate. If resting heart rate is R and maximum heart rate is M, then the formula that gives training heart rate is

$$T = R + \frac{3}{5}(M - R)$$

Use this formula along with the results of Problems 61 and 62 to fill in the following two tables.

63. For a 20-year-old person

Resting Heart Rate (beats per minute)	Training Heart Rate (beats per minute)
60	
65	
70	
75	
80	
85	

64. For a 40-year-old person

Resting Heart Rate (beats per minute)	Training Heart Rate (beats per minute)
60	
65	
70	
75	
80	
85	

B Use the rate equation $d = r \cdot t$ to solve Problems 65 and 66. [Example 5]

65. At 2:30 p.m. Shelly leaves her house and drives at an average speed of 55 miles per hour to her sister's house. She arrives at 6:30 p.m.

  a. How many hours was the drive to her sister's house?

  b. How many miles from her sister does Shelly live?

66. At 1:30 p.m. Cary leaves his house and drives at an average speed of 65 miles per hour to his brother's house. He arrives at 5:30 p.m.

  a. How many hours was the drive to his brother's house?

  b. How many miles from his brother's house does Cary live?

B Use the rate equation $r = \dfrac{d}{t}$ to solve Problems 67 and 68.

67. At 2:30 p.m. Brittney leaves her house and drives 260 miles to her sister's house. She arrives at 6:30 p.m.

  a. How many hours was the drive to her sister's house?

  b. What was Brittney's average speed?

68. At 8:30 a.m. Ethan leaves his house and drives 220 miles to his brother's house. He arrives at 12:30 p.m.

  a. How many hours was the drive to his brother's house?

  b. What was Ethan's average speed?

Find the complement and supplement of each angle. [Example 6]

69. 45°   **70.** 75°   **71.** 31°   **72.** 59°

Applying the Concepts

73. Digital Video On the internet, video is compressed so it is small enough for people to download. A formula for estimating the size, in kilobytes, of a compressed video is

$$S = \dfrac{height \cdot width \cdot fps \cdot time}{35{,}000}$$

where *height* and *width* are in pixels, *fps* is the number of frames per second the video is to play (television plays at 30 fps), and *time* is given in seconds.

  a. Estimate the size in kilobytes of a movie trailer that has a height of 480 pixels, has a width of 216 pixels, plays at 30 fps, and runs for 150 seconds.

  b. Estimate the size in kilobytes of a commercial trailer that has a height of 320 pixels, has a width of 144 pixels, plays at 15 fps, and runs for 150 seconds.

74. Vehicle Weight If you can measure the area that the tires on your car contact the ground, and you know the air pressure in the tires, then you can estimate the weight of your car, in pounds, with the following formula:

$$W = APN$$

where *W* is the vehicle's weight in pounds, *A* is the average tire contact area with a hard surface in square inches, *P* is the air pressure in the tires in pounds per square inch (psi, or lb/in²), and *N* is the number of tires.

  a. What is the approximate weight of a car if the average tire contact area is a rectangle 6 inches by 5 inches and if the air pressure in the tires is 30 psi?

  b. What is the approximate weight of a car if the average tire contact area is a rectangle 5 inches by 4 inches, and the tire pressure is 30 psi?

75. Temperature The chart shows the temperatures for some of the world's hottest places.

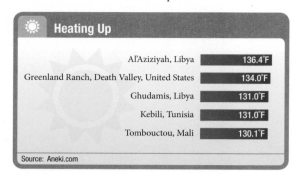

a. Use the formula $C = \frac{5}{9}(F - 32)$ to find the temperature in Celsius for Al'Aziziyah.

b. Use the formula $K = \frac{5}{9}(F - 32) + 273$ to find the temperature in Ghudamis, Libya, in Kelvin.

76. Fermat's Last Theorem Pierce deFermat was a French lawyer and an amateur mathematician. He is best known for his last theorem, which states that if n is an integer greater than 2, then there are no positive integers $x, y,$ and z that will make the formula $x^n + y^n = z^n$ true.

Use the formula $x^n + y^n = z^n$ to

a. Find x if $n = 1, y = 7,$ and $z = 15$.

b. Find y if $n = 1, x = 23,$ and $z = 37$.

Maintaining Your Skills

Simplify.

77. $\dfrac{\frac{3}{5}}{\frac{4}{5}}$

78. $\dfrac{\frac{5}{7}}{\frac{6}{7}}$

79. $\dfrac{1 + \frac{1}{2}}{1 - \frac{1}{2}}$

80. $\dfrac{1 + \frac{1}{3}}{1 - \frac{1}{3}}$

81. $\dfrac{\frac{1}{2} + \frac{1}{4}}{\frac{1}{4} + \frac{1}{8}}$

82. $\dfrac{\frac{1}{2} - \frac{1}{4}}{\frac{1}{4} - \frac{1}{8}}$

83. $\dfrac{\frac{3}{5} + \frac{3}{7}}{\frac{3}{5} - \frac{3}{7}}$

84. $\dfrac{\frac{5}{7} + \frac{5}{8}}{\frac{5}{7} - \frac{5}{8}}$

Chapter 8 Summary

Combining Similar Terms [8.1]

EXAMPLES

1. $7x + 2x = (7 + 2)x$
   $= 9x$

Two terms are *similar terms* if they have the same variable part. The expressions $7x$ and $2x$ are similar because the variable part in each is the same. Similar terms are combined by using the distributive property.

Finding the Value of an Algebraic Expression [8.1]

2. When $x = 5$, the expression $2x + 7$ becomes
   $2(5) + 7 = 10 + 7 = 17$

An *algebraic expression* is a mathematical expression that contains numbers and variables. Expressions that contain a variable will take on different values depending on the value of the variable.

The Solution to an Equation [8.2]

A *solution* to an equation is a number that, when used in place of the variable, makes the equation a true statement.

The Addition Property of Equality [8.2]

3. We solve $x - 4 = 9$ by adding 4 to each side.
   $x - 4 = 9$
   $x - 4 + \mathbf{4} = 9 + \mathbf{4}$
   $x + 0 = 13$
   $x = 13$

Let A, B, and C represent algebraic expressions.

If $\quad A = B$
then $\quad A + C = B + C$

In words: Adding the same quantity to both sides of an equation will not change the solution.

The Multiplication Property of Equality [8.3]

4. Solve $\frac{1}{3}x = 5$.
   $\frac{1}{3}x = 5$
   $\mathbf{3} \cdot \frac{1}{3}x = \mathbf{3} \cdot 5$
   $x = 15$

Let A, B, and C represent algebraic expressions, with C not equal to 0.

If $\quad A = B$
then $\quad AC = BC$

In words: Multiplying both sides of an equation by the same nonzero number will not change the solution to the equation. This property holds for division as well.

Steps Used to Solve a Linear Equation in One Variable [8.4]

5.
$2(x - 4) + 5 = -11$
$2x - 8 + 5 = -11$
$2x - 3 = -11$
$2x - 3 + \mathbf{3} = -11 + \mathbf{3}$
$2x = -8$
$\dfrac{2x}{2} = \dfrac{-8}{2}$
$x = -4$

Step 1 Simplify each side of the equation.

Step 2 Use the addition property of equality to get all variable terms on one side and all constant terms on the other side.

Step 3 Use the multiplication property of equality to get just one x isolated on either side of the equation.

Step 4 Check the solution in the original equation if necessary.

If the original equation contains fractions, you can begin by multiplying each side by the LCD for all fractions in the equation.

Evaluating Formulas [8.6]

6. When $w = 8$ and $l = 13$ the formula $P = 2w + 2l$ becomes
$P = 2 \cdot 8 + 2 \cdot 13$
$= 16 + 26$
$= 42$

In mathematics, a formula is an equation that contains more than one variable. For example, the formula for the perimeter of a rectangle is $P = 2l + 2w$. We evaluate a formula by substituting values for all but one of the variables and then solving the resulting equation for that variable.

🚫 COMMON MISTAKES

Before we end, we should mention a very common mistake made by students when they first begin to solve equations. It involves trying to subtract away the number in front of the variable, like this:

$7x = 21$
$7x - \mathbf{7} = 21 - \mathbf{7}$ Add -7 to both sides.
$x = 14$ ⟵ Mistake

The mistake is not in trying to subtract 7 from both sides of the equation. The mistake occurs when we say $7x - 7 = x$. It just isn't true. We can add and subtract only similar terms. The terms $7x$ and 7 are not similar, because one contains x and the other doesn't. The correct way to do the problem is like this:

$7x = 21$
$\dfrac{7x}{7} = \dfrac{21}{7}$ Divide both sides by 7.
$x = 3$

Chapter 8 Review

Simplify the expressions by combining similar terms. [8.1]

1. $10x + 7x$
2. $8x - 12x$
3. $2a + 9a + 3 - 6$
4. $4y - 7y + 8 - 10$
5. $6x - x + 4$
6. $-5a + a + 4 - 3$
7. $2a - 6 + 8a + 2$
8. $12y - 4 + 3y - 9$

Find the value of each expression when x is 4. [8.1]

9. $10x + 2$
10. $5x - 12$
11. $-2x + 9$
12. $-x + 8$

Find the value of each expression when x is -5. [8.1]

13. $3y + 6$
14. $9 - 2y$
15. $12 + y$
16. $-6y - 20$

17. Is $x = -3$ a solution to $5x - 2 = -17$? [8.2]
18. Is $x = 4$ a solution to $3x - 2 = 2x + 1$? [8.2]

Solve the equations. [8.2, 8.3, 8.4]

19. $x - 5 = 4$
20. $-x + 3 + 2x = 6 - 7$
21. $2x + 1 = 7$
22. $3x - 5 = 1$
23. $2x + 4 = 3x - 5$
24. $4x + 8 = 2x - 10$
25. $3(x - 2) = 9$
26. $4(x - 3) = -20$
27. $3(2x + 1) - 4 = -7$
28. $4(3x + 1) = -2(5x - 2)$
29. $5x + \dfrac{3}{8} = -\dfrac{1}{4}$
30. $\dfrac{7}{x} - \dfrac{2}{5} = 1$
31. $3(2x - 5) = 4x + 3$
32. $\dfrac{2}{3}(6x - 9) = 6x - 2$
33. $3x - 11 = 2(x - 2)$
34. $4(3x - 9) = 2(4x + 6)$
35. $5x + 9 = 4x - 7$
36. $4(x - 6) = 8$
37. $5x - 7 = 3$
38. $\dfrac{3}{x} - \dfrac{1}{4} = 2$

39. **Number Problem** The sum of a number and -3 is -5. Find the number. [8.5]

40. **Number Problem** Three times the sum of a number and 2 is -6. Find the number. [8.5]

41. **Geometry** The length of a rectangle is twice its width. If the perimeter is 42 meters, find the length and the width. [8.5]

42. **Geometry** Two angles are complementary angles. If one angle is 5 times larger than the other angle, find the two angles. [8.5]

43. **Geometry** The biggest angle in a triangle is 6 times bigger than the smallest angle. The third angle is half the largest angle. Find the three angles. [8.5]

In Problems 44–47, use the equation $3x + 2y = 6$ to find y. [8.6]

44. $x = -2$
45. $x = 6$
46. $x = 0$
47. $x = \dfrac{1}{3}$

In Problems 48–50, use the equation $3x + 2y = 6$ to find x when y has the given value. [8.6]

48. $y = 3$
49. $y = -3$
50. $y = 0$

In Problems 51–54, use the equation $y = \dfrac{2}{3}x - 4$ to find x when y has the given value. [8.6]

51. $y = 0$
52. $y = -6$
53. $y = 4$
54. $y = -10$

Chapter 8 Cumulative Review

Simplify.

1. $5{,}309 + 687$
2. $\dfrac{7}{11} + \dfrac{4}{5}$
3. $11.09 - 6.531$
4. $4\dfrac{1}{8} - 1\dfrac{3}{4}$
5. $2305(407)$
6. $0.002(230)$
7. $314\overline{)13{,}188}$
8. $\dfrac{6}{32} \div \dfrac{9}{48}$

9. Round the number 435,906 to the nearest ten thousand.
10. Write 0.48 as a fraction in lowest terms.
11. Change $\dfrac{76}{12}$ to a mixed number in lowest terms.
12. Find the difference of 0.45 and $\dfrac{2}{5}$.
13. Write the decimal 0.8 as a percent.

Use the table given in Chapter 6 to make the following conversion.

14. 7 kilograms to pounds

15. Write 124% as a fraction or mixed number in lowest terms.
16. What percent of 60 is 21?

Simplify.

17. $\left(\dfrac{1}{3}\right)^3 + \left(\dfrac{1}{9}\right)^2$
18. $8x + 9 - 9x - 14$
19. $-|-7|$
20. $\dfrac{-3(-8) + 4(-2)}{11 - 9}$
21. $19 - 5(7 - 4)$
22. $\sqrt{25} + \sqrt{16}$

Solve.

23. $\dfrac{3}{8}y = 21$
24. $-3(2x - 1) = 3(x + 5)$
25. $\dfrac{3.6}{4} = \dfrac{4.5}{x}$

26. Write the following ratio as a fraction in lowest terms: 0.04 to 0.32
27. Subtract -3 from 5.

28. **Surface Area** Find the surface area of a rectangular solid with length 7 inches, width 3 inches, and height 2 inches.

29. **Age** Ben is 8 years older than Ryan. In 6 years the sum of their ages will be 38. How old are they now?

30. **Gas Mileage** A truck travels 432 miles on 27 gallons of gas. What is the rate of gas mileage in miles per gallon?

31. **Discount** A surfboard that usually sells for $400 is marked down to $320. What is the discount? What is the discount rate?

32. **Geometry** Find the length of the hypotenuse of a right triangle with sides of 5 and 12 meters.

33. **Cost of Coffee** If coffee costs $6.40 per pound, how much will 2 lb 4 oz, cost?

34. **Interest** If $1,400 is invested at 6% simple interest for 90 days, how much interest is earned?

35. **Wildflower Seeds** C.J. works in a nursery, and one of his tasks is filling packets of wildflower seeds. If each packet is to contain $\dfrac{1}{4}$ pound of seeds, how many packets can be filled from 16 pounds of seeds?

36. **Commission** A car stereo salesperson receives a commission of 8% on all units he sells. If his total sales for March are $9,800, how much money in commission will he make?

37. **Volume** How many 8 fluid ounce glasses of water will it take to fill a 15-gallon aquarium?

Photo Messaging Suppose the graph shows the number of picture messages sent each month in millions. Use the information to answer the following questions.

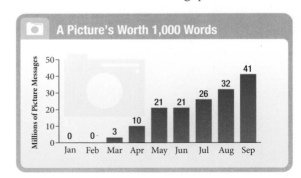

38. How many more messages were sent in September than July?

39. What is the ratio of messages sent in May to messages sent in March?

Chapter 8 Test

Simplify each expression by combining similar terms. [8.1]

1. $9x - 3x + 7 - 12$
2. $4b - 1 - b - 3$
3. $3(2x - 6) - 4x$
4. $4(5x - 6) - 6x + 12$

Find the value of each expression when $x = 3$. [8.1]

5. $3x - 12$
6. $-x + 12$

7. Is $x = -1$ a solution to $4x - 3 = -7$?

8. Is $x = 3$ a solution to $3x - 7 = -2$?

9. Use the equation $4x + 3y = 12$ to find y when $x = -3$.

10. Use the equation $y = \frac{4}{3}x + 7$ to find x when $y = -1$.

Solve each equation. [8.2, 8.3, 8.4]

11. $x - 7 = -3$
12. $a - 2.9 = -7.8$
13. $\frac{2}{3}y = 18$
14. $\frac{7}{x} - \frac{1}{6} = 1$
15. $3x - 7 = 5x + 1$
16. $2(x - 5) = -8$
17. $3(2x + 3) = -3(x - 5)$
18. $6(3x - 2) - 8 = 4x - 6$
19. $4(6x - 9) = 2(4x - 6)$
20. $3(x - 7) = -6$

21. **Number Problem** Twice the sum of a number and 3 is -10. Find the number. [8.5]

22. **Hot Air Balloon** The first successful crossing of the Atlantic in a hot air balloon was made in August 1978 by Maxie Anderson, Ben Abruzzo, and Larry Newman of the United States. The 3,100 mile trip took approximately 140 hours. Use the formula $r = \frac{d}{t}$ to find their average speed to the nearest whole number. [8.5]

23. **Geometry** The length of a rectangle is 4 centimeters longer than its width. If the perimeter is 28 centimeters, find the length and the width. [8.5]

24. **Age problem** Karen is 5 years younger than Susan. Three years ago, the sum of their ages was 11. How old are they now? [8.5]

25. **Geometry** The largest angle in a triangle is 3 times bigger than another angle, which is twice as big as the smallest angle. Find the three angles. [8.5]

26. **Coin Problem** A coin collection has seven more dimes than quarters. If the collection has a value of $2.10, how many quarters and dimes are in the collection? [8.5]

27. **Geometry** Two angles are complementary. If one angle is 3 times bigger than the other angle, find the two angles. [8.5]

28. **Perimeter of a Rectangle** If a rectangle has a length of 13 meters and a perimeter of 35 meters, find the width of the rectangle. [8.6]

29. **Google Earth** The length of the rectangular base of the Sphinx is 12 times longer than the width. If the perimeter of the base is 520 feet. Find the length and the width. [8.6]

Chapter 8 Projects

SOLVING EQUATIONS

Group Project

The Equation Game*

Number of People 2–5

Time Needed 30 minutes

Equipment Per group: deck of cards, timer or clock, pencil and paper, copy of rules.

Background The Equation Game is a fun way to practice working with equations.

Procedure Remove all the face cards from the deck. Aces will be 1's. The dealer deals four cards face up, a fifth card face down. Each player writes down the four numbers that are face up. Set the timer for 5 minutes, then flip the fifth card. Each player writes down equations that use the numbers on the first four cards to equal the number on the fifth card. When the five minutes are up, figure out the scores. An equation that uses

1 of the four cards scores 0 points
2 of the four cards scores 4 points
3 of the four cards scores 9 points
4 of the four cards scores 16 points

Check the other players' equations. If you find an error, you get 7 points. The person with the mistake gets no points for that equation.

Example The first four cards are a four, a nine, an ace, and a two. The fifth card is a seven. Here are some equations you could make:

$$9 - 2 = 7$$
$$4 + 2 + 1 = 7$$
$$\frac{9 - (4 - 2)}{1} = 7$$

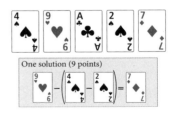

One solution (9 points)

*This project was adapted from www.exploratorium.edu/math_explorer/fantasticFour.html.

Research Project

Algebraic Symbolism

Algebra made a beginning as early as 1850 B.C. in Egypt. However, the symbols we use in algebra today took some time to develop. For example, the algebraic use of letters for numbers began much later during the Third century with Diophantus, a mathematician famous for studying *Diophantine equations.* In the early centuries, the full words *plus, minus, multiplied by, divided by,* and *equals* were written out. Imagine how much more difficult your homework would be if you had to write out all these words instead of using symbols. Algebraists began to come up with a system of symbols to make writing algebra easier. At first, not everyone agreed on the symbols to be used. For example, the present division sign ÷ was often used for subtraction. People in different countries used different symbols: the Italians preferred to use *p* and *m* for plus and minus, while the less traditional Germans were starting to use + and −.

Research the history of algebraic symbolism. Find out when the algebraic symbols we use today (such as letters to represent variables, +, −, ÷, ·, a/b and $\left(\frac{a}{b}\right)$ came into common use. Summarize your results in an essay.

Negative Exponents

Up to this point, all the exponents we have worked with have been positive numbers. We want to extend the type of numbers we can use for exponents to include negative numbers as well. The definition that follows allows us to work with exponents that are negative numbers.

A Negative Exponents

OBJECTIVES

A Simplify expressions containing negative exponents.

B Understand and apply the division property for exponents.

Definition

For **negative exponents**, if r is a positive integer and a is any number other than zero, then
$$a^{-r} = \frac{1}{a^r} \quad a \neq 0$$

NOTE
Because division by 0 is undefined, our definition includes the restriction $a \neq 0$.

The definition indicates that negative exponents give us reciprocals, as the following examples illustrate.

EXAMPLE 1   Simplify: 2^{-3}.

SOLUTION   The first step is to rewrite the expression with a positive exponent, by using our definition. After that, we simplify.

$$2^{-3} = \frac{1}{2^3} \quad \text{Definition of negative exponents}$$

$$= \frac{1}{8} \quad \text{The cube of 2 is 8.}$$

PRACTICE PROBLEMS

1. Simplify: 2^{-4}.

EXAMPLE 2   Simplify: 5^{-2}.

SOLUTION   First, we write the expression with a positive exponent, then we simplify.

$$5^{-2} = \frac{1}{5^2} \quad \text{Definition of negative exponents}$$

$$= \frac{1}{25} \quad \text{The square of 5 is 25.}$$

2. Simplify: 4^{-2}.

As you can see from our first two examples, when we apply the definition for negative exponents to expressions containing negative exponents, we end up with reciprocals.

EXAMPLE 3   Simplify: $(-3)^{-2}$.

SOLUTION   The fact that the base in this problem is negative does not change the procedure we use to simplify.

$$(-3)^{-2} = \frac{1}{(-3)^2} \quad \text{Definition of negative exponents}$$

$$= \frac{1}{9} \quad \text{The square of } -3 \text{ is 9.}$$

3. Simplify: $(-2)^{-3}$.

To check your understanding of negative exponents, look over the two lines below.

Answers
1. $\frac{1}{16}$
2. $\frac{1}{16}$
3. $-\frac{1}{8}$

Appendix A Negative Exponents

$$2^1 = 2 \qquad 2^2 = 4 \qquad 2^3 = 8 \qquad 2^4 = 16$$

$$2^{-1} = \frac{1}{2} \qquad 2^{-2} = \frac{1}{4} \qquad 2^{-3} = \frac{1}{8} \qquad 2^{-4} = \frac{1}{16}$$

The properties of exponents we have developed so far hold for negative exponents as well as positive exponents. For example, our multiplication property for exponents is still written as

$$a^r \cdot a^s = a^{r+s}$$

but now r and s can be negative numbers also.

EXAMPLE 4 Simplify: $2^5 \cdot 2^{-7}$.

SOLUTION This is multiplication with the same base, so we add exponents.

$$\begin{aligned} 2^5 \cdot 2^{-7} &= 2^{5+(-7)} & \text{Multiplication property for exponents} \\ &= 2^{-2} & \text{Add.} \\ &= \frac{1}{2^2} & \text{Definition of negative exponents} \\ &= \frac{1}{4} & \text{The square of 2 is 4.} \end{aligned}$$

When we simplify expressions containing negative exponents, let's agree that the final expression contains only positive exponents.

EXAMPLE 5 Simplify: $x^9 \cdot x^{-12}$.

SOLUTION Again, because we have the product of two expressions with the same base, we use the multiplication property for exponents to add exponents.

$$\begin{aligned} x^9 \cdot x^{-12} &= x^{9+(-12)} & \text{The multiplication property for exponents} \\ &= x^{-3} & \text{Add exponents.} \\ &= \frac{1}{x^3} & \text{Definition of negative exponents} \end{aligned}$$

B Division with Exponents

To develop our next property of exponents, we use the definition for positive exponents. Consider the expression $\frac{x^6}{x^4}$. We can simplify by expanding the numerator and denominator and then reducing to lowest terms by dividing out common factors.

$$\begin{aligned} \frac{x^6}{x^4} &= \frac{x \cdot x \cdot x \cdot x \cdot x \cdot x}{x \cdot x \cdot x \cdot x} & \text{Expand numerator and denominator.} \\ &= \frac{\cancel{x} \cdot \cancel{x} \cdot \cancel{x} \cdot \cancel{x} \cdot x \cdot x}{\cancel{x} \cdot \cancel{x} \cdot \cancel{x} \cdot \cancel{x}} & \text{Divide out common factors.} \\ &= x \cdot x \\ &= x^2 & \text{Write answer with exponent 2.} \end{aligned}$$

Note that the exponent in the answer is the difference of the exponents in the original problem. More specifically, if we subtract the exponent in the denominator from the exponent in the numerator, we obtain the exponent in the answer. This discussion leads us to another property of exponents.

4. Simplify: $3^4 \cdot 3^{-7}$.

5. Simplify: $x^6 \cdot x^{-10}$.

Answers
4. $\frac{1}{27}$
5. $\frac{1}{x^4}$

Appendix A Negative Exponents

> **Property Division Property for Exponents**
> If a is any number other than zero, and r and s are integers, then
> $$\frac{a^r}{a^s} = a^{r-s}$$
> *In words*: To divide two numbers with the same base, subtract the exponent in the denominator from the exponent in the numerator, and use the common base as the base in the answer.

EXAMPLE 6 Simplify: $\frac{2^5}{2^8}$.

SOLUTION Using our new property, we subtract the exponent in the denominator from the exponent in the numerator. The result is an expression containing a negative exponent.

$$\frac{2^5}{2^8} = 2^{5-8} \quad \text{Division property for exponents}$$
$$= 2^{-3} \quad \text{Subtract.}$$
$$= \frac{1}{2^3} \quad \text{Definition of negative exponents}$$
$$= \frac{1}{8} \quad \text{The cube of 2 is 8.}$$

6. Simplify: $\frac{2^6}{2^8}$.

To further justify our new property of exponents, we can rework the problem shown in Example 6, without using the division property for exponents, to see that we obtain the same answer.

$$\frac{2^5}{2^8} = \frac{2 \cdot 2 \cdot 2 \cdot 2 \cdot 2}{2 \cdot 2 \cdot 2 \cdot 2 \cdot 2 \cdot 2 \cdot 2 \cdot 2} \quad \text{Expand numerator and denominator.}$$
$$= \frac{\cancel{2} \cdot \cancel{2} \cdot \cancel{2} \cdot \cancel{2} \cdot \cancel{2}}{\cancel{2} \cdot \cancel{2} \cdot \cancel{2} \cdot \cancel{2} \cdot \cancel{2} \cdot 2 \cdot 2 \cdot 2}$$
$$= \frac{1}{2 \cdot 2 \cdot 2} \quad \text{Divide out common factors.}$$
$$= \frac{1}{8} \quad \text{Multiply.}$$

As you can see, the answer matches the answer obtained by using the division property for exponents.

EXAMPLE 7 Divide: $10^{-3} \div 10^5$.

SOLUTION We begin by writing the division problem in fractional form. Then we apply our division property.

$$10^{-3} \div 10^5 = \frac{10^{-3}}{10^5} \quad \text{Write problem in fractional form.}$$
$$= 10^{-3-5} \quad \text{Division property for exponents}$$
$$= 10^{-8} \quad \text{Subtract.}$$
$$= \frac{1}{10^8} \quad \text{Definition of negative exponents}$$

7. Divide: $10^{-5} \div 10^3$.

We can leave the answer in exponential form, as it is, or we can expand the denominator to obtain

$$= \frac{1}{100{,}000{,}000}$$

Answers
6. $\frac{1}{4}$
7. $\frac{1}{10^8}$

Appendix A Negative Exponents

8. Simplify: $\dfrac{10^{-6}}{10^{-8}}$.

EXAMPLE 8 Simplify: $\dfrac{10^{-8}}{10^{-6}}$.

SOLUTION Again, we are dividing expressions that have the same base. To find the exponent on the answer, we subtract the exponent in the denominator from the exponent in the numerator.

$$\dfrac{10^{-8}}{10^{-6}} = 10^{-8-(-6)} \quad \text{Division property for exponents}$$

$$= 10^{-2} \quad \text{Subtract.}$$

$$= \dfrac{1}{10^2} \quad \text{Definition of negative exponents}$$

$$= \dfrac{1}{100} \quad \text{Answer in expanded form} \quad \blacksquare$$

9. Simplify: $4x^{-3} \cdot 7x$.

EXAMPLE 9 Simplify: $3x^{-4} \cdot 5x$.

SOLUTION To begin to simplify this expression, we regroup using the commutative and associative properties. That way, the numbers 3 and 5 are grouped together, as are the powers of x. Note also how we write x as x^1, so we can see the exponent.

$$3x^{-4} \cdot 5x = 3x^{-4} \cdot 5x^1 \quad \text{Write } x \text{ as } x^1.$$

$$= (3 \cdot 5)(x^{-4} \cdot x^1) \quad \text{Commutative and associative properties}$$

$$= 15x^{-4+1} \quad \text{Multiply 3 and 5, then add exponents.}$$

$$= 15x^{-3} \quad \text{The sum of } -4 \text{ and } 1 \text{ is } -3.$$

$$= \dfrac{15}{x^3} \quad \text{Definition of negative exponents} \quad \blacksquare$$

10. Simplify: $\dfrac{x^{-5} \cdot x^2}{x^{-8}}$.

EXAMPLE 10 Simplify: $\dfrac{x^{-4} \cdot x^7}{x^{-2}}$.

SOLUTION We simplify the numerator first by adding exponents.

$$\dfrac{x^{-4} \cdot x^7}{x^{-2}} = \dfrac{x^{-4+7}}{x^{-2}} \quad \text{Multiplication property for exponents}$$

$$= \dfrac{x^3}{x^{-2}} \quad \text{The sum of } -4 \text{ and } 7 \text{ is } 3.$$

$$= x^{3-(-2)} \quad \text{Division property for exponents}$$

$$= x^5 \quad \text{Subtracting } -2 \text{ is equivalent to adding } +2. \quad \blacksquare$$

Answers
8. 100
9. $\dfrac{28}{x^2}$
10. x^5

Problem Set A

Write each expression with positive exponents, then simplify. [Examples 1–5]

1. 2^{-5}
2. 3^{-4}
3. 10^{-2}
4. 10^{-3}

5. x^{-3}
6. x^{-4}
7. $(-4)^{-2}$
8. $(-2)^{-4}$

9. $(-5)^{-3}$
10. $(-4)^{-3}$
11. $2^6 \cdot 2^{-8}$
12. $3^5 \cdot 3^{-8}$

13. $10^{-2} \cdot 10^5$
14. $10^{-3} \cdot 10^6$
15. $x^{-4} \cdot x^{-3}$
16. $x^{-3} \cdot x^{-7}$

17. $3^8 \cdot 3^{-6}$
18. $2^9 \cdot 2^{-6}$
19. $2 \cdot 2^{-4}$
20. $3^{-5} \cdot 3$

21. $10^{-3} \cdot 10^8 \cdot 10^{-2}$
22. $10^{-3} \cdot 10^9 \cdot 10^{-4}$
23. $x^{-5} \cdot x^{-4} \cdot x^{-3}$
24. $x^{-7} \cdot x^{-9} \cdot x^{-6}$

25. $2^{-3} \cdot 2 \cdot 2^3$
26. $3 \cdot 3^5 \cdot 3^{-5}$

Divide as indicated. Write your answer using only positive exponents. [Examples 6–8]

27. $\dfrac{2^9}{2^7}$
28. $\dfrac{3^8}{3^4}$
29. $\dfrac{2^7}{2^9}$
30. $\dfrac{3^5}{3^7}$

Appendix A Problem Set

31. $\dfrac{x^4}{x^3}$
32. $\dfrac{x^7}{x^5}$
33. $\dfrac{x^3}{x^4}$
34. $\dfrac{x^5}{x^7}$

35. $10^{-2} \div 10^5$
36. $10^{-4} \div 10^2$
37. $10^{-5} \div 10^2$
38. $10^{-3} \div 10$

39. $\dfrac{10^{-2}}{10^{-5}}$
40. $\dfrac{10^{-4}}{10^{-7}}$
41. $x^{12} \div x^{-12}$
42. $x^6 \div x^{-6}$

43. $2^{12} \div 2^{10}$
44. $3^{11} \div 3^9$
45. $\dfrac{10}{10^{-3}}$
46. $\dfrac{10^2}{10^{-5}}$

Simplify each expression. Write your answer using only positive exponents. [Examples 9, 10]

47. $3x^{-2} \cdot 5x^7$
48. $5x^{-3} \cdot 8x^6$
49. $2x^{-2} \cdot 3x^{-3}$
50. $4x^{-6} \cdot 5x^{-4}$

51. $(4x^{-4})(-3x^{-3})$
52. $(3x^2)(-4x^{-1})$
53. $7x \cdot 3x^{-4} \cdot 2x^5$
54. $6x \cdot 2x^{-3} \cdot 3x^4$

55. $\dfrac{x^{-3} \cdot x^7}{x^{-1}}$
56. $\dfrac{x^{-4} \cdot x^6}{x^{-2}}$
57. $\dfrac{x^{-2} \cdot x^{-8}}{x^{-12}}$
58. $\dfrac{x^{-5} \cdot x^{-7}}{x^{-15}}$

Scientific Notation

OBJECTIVES

A Write numbers in scientific notation.

B Convert numbers written in scientific notation to standard form.

There are many disciplines that deal with very large numbers and others that deal with very small numbers. For example, in astronomy, distances commonly are given in light-years. A light-year is the distance that light will travel in one year. It is approximately

$$5{,}880{,}000{,}000{,}000 \text{ miles}$$

It can be difficult to perform calculations with numbers in this form because of the number of zeros present. Scientific notation provides a way of writing very large, or very small, numbers in a more manageable form. Here is the formal definition:

A Scientific Notation

> **Definition**
>
> A number is in **scientific notation** when it is written as the product of a number between 1 and 10 and an integer power of 10. A number written in scientific notation has the form
>
> $$n \times 10^r$$
>
> where $1 \leq n < 10$ and $r =$ an integer.

EXAMPLE 1 The speed of light is 186,000 miles per second. Write 186,000 in scientific notation.

SOLUTION To write this number in scientific notation, we rewrite it as the product of a number between 1 and 10 and a power of 10. To do so, we move the decimal point 5 places to the left so that it appears between the 1 and the 8, giving us 1.86. Then we multiply this number by 10^5. The number that results has the same value as our original number but is written in scientific notation. Here is our result:

$$\text{Standard form} \longrightarrow 186{,}000 = 1.86 \times 10^5 \longleftarrow \text{Scientific notation}$$

Both numbers have exactly the same value. The number on the left is written in *standard form*, while the number on the right is written in scientific notation. ∎

B Converting to Standard Form

Lets now practice taking a number written in scientific notation and converting it to a number in standard form.

EXAMPLE 2 If your pulse rate is 60 beats per minute, then your heart will beat 8.64×10^4 times each day. Write 8.64×10^4 in standard form.

SOLUTION Because 10^4 is 10,000, we can think of this as simply a multiplication problem. That is,

$$8.64 \times 10^4 = 8.64 \times 10{,}000 = 86{,}400$$

PRACTICE PROBLEMS

1. Write 27,500 in scientific notation.

2. Write 7.89×10^5 in standard form.

Answers
1. 2.75×10^4
2. 789,000

Looking over our result, we can think of the exponent 4 as indicating the number of places we need to move the decimal point to write our number in standard form. Because our exponent is positive 4, we move the decimal point from its original position, between the 8 and the 6, four places to the right. If we need to add any zeros on the right we do so. The result is the standard form of our number, 86,400. ∎

Next, we turn our attention to writing small numbers in scientific notation. To do so, we use the negative exponents developed in the previous section. For example, the number 0.00075, when written in scientific notation, is equivalent to 7.5×10^{-4}. Here's why:

$$7.5 \times 10^{-4} = 7.5 \times \frac{1}{10^4} = 7.5 \times \frac{1}{10,000} = \frac{7.5}{10,000} = 0.00075$$

The table below lists some other numbers both in scientific notation and in standard form.

EXAMPLE 3 Each pair of numbers in the table below is equal.

Standard Form		Scientific Notation
376,000	=	3.76×10^5
49,500	=	4.95×10^4
3,200	=	3.2×10^3
591	=	5.91×10^2
46	=	4.6×10^1
8	=	8×10^0
0.47	=	4.7×10^{-1}
0.093	=	9.3×10^{-2}
0.00688	=	6.88×10^{-3}
0.0002	=	2×10^{-4}
0.000098	=	9.8×10^{-5}

As we read across the table, for each pair of numbers, notice how the decimal point in the number on the right is placed so that the number containing the decimal point is always a number between 1 and 10. Correspondingly, the exponent on 10 keeps track of how many places the decimal point was moved in converting from standard form to scientific notation. In general, when the exponent is positive, we are working with a large number. On the other hand, when the exponent is negative, we are working with a small number. (By small number, we mean a number that is less than 1, but larger than 0.)

We end this section with a diagram that shows two numbers, one large and one small, that are converted to scientific notation.

$$376,000 = 3.76 \times 10^5$$

Moved 5 places — Exponent keeps track of the 5 places we moved the decimal point.

Decimal point originally here

$$0.00688 = 6.88 \times 10^{-3}$$

Moved 3 places to right — Keeps track of the 3 places we moved the decimal point

---

3. Fill in the missing numbers in the table below.

	Standard Form		Scientific Notation
a.	24,500	=	
b.		=	5.6×10^5
c.	0.000789	=	
d.		=	4.8×10^{-3}

Answer
3. a. 2.45×10^4
 b. 560,000
 c. 7.89×10^{-4}
 d. 0.0048

Problem Set B

Write each number in scientific notation. [Examples 1, 3]

1. 425,000 **2.** 635,000 **3.** 6,780,000 **4.** 5,490,000

5. 11,000 **6.** 29,000 **7.** 89,000,000 **8.** 37,000,000

Write each number in standard form. [Examples 2, 3]

9. 3.84×10^4 **10.** 3.84×10^7 **11.** 5.71×10^7 **12.** 5.71×10^5

13. 3.3×10^3 **14.** 3.3×10^2 **15.** 8.913×10^7 **16.** 8.913×10^5

Write each number in scientific notation. [Examples 1, 3]

17. 0.00035 **18.** 0.0000035 **19.** 0.0007 **20.** 0.007

21. 0.06035 **22.** 0.0006035 **23.** 0.1276 **24.** 0.001276

Write each number in standard form. [Examples 2, 3]

25. 8.3×10^{-4} **26.** 8.3×10^{-7} **27.** 6.25×10^{-2} **28.** 7.83×10^{-4}

29. 3.125×10^{-1} **30.** 3.125×10^{-2} **31.** 5×10^{-3} **32.** 5×10^{-5}

Applying the Concepts

Super Bowl Advertising and Viewers The cost of a 30-second television ad along with the approximate number of viewers for four different Super Bowls is shown below. Complete the table by writing the ad cost in scientific notation, and the number of viewers in standard form.

	Year	Super Bowl	Ad Cost	Ad Cost in Scientific Notation	Viewers in Scientific Notation	Number of Viewers
33.	1967	I	$40,000		5.12×10^6	
34.	1982	XVI	$345,000		8.52×10^7	
35.	1997	XXXI	$1,200,000		8.79×10^7	
36.	2011	XLV	$3,000,000		1.11×10^8	

Galilean Moons The planet Jupiter has about 60 known moons. In the year 1610, Galileo first discovered the four largest moons of Jupiter: Io, Europa, Ganymede, and Callisto. These moons are known as the Galilean moons. Each moon has a unique period, or the time it takes to make a trip around Jupiter. Fill in the tables below.

37.

Jupiter's Moon	Period (seconds)
Io	153,000
Europa	3.07×10^5
Ganymede	618,000
Callisto	1.44×10^6

38.

Jupiter's Moon	Distance from Jupiter (kilometers)
Io	422,000
Europa	6.17×10^5
Ganymede	1,070,000
Callisto	1.88×10^6

Computer Science The smallest amount of data that a computer can hold is measured in bits. A byte is the next largest unit and is equal to 8, or 2^3, bits. Fill in the table below.

		Number of Bytes	
Unit	Exponential Form		Scientific Notation
39. Kilobyte	$2^{10} \approx 1,024$		
40. Megabyte	$2^{20} \approx 1,049,000$		
41. Gigabyte	$2^{30} \approx 1,074,000,000$		
42. Terabyte	$2^{40} \approx 1,099,500,000,000$		

Multiplication and Division with Scientific Notation

C

OBJECTIVES

A Multiply and divide numbers written in scientific notation.

A Multiplication and Division with Numbers Written in Scientific Notation

In this section, we extend our work with scientific notation to include multiplication and division with numbers written in scientific notation. To work the problems in this section, we use the material presented in the previous two sections, along with the commutative and associative properties of multiplication and the rule for multiplication with fractions. Here is our first example.

PRACTICE PROBLEMS

EXAMPLE 1 Multiply: $(3.5 \times 10^8)(2.2 \times 10^{-5})$.

1. Multiply: $(2.5 \times 10^6)(1.4 \times 10^2)$.

SOLUTION First we apply the commutative and associative properties to rearrange the numbers, so that the decimal numbers are grouped together and the powers of 10 are also.

$$(3.5 \times 10^8)(2.2 \times 10^{-5}) = (3.5)(2.2) \times (10^8)(10^{-5})$$

Next, we multiply the decimal numbers together and then the powers of ten. To multiply the powers of ten, we add exponents.

$$= 7.7 \times 10^{8+(-5)}$$
$$= 7.7 \times 10^3 \qquad \blacksquare$$

EXAMPLE 2 Find the product of 130,000,000 and 0.000005. Write your answer in scientific notation.

2. Find the product of 2,200,000 and 0.00015.

SOLUTION We begin by writing both numbers in scientific notation. Then we proceed as we did in Example 1: We group the numbers between 1 and 10 separately from the powers of 10.

$$(130{,}000{,}000)(0.000005) = (1.3 \times 10^8)(5 \times 10^{-6})$$
$$= (1.3)(5) \times (10^8)(10^{-6})$$
$$= 6.5 \times 10^2 \qquad \blacksquare$$

Our next examples involve division with numbers in scientific notation.

EXAMPLE 3 Divide: $\dfrac{8 \times 10^3}{4 \times 10^{-6}}$.

3. Divide: $\dfrac{6 \times 10^5}{2 \times 10^{-4}}$.

SOLUTION To separate the numbers between 1 and 10 from the powers of 10, we "undo" the multiplication and write the problem as the product of two fractions. Doing so looks like this:

$$\dfrac{8 \times 10^3}{4 \times 10^{-6}} = \dfrac{8}{4} \times \dfrac{10^3}{10^{-6}} \qquad \text{Write as two separate fractions.}$$

Next, we divide 8 by 4 to obtain 2. Then we divide 10^3 by 10^{-6} by subtracting exponents.

$$= 2 \times 10^{3-(-6)} \qquad \text{Divide.}$$
$$= 2 \times 10^9 \qquad \blacksquare$$

Answers
1. 3.5×10^8
2. 3.3×10^2
3. 3.0×10^9

Appendix C Multiplication and Division with Scientific Notation

4. Divide: $\dfrac{0.0038}{19,000,000}$.

EXAMPLE 4 Divide: $\dfrac{0.00045}{1,500,000}$.

SOLUTION To begin, write each number in scientific notation.

$$\dfrac{0.00045}{1,500,000} = \dfrac{4.5 \times 10^{-4}}{1.5 \times 10^{6}} \quad \text{Write numbers in scientific notation.}$$

Next, as in the previous example, we write the problem as two separate fractions in order to group the numbers between 1 and 10 together, as well as the powers of 10.

$$= \dfrac{4.5}{1.5} \times \dfrac{10^{-4}}{10^{6}} \quad \text{Write as two separate fractions.}$$

$$= 3 \times 10^{-4-6} \quad \text{Divide.}$$

$$= 3 \times 10^{-10} \quad -4 - 6 = -4 + (-6) = -10 \quad \blacksquare$$

5. Simplify: $\dfrac{(6.8 \times 10^{-4})(3.9 \times 10^{2})}{7.8 \times 10^{-6}}$.

EXAMPLE 5 Simplify: $\dfrac{(6.8 \times 10^{5})(3.9 \times 10^{-7})}{7.8 \times 10^{-4}}$.

SOLUTION We group the numbers between 1 and 10 separately from the powers of 10.

$$\dfrac{(6.8 \times 10^{5})(3.9 \times 10^{-7})}{7.8 \times 10^{-4}} = \dfrac{(6.8)(3.9)}{7.8} \times \dfrac{(10^{5})(10^{-7})}{10^{-4}}$$

$$= 3.4 \times 10^{5+(-7)-(-4)}$$

$$= 3.4 \times 10^{2} \quad \blacksquare$$

6. Simplify: $\dfrac{(0.000035)(45,000)}{0.000075}$.

EXAMPLE 6 Simplify: $\dfrac{(35,000)(0.0045)}{7,500,000}$.

SOLUTION We write each number in scientific notation, and then we proceed as we have in the examples above.

$$\dfrac{(35,000)(0.0045)}{7,500,000} = \dfrac{(3.5 \times 10^{4})(4.5 \times 10^{-3})}{7.5 \times 10^{6}}$$

$$= \dfrac{(3.5)(4.5)}{7.5} \times \dfrac{(10^{4})(10^{-3})}{10^{6}}$$

$$= 2.1 \times 10^{4+(-3)-6}$$

$$= 2.1 \times 10^{-5} \quad \blacksquare$$

Answers
4. 2.0×10^{-10}
5. 3.4×10^{4}
6. 2.1×10^{4}

Problem Set C

Find each product. Write all answers in scientific notation. [Examples 1, 2]

1. $(2 \times 10^4)(3 \times 10^6)$ **2.** $(3 \times 10^3)(1 \times 10^5)$ **3.** $(2.5 \times 10^7)(6 \times 10^3)$ **4.** $(3.8 \times 10^6)(5 \times 10^3)$

5. $(7.2 \times 10^3)(9.5 \times 10^{-6})$ **6.** $(8.5 \times 10^5)(4.2 \times 10^{-9})$ **7.** $(36{,}000)(450{,}000)$ **8.** $(25{,}000)(620{,}000)$

9. $(4{,}200)(0.00009)$ **10.** $(0.0000065)(86{,}000)$

Find each quotient. Write all answers in scientific notation. [Examples 3, 4]

11. $\dfrac{3.6 \times 10^5}{1.8 \times 10^2}$ **12.** $\dfrac{9.3 \times 10^{15}}{3.0 \times 10^5}$ **13.** $\dfrac{8.4 \times 10^{-6}}{2.1 \times 10^3}$ **14.** $\dfrac{6.0 \times 10^{-10}}{1.5 \times 10^3}$

15. $\dfrac{3.5 \times 10^5}{7.0 \times 10^{-10}}$ **16.** $\dfrac{1.6 \times 10^7}{8.0 \times 10^{-14}}$ **17.** $\dfrac{540{,}000}{9{,}000}$ **18.** $\dfrac{750{,}000{,}000}{250{,}000}$

19. $\dfrac{0.00092}{46{,}000}$ **20.** $\dfrac{0.00000047}{235{,}000}$

Simplify each expression, and write all answers in scientific notation. [Examples 5, 6]

21. $\dfrac{(3 \times 10^7)(8 \times 10^4)}{6 \times 10^5}$ **22.** $\dfrac{(4 \times 10^9)(6 \times 10^5)}{8 \times 10^3}$ **23.** $\dfrac{(2 \times 10^{-3})(6 \times 10^{-5})}{3 \times 10^{-4}}$ **24.** $\dfrac{(4 \times 10^{-5})(9 \times 10^{-10})}{6 \times 10^{-6}}$

25. $\dfrac{(3.5 \times 10^{-4})(4.2 \times 10^5)}{7 \times 10^3}$

26. $\dfrac{(2.4 \times 10^{-6})(3.6 \times 10^3)}{9 \times 10^5}$

27. $\dfrac{(0.00087)(40{,}000)}{1{,}160{,}000}$

28. $\dfrac{(0.0045)(24{,}000)}{270{,}000}$

29. $\dfrac{(525)(0.0000032)}{0.0025}$

30. $\dfrac{(465)(0.000004)}{0.0093}$

Applying the Concepts

Super Bowl Advertising and Viewers The cost of a 30-second television ad along with the approximate number of viewers for four different Super Bowls is shown below. Complete the table by finding the cost per viewer by dividing the ad cost by the number of viewers.

Year	Super Bowl	Ad Cost in Scientific Notation	Viewers in Scientific Notation	Ad Cost Per Viewer
31. 1967	I	$\$4.0 \times 10^4$	5.12×10^6	
32. 1982	XVI	$\$3.45 \times 10^5$	8.52×10^7	
33. 1997	XXXI	$\$1.2 \times 10^6$	8.79×10^7	
34. 2011	XLV	$\$3.0 \times 10^6$	1.11×10^8	

35. **Pyramids and Scientific Notation** The Great Pyramid at Giza is one of the largest and oldest man-made structures in the world. It weighs over 10^{10} kilograms. If each stone making up the pyramid weighs approximately 4,000 kilograms, how many stones make up the structure? Write your answer in scientific notation.

36. **Technology** Suppose a CD-ROM holds about 700 megabytes, or 7.0×10^8 bytes, of data. A four-and-a-half minute song downloaded from the Internet uses 5 megabytes, or 5.0×10^6 bytes, of storage space. How many four-and-a-half minute songs can fit on a CD?

Solutions to Selected Practice Problems

Solutions to all practice problems that require more than one step are shown here. Before you look back here to see where you have made a mistake, you should try the problem you are working on twice. If you do not get the correct answer the second time you work the problem, then the solution here should show you where you went wrong.

Chapter 1

Section 1.2

1. $\begin{array}{r} 63 \\ +25 \\ \hline 88 \end{array}$

2. $\begin{array}{r} 342 \\ +605 \\ \hline 947 \end{array}$

3. a. $\begin{array}{r} \overset{1}{375} \\ 121 \\ +473 \\ \hline 969 \end{array}$
    b. $\begin{array}{r} \overset{2\,2}{495} \\ 699 \\ +978 \\ \hline 2{,}172 \end{array}$

4. a. $\begin{array}{r} \overset{11\;11}{57{,}904} \\ 7{,}193 \\ 655 \\ \hline 65{,}752 \end{array}$
    b. $\begin{array}{r} \overset{1\,1\,2\,3}{68{,}495} \\ 7{,}236 \\ 878 \\ 29 \\ 5 \\ \hline 76{,}643 \end{array}$

7. a. $6 + 2 + 4 + 8 + 3 = (6 + 4) + (2 + 8) + 3 = 10 + 10 + 3 = 23$
    b. $24 + 17 + 36 + 13 = (24 + 36) + (17 + 13) = 60 + 30 = 90$

8. a. $n = 8$, since $8 + 9 = 17$
    b. $n = 8$, since $8 + 2 = 10$
    c. $n = 1$, since $8 + 1 = 9$
    d. $n = 6$, since $16 = 6 + 10$

9. a. $7 + 7 + 7 + 7 = 28$ ft   b. $88 + 88 + 33 + 33 = 242$ in.   c. $44 + 66 + 77 = 187$ yd

Section 1.3

5. a. Food $\$ 5{,}296$
        Car $4{,}847$
        Total $\$\overline{10{,}143}$ = $\$10{,}140$ to the nearest ten dollars

    b. Savings $\$2{,}149$
       Taxes $6{,}137$
       Total $\overline{\$8{,}286}$ = $\$8{,}300$ to the nearest hundred dollars

    c. House $\$10{,}200$
       Taxes $6{,}137$
       Misc. $6{,}142$
       Car $4{,}847$
       Savings $2{,}149$
       Total $\overline{\$29{,}475}$ = $\$29{,}000$ to the nearest thousand dollars

6. a. We round each of the four numbers in the sum to the nearest thousand, and then we add the rounded numbers.

5,287	rounds to	5,000
2,561	rounds to	3,000
888	rounds to	1,000
+4,898	rounds to	+ 5,000
		14,000

    We estimate the answer to this problem to be approximately 14,000. The actual answer, found by adding the original, unrounded numbers, is 13,634.

    b. We round each of the four numbers in the sum to the nearest thousand, and then we add the rounded numbers.

702	rounds to	1,000
3,944	rounds to	4,000
1,001	rounds to	1,000
+3,500	rounds to	+ 4,000
		10,000

    We estimate the answer to this problem to be approximately 10,000. The actual answer, found by adding the original, unrounded numbers, is 9,147.

7. We round each of the prices to the nearest dollar, and then we add the rounded numbers.

$2.69	rounds to	$3.00
$4.26	rounds to	$4.00
$1.99	rounds to	$2.00
$1.59	rounds to	$2.00
+ $5.19	rounds to	+ $5.00
		$16.00

    We estimate the total to be $16.00. Thus, $20.00 will be enough to pay for all of the groceries.

Section 1.4

1. a.
$$\begin{array}{r} 684 \\ -431 \\ \hline 253 \end{array}$$

b.
$$\begin{array}{r} 7{,}406 \\ -3{,}405 \\ \hline 4{,}001 \end{array}$$

2. a.
$$\begin{array}{r} 6{,}857 \\ -\ 405 \\ \hline 6{,}452 \end{array}$$

b.
$$\begin{array}{r} 345 \\ -\ 234 \\ \hline 111 \end{array}$$

3. a. $63 = 6 \text{ tens} + 3 \text{ ones} = 5 \text{ tens} + 13 \text{ ones}$
  $\underline{-47 = 4 \text{ tens} + 7 \text{ ones} = 4 \text{ tens} \quad\ 7 \text{ ones}}$
  $\phantom{-47 = 4 \text{ tens} + 7 \text{ ones} = }1 \text{ ten} +\ 6 \text{ ones}$
  Answer: 16

b. $532 = 5 \text{ hundreds} + 3 \text{ tens} + 2 \text{ ones} = 5 \text{ hundreds} + 2 \text{ tens} + 12 \text{ ones}$
  $\underline{-403 = 4 \text{ hundreds} + 0 \text{ tens} + 3 \text{ ones} = 4 \text{ hundreds}\quad 0 \text{ tens}\quad 3 \text{ ones}}$
  $\phantom{-403 = 4 \text{ hundreds} + 0 \text{ tens} + 3 \text{ ones} = }1 \text{ hundred} + 2 \text{ tens} + 9 \text{ ones}$
  Answer: 129

4. a.
$$\begin{array}{r} \overset{5\,15}{6\,\cancel{5}6} \\ -283 \\ \hline 373 \end{array}$$

b.
$$\begin{array}{r} \overset{2,\,16\,12}{3{,}7\,\cancel{2}\,\cancel{9}}, \\ -1{,}749 \\ \hline 1{,}980 \end{array}$$

Section 1.5

1. a. $4 \cdot 70 = 70 + 70 + 70 + 70 = 280$

b. $4 \cdot 700 = 700 + 700 + 700 + 700 = 2{,}800$

c. $4 \cdot 7{,}000 = 7{,}000 + 7{,}000 + 7{,}000 + 7{,}000 = 28{,}000$

4. a.
$$\begin{array}{r} \overset{5}{57} \\ \times\ 8 \\ \hline 456 \end{array}$$

b.
$$\begin{array}{r} \overset{5}{570} \\ \times\ 8 \\ \hline 4{,}560 \end{array}$$

5. a.
$$\begin{array}{r} 45 \\ \times 62 \\ \hline 90 \\ +2{,}700 \\ \hline 2{,}790 \end{array}$$
$\leftarrow 2(45) = 90$
$\leftarrow 60(45) = 2{,}700$

b.
$$\begin{array}{r} 620 \\ \times 45 \\ \hline 3{,}100 \\ +24{,}800 \\ \hline 27{,}900 \end{array}$$
$\leftarrow 5(620) = 3{,}100$
$\leftarrow 40(620) = 24{,}800$

6. a.
$$\begin{array}{r} 356 \\ \times 641 \\ \hline 356 \\ 14{,}240 \\ 213{,}600 \\ \hline 228{,}196 \end{array}$$
$\leftarrow 1(356) = 356$
$\leftarrow 40(356) = 14{,}240$
$\leftarrow 600(356) = 213{,}600$

b.
$$\begin{array}{r} 3{,}560 \\ \times 641 \\ \hline 3{,}560 \\ 142{,}400 \\ 2{,}136{,}000 \\ \hline 2{,}281{,}960 \end{array}$$
$\leftarrow 1(3{,}560) = 3{,}560$
$\leftarrow 40(3{,}560) = 142{,}400$
$\leftarrow 600(3{,}560) = 2{,}136{,}000$

10.
$$\begin{array}{r} 365 \\ \times\ 550 \\ \hline 18{,}250 \\ 182{,}500 \\ \hline 200{,}750 \text{ mg} \end{array}$$

11. $36(\$12) = \$432$ Total weekly earnings
$\$432 - \$109 = \$323$ Take-home pay

12. Fat: $3(10) = 30$ grams of fat; sodium: $3(160) = 480$ milligrams of sodium

13. Bowling for 3 hours burns $3(265) = 795$ calories. Eating two bags of chips means you are consuming $2(3)(160) = 960$ calories. No; bowling won't burn all the calories.

Section 1.6

1. a.
$$\begin{array}{r} 74 \\ 4\overline{)296} \\ \underline{28}\downarrow \\ 16 \\ \underline{16} \\ 0 \end{array}$$

b.
$$\begin{array}{r} 740 \\ 4\overline{)2{,}960} \\ \underline{28}\downarrow \\ 16 \\ \underline{16} \\ 00 \end{array}$$

2. a.
$$\begin{array}{r} 283 \\ 24\overline{)6{,}792} \\ \underline{48}\downarrow \\ 1\,99 \\ \underline{1\,92} \\ 72 \\ \underline{72} \\ 0 \end{array}$$

b.
$$\begin{array}{r} 2{,}830 \\ 24\overline{)67{,}920} \\ \underline{48}\downarrow \\ 19\,9 \\ \underline{19\,2} \\ 72 \\ \underline{72} \\ 00 \end{array}$$

3.
$$\begin{array}{r} 208 \\ 9\overline{)1{,}872} \\ \underline{18}\downarrow \\ 07 \\ \underline{0} \\ 72 \\ \underline{72} \\ 0 \end{array}$$

4. a. 69 R 20, or $69\frac{20}{27}$
$$\begin{array}{r} 27\overline{)1{,}883} \\ \underline{1\,62}\downarrow \\ 263 \\ \underline{243} \\ 20 \end{array}$$

b. 104 R 11, or $104\frac{11}{18}$
$$\begin{array}{r} 18\overline{)1{,}883} \\ \underline{18}\downarrow \\ 08 \\ \underline{0}\downarrow \\ 83 \\ \underline{72} \\ 11 \end{array}$$

5. 156 The family spent $156 per day
$$\begin{array}{r} 12\overline{)1{,}872} \\ \underline{1\,2}\downarrow \\ 67 \\ \underline{60}\downarrow \\ 72 \\ \underline{72} \\ 0 \end{array}$$

6. 537 They bought 537 basketballs
$$\begin{array}{r} 6\overline{)3{,}222} \\ \underline{3\,0}\downarrow \\ 22 \\ \underline{18}\downarrow \\ 42 \\ \underline{42} \\ 0 \end{array}$$

Section 1.7

1. Base 5, exponent 2; 5 to the second power, or 5 squared **2.** Base 2, exponent 3; 2 to the third power, or 2 cubed
3. Base 1, exponent 4; 1 to the fourth power **4.** $5^2 = 5 \cdot 5 = 25$ **5.** $9^2 = 9 \cdot 9 = 81$ **6.** $2^3 = 2 \cdot 2 \cdot 2 = 8$
7. $1^4 = 1 \cdot 1 \cdot 1 \cdot 1 = 1$ **8.** $2^5 = 2 \cdot 2 \cdot 2 \cdot 2 \cdot 2 = 32$ **9.** $7^1 = 7$ **10.** $4^1 = 4$ **11.** $9^0 = 1$ **12.** $1^0 = 1$

13. a. $5 \cdot 7 - 3 \cdot 6 = 35 - 18$ **b.** $5 \cdot 70 - 3 \cdot 60 = 350 - 180$ **14.** $7 + 3(6 + 4) = 7 + 3(10)$
$ = 17$ $ = 170$ $ = 7 + 30$
$ = 37$

15. a. $28 \div 7 - 3 = 4 - 3$ **b.** $6 \cdot 3^2 + 64 \div 2^4 - 2 = 6 \cdot 9 + 64 \div 16 - 2$
$ = 1$ $ = 54 + 4 - 2$
$ = 58 - 2$
$ = 56$

16. a. $5 + 3[24 - 5(6 - 2)] = 5 + 3[24 - 5(4)]$ **b.** $50 + 30[240 - 50(6 - 2)] = 50 + 30[240 - 50(4)]$
$ = 5 + 3[24 - 20]$ $ = 50 + 30[240 - 200]$
$ = 5 + 3[4]$ $ = 50 + 30(40)$
$ = 5 + 12$ $ = 50 + 1,200$
$ = 17$ $ = 1,250$

17. Mean $= \dfrac{187 + 273 + 150 + 173 + 227}{5} = \dfrac{1010}{5} = 202$ miles **18.** 150 173 187 227 273 $= 187$ miles
 ↑
 median

19. $40,770 $42,635 $44,475 $46,320
 median

$\dfrac{\$42,635 + \$44,475}{2} = \$43,555$

Section 1.8

1. $A = bh = 3 \cdot 2 = 6$ cm²   **4.** $V = 15 \cdot 12 \cdot 8 = 1,440$ ft³
2. $A = lw = 70 \cdot 35 = 2,450$ mm²   **5. a.** Surface area $= 2(15 \cdot 8) + 2(8 \cdot 12) + (15 \cdot 12) = 612$ ft²
3. $A = 38 \cdot 13 + 24 \cdot 27$ **b.** Two gallons will cover it, with some paint left over.
$ = 494 + 648$
$ = 1142$ ft²

Chapter 2

Section 2.1

6. $\dfrac{2}{3} = \dfrac{2 \cdot 4}{3 \cdot 4} = \dfrac{8}{12}$ **7.** $\dfrac{15}{20} = \dfrac{15 \div 5}{20 \div 5} = \dfrac{3}{4}$ **9.** $\dfrac{1}{3} \cdot \dfrac{4}{4} = \dfrac{4}{12}; \dfrac{1}{6} \cdot \dfrac{2}{2} = \dfrac{2}{12}; \dfrac{1}{4} \cdot \dfrac{3}{3} = \dfrac{3}{12}; \dfrac{2}{12}, \dfrac{3}{12}, \dfrac{4}{12}, \dfrac{5}{12}$

Section 2.2

1. 37 and 59 are prime numbers; 39 is divisible by 3 and 13; 51 is divisible by 3 and 17.

2. a. $90 = 9 \cdot 10$ **b.** $900 = 9 \cdot 100$ **4.** $\dfrac{12}{18} = \dfrac{12 \div 6}{18 \div 6} = \dfrac{2}{3}$ **5.** $\dfrac{15}{20} = \dfrac{3 \cdot 5}{2 \cdot 2 \cdot 5} = \dfrac{3}{4}$
$ = 3 \cdot 3 \cdot 2 \cdot 5$ $ = 3 \cdot 3 \cdot 25 \cdot 4$
$ = 2 \cdot 3^2 \cdot 5$ $ = 3 \cdot 3 \cdot 5 \cdot 5 \cdot 2 \cdot 2$
$ = 2^2 \cdot 3^2 \cdot 5^2$

6. a. $\dfrac{8}{72} = \dfrac{2 \cdot 2 \cdot 2 \cdot 1}{2 \cdot 2 \cdot 2 \cdot 3 \cdot 3} = \dfrac{1}{9}$ **b.** $\dfrac{16}{144} = \dfrac{2 \cdot 2 \cdot 2 \cdot 2}{2 \cdot 2 \cdot 2 \cdot 2 \cdot 3 \cdot 3} = \dfrac{1}{9}$ **7.** $\dfrac{5}{50} = \dfrac{5 \cdot 1}{5 \cdot 2 \cdot 5} = \dfrac{1}{10}$ **8.** $\dfrac{120}{25} = \dfrac{2 \cdot 2 \cdot 2 \cdot 3 \cdot 5}{5 \cdot 5} = \dfrac{24}{5}$

9. a. $\dfrac{30}{35} = \dfrac{2 \cdot 3 \cdot 5}{5 \cdot 7} = \dfrac{6}{7}$ **b.** $\dfrac{300}{350} = \dfrac{2 \cdot 2 \cdot 3 \cdot 5 \cdot 5}{2 \cdot 5 \cdot 5 \cdot 7} = \dfrac{6}{7}$

Section 2.3

1. $\dfrac{2}{3} \cdot \dfrac{5}{9} = \dfrac{10}{27}$

2. $\dfrac{2}{5} \cdot 7 = \dfrac{2}{5} \cdot \dfrac{7}{1}$
$= \dfrac{14}{5}$

3. $\dfrac{1}{3}\left(\dfrac{4}{5} \cdot \dfrac{1}{3}\right) = \dfrac{1}{3}\left(\dfrac{4}{15}\right)$
$= \dfrac{4}{45}$

4. $\dfrac{1}{4}\left(\dfrac{2}{3} \cdot \dfrac{1}{2}\right) = \left(\dfrac{1}{4} \cdot \dfrac{2}{3}\right)\dfrac{1}{2}$
$= \dfrac{2}{12} \cdot \dfrac{1}{2}$
$= \dfrac{2}{24}$
$= \dfrac{1}{12}$

5. a. $\dfrac{12}{25} \cdot \dfrac{5}{6} = \dfrac{12 \cdot 5}{25 \cdot 6}$
$= \dfrac{(2 \cdot 2 \cdot 3) \cdot 5}{(5 \cdot 5) \cdot (2 \cdot 3)}$
$= \dfrac{2}{5}$

b. $\dfrac{12}{25} \cdot \dfrac{50}{60} = \dfrac{12 \cdot 50}{25 \cdot 60}$
$= \dfrac{(2 \cdot 2 \cdot 3) \cdot (2 \cdot 5 \cdot 5)}{(5 \cdot 5) \cdot (2 \cdot 2 \cdot 3 \cdot 5)}$
$= \dfrac{2}{5}$

6. a. $\dfrac{8}{3} \cdot \dfrac{9}{24} = \dfrac{8 \cdot 9}{3 \cdot 24}$
$= \dfrac{(2 \cdot 2 \cdot 2) \cdot (3 \cdot 3)}{3 \cdot (2 \cdot 2 \cdot 2 \cdot 3)}$
$= \dfrac{1}{1}$
$= 1$

b. $\dfrac{8}{30} \cdot \dfrac{90}{24} = \dfrac{8 \cdot 90}{30 \cdot 24}$
$= \dfrac{(2 \cdot 2 \cdot 2) \cdot (2 \cdot 3 \cdot 3 \cdot 5)}{(2 \cdot 3 \cdot 5) \cdot (2 \cdot 2 \cdot 2 \cdot 3)}$
$= \dfrac{1}{1}$
$= 1$

7. $\dfrac{3}{4} \cdot \dfrac{8}{3} \cdot \dfrac{1}{6} = \dfrac{3 \cdot 8 \cdot 1}{4 \cdot 3 \cdot 6}$
$= \dfrac{(3 \cdot 2 \cdot 2 \cdot 2) \cdot 1}{(2 \cdot 2) \cdot 3 \cdot (2 \cdot 3)}$
$= \dfrac{1}{3}$

8. $\left(\dfrac{2}{3}\right)^2 = \dfrac{2}{3} \cdot \dfrac{2}{3}$
$= \dfrac{4}{9}$

9. a. $\left(\dfrac{3}{4}\right)^2 \cdot \dfrac{1}{2} = \dfrac{3}{4} \cdot \dfrac{3}{4} \cdot \dfrac{1}{2}$
$= \dfrac{9}{32}$

b. $\left(\dfrac{2}{3}\right)^3 \cdot \dfrac{9}{8} = \dfrac{2}{3} \cdot \dfrac{2}{3} \cdot \dfrac{2}{3} \cdot \dfrac{9}{8}$
$= \dfrac{2 \cdot 2 \cdot 2 \cdot 9}{3 \cdot 3 \cdot 3 \cdot 8}$
$= \dfrac{2 \cdot 2 \cdot 2 \cdot (3 \cdot 3)}{3 \cdot 3 \cdot 3 \cdot (2 \cdot 2 \cdot 2)}$
$= \dfrac{2 \cdot 2 \cdot 2 \cdot (3 \cdot 3)}{3 \cdot 3 \cdot 3 \cdot (2 \cdot 2 \cdot 2)}$
$= \dfrac{1}{3}$

10. a. $\dfrac{2}{3} \cdot \dfrac{1}{2} = \dfrac{2 \cdot 1}{3 \cdot 2}$
$= \dfrac{1}{3}$

b. $\dfrac{3}{5}(15) = \dfrac{3}{5}\left(\dfrac{15}{1}\right)$
$= \dfrac{3 \cdot 3 \cdot 5}{5}$
$= \dfrac{9}{1}$
$= 9$

11. a. $\dfrac{2}{3}(12) = \dfrac{2}{3}\left(\dfrac{12}{1}\right)$
$= \dfrac{2 \cdot 2 \cdot 2 \cdot 3}{3 \cdot 1}$
$= \dfrac{8}{1}$
$= 8$

b. $\dfrac{2}{3}(120) = \dfrac{2}{3}\left(\dfrac{120}{1}\right)$
$= \dfrac{2 \cdot 2 \cdot 2 \cdot 2 \cdot 3 \cdot 5}{3}$
$= \dfrac{80}{1}$
$= 80$

12. $A = \dfrac{1}{2}(7)(10)$
$= 35 \text{ in}^2$

13.
$A = 4 \times 4 = 16 \text{ ft}^2$
$A = \dfrac{1}{2} \times 2 \times 2 = 2 \text{ ft}^2$
$A = 8 \times 2 = 16 \text{ ft}^2$
Total area $= 2 + 16 + 16 = 34 \text{ ft}^2$

Section 2.4

1. a. $\dfrac{1}{3} \div \dfrac{1}{6} = \dfrac{1}{3} \cdot \dfrac{6}{1}$
$= \dfrac{6}{3}$
$= 2$

b. $\dfrac{1}{30} \div \dfrac{1}{60} = \dfrac{1}{30} \cdot \dfrac{60}{1}$
$= \dfrac{1 \cdot 2 \cdot 2 \cdot 3 \cdot 5}{2 \cdot 3 \cdot 5 \cdot 1}$
$= \dfrac{2}{1}$
$= 2$

2. $\dfrac{5}{9} \div \dfrac{10}{3} = \dfrac{5}{9} \cdot \dfrac{3}{10}$
$= \dfrac{5 \cdot 3}{3 \cdot 3 \cdot 2 \cdot 5}$
$= \dfrac{1}{6}$

3. a. $\dfrac{3}{4} \div 3 = \dfrac{3}{4} \cdot \dfrac{1}{3}$
$= \dfrac{1}{4}$

b. $\dfrac{3}{5} \div 3 = \dfrac{3}{5} \cdot \dfrac{1}{3}$
$= \dfrac{1}{5}$

c. $\dfrac{3}{7} \div 3 = \dfrac{3}{7} \cdot \dfrac{1}{3}$
$= \dfrac{1}{7}$

4. $4 \div \left(\dfrac{1}{5}\right) = 4(5)$
$= 20$

5. a. $\dfrac{5}{32} \div \dfrac{10}{42} = \dfrac{5}{32} \cdot \dfrac{42}{10}$
$= \dfrac{5 \cdot (2 \cdot 3 \cdot 7)}{2 \cdot 2 \cdot 2 \cdot 2 \cdot 2 \cdot 2 \cdot 5}$
$= \dfrac{21}{32}$

b. $\dfrac{15}{32} \div \dfrac{30}{42} = \dfrac{15}{32} \cdot \dfrac{42}{30}$
$= \dfrac{(3 \cdot 5) \cdot (2 \cdot 3 \cdot 7)}{(2 \cdot 2 \cdot 2 \cdot 2 \cdot 2) \cdot (2 \cdot 3 \cdot 5)}$
$= \dfrac{21}{32}$

6. $\dfrac{12}{25} \div 6 = \dfrac{12}{25} \cdot \dfrac{1}{6}$
$= \dfrac{2 \cdot 2 \cdot 3 \cdot 1}{5 \cdot 5 \cdot 2 \cdot 3}$
$= \dfrac{2}{25}$

b. $\dfrac{24}{25} \div 6 = \dfrac{24}{25} \cdot \dfrac{1}{6}$
$= \dfrac{2 \cdot 2 \cdot 2 \cdot 3 \cdot 1}{(5 \cdot 5) \cdot (2 \cdot 3)}$
$= \dfrac{4}{25}$

7. a. $12 \div \dfrac{4}{3} = 12\left(\dfrac{3}{4}\right)$
$= 9$

b. $12 \div \dfrac{4}{5} = 12\left(\dfrac{5}{4}\right)$
$= 15$

c. $12 \div \dfrac{4}{7} = 12\left(\dfrac{7}{4}\right)$
$= 21$

8. $\dfrac{5}{4} \div \dfrac{1}{8} + 8 = \dfrac{5}{4} \cdot \dfrac{8}{1} + 8$
$= 10 + 8$
$= 18$

9. $18 \div \left(\dfrac{3}{5}\right)^2 + 48 \div \left(\dfrac{2}{5}\right)^2 = 18 \div \dfrac{9}{25} + 48 \div \dfrac{4}{25}$
$= 18 \cdot \dfrac{25}{9} + 48 \cdot \dfrac{25}{4}$
$= 50 + 300$
$= 350$

10. $12 \div \dfrac{3}{4} = 12 \cdot \dfrac{4}{3}$
$= 4 \cdot 4$
$= 16$ blankets

Section 2.5

1. $\dfrac{3}{10} + \dfrac{1}{10} = \dfrac{3+1}{10}$
$= \dfrac{4}{10}$
$= \dfrac{2}{5}$

2. $\dfrac{a+5}{12} + \dfrac{3}{12} = \dfrac{a+5+3}{12}$
$= \dfrac{a+8}{12}$

3. $\dfrac{8}{7} - \dfrac{5}{7} = \dfrac{8-5}{7}$
$= \dfrac{3}{7}$

4. $\dfrac{5}{9} + \dfrac{8}{9} + \dfrac{5}{9} = \dfrac{5+8+5}{9}$
$= \dfrac{18}{9}$
$= 2$

5. a. $\left.\begin{array}{l}18 = 2 \cdot 3 \cdot 3 \\ 14 = 2 \cdot 7\end{array}\right\}$ LCD $= 2 \cdot 3 \cdot 3 \cdot 7$
$= 126$

b. $\left.\begin{array}{l}36 = 2 \cdot 2 \cdot 3 \cdot 3 \\ 28 = 2 \cdot 2 \cdot 7\end{array}\right\}$ LCD $= 2 \cdot 2 \cdot 3 \cdot 3 \cdot 7$
$= 252$

6. a. LCD $= 126$; $\dfrac{5}{18} + \dfrac{3}{14} = \dfrac{5 \cdot 7}{18 \cdot 7} + \dfrac{3 \cdot 9}{14 \cdot 9}$
$= \dfrac{35}{126} + \dfrac{27}{126}$
$= \dfrac{62}{126}$
$= \dfrac{31}{63}$

b. LCD $= 252$; $\dfrac{5}{36} + \dfrac{3}{28} = \dfrac{5 \cdot 7}{36 \cdot 7} + \dfrac{3 \cdot 9}{28 \cdot 9}$
$= \dfrac{35}{252} + \dfrac{27}{252}$
$= \dfrac{62}{252}$
$= \dfrac{31}{126}$

8. a. $\dfrac{2}{9} + \dfrac{4}{15} = \dfrac{2 \cdot 5}{9 \cdot 5} + \dfrac{4 \cdot 3}{15 \cdot 3}$
$= \dfrac{10}{45} + \dfrac{12}{45}$
$= \dfrac{22}{45}$

b. $\dfrac{2}{27} + \dfrac{4}{45} = \dfrac{2 \cdot 5}{27 \cdot 5} + \dfrac{4 \cdot 3}{45 \cdot 3}$
$= \dfrac{10}{135} + \dfrac{12}{135}$
$= \dfrac{22}{135}$

9. LCD $= 100$; $\dfrac{8}{25} - \dfrac{3}{20} = \dfrac{8 \cdot 4}{25 \cdot 4} - \dfrac{3 \cdot 5}{20 \cdot 5}$
$= \dfrac{32}{100} - \dfrac{15}{100}$
$= \dfrac{17}{100}$

10. LCD $= 20$; $\dfrac{3}{4} - \dfrac{1}{5} = \dfrac{3 \cdot 5}{4 \cdot 5} - \dfrac{1 \cdot 4}{5 \cdot 4}$
$= \dfrac{15}{20} - \dfrac{4}{20}$
$= \dfrac{11}{20}$

11. a. LCD $= 36$;
$\dfrac{1}{9} + \dfrac{1}{4} + \dfrac{1}{6} = \dfrac{1 \cdot 4}{9 \cdot 4} + \dfrac{1 \cdot 9}{4 \cdot 9} + \dfrac{1 \cdot 6}{6 \cdot 6}$
$= \dfrac{4}{36} + \dfrac{9}{36} + \dfrac{6}{36}$
$= \dfrac{19}{36}$

b. LCD $= 360$;
$\dfrac{1}{90} + \dfrac{1}{40} + \dfrac{1}{60} = \dfrac{1 \cdot 4}{90 \cdot 4} + \dfrac{1 \cdot 9}{40 \cdot 9} + \dfrac{1 \cdot 6}{60 \cdot 6}$
$= \dfrac{4}{360} + \dfrac{9}{360} + \dfrac{6}{360}$
$= \dfrac{19}{360}$

12. LCD $= 4$; $2 - \dfrac{3}{4} = \dfrac{2}{1} - \dfrac{3}{4}$
$= \dfrac{2 \cdot 4}{1 \cdot 4} - \dfrac{3}{4}$
$= \dfrac{8}{4} - \dfrac{3}{4}$
$= \dfrac{5}{4}$

Section 2.6

1. $5\frac{2}{3} = 5 + \frac{2}{3}$
$= \frac{5}{1} + \frac{2}{3}$
$= \frac{5 \cdot 3}{1 \cdot 3} + \frac{2}{3}$
$= \frac{15}{3} + \frac{2}{3}$
$= \frac{17}{3}$

2. $3\frac{1}{6} = 3 + \frac{1}{6}$
$= \frac{3}{1} + \frac{1}{6}$
$= \frac{3 \cdot 6}{1 \cdot 6} + \frac{1}{6}$
$= \frac{18}{6} + \frac{1}{6}$
$= \frac{19}{6}$

3. $5\frac{2}{3} = \frac{(3 \cdot 5) + 2}{3}$
$= \frac{17}{3}$

4. $6\frac{4}{9} = \frac{(9 \cdot 6) + 4}{9}$
$= \frac{58}{9}$

5. $3\overline{)11}\;\;\frac{3}{9}$ so $\frac{11}{3} = 3\frac{2}{3}$

6. $5\overline{)14}\;\;\frac{2}{10}\;\frac{4}{4}$ so $\frac{14}{5} = 2\frac{4}{5}$

7. $26\overline{)207}\;\;\frac{7}{182}\;\frac{25}{}$ so $\frac{207}{26} = 7\frac{25}{26}$

Section 2.7

1. $2\frac{3}{4} \cdot 4\frac{1}{3} = \frac{11}{4} \cdot \frac{13}{3}$
$= \frac{143}{12}$
$= 11\frac{11}{12}$

2. $2 \cdot 3\frac{5}{8} = \frac{2}{1} \cdot \frac{29}{8}$
$= \frac{58}{8}$
$= 7\frac{2}{8}$
$= 7\frac{1}{4}$

3. $1\frac{3}{5} \div 3\frac{2}{5} = \frac{8}{5} \div \frac{17}{5}$
$= \frac{8}{5} \cdot \frac{5}{17}$
$= \frac{8}{17}$

Section 2.8

1. $3\frac{2}{3} + 2\frac{1}{4} = 3 + \frac{2}{3} + 2 + \frac{1}{4}$
$= (3 + 2) + \left(\frac{2}{3} + \frac{1}{4}\right)$
$= 5 + \left(\frac{2 \cdot 4}{3 \cdot 4} + \frac{1 \cdot 3}{4 \cdot 3}\right)$
$= 5 + \left(\frac{8}{12} + \frac{3}{12}\right)$
$= 5 + \frac{11}{12} = 5\frac{11}{12}$

2. $5\frac{3}{4} = 5\frac{3 \cdot 5}{4 \cdot 5} = 5\frac{15}{20}$
$+ 6\frac{4}{5} = 6\frac{4 \cdot 4}{5 \cdot 4} = 6\frac{16}{20}$
$11\frac{31}{20} = 11 + 1\frac{11}{20} = 12\frac{11}{20}$

3. $6\frac{3}{4} = 6\frac{3 \cdot 2}{4 \cdot 2} = 6\frac{6}{8}$
$+ 2\frac{7}{8} = 2\frac{7}{8} = 2\frac{7}{8}$
$8\frac{13}{8} = 9\frac{5}{8}$

4. $2\frac{1}{3} = 2\frac{1 \cdot 4}{3 \cdot 4} = 2\frac{4}{12}$
$1\frac{1}{4} = 1\frac{1 \cdot 3}{4 \cdot 3} = 1\frac{3}{12}$
$+ 3\frac{11}{12} = 3\frac{11}{12} = 3\frac{11}{12}$
$6\frac{18}{12} = 7\frac{6}{12} = 7\frac{1}{2}$

5. $4\frac{7}{8}$
$-1\frac{5}{8}$
$3\frac{2}{8} = 3\frac{1}{4}$

6. $12\frac{7}{10} = 12\frac{7}{10} = 12\frac{7}{10}$
$-7\frac{2}{5} = -7\frac{2 \cdot 2}{5 \cdot 2} = -7\frac{4}{10}$
$5\frac{3}{10}$

7. $10 = 9\frac{7}{7}$
$-5\frac{4}{7} = -5\frac{4}{7}$
$4\frac{3}{7}$

8. $6\frac{1}{3} = \left(5 + \frac{3}{3}\right) + \frac{1}{3} = 5\frac{4}{3}$
$-2\frac{2}{3} = \;\;\;\;\;\;\;\;\;\;\;\;\;\;\; -2\frac{2}{3} = -2\frac{2}{3}$
$3\frac{2}{3}$

9. $6\frac{3}{4} = 6\frac{3 \cdot 3}{4 \cdot 3} = 6\frac{9}{12} = 5\frac{21}{12}$
$-2\frac{5}{6} = -2\frac{5 \cdot 2}{6 \cdot 2} = -2\frac{10}{12} = -2\frac{10}{12}$
$3\frac{11}{12}$

Section 2.9

1. $4 + \left(1\frac{1}{2}\right)\left(2\frac{3}{4}\right) = 4 + \left(\frac{3}{2}\right)\left(\frac{11}{4}\right)$
$= 4 + \frac{33}{8}$
$= \frac{32}{8} + \frac{33}{8}$
$= \frac{65}{8}$
$= 8\frac{1}{8}$

2. $\left(\frac{2}{3} + \frac{1}{6}\right)\left(2\frac{5}{6} + 1\frac{1}{3}\right) = \left(\frac{5}{6}\right)\left(4\frac{1}{6}\right)$
$= \frac{5}{6}\left(\frac{25}{6}\right)$
$= \frac{125}{36}$
$= 3\frac{17}{36}$

3. $\frac{3}{7} + \frac{1}{3}\left(1\frac{1}{2} + 4\frac{1}{2}\right)^2 = \frac{3}{7} + \frac{1}{3}(6)^2$
$= \frac{3}{7} + \frac{1}{3}(36)$
$= \frac{3}{7} + 12$
$= 12\frac{3}{7}$

4. $\dfrac{\frac{2}{3}}{\frac{5}{9}} = \dfrac{2}{3} \div \dfrac{5}{9}$

$= \dfrac{2}{3} \cdot \dfrac{9}{5}$

$= \dfrac{18}{15}$

$= \dfrac{6}{5} = 1\dfrac{1}{5}$

5. $\dfrac{\frac{1}{2} + \frac{3}{4}}{\frac{2}{3} - \frac{1}{4}} = \dfrac{12\left(\frac{1}{2} + \frac{3}{4}\right)}{12\left(\frac{2}{3} - \frac{1}{4}\right)}$

$= \dfrac{12 \cdot \frac{1}{2} + 12 \cdot \frac{3}{4}}{12 \cdot \frac{2}{3} - 12 \cdot \frac{1}{4}}$

$= \dfrac{6 + 9}{8 - 3}$

$= \dfrac{15}{5} = 3$

6. $\dfrac{4 + \frac{2}{3}}{3 - \frac{1}{4}} = \dfrac{12\left(4 + \frac{2}{3}\right)}{12\left(3 - \frac{1}{4}\right)}$

$= \dfrac{12 \cdot 4 + 12 \cdot \frac{2}{3}}{12 \cdot 3 - 12 \cdot \frac{1}{4}}$

$= \dfrac{48 + 8}{36 - 3}$

$= \dfrac{56}{33} = 1\dfrac{23}{33}$

Chapter 3

Section 3.1

1. $700 + 80 + 5 + \dfrac{4}{10} + \dfrac{6}{100} + \dfrac{2}{1,000}$ **2. a.** Six hundredths **b.** Seven tenths **c.** Eight thousandths

3. a. Five and six hundredths **b.** Four and seven tenths **c.** Three and eight thousandths
4. a. Five and ninety-eight hundredths **b.** Five and ninety-eight thousandths   **5.** Three hundred five and four hundred six thousandths

Section 3.2

1. a. $38.45 = 38\dfrac{45}{100} = 38\dfrac{450}{1,000}$
$+456.073 = 456\dfrac{73}{1,000} = 456\dfrac{73}{1,000}$
$\overline{\phantom{+456.073 = 456\dfrac{73}{1,000}}}$
$494\dfrac{523}{1,000} = 494.523$

b. $38.045 = 38\dfrac{45}{1,000} = 38\dfrac{45}{1,000}$
$+456.73 = 456\dfrac{73}{100} = 456\dfrac{730}{1,000}$
$\overline{\phantom{+456.73 = 456\dfrac{73}{100}}}$
$494\dfrac{775}{1,000} = 494.775$

2. 78.674
$\underline{-23.431}$
55.243

3. 16.000
0.033
4.600
$\underline{+0.080}$
20.713

4. a. 6.70
$\underline{-2.05}$
4.65

b. 6.7000
$\underline{-2.0563}$
4.6437

5. 7.000 10.567
$\underline{+3.567}$ and $\underline{-5.890}$
10.567 4.677

6. $\$10.00$ 1 quarter + 2 dimes + 4 pennies = 0.25 + 0.20 + 0.04 = 0.49, which is too much change.   **7. a.** $P = 1.38 + 1.38 + 1.38 + 1.38 = 5.52$ in.
$\underline{-9.56}$ One of the dimes should be a nickel. Tell the clerk that you have been given too much change.
$\$.44$

b. $P = 6.6 + 4.7 + 4.7 = 16.0$ cm

Section 3.3

1. a. $0.4 \times 0.6 = \dfrac{4}{10} \times \dfrac{6}{10}$
$= \dfrac{24}{100}$
$= 0.24$

b. $0.04 \times 0.06 = \dfrac{4}{100} \times \dfrac{6}{100}$
$= \dfrac{24}{10,000}$
$= 0.0024$

2. a. $0.5 \times 0.007 = \dfrac{5}{10} \times \dfrac{7}{1,000}$
$= \dfrac{35}{10,000}$
$= 0.0035$

b. $0.05 \times 0.07 = \dfrac{5}{100} \times \dfrac{7}{100}$
$= \dfrac{35}{10,000}$
$= 0.0035$

3. a. $3.5 \times 0.04 = 3\dfrac{5}{10} \times \dfrac{4}{100}$
$= \dfrac{35}{10} \times \dfrac{4}{100}$
$= \dfrac{140}{1,000}$
$= \dfrac{14}{100}$
$= 0.14$

b. $0.35 \times 0.4 = \dfrac{35}{100} \times \dfrac{4}{10}$
$= \dfrac{140}{1,000}$
$= \dfrac{14}{100}$
$= 0.14$

4. a. $3 + 2 = 5$ digits to the right   **b.** $2 + 4 = 6$ digits to the right

5. a. 4.03
$\underline{\times5.22}$
806
8060
$\underline{201500}$
21.0366

b. 40.3
$\underline{\times0.522}$
806
8060
$\underline{201500}$
21.0366

Solutions to Selected Practice Problems

6. a. $80 \times 6 = 480$
b. $40 \times 180 = 7{,}200$
c. $8^2 = 64$

7. a. $0.03(5.5 + 0.02) = 0.03(5.52)$
$= 0.1656$

b. $0.03(0.55 + 0.002) = 0.03(0.552)$
$= 0.01656$

8. a. $5.7 + 14(2.4)^2 = 5.7 + 14(5.76)$
$= 5.7 + 80.64$
$= 86.34$

b. $0.57 + 1.4(2.4)^2 = 0.57 + 1.4(5.76)$
$= 0.57 + 8.064$
$= 8.634$

9. a. $A = s^2 = 1.38^2 = 1.90 \text{ in}^2$
b. $A = lw = (39.6)(25.1) = 993.96 \text{ mm}^2$

10. $6.82(36) + 10.23(14) = 245.52 + 143.22$
$= \$388.74$

11. $C = 3.14(3)$
$= 9.42 \text{ cm}$

12. a. $C = \pi d = (3.14)(0.92) = 2.89 \text{ in.}$
b. $C = 2\pi r = 2(3.14)(13.20) = 82.90 \text{ mm}$

13. Radius $= \frac{1}{2}(20) = 10 \text{ ft}$
$A = \pi r^2 = (3.14)(10)^2$
$= 314 \text{ ft}^2$

14. Radius $= 2(0.125) = 0.250$
$V = \pi r^2 h$
$= (3.14)(0.250)^2(6)$
$= 1.178 \text{ in}^3$

Section 3.4

1.
```
       154.2
30)4,626.0
    3 0
    1 62
    1 50
     126
     120
       6 0
       6 0
         0
```

2. a.
```
      6.7
5) 33.5
   30
    3 5
    3 5
      0
```

b.
```
      6.9
5)34.5
  30
   4 5
   4 5
     0
```

c.
```
      7.1
5)35.5
  35
   0 5
     5
     0
```

3. a.
```
       2.636
18)47.448
   36
   11 4
   10 8
      64
      54
      108
      108
        0
```

b.
```
       26.36
18)474.48
   36
   114
   108
     64
     54
     1 08
     1 08
        0
```

4. a.
```
        45.54
25)1,138.50
   1 00
     138
     125
      13 5
      12 5
       1 00
       1 00
          0
```

b.
```
        4.554
25)113.850
   100
    13 8
    12 5
     1 35
     1 25
       100
       100
         0
```

5. a.
```
          3.15
4.2.)13.2.30
     126
       6 3
       4 2
       2 10
       2 10
          0
```

b.
```
          31.5
0.42.)13.23.0
      12 6
         63
         42
        21 0
        21 0
           0
```

6.
```
          1.422
0.32.)0.45.530     Answer to nearest
       32           hundredth is 1.42
       13 5
       12 8
          73
          64
          90
          64
          26
```

7. a.
```
         3 16.66    Answer to nearest
0.06.)19.00.00      tenth is 316.7
      18
      10
       6
       40
       36
        40
        36
         40
```

b.
```
         31.66     Answer to nearest
0.06.)1.90.00       tenth is 31.7
      18
      10
       6
       40
       36
        40
        36
         4
```

8.
```
         28.5 hours
6.54.)186.39.0
      130 8
       55 59
       52 32
        3 27 0
        3 27 0
             0
```

9. $\dfrac{4.39 - 0.43}{0.33} = \dfrac{3.96}{0.33}$

$= 12$ additional minutes

The call was 13 minutes long.

Section 3.5

1. a. $5\overline{)2.0}$ gives $.4$, $\underline{2\,0}$, 0 so $\dfrac{2}{5} = 0.4$ **b.** $5\overline{)3.0}$ gives 0.6, $\underline{3\,0}$, 0 so $\dfrac{3}{5} = 0.6$ **c.** $5\overline{)4.0}$ gives $.8$, $\underline{4\,0}$, 0 so $\dfrac{4}{5} = 0.8$

2. a. $12\overline{)11.0000}$ gives 0.9166 ... so $\dfrac{11}{12} = 0.917$ to the nearest thousandth   **b.** $13\overline{)12.0000}$ gives $.9230$ ... so $\dfrac{12}{13} = 0.923$ to the nearest thousandth

3. $11\overline{)5.0000}$ gives 0.4545 ... so $\dfrac{5}{11} = 0.\overline{45}$

4. a. $0.48 = \dfrac{48}{100} = \dfrac{12}{25}$ **b.** $0.048 = \dfrac{48}{1{,}000} = \dfrac{6}{125}$

5. $0.025 = \dfrac{25}{1{,}000} = \dfrac{1}{40}$

6. $12.8 = 12\dfrac{8}{10} = 12\dfrac{4}{5}$

7. $\dfrac{14}{25}(2.43 + 0.27)$
$= 0.56(2.43 + 0.27)$
$= 0.56(2.70)$
$= 1.512$

8. $\dfrac{1}{4} + 0.25\left(\dfrac{3}{5}\right) = \dfrac{1}{4} + \dfrac{1}{4}\left(\dfrac{3}{5}\right)$
$= \dfrac{1}{4} + \dfrac{3}{20}$
$= \dfrac{5}{20} + \dfrac{3}{20}$
$= \dfrac{8}{20}$
$= \dfrac{2}{5}$ or 0.4

9. $\left(\dfrac{1}{3}\right)^3(5.4) + \left(\dfrac{1}{5}\right)^2(2.5) = \dfrac{1}{27}(5.4) + \dfrac{1}{25}(2.5) = 0.2 + 0.1 = 0.3$

10. $35.50 - \dfrac{1}{4}(35.50) = \dfrac{3}{4}(35.50) = 26.625 = \$26.63$ to the nearest cent

11. $A = \dfrac{1}{2}bh = \dfrac{1}{2}(6.6)(3.3) = 10.89$ cm²

12. $V = \pi r^2 h + \dfrac{1}{2} \cdot \dfrac{4}{3}\pi r^3 = (3.14)(10)^2(10) + \dfrac{1}{2} \cdot \dfrac{4}{3}(3.14)(10)^3$
$= 3{,}140 + \dfrac{2}{3}(3{,}140)$
$= 3{,}140 + 2{,}093.3$
$= 5{,}233.3$ in³

Section 3.6

1. $4\sqrt{25} = 4 \cdot 5$
$= 20$

2. $\sqrt{36} + \sqrt{4} = 6 + 2$
$= 8$

3. $\sqrt{\dfrac{36}{100}} = \dfrac{6}{10}$
$= \dfrac{3}{5}$

4. $14\sqrt{36} = 14 \cdot 6$
$= 84$

5. $\sqrt{81} - \sqrt{25} = 9 - 5$
$= 4$

6. $\sqrt{\dfrac{64}{121}} = \dfrac{8}{11}$

7. $5\sqrt{14} \approx 5(3.7416574)$
$= 18.708287$
$= 18.7083$ to the nearest ten thousandth

8. $\sqrt{405} + \sqrt{147} \approx 20.124612 + 12.124356$
$= 32.248968$
$= 32.25$ to the nearest hundredth

9. $\sqrt{\dfrac{7}{12}} \approx \sqrt{0.5833333}$
$= 0.7637626$
$= 0.764$ to the nearest thousandth

10. a. $c = \sqrt{5^2 + 5^2}$
$= \sqrt{25 + 25}$
$= \sqrt{50}$
$= 7.07$ ft to the nearest hundredth

b. $c = \sqrt{16^2 + 12^2}$
$= \sqrt{256 + 144}$
$= \sqrt{400}$
$= 20$ cm

11. $c = \sqrt{12^2 + 5^2}$
$= \sqrt{144 + 25}$
$= \sqrt{169}$
$= 13$ ft

Chapter 4

Section 4.1

1. a. $\dfrac{32}{48} = \dfrac{2}{3}$ b. $\dfrac{3.2}{4.8} = \dfrac{3.2 \times 10}{4.8 \times 10}$ c. $\dfrac{0.32}{0.48} = \dfrac{0.32 \times 100}{0.48 \times 100}$ 2. a. $\dfrac{\frac{3}{5}}{\frac{9}{10}} = \dfrac{3}{5} \cdot \dfrac{10}{9} = \dfrac{2}{3}$ b. $\dfrac{0.6}{0.9} = \dfrac{0.6 \times 10}{0.9 \times 10}$

$\phantom{1. a. \dfrac{32}{48} = \dfrac{2}{3}\ \ \ \ }= \dfrac{32}{48}= \dfrac{32}{48}= \dfrac{6}{9}$

$\phantom{1. a. \dfrac{32}{48} = \dfrac{2}{3}\ \ \ \ }= \dfrac{2}{3}= \dfrac{2}{3}= \dfrac{2}{3}$

3. a. $\dfrac{0.06}{0.12} = \dfrac{0.06 \times 100}{0.12 \times 100} = \dfrac{6}{12} = \dfrac{1}{2}$ b. $\dfrac{600}{1{,}200} = \dfrac{1}{2}$ 4. $\dfrac{12}{16} = \dfrac{3}{4}$

5. Alcohol to water: $\dfrac{4}{12} = \dfrac{1}{3}$; water to alcohol: $\dfrac{12}{4} = \dfrac{3}{1}$; water to total solution: $\dfrac{12}{16} = \dfrac{3}{4}$

Section 4.2

1. $\dfrac{107 \text{ miles}}{2 \text{ hours}} = 53.5$ miles/hour 2. $\dfrac{192 \text{ miles}}{6 \text{ gallons}} = 32$ miles/gallon

3. $\dfrac{48¢}{5.5 \text{ ounces}} = 8.7¢/\text{ounce}$; $\dfrac{75¢}{11.5 \text{ ounces}} = 6.5¢/\text{ounce}$; $\dfrac{219¢}{46 \text{ ounces}} = 4.8¢/\text{ounce}$

(Answers are rounded to the nearest tenth.)

Section 4.3

1. $\dfrac{8 \cdot n}{8} = n$ 2. $\dfrac{3 \cdot y}{3} = y$ 3. $\dfrac{8 \cdot n}{8} = \dfrac{40}{8}$ 4. $\dfrac{35}{7} = \dfrac{7 \cdot a}{7}$

$n = 5 5 = a$

Section 4.4

1. First term = 2, second term = 3, third term = 6, fourth term = 9, means: 3 and 6; extremes: 2 and 9

2. a. $5 \cdot 18 = 90$ 3. a. $3 \cdot x = 4 \cdot 9$ 4. $2 \cdot 19 = 8 \cdot y$ 5. a. $15 \cdot n = 6(0.3)$ 6. a. $\dfrac{3}{4} \cdot 8 = 7 \cdot x$ 7. $\dfrac{b}{18} = \dfrac{0.5}{1}$

$6 \cdot 15 = 90 3 \cdot x = 36 38 = 8 \cdot y 15 \cdot n = 1.8 6 = 7 \cdot x b \cdot 1 = 18(0.5)$

b. $13 \cdot 3 = 39$ $\dfrac{3 \cdot x}{3} = \dfrac{36}{3}$ $\dfrac{38}{8} = \dfrac{8 \cdot y}{8}$ $\dfrac{\cancel{15} \cdot n}{\cancel{15}} = \dfrac{1.8}{15}$ $\dfrac{6}{7} = \dfrac{7 \cdot x}{7}$ $b = 9$

$39 \cdot 1 = 39 x = 12 4.75 = y n = 0.12 \dfrac{6}{7} = x$

b. $5 \cdot x = 8 \cdot 3$ b. $0.35(100) = n \cdot 7$

c. $\dfrac{2}{3} \cdot 5 = \dfrac{10}{3}$ $5 \cdot x = 24$ $35 = n \cdot 7$ b. $6 \cdot x = \dfrac{3}{5} \cdot 15$

$\dfrac{5}{3} \cdot 2 = \dfrac{10}{3}$ $\dfrac{5 \cdot x}{5} = \dfrac{24}{5}$ $\dfrac{35}{7} = \dfrac{n \cdot 7}{7}$ $6 \cdot x = 9$

d. $0.12(3) = 0.36$ $x = 4.8$ $5 = n$ $\dfrac{6 \cdot x}{6} = \dfrac{9}{6}$

$0.18(2) = 0.36$ $x = \dfrac{3}{2}$

Section 4.5

1. a. $\dfrac{x}{54} = \dfrac{10}{18}$
$x \cdot 18 = 54 \cdot 10$
$x \cdot 18 = 540$
$\dfrac{x \cdot \cancel{18}}{\cancel{18}} = \dfrac{540}{18}$
$x = 30$ hits

b. $\dfrac{x}{27} = \dfrac{10}{18}$
$x \cdot 18 = 27 \cdot 10$
$x \cdot 18 = 270$
$\dfrac{x \cdot \cancel{18}}{\cancel{18}} = \dfrac{270}{18}$
$x = 15$ hits

2. a. $\dfrac{x}{10} = \dfrac{288}{6}$
$x \cdot 6 = 10 \cdot 288$
$x \cdot 6 = 2{,}880$
$\dfrac{x \cdot 6}{6} = \dfrac{2{,}880}{6}$
$x = 480$ miles

b. $\dfrac{x}{11} = \dfrac{288}{6}$
$x \cdot 6 = 11 \cdot 288$
$x \cdot 6 = 3{,}168$
$\dfrac{x \cdot 6}{6} = \dfrac{3{,}168}{6}$
$x = 528$ miles

3. $\dfrac{x}{4.75} = \dfrac{105}{1}$
$x \cdot 1 = 4.75(105)$
$x = 498.75$ miles

Section 4.6

1. $\dfrac{x}{14} = \dfrac{25}{10}$
$10x = 350$
$x = 35$

2. $\dfrac{h}{120} = \dfrac{360}{160}$
$\dfrac{h}{120} = \dfrac{9}{4}$ reduce to lowest terms
$4h = 1080$
$h = 270$ pixels

3. We see AC has a length of 3 units and BC has a length of 4 units. Since AC is proportional to GI, which has a length of 9 units, we set up a proportion to find the length of a new side, HI.

$\dfrac{HI}{BC} = \dfrac{GI}{AC}$

$\dfrac{HI}{4} = \dfrac{9}{3}$

$\dfrac{HI}{4} = 3$

$HI = 12$

Now we can draw HI with length 12 units, and complete the triangle by drawing line GH

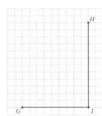 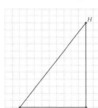

4. $\dfrac{x}{18} = \dfrac{42}{12}$
$12x = 756$
$x = 63$ ft

5. $\dfrac{b}{15} = \dfrac{32}{24}$
$24b = 480$
$b = 20$ in.

Chapter 5

Section 5.1

21. $\dfrac{82}{100} = \dfrac{41}{50}$

22. $\dfrac{6.5}{100} = \dfrac{65}{1000} = \dfrac{13}{200}$

23. $\dfrac{42\frac{1}{2}}{100} = \dfrac{\frac{85}{2}}{100}$
$= \dfrac{85}{2} \cdot \dfrac{1}{100}$
$= \dfrac{85}{200}$
$= \dfrac{17}{40}$

24. a. $\dfrac{9}{10} = 0.9 = 90\%$

b. $\dfrac{9}{20} = 0.45 = 45\%$

25. a. $\dfrac{5}{8} = 0.625 = 62.5\%$

b. $\dfrac{9}{8} = 1.125 = 112.5\%$

26. a. $\dfrac{7}{12} = 0.58\overline{3} = 58\dfrac{1}{3}\%$ or $\approx 58.3\%$

b. $\dfrac{13}{12} = 1.08\overline{3} = 108\dfrac{1}{3}\%$ or $\approx 108.3\%$

Section 5.2

1. a. $n = 0.25(74)$
$= 18.5$

b. $n = 0.50(74)$
$= 37$

2. a. $n \cdot 84 = 21$
$\dfrac{n \cdot 84}{84} = \dfrac{21}{84}$
$n = 0.25$
$n = 25\%$

b. $n \cdot 84 = 42$
$\dfrac{n \cdot 84}{84} = \dfrac{42}{84}$
$n = 0.50$
$n = 50\%$

3. a. $35 = 0.40 \cdot n$
$\dfrac{35}{0.40} = \dfrac{0.40 \cdot n}{0.40}$
$87.5 = n$

b. $70 = 0.40 \cdot n$
$\dfrac{70}{0.40} = \dfrac{0.40 \cdot n}{0.40}$
$175 = n$

4. $n = 0.635(45)$
$n \approx 28.6$

5. $n \cdot 85 = 11.9$
$\dfrac{n \cdot 85}{85} = \dfrac{11.9}{85}$
$n = 0.14$
$n = 14\%$

6. $62 = 0.39 \cdot n$
$\dfrac{62}{0.39} = \dfrac{0.39 \cdot n}{0.39}$
$159.0 \approx n$

7. 25 is what percent of 160?
$25 = n \cdot 160$
$\dfrac{25}{160} = n$
$n = 0.156 = 15.6\%$ to the nearest tenth of a percent

8. a. $\dfrac{25}{100} = \dfrac{n}{74}$
$25 \cdot 74 = 100 \cdot n$
$1850 = 100 \cdot n$
$18.5 = n$

b. $\dfrac{50}{100} = \dfrac{n}{74}$
$50 \cdot 74 = 100 \cdot n$
$3{,}700 = 100 \cdot n$
$37 = n$

9. a. $\dfrac{n}{100} = \dfrac{21}{84}$
$84 \cdot n = 21 \cdot 100$
$84 \cdot n = 2100$
$n = 25$
25%

b. $\dfrac{n}{100} = \dfrac{42}{84}$
$84 \cdot n = 42 \cdot 100$
$84 \cdot n = 4{,}200$
$n = 50$
50%

10. a. $\dfrac{40}{100} = \dfrac{35}{n}$
$40 \cdot n = 35 \cdot 100$
$40 \cdot n = 3500$
$n = 87.5$

b. $\dfrac{40}{100} = \dfrac{70}{n}$
$40 \cdot n = 70 \cdot 100$
$40 \cdot n = 7{,}000$
$n = 175$

Section 5.3

1. 114 is what percent of 150?
$114 = n \cdot 150$
$\dfrac{114}{150} = \dfrac{n \cdot 150}{150}$
$n = 0.76$
$n = 76\%$

2. What is 75% of 40?
$n = 0.75(40)$
$n = 30$ milliliters HCl

3. 3,360 is 42% of what number?
$3{,}360 = 0.42 \cdot n$
$\dfrac{3{,}360}{0.42} = \dfrac{0.42 \cdot n}{0.42}$
$n = 8{,}000$ students

4. What is 35% of 300?
$n = 0.35(300)$
$= 105$ students

Section 5.4

1. What is 6% of $625?
$n = 0.06(625)$
$n = \$37.50$

2. $4.35 is 3% of what number?
$4.35 = 0.03 \cdot n$
$n = \$145$ Purchase price
Total price $= \$145 + \$4.35$
$= \$149.35$

3. $11.82 is what percent of $197?
$11.82 = n \cdot 197$
$n = 0.06$
$= 6\%$

4. What is 3% of 115,000?
$n = 0.03(115{,}000)$
$n = \$3{,}450$

5. 10% of what number is $115?
$0.10 \cdot n = 115$
$n = \$1{,}150$

6. $105 is what percent of $750?
$105 = n \cdot 750$
$n = 0.14$
$= 14\%$

Section 5.5

1. $0.07(18{,}000) = 1{,}260$

$18,000	Old salary
+ 1,260	Raise
$19,260	New salary

2. $0.044(11{,}180) = 492$

11,180	Drunk drivers in 1997
− 492	Decrease
10,688 ≈ 10,700 to the nearest hundred	

3. $35 − \$28 = \$7$ Decrease
$7 is what percent of $35?
$7 = n \cdot 35$
$n = 0.20 = 20\%$ Decrease

4. What is 15% of $550?
$n = 0.15(550)$
$n = \$82.50$ Discount

$550.00	Original price
− 82.50	Less discount
$467.50	Sale price

5. What is 15% of $45?
$n = 0.15(45)$
$n = \$6.75$ Discount

$45.00	Original price
− 6.75	Less discount
$38.25	Sale price

What is 5% of $38.25?
$n = 0.05(38.25)$
$n = \$1.91$ to the nearest cent

$38.25	Sale price
+ 1.91	Sales tax
$40.16	Total price

Section 5.6

1. Interest = 0.08($3,000)
       = $240
   $3,000 Principal
   + 240 Interest
   $3,240 Amount after 1 year

2. Interest = 0.12($7,500)
       = $900
   $7,500 Principal
   + 900 Interest
   $8,400 Total amount to pay back

3. 0.024(5,243.56) = $125.85 finance charge

4. $I = P \cdot R \cdot T$
   $I = 700 \times 0.04 \times \frac{90}{360}$
   $I = 700 \times 0.04 \times \frac{1}{4}$
   $I = \$7$ Interest

5. $I = P \cdot R \cdot T$
   $I = 1{,}200 \times 0.095 \times \frac{120}{360}$
   $I = 1{,}200 \times 0.095 \times \frac{1}{3}$
   $I = \$38$ Interest
   $1,200 Principal
   +   38 Interest
   $1,238 Total amount withdrawn

6. Interest after 1 year is
   0.06($5,000) = $300
   Total in account after 1 year is
   $5,000 Principal
   + 300 Interest
   $5,300

   Interest paid the second year is
   0.06 ($5,300) = $318
   Total in account after 2 years is
   $5,300 Principal
   + 318 Interest
   $5,618

7. Interest at the end of first quarter
   $I = \$20{,}000 \times 0.08 \times \frac{1}{4} = \$400$
   Total in account at end of first quarter
   $20,000 + $400 = $20,400
   Interest for the second quarter
   $I = \$20{,}400 \times 0.08 \times \frac{1}{4} = \$408$
   Total in account at end of second quarter
   $20,400 + $408 = $20,808

   Interest for the third quarter
   $I = \$20{,}808 \times 0.08 \times \frac{1}{4} = \$416.16$
   Total in account at the end of third quarter
   $20,808 + $416.16 = $21,224.16
   Interest for the fourth quarter
   $I = \$21{,}224.16 \times 0.08 \times \frac{1}{4} = \$424.48$ to the nearest cent
   Total in account at end of 1 year
   $21,224.16
   +   424.48
   $21,648.64

Section 5.7

1. a. 9 + 11 + 16 + 10 = 46

2. a. 0.76(600) = 456 people
   b. 0.24(600) = 144 people

3.

Chapter 6

Section 6.1

1. 8 ft = 8 × 12 in.
   = 96 in.

2. 26 ft = 26 ft × $\frac{1 \text{ yd}}{3 \text{ ft}}$
   = $\frac{26}{3}$ yd
   = $8\frac{2}{3}$ yd, or 8.67 yd, rounded to the nearest hundredth

3. 220 yd = 220 yd × $\frac{3 \text{ ft}}{1 \text{ yd}}$ × $\frac{12 \text{ in.}}{1 \text{ ft}}$
   = 220 × 3 × 12 in.
   = 7,920 in.

4. 67 cm = 67 cm × $\frac{1 \text{ m}}{100 \text{ cm}}$
   = $\frac{67 \text{ m}}{100}$
   = 0.67 m

5. 78.4 mm = 78.4 mm × $\frac{1 \text{ m}}{1{,}000 \text{ mm}}$ × $\frac{10 \text{ dm}}{1 \text{ m}}$
   = $\frac{78.4 \times 10}{1{,}000}$ dm
   = 0.784 dm

6. 6 pens = 6 pens × $\frac{28 \text{ feet of fencing}}{1 \text{ pen}}$ × $\frac{1.72 \text{ dollars}}{1 \text{ foot of fencing}}$
   = 6 × 28 × 1.72 dollars
   = $288.96

7. 1,100 feet per minute = $\frac{1{,}100 \text{ feet}}{1 \text{ minute}} \cdot \frac{1 \text{ mile}}{5{,}280 \text{ feet}} \cdot \frac{60 \text{ minutes}}{1 \text{ hour}}$
   = $1{,}100 \cdot \frac{60 \text{ miles}}{5{,}280 \text{ hours}}$
   = 12.5 miles per hour, which is a reasonable speed for a chair lift.

Section 6.2

1. $1 \text{ yd}^2 = 1 \text{ yd} \times \text{yd} \times \dfrac{3 \text{ ft}}{1 \text{ yd}} \times \dfrac{3 \text{ ft}}{1 \text{ yd}} = 3 \text{ ft} \times 3 \text{ ft} = 9 \text{ ft}^2$

2. Length = 36 in. + 12 in. = 48 in.; Width = 24 in. + 12 in. = 36 in.;
   Area = 48 in. × 36 in. = 1,728 in²
   Area in square feet = $1{,}728 \text{ in}^2 \times \dfrac{1 \text{ ft}^2}{144 \text{ in}^2} = \dfrac{1{,}728}{144} \text{ ft}^2 = 12 \text{ ft}^2$

3. $A = 1.5 \times 45$
   $= 67.5 \text{ yd}^2$
   $= 67.5 \text{ yd}^2 \times \dfrac{9 \text{ ft}^2}{1 \text{ yd}^2}$
   $= 607.5 \text{ ft}^2$

4. $55 \text{ acres} = 55 \text{ acres} \times \dfrac{43{,}560 \text{ ft}^2}{1 \text{ acre}}$
   $= 55 \times 43{,}560 \text{ ft}^2$
   $= 2{,}395{,}800 \text{ ft}^2$

5. $960 \text{ acres} = 960 \text{ acres} \times \dfrac{1 \text{ mi}^2}{640 \text{ acres}}$
   $= \dfrac{960}{640} \text{ mi}^2$
   $= 1.5 \text{ mi}^2$

6. $1 \text{ m}^2 = 1 \text{ m}^2 \times \dfrac{100 \text{ dm}^2}{1 \text{ m}^2} \times \dfrac{100 \text{ cm}^2}{1 \text{ dm}^2}$
   $= 10{,}000 \text{ cm}^2$

7. $5 \text{ gal} = 5 \text{ gal} \times \dfrac{4 \text{ qt}}{1 \text{ gal}} \times \dfrac{2 \text{ pt}}{1 \text{ qt}}$
   $= 5 \times 4 \times 2 \text{ qt}$
   $= 40 \text{ pt}$

8. $2{,}000 \text{ qt} = 2{,}000 \text{ qt} \times \dfrac{1 \text{ gal}}{4 \text{ qt}}$
   $= \dfrac{2{,}000}{4} \text{ gal}$
   $= 500 \text{ gal}$
   The number of 10-gal containers in 500 gal is $\dfrac{500}{10} = 50$ containers.

9. $3.5 \text{ liters} = 3.5 \text{ liters} \times \dfrac{1{,}000 \text{ mL}}{1 \text{ liter}}$
   $= 3.5 \times 1{,}000 \text{ mL}$
   $= 3{,}500 \text{ mL}$

Section 6.3

1. $15 \text{ lb} = 15 \text{ lb} \times \dfrac{16 \text{ oz}}{1 \text{ lb}}$
   $= 15 \times 16 \text{ oz}$
   $= 240 \text{ oz}$

2. $5 \text{ T} = 5 \text{ T} \times \dfrac{2{,}000 \text{ lb}}{1 \text{ T}}$
   $= 10{,}000 \text{ lb}$
   10,000 lb is the equivalent of 5 tons.

3. $5 \text{ kg} = 5 \text{ kg} \times \dfrac{1{,}000 \text{ g}}{1 \text{ kg}} \times \dfrac{1{,}000 \text{ mg}}{1 \text{ g}}$
   $= 5 \times 1{,}000 \times 1{,}000 \text{ mg}$
   $= 5{,}000{,}000 \text{ mg}$

Section 6.4

1. $10 \text{ in.} = 10 \text{ in.} \times \dfrac{2.54 \text{ cm}}{1 \text{ in.}}$
   $= 10 \times 2.54 \text{ cm}$
   $= 25.4 \text{ cm}$

2. $16.4 \text{ ft} = 16.4 \text{ ft} \times \dfrac{1 \text{ m}}{3.28 \text{ ft}}$
   $= \dfrac{16.4}{3.28} \text{ m}$
   $= 5 \text{ m}$

3. $10 \text{ liters} = 10 \text{ liters} \times \dfrac{1 \text{ gal}}{3.79 \text{ liters}}$
   $= \dfrac{10}{3.79} \text{ gal}$
   $= 2.64 \text{ gal}$ (rounded to the nearest hundredth)

4. $2.2 \text{ liters} = 2.2 \text{ liters} \times \dfrac{1{,}000 \text{ mL}}{1 \text{ liter}} \times \dfrac{1 \text{ in}^3}{16.39 \text{ mL}}$
   $= \dfrac{2.2 \times 1{,}000}{16.39} \text{ in}^3$
   $= 134 \text{ in}^3$ (rounded to the nearest cubic inch)

5. $165 \text{ lb} = 165 \text{ lb} \times \dfrac{1 \text{ kg}}{2.20 \text{ lb}}$
   $= \dfrac{165}{2.20} \text{ kg}$
   $= 75 \text{ kg}$

6. $F = \dfrac{9}{5}(40) + 32$
   $= 72 + 32$
   $= 104°F$

7. $C = \dfrac{5(101.6 - 32)}{9}$
   $= \dfrac{5(69.6)}{9}$
   $= 38.7°C$ (rounded to the nearest tenth)

Section 6.5

1. a. $2 \text{ hr } 45 \text{ min} = 2 \text{ hr} \times \dfrac{60 \text{ min}}{1 \text{ hr}} + 45 \text{ min}$
$= 120 \text{ min} + 45 \text{ min}$
$= 165 \text{ min}$

b. $2 \text{ hr } 45 \text{ min} = 2 \text{ hr} + 45 \text{ min} \times \dfrac{1 \text{ hr}}{60 \text{ min}}$
$= 2 \text{ hr} + 0.75 \text{ hr}$
$= 2.75 \text{ hr}$

2.
```
   4 min   27 sec
 + 8 min   46 sec
  12 min   73 sec
```
Since there are 60 seconds in every minute, we write 73 seconds as 1 minute 13 seconds. We have
12 min 73 sec = 12 min + 1 min 13 sec
= 13 min 13 sec

3.
```
  6 hr  25 min      5 hr  85 min
 −      42 min  =  −      42 min
                    5 hr  43 min
```

4. First, we multiply each unit by 4:
```
  3 lb   8 oz
 ×       4
 12 lb  32 oz
```
To convert the 32 ounces to pounds, we multiply the ounces by the conversion factor
$12 \text{ lb } 32 \text{ oz} = 12 \text{ lb} + 32 \text{ oz} \times \dfrac{1 \text{ lb}}{16 \text{ oz}}$
$= 12 \text{ lb} + 2 \text{ lbs}$
$= 14 \text{ lb}$
Finally, we multiply the 14 lb and $5.00 for a total price of $70.00.

5. First, we multiply each unit by 3:
```
  5 ft   8 in.
 ×       3
 15 ft  24 in.
```
To convert the 24 inches to feet, we multiply the inches by the conversion factor
$15 \text{ ft } 24 \text{ in.} = 15 \text{ ft} + 12 \text{ in.} \times \dfrac{1 \text{ ft}}{12 \text{ in}}$
$= 15 \text{ ft} + 2 \text{ ft}$
$= 17 \text{ ft}$

Chapter 7

Section 7.2

1. $2 + (-5) = -3$

2. $-2 + 5 = 3$

3. $-2 + (-5) = -7$

4. $2 + 6 = 8$

5. $2 + (-6) = -4$

6. $-2 + 6 = 4$

7. $-2 + (-6) = -8$

8. $15 + 12 = 27$
$15 + (-12) = 3$
$-15 + 12 = -3$
$-15 + (-12) = -27$

9. $12 + (-3) + (-7) + 5 = 9 + (-7) + 5$
$= 2 + 5$
$= 7$

10. $[-2 + (-12)] + [7 + (-5)] = [-14] + [2]$
$= -12$

11. $-5.76 + (-3.24) = -9.00$

12. $6.88 + (-8.55) = -1.67$

13. $\dfrac{5}{6} + \left(-\dfrac{2}{6}\right) = \dfrac{3}{6} = \dfrac{1}{2}$

14. $\dfrac{1}{2} + \left(-\dfrac{3}{4}\right) + \left(-\dfrac{5}{8}\right) = \dfrac{1 \cdot 4}{2 \cdot 4} + \left(-\dfrac{3 \cdot 2}{4 \cdot 2}\right) + \left(-\dfrac{5}{8}\right)$
$= \dfrac{4}{8} + \left(-\dfrac{6}{8}\right) + \left(-\dfrac{5}{8}\right)$
$= -\dfrac{2}{8} + \left(-\dfrac{5}{8}\right)$
$= -\dfrac{7}{8}$

Section 7.3

1. $7 - 3 = 7 + (-3)$
$= 4$

2. $-7 - 3 = -7 + (-3)$
$= -10$

3. $-8 - 6 = -8 + (-6)$
$= -14$

4. $10 - (-6) = 10 + 6$
$= 16$

5. $-10 - (-15) = -10 + 15$
$= 5$

6. a. $8 - 5 = 8 + (-5)$
$= 3$

b. $-8 - 5 = -8 + (-5)$
$= -13$

c. $8 - (-5) = 8 + 5$
$= 13$

d. $-8 - (-5) = -8 + 5$
$= -3$

e. $12 - 10 = 12 + (-10)$
$= 2$

f. $-12 - 10 = -12 + (-10)$
$= -22$

g. $12 - (-10) = 12 + 10$
$= 22$

h. $-12 - (-10) = -12 + 10$
$= -2$

7. $-4 + 6 - 7 = -4 + 6 + (-7)$
$= 2 + (-7)$
$= -5$

8. $15 - (-5) - 8 = 15 + 5 + (-8)$
$= 20 + (-8)$
$= 12$

9. $-8 - 2 = -8 + (-2)$
$= -10$

10. $7 - (-5) = 7 + 5$
$= 12$

11. $-8 - (-6) = -8 + 6$
$= -2$

12. $-57.8 - 70.4 = -57.8 + (-70.4)$
$= -128.2$

13. $-\dfrac{5}{8} - \dfrac{3}{8} = -\dfrac{5}{8} + \left(-\dfrac{3}{8}\right)$
$= -\dfrac{8}{8}$
$= -1$

Section 7.4

1. $2(-6) = (-6) + (-6)$
$= -12$

2. $-2(6) = 6(-2) = (-2) + (-2) + (-2) + (-2) + (-2) + (-2)$
$= -12$

3. $-2(-6) = 12$

10. $-5(2)(-4) = -10(-4)$
$= 40$

11. $\dfrac{3}{4}\left(-\dfrac{4}{7}\right) = -\dfrac{3 \cdot 4}{4 \cdot 7} = -\dfrac{3}{7}$

12. $\left(-\dfrac{5}{6}\right)\left(-\dfrac{9}{20}\right) = \dfrac{5 \cdot 9}{6 \cdot 20} = \dfrac{5 \cdot 3 \cdot 3}{2 \cdot 3 \cdot 4 \cdot 5} = \dfrac{3}{8}$

13. $(-3)(6.7) = -20.1$

14. $(-0.6)(-0.5) = 0.30$

15. a. $(-8)^2 = (-8)(-8) = 64$
b. $-8^2 = -8 \cdot 8 = -64$
c. $(-3)^3 = (-3)(-3)(-3) = -27$
d. $-3^3 = -3 \cdot 3 \cdot 3 = -27$

16. $-2[5 + (-8)] = -2[-3]$
$= 6$

17. $-3 + 4(-7 + 3) = -3 + 4(-4)$
$= -3 + (-16)$
$= -19$

18. $-3(5) + 4(-4) = -15 + (-16)$
$= -31$

19. $-2(3 - 5) - 7(-2 - 4) = -2(-2) - 7(-6)$
$= 4 - (-42)$
$= 4 + 42$
$= 46$

20. $(-6 - 1)(4 - 9) = (-7)(-5)$
$= 35$

Section 7.5

6. $\dfrac{8(-5)}{-4} = \dfrac{-40}{-4}$
$= 10$

7. $\dfrac{-20 + 6(-2)}{7 - 11} = \dfrac{-20 + (-12)}{-4}$
$= \dfrac{-32}{-4}$
$= 8$

8. $-3(4^2) + 10 \div (-5) = -3(16) + 10 \div (-5)$
$= -48 + (-2)$
$= -50$

9. $-80 \div 2 \div 10 = -40 \div 10$
$= -4$

Section 7.6

1. $5(7a) = (5 \cdot 7)a$
$= 35a$

2. $-3(9x) = (-3 \cdot 9)x$
$= -27x$

3. $5(-8y) = [5(-8)]y$
$= -40y$

4. $6 + (9 + x) = (6 + 9) + x$
$= 15 + x$

5. $(3x + 7) + 4 = 3x + (7 + 4)$
$= 3x + 11$

6. $6(x + 4) = 6(x) + 6(4)$
$= 6x + 24$

7. $7(a - 5) = 7(a) - 7(5)$
$= 7a - 35$

8. $6(4x + 5) = 6(4x) + 6(5)$
$= (6 \cdot 4)x + 6(5)$
$= 24x + 30$

9. $3(8a - 4) = 3(8a) - 3(4)$
$= 24a - 12$

10. $8(3x + 4y) = 8(3x) + 8(4y)$
$= 24x + 32y$

11. $-3b + 5b = (-3 + 5)b$
$= 2b$

12. $5a - 4 - 3a$
$= 5a - 3a - 4$
$= (5 - 3)a - 4$
$= 2a - 4$

13. $A = s^2 = 12^2 = 144 \text{ ft}^2$
$P = 4s = 4(12) = 48 \text{ ft}$

14. $A = lw = 100(53) = 5{,}300 \text{ yd}^2$
$P = 2l + 2w = 2(100) + 2(53) = 200 + 106 = 306 \text{ yd}$

Chapter 8

Section 8.1

1. $6(x + 4) = 6(x) + 6(4)$
$= 6x + 24$

2. $-3(2x + 4) = -3(2x) + (-3)(4)$
$= -6x + (-12)$
$= -6x - 12$

3. $\frac{1}{2}(2x - 4) = \frac{1}{2}(2x) - \frac{1}{2}(4)$
$= x - 2$

4. $6x - 2 + 3x + 8 = 6x + 3x + (-2) + 8$
$= 9x + 6$

5. $2(4x + 3) + 7 = 2(4x) + 2(3) + 7$
$= 8x + 6 + 7$
$= 8x + 13$

6. $3(2x + 1) + 5(4x - 3) = 3(2x) + 3(1) + 5(4x) - 5(3)$
$= 6x + 3 + 20x - 15$
$= 26x - 12$

7 a. When → $x = 3$
the expression → $4x - 7$
becomes → $4(3) - 7 = 12 - 7$
$= 5$

b. When → $x = -2$
the expression → $2x - 5 + 6x$
becomes → $2(-2) - 5 + 6(-2) = -4 - 5 - 12$
$= -21$

c. When → $y = -2$
the expression → $y^2 - 10y + 25$
becomes → $(-2)^2 - 10(-2) + 25$
$= 4 + 20 + 25$
$= 49$

8. $A = lw$
$= 25(8 + 2x)$
$= 25(8) + 25(2x)$
$= 200 + 50x \text{ cm}^2$

9. a. $x = 90° - 45° = 45°$
b. $x = 180° - 60° = 120°$

Section 8.2

1. When → $x = 3$
the equation → $5x - 4 = 11$
becomes → $5(3) - 4 = 11$
$15 - 4 = 11$
$11 = 11$

2. When → $a = -3$
the equation → $6a - 3 = 2a + 4$
becomes → $6(-3) - 3 = 2(-3) + 4$
$-18 - 3 = -6 + 4$
$-21 = -2$
This is a false statement, so $a = -3$ is not a solution.

3. $x + 5 = -2$
$x + 5 + (-5) = -2 + (-5)$
$x + 0 = -7$
$x = -7$

4. $a - 2 = 7$
$a - 2 + 2 = 7 + 2$
$a + 0 = 9$
$a = 9$

5. $y + 6 - 2 = 8 - 9$
$y + 4 = -1$
$y + 4 + (-4) = -1 + (-4)$
$y + 0 = -5$
$y = -5$

6. $5x - 3 - 4x = 4 - 7$
$x - 3 = -3$
$x - 3 + 3 = -3 + 3$
$x + 0 = 0$
$x = 0$

7. $-5 - 7 = x + 2$
$-12 = x + 2$
$-12 + (-2) = x + 2 + (-2)$
$-14 = x + 0$
$-14 = x$

8. $a - \frac{2}{3} = \frac{5}{6}$
$a - \frac{2}{3} + \frac{2}{3} = \frac{5}{6} + \frac{2}{3}$
$a = \frac{9}{6} = \frac{3}{2}$

9. $5(3a - 4) - 14a = 25$
$15a - 20 - 14a = 25$
$a - 20 = 25$
$a - 20 + 20 = 25 + 20$
$a = 45$

Section 8.3

1. $\frac{1}{3}x = 5$
$3 \cdot \frac{1}{3}x = 3 \cdot 5$
$x = 15$

2. $\frac{1}{5}a + 3 = 7$
$\frac{1}{5}a + 3 + (-3) = 7 + (-3)$
$\frac{1}{5}a = 4$
$5 \cdot \frac{1}{5}a = 5 \cdot 4$
$a = 20$

3. $\frac{3}{5}y = 6$
$\frac{5}{3} \cdot \frac{3}{5}y = \frac{5}{3} \cdot 6$
$y = 10$

4. $-\frac{3}{4}x = \frac{6}{5}$
$-\frac{4}{3}\left(-\frac{3}{4}x\right) = -\frac{4}{3} \cdot \frac{6}{5}$
$x = -\frac{8}{5}$

5. $6x = -42$
$\frac{6x}{6} = \frac{-42}{6}$
$x = -7$

6. $-5x + 6 = -14$
$-5x + 6 + (-6) = -14 + (-6)$
$-5x = -20$
$\frac{-5x}{-5} = \frac{-20}{-5}$
$x = 4$

7. $3x - 7x + 5 = 3 - 18$
$-4x + 5 = -15$
$-4x + 5 + (-5) = -15 + (-5)$
$-4x = -20$
$\frac{-4x}{-4} = \frac{-20}{-4}$
$x = 5$

8. $-5 + 4 = 2x - 11 + 3x$
$-1 = 5x - 11$
$-1 + 11 = 5x - 11 + 11$
$10 = 5x$
$\frac{10}{5} = \frac{5x}{5}$
$2 = x$

Section 8.4

1.
$4(x + 3) = -8$
$4x + 12 = -8$
$4x + 12 + (-12) = -8 + (-12)$
$4x = -20$
$\dfrac{4x}{4} = \dfrac{-20}{4}$
$x = -5$

2.
$6a + 7 = 4a - 3$
$6a + (-4a) + 7 = 4a + (-4a) - 3$
$2a + 7 = -3$
$2a + 7 + (-7) = -3 + (-7)$
$2a = -10$
$\dfrac{2a}{2} = \dfrac{-10}{2}$
$a = -5$

3.
$5(x - 2) + 3 = -12$
$5x - 10 + 3 = -12$
$5x - 7 = -12$
$5x - 7 + 7 = -12 + 7$
$5x = -5$
$\dfrac{5x}{5} = \dfrac{-5}{5}$
$x = -1$

4.
$3(4x - 5) + 6 = 3x + 9$
$12x - 15 + 6 = 3x + 9$
$12x - 9 = 3x + 9$
$12x + (-3x) - 9 = 3x + (-3x) + 9$
$9x - 9 = 9$
$9x - 9 + 9 = 9 + 9$
$9x = 18$
$\dfrac{9x}{9} = \dfrac{18}{9}$
$x = 2$

5.
$\dfrac{x}{3} + \dfrac{x}{6} = 9$
$6\left(\dfrac{x}{3} + \dfrac{x}{6}\right) = 6(9)$
$6\left(\dfrac{x}{3}\right) + 6\left(\dfrac{x}{6}\right) = 6(9)$
$2x + x = 54$
$3x = 54$
$x = 18$

6.
$3x + \dfrac{1}{4} = \dfrac{5}{8}$
$8\left(3x + \dfrac{1}{4}\right) = 8\left(\dfrac{5}{8}\right)$
$8(3x) + 8\left(\dfrac{1}{4}\right) = 8\left(\dfrac{5}{8}\right)$
$24x + 2 = 5$
$24x = 3$
$x = \dfrac{1}{8}$

7.
$\dfrac{4}{x} + 3 = \dfrac{11}{5}$
$5x\left(\dfrac{4}{x} + 3\right) = 5x\left(\dfrac{11}{5}\right)$
$5x\left(\dfrac{4}{x}\right) + 5x(3) = 5x\left(\dfrac{11}{5}\right)$
$20 + 15x = 11x$
$20 = -4x$
$-5 = x$

8.
$\dfrac{1}{5}x - 2.4 = 8.3$
$\dfrac{1}{5}x - 2.4 + 2.4 = 8.3 + 2.4$
$\dfrac{1}{5}x = 10.7$
$5\left(\dfrac{1}{5}x\right) = 5(10.7)$
$x = 53.5$

9.
$7a - 0.18 = 2a + 0.77$
$7a + (-2a) - 0.18 = 2a + (-2a) + 0.77$
$5a - 0.18 = 0.77$
$5a - 0.18 + 0.18 = 0.77 + 0.18$
$5a = 0.95$
$\dfrac{5a}{5} = \dfrac{0.95}{5}$
$a = 0.19$

Section 8.5

1. Step 1: Read and list.
   Known items: The numbers 3 and 10
   Unknown item: The number in question

Step 2: Assign a variable and translate the information.
   Let x = the number asked for in the problem.
   Then "The sum of a number and 3" translates to $x + 3$.

Step 3: Reread and write an equation.
   The sum of x and 3 is 10.
   $x + 3 = 10$

Step 4: Solve the equation.
   $x + 3 = 10$
   $x = 7$

Step 5: Write your answer.
   The number is 7.

Step 6: Reread and check.
   The sum of 7 and 3 is 10.

2. Step 1: Read and list.
   Known items: The numbers 4 and 34, twice a number, and three times a number
   Unknown item: The number in question

Step 2: Assign a variable and translate the information.
   Let x = the number asked for in the problem.
   Then "The sum of twice a number and three times the number" translates to $2x + 3x$.

Step 3: Reread and write an equation.

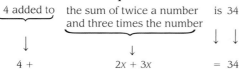

$$4 + 2x + 3x = 34$$

Step 4: Solve the equation.
$$4 + 2x + 3x = 34$$
$$5x + 4 = 34$$
$$5x = 30$$
$$x = 6$$

Step 5: Write your answer.
The number is 6.

Step 6: Reread and check.
Twice 6 is 12 and three times 6 is 18. Their sum is $12 + 18 = 30$. Four added to this is 34. Therefore, 4 added to the sum of twice 6 and three times 6 is 34.

3. Step 1: Read and list.
Known items: Length is twice width; perimeter 42 cm
Unknown items: The length and the width

Step 2: Assign a variable and translate the information.
Let $x =$ the width. Since the length is twice the width, the length must be $2x$. Here is a picture.

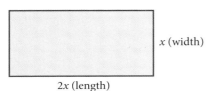

Step 3 Reread and write an equation.
The perimeter is the sum of the sides, and is given as 42; therefore,
$$x + x + 2x + 2x = 42$$

Step 4: Solve the equation.
$$x + x + 2x + 2x = 42$$
$$6x = 42$$
$$x = 7$$

Step 5: Write your answer.
The width is 7 centimeters and the length is
$2(7) = 14$ centimeters

Step 6: Reread and check.
The length, 14, is twice the width, 7. The perimeter is $7 + 7 + 14 + 14 = 42$ centimeters.

4. Step 1: Read and list.
Known items: Three angles are in a triangle. One is 3 times the smallest. The largest is 5 times the smallest.
Unknown item: The three angles

Step 2: Assign a variable and translate the information.
Let $x =$ the smallest angle. The other two angles are $3x$ and $5x$.

Step 3: Reread and write an equation.
The three angles must add up to $180°$, so
$$x + 3x + 5x = 180°$$

Step 4: Solve the equation.
$$x + 3x + 5x = 180°$$
$$9x = 180° \quad \text{Add similar terms on left side}$$
$$x = 20° \quad \text{Divide each side by 9}$$

Step 5: Write the answer.
The three angles are, 20°, 3(20°) = 60°, and 5(20°) = 100°.

Step 6: Reread and check.
The sum of the three angles is 20° + 60° + 100° = 180°. One angle is 3 times the smallest, while the largest is 5 times the smallest.

5. Step 1: Read and list.
Known items: Joyce is 21 years older than Travis. Six years from now their ages will add to 49.
Unknown items: Their ages now

Step 2: Assign a variable and translate the information.
Let x = Travis's age now; since Joyce is 21 years older than that, she is presently $x + 21$ years old.

Step 3: Reread and write an equation.

	Now	in 6 years
Joyce	$x + 21$	$x + 27$
Travis	x	$x + 6$

$x + 27 + x + 6 = 49$

Step 4: Solve the equation.
$$x + 27 + x + 6 = 49$$
$$2x + 33 = 49$$
$$2x = 16$$
$$x = 8$$

Step 5: Write your answer.
Travis is now 8 years old, and Joyce is 8 + 21 = 29 years old.

Step 6: Reread and check.
Joyce is 21 years older than Travis. In six years, Joyce will be 35 years old and Travis will be 14 years old. At that time, the sum of their ages will be
35 + 14 = 49.

6. Step 1: Read and list.
Known items: Charges are $11 per day and 16 cents per mile. Car is rented for 2 days. Total charge is $41.
Unknown items: How many miles the car was driven

Step 2: Assign a variable and translate the information.
Let x = the number of miles the car was driven. Two days rental is 2(11). The cost for driving x miles is 0.16x.

Step 3: Reread and write an equation.
The total cost is the two days' rental plus the mileage cost. It must add up to 41. The equation is
2(11) + 0.16x = 41

Step 4: Solve the equation.
$$2(11) + 0.16x = 41$$
$$22 + 0.16x = 41$$
$$22 + (-22) + 0.16x = 41 + (-22)$$
$$0.16x = 19$$
$$\frac{0.16x}{0.16} = \frac{19}{0.16}$$
$$x = 118.75$$

Step 5: Write your answer.
The car was driven 118.75 miles.

Step 6: Reread and check.
The charge for two days is 2(11) = $22. The 118.75 miles adds 118.75(0.16) = $19. The total is $22 + $19 = $41, which checks with the total charge given in the problem.

7. Step 1 Read and list.

Known items: We have dimes and quarters. There are 7 more dimes than quarters, and the total value of the coins is $1.75.

Unknown items: How many of each type of coin Amy has

Step 2: Assign a variable and translate the information.

Let x = the number of quarters. Here is a table that summarizes the information in the problem.

	Quarters	Dimes
Number of	x	$x + 7$
Value of	$0.25x$	$0.10(x + 7)$

Step 3: Reread and write an equation.

The value of the quarters plus the value of the dimes must add to 1.75. Therefore, our equation is

$0.25x + 0.10(x + 7) = 1.75$

Step 4: Solve the equation.

$$0.25x + 0.10(x + 7) = 1.75$$
$$0.25x + 0.10x + 0.70 = 1.75$$
$$0.35x + 0.70 = 1.75$$
$$0.35x = 1.05$$
$$x = 3$$

Step 5: Write your answer.

She has 3 quarters. The number of dimes is 7 more than that, which is 10.

Step 6: Reread and check.

3 quarters are worth $0.75. 10 dimes are worth $1.00. The total of the two is $1.75, which checks with the information in the problem.

Section 8.6

1. When $P = 80$ and $w = 6$
the formula → $P = 2l + 2w$
becomes → $80 = 2l + 2(6)$
$80 = 2l + 12$
$68 = 2l$
$34 = l$
The length is 34 feet.

2. When $F = 77$
the formula → $C = \frac{5}{9}(F - 32)$
becomes → $C = \frac{5}{9}(77 - 32)$
$= \frac{5}{9}(45)$
$= \frac{5}{9} \cdot \frac{45}{1}$
$= \frac{225}{9}$
$= 25$ degrees Celsius

3. When $x = 0$
the formula → $y = 2x + 6$
becomes → $y = 2 \cdot 0 + 6$
$= 0 + 6$
$= 6$

4. When $x = -3$
the formula → $2x + 3y = 4$
becomes → $2(-3) + 3y = 4$
$-6 + 3y = 4$
$3y = 10$
$y = \frac{10}{3}$

5. a. $11 - 9 = 2$ hr
b. $d = 60$ mi/hr $\cdot$ 2 hr
$d = 60 \cdot 2$
$d = 120$ miles

6. With $x = 35°$ we use the formulas for finding the complement and the supplement of an angle:
The complement of $35°$ is $90° - 35° = 55°$
The supplement of $35°$ is $180° - 35° = 145°$

65.

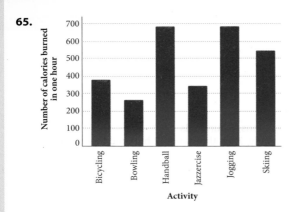

Problem Set 1.4

1. The difference of 10 and 2 **3.** The difference of a and 6 **5.** The difference of 8 and 2 is 6. **7.** 3 **9.** 8 **11.** 23

13.

First Number a	Second Number b	The Difference of a and b $a-b$
25	15	10
24	16	8
23	17	6
22	18	4

15.

First Number a	Second Number b	The Difference of a and b $a-b$
400	256	144
400	144	256
225	144	81
225	81	144

17. $8-3$ **19.** $y-9$ **21.** $3-2=1$ **23.** $37-9x=10$ **25.** $2y-15x=24$ **27.** $(x+2)-(x+1)=1$
29. 32 **31.** 22 **33.** 10 **35.** 111 **37.** 312 **39.** 403 **41.** 1,111 **43.** 4,544 **45.** 15 **47.** 33 **49.** 5 **51.** 33
53. 95 **55.** 152 **57.** 274 **59.** 488 **61.** 538 **63.** 163 **65.** 1,610 **67.** 46,083 **69.** $255
71. $172,500 **73.** 91 mph **75.** 33 GB **77. a.**

Country	Production)
United States	10,781,000
France	3,019,000
Germany	6,213,000
South Korea	4,086,000
Japan	11,596,000

b. 2,127,000 more **79.** $3,500

Problem Set 1.5

1. 300 **3.** 600 **5.** 3,000 **7.** 5,000 **9.** 21,000 **11.** 81,000

13.

First Number a	Second Number b	Their Product ab
11	11	121
11	22	242
22	22	484
22	44	968

15.

First Number a	Second Number b	Their Product ab
25	10	250
25	100	2,500
25	1,000	25,000
25	10,000	250,000

17.

First Number a	Second Number b	Their Product ab
12	20	240
36	20	720
12	40	480
36	40	1,440

7. Step 1 Read and list.
  Known items: We have dimes and quarters. There are 7 more dimes than quarters, and the total value of the coins is $1.75.
  Unknown items: How many of each type of coin Amy has

Step 2: Assign a variable and translate the information.
  Let x = the number of quarters. Here is a table that summarizes the information in the problem.

	Quarters	Dimes
Number of	x	$x + 7$
Value of	$0.25x$	$0.10(x + 7)$

Step 3: Reread and write an equation.
  The value of the quarters plus the value of the dimes must add to 1.75. Therefore, our equation is
  $0.25x + 0.10(x + 7) = 1.75$

Step 4: Solve the equation.
$$0.25x + 0.10(x + 7) = 1.75$$
$$0.25x + 0.10x + 0.70 = 1.75$$
$$0.35x + 0.70 = 1.75$$
$$0.35x = 1.05$$
$$x = 3$$

Step 5: Write your answer.
  She has 3 quarters. The number of dimes is 7 more than that, which is 10.

Step 6: Reread and check.
  3 quarters are worth $0.75. 10 dimes are worth $1.00. The total of the two is $1.75, which checks with the information in the problem.

Section 8.6

1. When → $P = 80$ and $w = 6$
the formula → $P = 2l + 2w$
becomes → $80 = 2l + 2(6)$
$80 = 2l + 12$
$68 = 2l$
$34 = l$
The length is 34 feet.

2. When → $F = 77$
the formula → $C = \dfrac{5}{9}(F - 32)$
becomes → $C = \dfrac{5}{9}(77 - 32)$
$= \dfrac{5}{9}(45)$
$= \dfrac{5}{9} \cdot \dfrac{45}{1}$
$= \dfrac{225}{9}$
$= 25$ degrees Celsius

3. When → $x = 0$
the formula → $y = 2x + 6$
becomes → $y = 2 \cdot 0 + 6$
$= 0 + 6$
$= 6$

4. When → $x = -3$
the formula → $2x + 3y = 4$
becomes → $2(-3) + 3y = 4$
$-6 + 3y = 4$
$3y = 10$
$y = \dfrac{10}{3}$

5. a. $11 - 9 = 2$ hr
b. $d = 60$ mi/hr $\cdot 2$ hr
$d = 60 \cdot 2$
$d = 120$ miles

6. With $x = 35°$ we use the formulas for finding the complement and the supplement of an angle:
The complement of $35°$ is $90° - 35° = 55°$
The supplement of $35°$ is $180° - 35° = 145°$

Appendices

Appendix A

1. $2^{-4} = \dfrac{1}{2^4}$
$= \dfrac{1}{16}$

2. $4^{-2} = \dfrac{1}{4^2}$
$= \dfrac{1}{16}$

3. $(-2)^{-3} = \dfrac{1}{(-2)^3}$
$= \dfrac{1}{-8}$
$= -\dfrac{1}{8}$

4. $3^4 \cdot 3^{-7} = 3^{4+(-7)}$
$= 3^{-3}$
$= \dfrac{1}{3^3}$
$= \dfrac{1}{27}$

5. $x^6 \cdot x^{-10} = x^{6+(-10)}$
$= x^{-4}$
$= \dfrac{1}{x^4}$

6. $\dfrac{2^6}{2^8} = 2^{6-8}$
$= 2^{-2}$
$= \dfrac{1}{2^2}$
$= \dfrac{1}{4}$

7. $\dfrac{10^{-5}}{10^3} = 10^{-5-3}$
$= 10^{-8}$
$= \dfrac{1}{10^8}$

8. $\dfrac{10^{-6}}{10^{-8}} = 10^{-6-(-8)}$
$= 10^{-6+8}$
$= 10^2$
$= 100$

9. $4x^{-3} \cdot 7x = 4x^{-3} \cdot 7x^1$
$= (4 \cdot 7)(x^{-3} \cdot x^1)$
$= 28x^{-3+1}$
$= 28x^{-2}$
$= \dfrac{28}{x^2}$

10. $\dfrac{x^{-5} \cdot x^2}{x^{-8}} = \dfrac{x^{-5+2}}{x^{-8}}$
$= \dfrac{x^{-3}}{x^{-8}}$
$= x^{-3-(-8)}$
$= x^{-3+8}$
$= x^5$

Appendix C

1. $(2.5 \times 10^6)(1.4 \times 10^2) = (2.5)(1.4) \times (10^6)(10^2)$
$= 3.5 \times 10^{6+2}$
$= 3.5 \times 10^8$

2. $(2{,}200{,}000)(0.00015) = (2.2 \times 10^6)(1.5 \times 10^{-4})$
$= (2.2)(1.5) \times (10^6)(10^{-4})$
$= 3.3 \times 10^{6+(-4)}$
$= 3.3 \times 10^2$

3. $\dfrac{6 \times 10^5}{2 \times 10^{-4}} = \dfrac{6}{2} \times \dfrac{10^5}{10^{-4}}$
$= 3.0 \times 10^{5-(-4)}$
$= 3.0 \times 10^{5+4}$
$= 3.0 \times 10^9$

4. $\dfrac{0.0038}{19{,}000{,}000} = \dfrac{3.8 \times 10^{-3}}{1.9 \times 10^7}$
$= \dfrac{3.8}{1.9} \times \dfrac{10^{-3}}{10^7}$
$= 2.0 \times 10^{-3-7}$
$= 2.0 \times 10^{-10}$

5. $\dfrac{(6.8 \times 10^{-4})(3.9 \times 10^2)}{7.8 \times 10^{-6}} = \dfrac{(6.8)(3.9)}{(7.8)} \times \dfrac{(10^{-4})(10^2)}{10^{-6}}$
$= 3.4 \times 10^{-4+2-(-6)}$
$= 3.4 \times 10^4$

6. $\dfrac{(0.000035)(45{,}000)}{0.000075} = \dfrac{(3.5 \times 10^{-5})(4.5 \times 10^4)}{7.5 \times 10^{-5}}$
$= \dfrac{(3.5)(4.5)}{(7.5)} \times \dfrac{(10^{-5})(10^4)}{10^{-5}}$
$= 2.1 \times 10^{-5+4-(-5)}$
$= 2.1 \times 10^4$

Answers to Odd-Numbered Problems

Chapter 1

Problem Set 1.1
1. 8 ones, 7 tens   **3.** 5 ones, 4 tens   **5.** 8 ones, 4 tens, 3 hundreds   **7.** 8 ones, 0 tens, 6 hundreds
9. 8 ones, 7 tens, 3 hundreds, 2 thousands   **11.** 9 ones, 6 tens, 5 hundreds, 3 thousands, 7 ten thousands, 2 hundred thousands
13. Ten thousands   **15.** Hundred millions   **17.** Ones   **19.** Hundred thousands   **21.** 600 + 50 + 8   **23.** 60 + 8
25. 4,000 + 500 + 80 + 7   **27.** 30,000 + 2,000 + 600 + 70 + 4   **29.** 3,000,000 + 400,000 + 60,000 + 2,000 + 500 + 70 + 7
31. 400 + 7   **33.** 30,000 + 60 + 8   **35.** 3,000,000 + 4,000 + 8   **37.** Twenty-nine   **39.** Forty
41. Five hundred seventy-three   **43.** Seven hundred seven   **45.** Seven hundred seventy
47. Twenty-three thousand, five hundred forty   **49.** Three thousand, four   **51.** Three thousand, forty
53. One hundred four million, sixty-five thousand, seven hundred eighty   **55.** Five billion, three million, forty thousand, eight
57. Two million, five hundred forty-six thousand, seven hundred thirty-one   **59.** 325   **61.** 5,432   **63.** 86,762   **65.** 2,000,200
67. 2,002,200   **69. a.** Twenty-eight thousand, six hundred thirty-one   **b.** Ninety-three thousand, three hundred thirty-three
71. Ten thousands   **73.** Three million, three hundred five thousand, three hundred ninety-three
75. Four thousand, three hundred twenty   **77.** Seven hundred fifty dollars and no cents   **79.** 310,000,000
81. One hundred twenty-seven million   **83.** 616,000   **85.** Three million, eight hundred thirty-one thousand

Problem Set 1.2
1. 15   **3.** 14   **5.** 24   **7.** 15   **9.** 20   **11.** 68   **13.** 98   **15.** 7,297   **17.** 6,487   **19.** 96   **21.** 7,449   **23.** 65
25. 102   **27.** 875   **29.** 829   **31.** 10,391   **33.** 16,204   **35.** 155,554   **37.** 111,110   **39.** 17,391   **41.** 14,892
43. 180   **45.** 2,220   **47.** 18,285   **49.**

First Number a	Second Number b	Their Sum $a+b$
61	38	99
63	36	99
65	34	99
67	32	99

51.

First Number a	Second Number b	Their Sum $a+b$
9	16	25
36	64	100
81	144	225
144	256	400

53. The sum of 4 and 9   **55.** The sum of 8 and 1   **57.** The sum of 2 and 3 is 5.   **59. a.** 5 + 2   **b.** 8 + 3
61. a. $m + 1$ **b.** $m + n$ **63.** 9 + 5   **65.** 8 + 3   **67.** 4 + 6   **69.** 1 + (2 + 3)   **71.** 2 + (1 + 6)   **73.** (1 + 9) + 1
75. 4 + (n + 1)   **77.** n = 4   **79.** n = 5   **81.** n = 8   **83.** n = 8   **85.** 12 in.   **87.** 16 ft   **89.** 26 yd   **91.** 18 in.
93. a. 101 million more users   **b.** 504 million more users   **95.** $349   **97.** $198,365

Problem Set 1.3
1. 40   **3.** 50   **5.** 50   **7.** 80   **9.** 460   **11.** 470   **13.** 56,780   **15.** 4,500   **17.** 500   **19.** 800   **21.** 900   **23.** 1,100
25. 5,000   **27.** 39,600   **29.** 5,000   **31.** 10,000   **33.** 1,000   **35.** 658,000   **37.** 510,000   **39.** 3,789,000

Original Number	Rounded to the Nearest		
	Ten	Hundred	Thousand
41. 7,821	7,820	7,800	8,000
43. 5,999	6,000	6,000	6,000
45. 10,985	10,990	11,000	11,000
47. 99,999	100,000	100,000	100,000

49. 1,200   **51.** 1,900   **53.** 58,000   **55.** 33,400   **57.** 190,000   **59.** 81,400
61. a. 35,695,000 cars   **b.** Yes, to the thousands   **c.** 9,232,000 cars   **d.** 15,682,000 cars
63. a. Yes, to the thousands   **b.** Yes, to the tens   **c.** Yes, to the nearest foot

65.

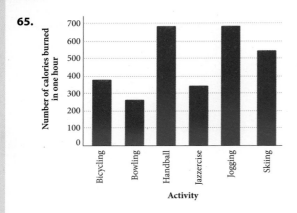

Problem Set 1.4

1. The difference of 10 and 2   **3.** The difference of a and 6   **5.** The difference of 8 and 2 is 6.   **7.** 3   **9.** 8   **11.** 23

13.

First Number a	Second Number b	The Difference of a and b $a-b$
25	15	10
24	16	8
23	17	6
22	18	4

15.

First Number a	Second Number b	The Difference of a and b $a-b$
400	256	144
400	144	256
225	144	81
225	81	144

17. $8 - 3$ **19.** $y - 9$ **21.** $3 - 2 = 1$ **23.** $37 - 9x = 10$ **25.** $2y - 15x = 24$ **27.** $(x + 2) - (x + 1) = 1$
29. 32   **31.** 22   **33.** 10   **35.** 111   **37.** 312   **39.** 403   **41.** 1,111   **43.** 4,544   **45.** 15   **47.** 33   **49.** 5   **51.** 33
53. 95   **55.** 152   **57.** 274   **59.** 488   **61.** 538   **63.** 163   **65.** 1,610   **67.** 46,083   **69.** $255
71. $172,500   **73.** 91 mph   **75.** 33 GB   **77. a.**

Country	Production)
United States	10,781,000
France	3,019,000
Germany	6,213,000
South Korea	4,086,000
Japan	11,596,000

b. 2,127,000 more   **79.** $3,500

Problem Set 1.5

1. 300   **3.** 600   **5.** 3,000   **7.** 5,000   **9.** 21,000   **11.** 81,000

13.

First Number a	Second Number b	Their Product ab
11	11	121
11	22	242
22	22	484
22	44	968

15.

First Number a	Second Number b	Their Product ab
25	10	250
25	100	2,500
25	1,000	25,000
25	10,000	250,000

17.

First Number a	Second Number b	Their Product ab
12	20	240
36	20	720
12	40	480
36	40	1,440

19. The product of 6 and 7   **21.** The product of 2 and n **23.** The product of 9 and 7 is 63.   **25.** $7 \cdot n$ **27.** $6 \cdot 7 = 42$
29. $0 \cdot 6 = 0$ **31.** Products: $9 \cdot 7$ and 63   **33.** Products: 4(4) and 16   **35.** Factors: 2, 3, and 4   **37.** Factors: 2, 2, and 3
39. 100   **41.** 228   **43.** 36   **45.** 1,440   **47.** 950   **49.** 1,725   **51.** 121   **53.** 1,552   **55.** 4,200   **57.** 66,248
59. 279,200   **61.** 12,321   **63.** 106,400   **65.** 198,592   **67.** 612,928   **69.** 333,180   **71.** 18,053,805   **73.** 263,646,976
75. 9(5)   **77.** $7 \cdot 6$ **79.** $(2 \cdot 7) \cdot 6$ **81.** $(3 \times 9) \times 1$ **83.** $7(2) + 7(3) = 35$ **85.** $9(4) + 9(7) = 99$ **87.** $3x + 3$
89. $2x + 10$ **91.** $n = 3$ **93.** $n = 9$ **95.** $n = 0$ **97.** 2,860 mi   **99.** $7.00   **101.** 148,800 jets   **103.** 2,081 calories
105. 280 calories   **107.** Yes   **109.** 8,000   **111.** 1,500,000   **113.** 1,400,000   **115.** 40   **117.** 54

Problem Set 1.6

1. $6 \div 3$ **3.** $45 \div 9$ **5.** $r \div s$ **7.** $20 \div 4 = 5$ **9.** $2 \cdot 3 = 6$ **11.** $9 \cdot 4 = 36$ **13.** $6 \cdot 8 = 48$ **15.** $7 \cdot 4 = 28$ **17.** 5
19. 8   **21.** Undefined   **23.** 45   **25.** 23   **27.** 1,530   **29.** 1,350   **31.** 18,000   **33.** 16,680   **35.** a   **37.** b   **39.** 1
41. 2   **43.** 4   **45.** 6   **47.** 45   **49.** 49   **51.** 432   **53.** 1,438   **55.** 705   **57.** 3,020
59.

First Number a	Second Number b	The Quotient of a and b $\dfrac{a}{b}$
100	25	4
100	26	3 R 22
100	27	3 R 19
100	28	3 R 16

61. 61 R 4   **63.** 90 R 1   **65.** 13 R 7   **67.** 234 R 6

69. 402 R 4   **71.** 35 R 35   **73.** $3,525   **75.** $8.00   **77.** 102 basketballs   **79.** 6 glasses, with 2 ounces left over
81. $4,325,000   **83.** $4.00

Problem Set 1.7

1. Base 4; exponent 5   **3.** Base 3; exponent 6   **5.** Base 8; exponent 2   **7.** Base 9; exponent 1   **9.** Base 4; exponent 0   **11.** 36
13. 8   **15.** 1   **17.** 1   **19.** 81   **21.** 10   **23.** 12   **25.** 1   **27.** 12   **29.** 100   **31.** 4   **33.** 43   **35.** 16   **37.** 84
39. 14   **41.** 74   **43.** 12,768   **45.** 104   **47.** 416   **49.** 66   **51.** 21   **53.** 7   **55.** 16   **57.** 84   **59.** 40   **61.** 41
63. 18   **65.** 405   **67.** 124   **69.** 11   **71.** 91   **73.** 7   **75.** $8(4 + 2) = 48$ **77.** $2(10 + 3) = 26$ **79.** $3(3 + 4) + 4 = 25$
81. $(20 \div 2) - 9 = 1$ **83.** $(8 \cdot 5) + (5 \cdot 4) = 60$ **85.** mean = 3; range = 4   **87.** mean = 6; range = 10
89. median = 11; range = 10   **91.** median = 50; range = 90   **93.** mode = 18; range = 59   **95.** Big Mac has twice the calories
97. a. 78   **b.** 76   **c.** 76   **d.** 47   **99.** mean = 6,881 students; range = 819 students   **101. a.** 126   **b.** 126.5   **c.** 130   **d.** 28
103. a. $4.109   **b.** $4.106   **c.** $0.303   **105.** 4   **107.** 30

Problem Set 1.8

1. 25 cm^2 **3.** 336 m^2 **5.** 60 ft^2 **7.** 45 m^2 **9.** 16 cm^2 **11.** 2,200 ft^2 **13.** 945 cm^2 **15.** 100 in^2
17. Volume = 64 cm^3; surface area = 96 cm^2 **19.** Volume = 72 ft^3; surface area = 108 ft^2 **21. a.** 5.4 in^2 **b.** 0.864 in^2
23. 124 tiles   **25.** 6,600 yd^2 **27.** The area increases from 25 ft^2 to 49 ft^2, which is an increase of 24 ft^2.   **29.** 8,509 mm^2
31. 2,106 ft^2 **33.** 1,352 ft^3 **35.** 7 ft   **37.** 9 ft

Chapter 1 Review

1. One thousand, six hundred fifty-six   **2.** Three million, seven hundred sixty-five thousand, eight hundred seven
3. 5,245,652   **4.** 12,012,012   **5.** 1,000,000 + 300,000 + 90,000 + 4,000 + 800 + 10 + 2   **6.** 200,000 + 1,000 + 700 + 20 + 7
7. d   **8.** f   **9.** c   **10.** a   **11.** b   **12.** e   **13.** g   **14.** d   **15.** 749   **16.** 1,382   **17.** 8,272   **18.** 9,141
19. 3,781,090   **20.** 3,781,100   **21.** 3,800,000   **22.** 4,000,000   **23.** 314   **24.** 614   **25.** 3,149   **26.** 1,738   **27.** 584
28. 6,888   **29.** 3,717   **30.** 42,924   **31.** 173   **32.** 68   **33.** 428   **34.** 76   **35.** 79   **36.** 183   **37.** 222   **38.** 3
39. 8   **40.** 10   **41.** 32   **42.** 36   **43.** $3(4 + 6) = 30$ **44.** $9(5 - 3) = 18$ **45.** $2(17 - 5) = 24$ **46.** $5(8 + 2) = 50$
47. mean $3,411; median $3,207; range $2,625   **48.** $P = 22$ m; $A = 24$ m^2 **49.** $P = 38$ ft; $A = 84$ ft^2

Chapter 1 Test

1. Twenty thousand, three hundred forty-seven   **2.** 2,045,006   **3.** 100,000 + 20,000 + 3,000 + 400 + 7   **4.** f   **5.** c   **6.** a
7. e   **8.** 876   **9.** 16,383   **10.** 520,000   **11.** 524   **12.** 3,085   **13.** 1,674   **14.** 22,258   **15.** 85   **16.** 21   **17.** 11
18. 4   **19.** 107   **20.** $3x - 6$ **21.** $2(11 + 7) = 36$ **22.** $(20 \div 5) + 9 = 13$ **23.** mean $573.34; median $586.95; range $470.00
24. $P = 46$ m; $A = 91$ m^2 **25.** 73,000 MW   **26.** 173,000 MW

A Glimpse of Algebra

1. $36x^2$ **3.** $16x^2$ **5.** $27a^3$ **7.** $8a^3b^3$ **9.** $81x^2y^2$ **11.** $25x^2y^2z^2$ **13.** $1,296x^4y^2$ **15.** $576x^6$ **17.** $200x^5$ **19.** $7,200a^7$
21. $432x^5y^5$ **23.** $400x^6y^6z^6$ **25.** $x^5y^7z^6$ **27.** $324a^{14}b^{16}$ **29.** $40x^{11}y^{12}$

Chapter 2

Problem Set 2.1

1. 1 **3.** 2 **5.** x **7.** a **9.** 5 **11.** 1 **13.** 12

15.

Numerator a	Denominator b	Fraction $\frac{a}{b}$
3	5	$\frac{3}{5}$
1	7	$\frac{1}{7}$
x	y	$\frac{x}{y}$
$x+1$	x	$\frac{x+1}{x}$

17. $\frac{3}{4}, \frac{1}{2}, \frac{9}{10}$ **19.** True **21.** False **23.** $\frac{3}{4}$ **25.** $\frac{43}{47}$
27. $\frac{4}{3}$ **29.** $\frac{13}{17}$ **31.** $\frac{4}{6}$ **33.** $\frac{5}{6}$ **35.** $\frac{8}{12}$ **37.** $\frac{8}{12}$
39. Answers will vary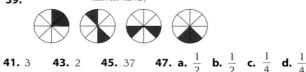
41. 3 **43.** 2 **45.** 37 **47.** a. $\frac{1}{2}$ b. $\frac{1}{2}$ c. $\frac{1}{4}$ d. $\frac{1}{4}$

49–57. $\frac{1}{16} \; \frac{1}{8} \; \frac{1}{4} \quad \frac{5}{8} \; \frac{3}{4} \; \frac{15}{16} \quad \frac{5}{4} \; \frac{3}{2} \quad \frac{15}{8} \; \frac{31}{16}$ on number line from 0 to 2

59. $\frac{1}{20} < \frac{4}{25} < \frac{3}{10} < \frac{2}{5}$ **61.** $\frac{5}{100}$ inch

63.

How Often Workers Send Non-Work-Related E-mail from the Office	Fraction of Respondents Saying Yes
Never	$\frac{4}{25}$
1 to 5 times a day	$\frac{47}{100}$
5 to 10 times a day	$\frac{8}{25}$
More than 10 times a day	$\frac{1}{20}$

65. $\frac{4}{5}$ **67.** $\frac{311}{500}$ **69.** $\frac{19}{33}$ **71.** a. $\frac{90}{360}$ b. $\frac{45}{360}$ c. $\frac{180}{360}$ d. $\frac{270}{360}$
73. d **75.** a **77.** 108 **79.** 60 **81.** 4 **83.** 5
85. 7 **87.** 51 **89.** 23 **91.** 32 **93.** 16 **95.** 18

Problem Set 2.2

1. Prime **3.** Composite; 3, 5, and 7 are factors **5.** Composite; 3 is a factor **7.** Prime **9.** $2^2 \cdot 3$ **11.** 3^4 **13.** $5 \cdot 43$
15. $3 \cdot 5$ **17.** $\frac{1}{2}$ **19.** $\frac{2}{3}$ **21.** $\frac{4}{5}$ **23.** $\frac{9}{5}$ **25.** $\frac{7}{11}$ **27.** $\frac{3}{5}$ **29.** $\frac{1}{7}$ **31.** $\frac{7}{9}$ **33.** $\frac{7}{5}$ **35.** $\frac{1}{5}$ **37.** $\frac{11}{7}$
39. $\frac{5}{3}$ **41.** $\frac{8}{9}$ **43.** $\frac{42}{55}$ **45.** $\frac{17}{19}$ **47.** $\frac{14}{33}$ **49.** a. $\frac{2}{17}$ b. $\frac{3}{26}$ c. $\frac{1}{9}$ d. $\frac{3}{28}$ e. $\frac{2}{19}$
51. a. $\frac{1}{45}$ b. $\frac{1}{30}$ c. $\frac{1}{18}$ d. $\frac{1}{15}$ e. $\frac{1}{10}$ **53.** a. $\frac{1}{3}$ b. $\frac{5}{6}$ c. $\frac{1}{5}$ **55.** $\frac{9}{16}$ **57–59.** $\frac{1}{2} = \frac{2}{4} = \frac{4}{8} = \frac{8}{16}$ $\frac{3}{2} = \frac{6}{4} = \frac{12}{8} = \frac{24}{16}$ on number line from 0 to 2, with $\frac{1}{4} = \frac{2}{8} = \frac{4}{16}$ and $\frac{5}{4} = \frac{10}{8} = \frac{20}{16}$

61. $\frac{106}{115}$ **63.** $\frac{1}{3}$ **65.** $\frac{8}{25}$ **67.** $\frac{3}{4}$ **69.** $\frac{37}{70}$ **71.** $\frac{1}{3}$ **73.** $\frac{1}{8}$ **75.** 3 **77.** 45 **79.** 25 **81.** $2^2 \cdot 5 \cdot 3$
83. $2^2 \cdot 5 \cdot 3$ **85.** 9 **87.** 25 **89.** 12 **91.** 18 **93.** 42 **95.** 53

Problem Set 2.3

1. $\frac{8}{15}$ **3.** $\frac{7}{8}$ **5.** 1 **7.** $\frac{27}{4}$ **9.** 1 **11.** $\frac{1}{24}$ **13.** $\frac{24}{125}$ **15.** $\frac{105}{8}$

17.

First Number x	Second Number y	Their Product xy
$\frac{1}{2}$	$\frac{2}{3}$	$\frac{1}{3}$
$\frac{2}{3}$	$\frac{3}{4}$	$\frac{1}{2}$
$\frac{3}{4}$	$\frac{4}{5}$	$\frac{3}{5}$
$\frac{5}{a}$	$\frac{a}{6}$	$\frac{5}{6}$

19.

First Number x	Second Number y	Their Product xy
$\frac{1}{2}$	30	15
$\frac{1}{5}$	30	6
$\frac{1}{6}$	30	5
$\frac{1}{15}$	30	2

21. $\frac{3}{5}$ **23.** 9 **25.** 1 **27.** 8 **29.** $\frac{1}{15}$ **31.** $\frac{4}{9}$ **33.** $\frac{9}{16}$ **35.** $\frac{1}{4}$ **37.** $\frac{8}{27}$ **39.** $\frac{1}{2}$ **41.** $\frac{9}{100}$ **43.** 3 **45.** 24
47. 4 **49.** 9 **51.** $\frac{3}{10}$; numerator should be 3, not 4.
53. a.

Number x	Square x^2
1	1
2	4
3	9
4	16
5	25
6	36
7	49
8	64

b. Either *larger* or *greater* will work.   **55.** 133 in²

57. $\frac{4}{9}$ ft² **59.** 3 yd² **61.** 138 in² **63. a.** $\frac{9}{10}$ inch **b.** $\frac{49}{50}$ inch **c.** 1 inch **65.** 126,500 ft³ **67.** About 8 million
69. 846 adults **71.** Canada: ≈ 1,668,000 barrels Venezuela: ≈ 1,251,000 barrels Iraq: ≈ 834,000 barrels **73.** $\frac{1}{27}$ **75.** $\frac{8}{27}$
77. 2 **79.** 3 **81.** 2 **83.** 5 **85.** 3 **87.** $\frac{4}{3}$ **89.** 3 **91.** $\frac{1}{7}$ **93.** 100 **95.** 9 **97.** 18 **99.** 8

Problem Set 2.4

1. $\frac{15}{4}$ **3.** $\frac{4}{3}$ **5.** 9 **7.** 200 **9.** $\frac{3}{8}$ **11.** 1 **13.** $\frac{49}{64}$ **15.** $\frac{3}{4}$ **17.** $\frac{15}{16}$ **19.** $\frac{1}{6}$ **21.** 6 **23.** $\frac{5}{18}$ **25.** $\frac{9}{2}$
27. $\frac{2}{9}$ **29.** 9 **31.** $\frac{4}{5}$ **33.** $\frac{15}{22}$ **35.** 40 **37.** $\frac{7}{10}$ **39.** 13 **41.** 12 **43.** 186 **45.** 646 **47.** $\frac{3}{5}$ **49.** 40
51. $3 \cdot 5 = 15$; $3 \div \frac{1}{5} = 3 \cdot \frac{5}{1} = 15$ **53.** 490 feet **55.** 14 blankets **57.** 6 **59.** 28 cartons **61.** 20 lots
63. $\frac{3}{6}$ **65.** $\frac{9}{6}$ **67.** $\frac{4}{12}$ **69.** $\frac{8}{12}$ **71.** $\frac{14}{30}$ **73.** $\frac{18}{30}$ **75.** $\frac{12}{24}$ **77.** $\frac{4}{24}$ **79.** $\frac{15}{36}$ **81.** $\frac{9}{36}$

83.

	Rounded to the Nearest Number		
	Ten	Hundred	Thousand
74	70	100	0
747	750	700	1,000
474	470	500	0

85. b

Problem Set 2.5

1. $\frac{2}{3}$ **3.** $\frac{1}{4}$ **5.** $\frac{1}{2}$ **7.** $\frac{1}{3}$ **9.** $\frac{3}{2}$ **11.** $\frac{x+6}{2}$ **13.** $\frac{4}{5}$ **15.** $\frac{10}{3}$

17.

First Number a	Second Number b	The Sum of a and b $a+b$
$\frac{1}{2}$	$\frac{1}{3}$	$\frac{5}{6}$
$\frac{1}{3}$	$\frac{1}{4}$	$\frac{7}{12}$
$\frac{1}{4}$	$\frac{1}{5}$	$\frac{9}{20}$
$\frac{1}{5}$	$\frac{1}{6}$	$\frac{11}{30}$

19.

First Number a	Second Number b	The Sum of a and b $a+b$
$\frac{1}{12}$	$\frac{1}{2}$	$\frac{7}{12}$
$\frac{1}{12}$	$\frac{1}{3}$	$\frac{5}{12}$
$\frac{1}{12}$	$\frac{1}{4}$	$\frac{4}{12} = \frac{1}{3}$
$\frac{1}{12}$	$\frac{1}{6}$	$\frac{3}{12} = \frac{1}{4}$

21. $\frac{7}{9}$ **23.** $\frac{7}{3}$ **25.** $\frac{7}{4}$ **27.** $\frac{7}{6}$ **29.** $\frac{9}{20}$ **31.** $\frac{7}{10}$ **33.** $\frac{19}{24}$ **35.** $\frac{13}{60}$ **37.** $\frac{31}{100}$ **39.** $\frac{67}{144}$ **41.** $\frac{29}{35}$ **43.** $\frac{949}{1,260}$
45. $\frac{13}{420}$ **47.** $\frac{41}{24}$ **49.** $\frac{53}{60}$ **51.** $\frac{5}{4}$ **53.** $\frac{88}{9}$ **55.** $\frac{3}{4}$ **57.** $\frac{1}{4}$ **59.** 19 **61.** 3 **63.** $\frac{160}{63}$ **65.** $\frac{5}{8}$ **67.** $\frac{1}{4}$ inch
69. $\frac{9}{2}$ pints **71.** $\frac{61}{400}$ **73.** $\frac{5}{3}$ hours **75.** $\frac{1}{3}$ **77.** 10 lots **79.** $\frac{3}{2}$ in. **81.** $\frac{9}{5}$ ft **83.** $\frac{7}{3}$ **85.** 3 **87.** 59 **89.** $\frac{16}{8}$
91. $\frac{8}{8}$ **93.** $\frac{11}{4}$ **95.** $\frac{17}{8}$ **97.** $\frac{9}{8}$ **99.** 2 R 3 **101.** 8 R 16 **103.** $\frac{9}{10}$ **105.** 8 **107.** 3 **109.** 2 **111.** $\frac{2}{7}$
113. $\frac{15}{22}$

Problem Set 2.6

1. $\dfrac{14}{3}$ 3. $\dfrac{21}{4}$ 5. $\dfrac{13}{8}$ 7. $\dfrac{47}{3}$ 9. $\dfrac{104}{21}$ 11. $\dfrac{427}{33}$ 13. $1\dfrac{1}{8}$ 15. $4\dfrac{3}{4}$ 17. $4\dfrac{5}{6}$ 19. $3\dfrac{1}{4}$ 21. $4\dfrac{1}{27}$ 23. $28\dfrac{8}{15}$
25. $\dfrac{11}{4}$ 27. $\dfrac{37}{8}$ 29. $\dfrac{14}{5}$ 31. $\dfrac{9}{40}$ 33. $\dfrac{3}{8}$ 35. $\dfrac{32}{35}$ 37. $\dfrac{4}{7}$ 39. $\dfrac{4}{5}$ 41. 9 43. 98

Problem Set 2.7

1. $5\dfrac{1}{10}$ 3. $13\dfrac{2}{3}$ 5. $6\dfrac{93}{100}$ 7. $5\dfrac{5}{6}$ 9. $9\dfrac{3}{4}$ 11. $3\dfrac{1}{5}$ 13. $12\dfrac{1}{2}$ 15. $9\dfrac{9}{20}$ 17. $\dfrac{32}{45}$ 19. $1\dfrac{2}{3}$ 21. 4 23. $4\dfrac{3}{10}$
25. $\dfrac{1}{10}$ 27. $3\dfrac{1}{5}$ 29. $2\dfrac{1}{8}$ 31. $7\dfrac{1}{2}$ 33. $\dfrac{11}{13}$ 35. $5\dfrac{1}{2}$ cups 37. $1\dfrac{1}{3}$ 39. $2{,}687\dfrac{1}{5}$ cents 41. $163\dfrac{3}{4}$ mi
43. $6\dfrac{758}{1{,}207}$; 6 shares 45. $\$2{,}516\dfrac{2}{3}$ 47. $182\dfrac{9}{16}$ ft² 49. $3\dfrac{7}{16}$ mi²
51. Can 1 contains $157\dfrac{1}{2}$ calories, whereas Can 2 contains $87\dfrac{1}{2}$ calories. Can 1 contains 70 more calories than Can 2.
53. Can 1 contains 1,960 milligrams of sodium, whereas Can 2 contains 1,050 milligrams of sodium. Can 1 contains 910 more milligrams of sodium than Can 2.
55. a. $\dfrac{10}{15}$ b. $\dfrac{3}{15}$ c. $\dfrac{9}{15}$ d. $\dfrac{5}{15}$ 57. a. $\dfrac{5}{20}$ b. $\dfrac{12}{20}$ c. $\dfrac{18}{20}$ d. $\dfrac{2}{20}$ 59. $\dfrac{13}{15}$ 61. $\dfrac{14}{9} = 1\dfrac{5}{9}$ 63. $\dfrac{3}{5}$ 65. $\dfrac{3}{14}$ 67. $2\dfrac{1}{4}$
69. $3\dfrac{1}{16}$

Problem Set 2.8

1. $5\dfrac{4}{5}$ 3. $12\dfrac{2}{5}$ 5. $3\dfrac{4}{9}$ 7. 12 9. $1\dfrac{3}{8}$ 11. $14\dfrac{1}{6}$ 13. $4\dfrac{1}{12}$ 15. $2\dfrac{1}{12}$ 17. $26\dfrac{7}{12}$ 19. 12 21. $2\dfrac{1}{2}$ 23. $8\dfrac{6}{7}$
25. $3\dfrac{3}{8}$ 27. $10\dfrac{4}{15}$ 29. $2\dfrac{1}{15}$ 31. $21\dfrac{17}{20}$ 33. 9 35. $18\dfrac{1}{10}$ 37. 14 39. 17 41. $24\dfrac{1}{24}$ 43. $27\dfrac{6}{7}$ 45. $6\dfrac{1}{4}$
47. $9\dfrac{7}{10}$ 49. $5\dfrac{1}{2}$ 51. $\dfrac{2}{3}$ 53. $1\dfrac{11}{12}$ 55. $3\dfrac{11}{12}$ 57. $5\dfrac{19}{20}$ 59. $5\dfrac{1}{2}$ 61. $\dfrac{13}{24}$ 63. $3\dfrac{1}{2}$ 65. $\$2\dfrac{1}{2}$ 67. $\$250$
69. $\$300$ 71. a. NFL: $P = 306\dfrac{2}{3}$ yd; Canadian: $P = 350$ yd; Arena: $P = 156\dfrac{2}{3}$ yd
b. NFL: $A = 5{,}333\dfrac{1}{3}$ sq yd; Canadian: $A = 7{,}150$ sq yd; Arena: $A = 1{,}416\dfrac{2}{3}$ sq yd
73. $31\dfrac{1}{6}$ in. 75. $85\dfrac{9}{10}$ in. 77. $4\dfrac{63}{64}$ 79. 2 81. $\dfrac{11}{8} = 1\dfrac{3}{8}$ 83. $3\dfrac{5}{8}$ 85. 17 87. 14 89. 104 91. 96
93. 40 95. $3\dfrac{1}{2}$ 97. $5\dfrac{29}{40}$ 99. $34\dfrac{4}{5}$ sec.

Problem Set 2.9

1. 7 3. 7 5. 2 7. 35 9. $\dfrac{7}{8}$ 11. $8\dfrac{1}{3}$ 13. $\dfrac{11}{36}$ 15. $3\dfrac{2}{3}$ 17. $6\dfrac{3}{8}$ 19. $4\dfrac{5}{12}$ 21. $\dfrac{8}{9}$ 23. $\dfrac{1}{2}$ 25. $1\dfrac{1}{10}$
27. 5 29. $\dfrac{3}{5}$ 31. $\dfrac{7}{11}$ 33. 5 35. $\dfrac{17}{28}$ 37. $1\dfrac{7}{16}$ 39. $\dfrac{13}{22}$ 41. $\dfrac{5}{22}$ 43. $\dfrac{15}{16}$ 45. $1\dfrac{5}{17}$ 47. $\dfrac{3}{29}$ 49. $1\dfrac{34}{67}$
51. $\dfrac{346}{441}$ 53. $5\dfrac{2}{5}$ 55. 8 57. 5 miles² 59. $89\dfrac{27}{32}$ 61. $123\dfrac{45}{64}$ ft² 63. $\dfrac{2}{3}$ 65. $\dfrac{1}{6}$ 67. $\dfrac{1}{7}$ 69. $9\dfrac{7}{9}$

Chapter 2 Review

1. $\dfrac{3}{4}$ 2. $\dfrac{1}{3}$ 3. $\dfrac{11}{7}$ 4. $\dfrac{3}{5}$ 5. 1 6. $\dfrac{4}{9}$ 7. $\dfrac{8}{21}$ 8. 30 9. $\dfrac{2}{3}$ 10. $\dfrac{3}{10}$ 11. $\dfrac{3}{8}$ 12. $\dfrac{2}{7}$ 13. $\dfrac{1}{2}$ 14. 2
15. $\dfrac{7}{2}$ 16. $\dfrac{37}{156}$ 17. $\dfrac{1}{36}$ 18. $1\dfrac{33}{100}$ 19. $\dfrac{29}{8}$ 20. $\dfrac{23}{3}$ 21. $3\dfrac{3}{4}$ 22. $13\dfrac{3}{4}$ 23. $\dfrac{8}{13}$ 24. $1\dfrac{7}{8}$ 25. 12
26. $7\dfrac{3}{5}$ 27. $17\dfrac{11}{12}$ 28. $2\dfrac{4}{9}$ 29. $11\dfrac{2}{3}$ 30. $5\dfrac{11}{16}$ 31. 5 32. $\dfrac{3}{5}$ 33. $\dfrac{1}{2}$ 34. $\dfrac{131}{154}$ 35. 20 items 36. 60 students
37. 9 38. $\dfrac{1}{6}$ 39. $1\dfrac{7}{8}$ cups 40. $1\dfrac{19}{24}$ in. 41. $10\dfrac{1}{2}$ tablespoons 42. $32\dfrac{1}{2}$ lb 43. Area $= 25\dfrac{1}{5}$ ft²; perimeter $= 28$ ft

Chapter 2 Cumulative Review

1. 6,317 2. 73,440 3. $\dfrac{81}{169}$ 4. 10 5. $\dfrac{1}{4}$ 6. $8\dfrac{19}{48}$ 7. $\dfrac{4}{11}$ 8. 60 9. 24 10. 2 11. $4\dfrac{8}{21}$ 12. $\dfrac{5}{7}$
13. 625 14. 6 15. $\dfrac{11}{12}$ 16. 622,120 17. $\dfrac{4}{9}$ 18. 20 19. 9 20. $\dfrac{1}{18}$ 21. 2,274 22. $\dfrac{8}{3} = 2\dfrac{2}{3}$ 23. $\dfrac{81}{125}$
24. $6\dfrac{6}{11}$ 25. 1090 26. $\dfrac{55}{36} = 1\dfrac{19}{36}$ 27. $\dfrac{9}{39}$ 28. $\dfrac{2}{7}$ 29. $\dfrac{46}{45} = 1\dfrac{1}{45}$
30. Thirty thousand, seven hundred sixty; $30{,}000 + 700 + 60$ 31. 1,506,421 babies

Chapter 2 Test

1. $\dfrac{2}{5}$ 2. $\dfrac{3}{8}$ 3. $\dfrac{23}{5}$ 4. $\dfrac{7}{10}$ 5. $1\dfrac{11}{36}$ 6. $9\dfrac{17}{24}$ 7. $3\dfrac{2}{3}$ 8. $5\dfrac{2}{5}$ 9. $\dfrac{1}{3}$ 10. $\dfrac{1}{9}$ 11. $\dfrac{5}{16}$ 12. $\dfrac{5}{6}$
13. $2\dfrac{1}{2}$ 14. $\dfrac{1}{4}$ 15. $\dfrac{3}{8}$ 16. $4\dfrac{1}{2}$ 17. $6\dfrac{2}{3}$ 18. $2\dfrac{13}{18}$ 19. $16\dfrac{3}{4}$ 20. $9\dfrac{11}{12}$ 21. $\dfrac{1}{2}$ 22. $1\dfrac{2}{3}$ 23. $12\dfrac{3}{8}$
24. $16\dfrac{3}{4}$ 25. $\dfrac{1}{6}$ 26. $3\dfrac{3}{4}$ 27. $\dfrac{1}{10}$ 28. $\dfrac{7}{10}$ 29. $27\dfrac{1}{2}$ in. 30. Perimeter $= 31$ ft 31. $8\dfrac{13}{30}$ in. 32. $\dfrac{14}{15}$ in.

Chapter 2 Glimpse of Algebra

1. $\dfrac{x+3}{4}$ 3. $\dfrac{x+2}{5}$ 5. $\dfrac{7x}{8}$ 7. $\dfrac{2}{x}$ 9. $\dfrac{3x+2}{6}$ 11. $\dfrac{2x-1}{4}$ 13. $\dfrac{12+3x}{4x}$ 15. $\dfrac{4x-5}{5x}$ 17. $\dfrac{7x+24}{12x}$ 19. $\dfrac{3x+4}{4x}$
21. $\dfrac{12+x}{4}$ 23. $\dfrac{10-x}{2}$ 25. $\dfrac{28-a}{7}$ 27. $\dfrac{5+2a}{5}$ 29. $\dfrac{8x+3}{x}$ 31. $\dfrac{2x-5}{x}$ 33. $\dfrac{4x+3}{4}$ 35. $\dfrac{11x}{9}$ 37. $\dfrac{3a}{7}$
39. $\dfrac{5x}{4}$

Chapter 3

Problem Set 3.1

1. Tens   3. Tenths   5. Hundred thousandths   7. Ones   9. Hundreds   11. Three tenths   13. Fifteen thousandths
15. Three and four tenths   17. Fifty-two and seven tenths   19. 0.55   21. 6.9   23. 11.11   25. 100.02   27. 3,000.003
29. $405\dfrac{36}{100}$ 31. $9\dfrac{9}{1,000}$ 33. $1\dfrac{234}{1,000}$ 35. $\dfrac{305}{100,000}$ 37. a. < b. >   39. 0.002 0.005 0.02 0.025 0.05 0.052
41. 7.451, 7.54   43. $\dfrac{1}{4}$ 45. $\dfrac{1}{8}$ 47. $\dfrac{5}{8}$ 49. $\dfrac{7}{8}$ 51. 9.99   53. 10.05   55. 0.05   57. 0.01
59. Hundredths   61. Three and eleven hundredths grams; two and five tenths grams
63. Fifteen hundredths   65.   67. $6\dfrac{31}{100}$ 69. $6\dfrac{23}{50}$ 71. $18\dfrac{123}{1,000}$
73. $\dfrac{3}{16}<\dfrac{3}{10}<\dfrac{3}{8}<\dfrac{3}{4}$ 75. <   77. >

PRICE OF 1 GALLON OF REGULAR GASOLINE	
Date	Price (Dollars)
4/4/11	4.057
4/11/11	4.161
4/18/11	4.205
4/25/11	4.217

Problem Set 3.2

1. 6.19   3. 1.13   5. 6.29   7. 9.042   9. 8.021   11. 11.7843   13. 24.343   15. 24.111   17. 258.5414   19. 666.66
21. 11.11   23. 3.57   25. 4.22   27. 120.41   29. 44.933   31. 7.673   33. 530.865   35. 27.89   37. 35.64
39. 411.438   41. 6   43. 1   45. 3.1   47. 5.9   49. 3.272   51. 4.001   53. 0.99 seconds   55. $1,571.10   57. 4.5 in.
59. $5.43   61. 0.4 seconds   63. 2 in.   65. $3.25; three $1 bills and a quarter   67. 3.25   69. $\dfrac{3}{100}$ 71. $\dfrac{51}{10,000}$ 73. $1\dfrac{1}{2}$
75. 1,400   77. $\dfrac{3}{20}$ 79. $\dfrac{147}{1,000}$ 81. 132,980   83. 2,115   85. 12   87. 16   89. 20   91. 68

Problem Set 3.3

1. 0.28   3. 0.028   5. 0.0027   7. 0.78   9. 0.792   11. 0.0156   13. 24.29821   15. 0.03   17. 187.85   19. 0.002
21. 27.96   23. 0.43   25. 49,940   27. 9,876,540   29. 1.89   31. 0.0025   33. 5.1106   35. 7.3485   37. 4.4
39. 2.074   41. 3.58   43. 187.4   45. 116.64   47. 20.75   49. 371.34 meters   51. 0.126
53. Moves it two places to the right   55. $1,381.38   57. a. 83.21 mm b. 551.27 mm^2 c. 1,102.53 mm^3 59. 1.18 in^2
61. $C = 25.12$ in.; $A = 50.24$ in^2 63. $C = 24,492$ mi   65. 168 in.   67. 100.48 ft^3 69. 50.24 ft^3 71. 1,879   73. 1,516 R 4
75. 298   77. 34.8   79. 49.896   81. 825   83. $\dfrac{3}{2}<1\dfrac{2}{3}<1\dfrac{5}{6}<\dfrac{25}{12}$ 85. No

Problem Set 3.4

1. 19.7   3. 6.2   5. 5.2   7. 11.04   9. 4.8   11. 9.7   13. 2.63   15. 42.24   17. 2.55   19. 1.35   21. 6.5   23. 9.9
25. 0.05   27. 89   29. 2.2   31. 1.35   33. 16.97   35. 0.25   37. 2.71   39. 11.69   41. 3.98   43. 5.98
45. 0.77778   47. 307.20607   49. 0.70945   51. 3,472 square miles   53. 7.5 mi   55. 25.74 seconds   57. 22.4 mi
59. 5 hr   61. 7 min   63.

Rank	Name	Number of Events	Total Earnings	Average per Event
1.	Na Yeon Choi	23	$1,871,165.50	$81,355
2.	Jiyai Shin	19	$1,783,127.00	$93,849
3.	Cristie Kerr	21	$1,601,551.75	$76,264
4.	Yani Tseng	19	$1,573,529.00	$82,817
5.	Suzann Pettersen	19	$1,557,174.50	$81,957

65. $\dfrac{3}{4}$ 67. $\dfrac{2}{3}$ 69. $\dfrac{3}{8}$ 71. $\dfrac{19}{50}$ 73. $\dfrac{6}{10}$ 75. $\dfrac{60}{100}$ 77. $\dfrac{12}{15}$ 79. $\dfrac{60}{15}$ 81. $\dfrac{18}{15}$ 83. 0.75
85. 0.875   87. 19   89. 3

Problem Set 3.5

1. 0.125 **3.** 0.625

5.
Fraction	$\frac{1}{4}$	$\frac{2}{4}$	$\frac{3}{4}$	$\frac{4}{4}$
Decimal	0.25	0.5	0.75	1

7.
Fraction	$\frac{1}{6}$	$\frac{2}{6}$	$\frac{3}{6}$	$\frac{4}{6}$	$\frac{5}{6}$	$\frac{6}{6}$
Decimal	$0.1\overline{6}$	$0.\overline{3}$	0.5	$0.\overline{6}$	$0.8\overline{3}$	1

9. 0.48 **11.** 0.4375 **13.** 0.92

15. 0.27 **17.** 0.09 **19.** 0.28

21.
Decimal	0.125	0.250	0.375	0.500	0.625	0.750	0.875
Fraction	$\frac{1}{8}$	$\frac{1}{4}$	$\frac{3}{8}$	$\frac{1}{2}$	$\frac{5}{8}$	$\frac{3}{4}$	$\frac{7}{8}$

23. $\frac{3}{20}$ **25.** $\frac{2}{25}$ **27.** $\frac{3}{8}$ **29.** $5\frac{3}{5}$ **31.** $5\frac{3}{50}$ **33.** $1\frac{11}{50}$ **35.** 2.4 **37.** 3.98 **39.** 3.02 **41.** 0.3 **43.** 0.072
45. 0.8 **47.** 1 **49.** 0.25 **51.** $7\frac{1}{5}$ miles **53.** $16.22 **55.** $52.66 **57.** 9 in.

59.
CHANGE IN STOCK PRICE		
Date	Gain ($)	As a Decimal ($) (To the Nearest hundredth)
April 25, 2011	$\frac{1}{20}$	0.05
April 26, 2011	$\frac{9}{20}$	0.45
April 27, 2011	$\frac{2}{25}$	0.08
April 28, 2011	$\frac{1}{4}$	0.25
April 29, 2011	$\frac{63}{100}$	0.63

61. 104.625 calories **63.** $10.38

65. Yes **67.** 33.49 m³ **69.** 226.1 ft³ **71.** 22.28 in² **73.** 2,128 **75.** 1,866 R 4 **77.** 36.945 **79.** 563
81. $\frac{3}{10} < \frac{2}{5} < \frac{1}{2} < \frac{4}{5}$ **83.** 852 **85.** 20,675

Problem Set 3.6

1. 8 **3.** 9 **5.** 6 **7.** 5 **9.** 15 **11.** 48 **13.** 45 **15.** 48 **17.** 15 **19.** 1 **21.** 78 **23.** 9 **25.** $\frac{4}{7}$ **27.** $\frac{3}{4}$
29. False **31.** True **33.** 1.1180 **35.** 11.1803 **37.** 3.46 **39.** 11.18 **41.** 0.58 **43.** 0.58 **45.** 12.124 **47.** 9.327
49. 12.124 **51.** 12.124 **53.** 10 in. **55.** 13 ft **57.** 6.40 in. **59.** 6.7 miles **61.** 30 ft **63.** 25 ft **65.** 5 miles
67. 16 · 2 **69.** 25 · 3 **71.** 25 · 2 **73.** 4 · 10 **75.** 16 · 1 **77.** 49 · 2 **79.** 16 · 3 **81.** $\frac{2}{5}$ **83.** $9\frac{8}{25}$ **85.** 8 **87.** $2\frac{4}{5}$
89. $\frac{7}{16}$ **91.** 19

Chapter 3 Review

1. Thousandths **2.** Hundredths **3.** 37.0042 **4.** 100.00202 **5.** 98.77 **6.** 100 **7.** 5.816 **8.** 7.779 **9.** 36.381
10. 21.5736 **11.** 7.65 **12.** 1.23 **13.** 7.09 **14.** 0.015625 **15.** 5.729 **16.** 13.92 **17.** 1.008 **18.** 12.5
19. 34.1 **20.** 0.1764 **21.** 4.16 **22.** 38 **23.** 0.875 **24.** 0.1875 **25.** $\frac{141}{200}$ **26.** $\frac{123}{500}$ **27.** $14\frac{1}{8}$ **28.** $5\frac{1}{20}$
29. 2.42 **30.** 48.585 **31.** 79 **32.** 0.7 **33.** 1.224 **34.** 3.24 **35.** 0.4 **36.** 13 **37.** 11 **38.** 14.21 cm

Chapter 3 Cumulative Review

1. 4,079 **2.** 327 **3.** 16,072 **4.** 0.316 **5.** 6.22 **6.** $\frac{31}{35}$ **7.** 55.728 **8.** 76 **9.** $\frac{1}{2}$ **10.** 464,000 **11.** $15\frac{3}{4}$
12. $\frac{11}{5}$ **13.** 20

14.
Decimal	Fraction
0.125	$\frac{1}{8}$
0.250	$\frac{1}{4}$
0.375	$\frac{3}{8}$
0.500	$\frac{1}{2}$
0.625	$\frac{5}{8}$
0.750	$\frac{3}{4}$
0.875	$\frac{7}{8}$
1	$\frac{8}{8}$

15. 9
16. Commutative and associative properties of multiplication
17. 3(13 + 4) = 51 **18.** $\frac{12}{7}$ **19.** False **20.** 32
21. $\frac{7}{9}$ **22.** 9 **23.** 0.94 **24.** $\frac{3}{16}$ **25.** $14\frac{7}{8}$
26. 84 **27.** $8\frac{1}{4}$ cups **28.** $9.60
29. 401.67 yds **30.** 424.61 yds

Chapter 3 Test

1. Five and fifty-three thousandths **2.** Thousandths **3.** 17.0406 **4.** 46.75 **5.** 8.18 **6.** 6.056 **7.** 35.568 **8.** 8.72 **9.** 11.36
10. 8.907 **11.** 1.568 **12.** 0.24 **13.** 0.92 **14.** $\frac{14}{25}$ **15.** 14.664 **16.** 4.69 **17.** 17.129 **18.** 0.26 **19.** 6.48
20. 10.784 **21.** 6 **22.** 9 **23.** $11.53 **24.** $24.47 **25.** $6.55 **26.** 14.3 ft **27.** 10.1 ft **28.** 19.21 inches

A Glimpse of Algebra

1. $6x^2 + 9x + 8$ **3.** $5a + 8$ **5.** $9x + 6$ **7.** $5y^3 + 5y^2 + 9y + 7$ **9.** $6x^2 + 4x + 5$ **11.** $3a^2 + 7a + 6$ **13.** $5x^3 + 7x^2 + 16x + 13$

Chapter 4

Problem Set 4.1

1. $\frac{4}{3}$ **3.** $\frac{16}{3}$ **5.** $\frac{2}{5}$ **7.** $\frac{1}{2}$ **9.** $\frac{3}{1}$ **11.** $\frac{7}{6}$ **13.** $\frac{7}{5}$ **15.** $\frac{5}{7}$ **17.** $\frac{8}{5}$ **19.** $\frac{1}{3}$ **21.** $\frac{1}{10}$ **23.** $\frac{3}{25}$
25. a. $\frac{1}{2}$ **b.** $\frac{1}{3}$ **c.** $\frac{2}{3}$ **27.** $\frac{5}{3}$ **29.** $\frac{20}{1}$ **31.** $\frac{75}{2}$ **33. a.** $\frac{13}{8}$ **b.** $\frac{1}{4}$ **c.** $\frac{3}{8}$ **d.** $\frac{13}{3}$
35. a. $\frac{6}{1}$ **b.** $\frac{5}{1}$ **c.** $\frac{4}{1}$ **d.** $\frac{3}{1}$ **e.** $\frac{4}{1}$ **37. a.** $\frac{43}{38}$ **b.** $\frac{1}{1}$ **c.** $\frac{19}{3}$ **39.** 18 **41.** 62.5 **43.** 0.615 **45.** 176 **47.** 184
49. 0.087 **51.** 0.048 **53.** $\frac{5}{8}$ **55.** $\frac{11}{2} = 5\frac{1}{2}$ **57.** $\frac{5}{14}$ **59.** 96

Problem Set 4.2

1. 55 mi/hr **3.** 84 km/hr **5.** 0.2 gal/sec **7.** 14 L/min **9.** 19 mi/gal **11.** $4.\overline{3}$ mi/L **13.** Rocky Mountains
15. 480 mL/hr **17.** 4.95¢ per oz **19.** 34.7¢ per diaper, 31.6¢ per diaper; Happy Baby **21.** $1.00 = 0.674 Euros
23. $1.00 = 0.606 pounds **25.** The 18 oz box is the best buy at $0.277 per ounce. **27.** 54.03 mi/hr **29.** 9.3 mi/gal **31.** $64
33. $16,000 **35.** $n = 6$ **37.** $n = 4$ **39.** $n = 4$ **41.** $n = 65$ **43.** $\frac{7}{8}$ **45.** $\frac{1}{40}$ **47.** $\frac{1}{60}$ **49.** $\frac{11}{6} = 1\frac{5}{6}$
51. The 100 ounce size is the better value.

Problem Set 4.3

1. 35 **3.** 18 **5.** 14 **7.** n **9.** y **11.** $n = 2$ **13.** $x = 7$ **15.** $y = 7$ **17.** $n = 8$ **19.** $a = 8$ **21.** $x = 2$ **23.** $y = 1$
25. $a = 6$ **27.** $n = 5$ **29.** $x = 3$ **31.** $n = 7$ **33.** $y = 1$ **35.** $y = 9$ **37.** $n = 3\frac{1}{2}$ **39.** $x = 3\frac{1}{2}$ **41.** $a = 2\frac{2}{5}$
43. $y = \frac{4}{7}$ **45.** $y = \frac{10}{13}$ **47.** $x = 2\frac{1}{2}$ **49.** $n = 1\frac{1}{2}$ **51.** $9 **53.** $\frac{3}{4}$ **55.** 1.2 **57.** 6.5 **59.** 0.75 **61.** 5.5
63. 0.03 **65.** 0.375 **67.** $\frac{17}{50}$ **69.** $2\frac{2}{5}$ **71.** $1\frac{3}{4}$ **73.** $\frac{7}{8}$

Problem Set 4.4

1. Means: 3, 5; extremes: 1, 15; products: 15 **3.** Means: 25, 2; extremes: 10, 5; products: 50 **5.** Means: $\frac{1}{2}$, 4; extremes: $\frac{1}{3}$, 6; products: 2
7. Means: 5, 1; extremes: 0.5, 10; products: 5 **9.** 10 **11.** $\frac{12}{5}$ **13.** $\frac{3}{2}$ **15.** $\frac{10}{9}$ **17.** 7 **19.** 14 **21.** 18 **23.** 6
25. 40 **27.** 50 **29.** 108 **31.** 3 **33.** $\frac{2}{5}$ **35.** $\frac{1}{4}$ **37.** 108 **39.** 65 **41.** 41 **43.** 108 **45.** 2,560
47. 20 **49.** 297.5 **51.** 450 **53.** 5 **55.** Tens **57.** 26.516 **59.** 0.39

Problem Set 4.5

1. 329 mi **3.** 360 points **5.** 15 pt **7.** 427.5 mi **9.** 900 eggs **11.** 35.5 mg/pill **13.** 5.5 mL **15.** 2.5 tablets
17. 435 in. = 36.25 ft **19.** $313.50 **21.** 265 g **23.** 91.3 L **25.** 60,113 people **27.** 78 teachers **29.** 850 meters
31. 15 mL **33.** 900 mg/day **35.** 2 **37.** 147 **39.** 20 **41.** 147 **43.** 1.35 **45.** 3.816 **47.** 4 **49.** 160 **51.** 183.79

Problem Set 4.6

1. 9 **3.** 14 **5.** 12 **7.** 25 **9.** 32
11. 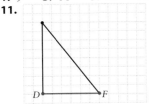 **13.** **15.** **17.**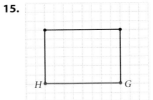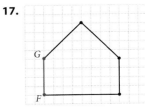

19. 45 in. **21.** 960 pixels **23.** 1,200 pixels **25.** 57 ft **27.** 177 ft **29.** 13.99 **31.** 40.999 **33.** 0.10545 **35.** 18
37. $4\frac{1}{6}$ **39.** 4 **41. a.** $\frac{1}{2}$ **b.** 18 rectangles should be shaded. **c.** $\frac{1}{3}$

Chapter 4 Review

1. $\frac{3}{10}$ **2.** $\frac{10}{3}$ **3.** $\frac{3}{4}$ **4.** $\frac{9}{5}$ **5.** $\frac{7}{5}$ **6.** $\frac{12}{11}$ **7.** $\frac{1}{2}$ **8.** $\frac{1}{8}$ **9.** $\frac{1}{3}$ **10.** $\frac{2}{3}$ **11.** $\frac{3}{17}$ **12.** $\frac{3}{5}$ **13.** 19 mi/gal
14. 1,100 ft/sec **15.** 9.05¢/ounce; 8.41¢/ounce; 32-ounce carton is the better buy.
16. 20.8¢/soda, 25.0¢/soda; the store brand is less expensive **17.** $x = 2$ **18.** $x = 3\frac{1}{5}$ **19.** 49 **20.** 6 **21.** $\frac{1}{10}$
22. 6 **23.** 1,500 mL **24.** 24 mg **25.** 3 tablets **26.** 150 mg **27.** 9 **28.** $x = 15$ cm

Chapter 4 Cumulative Review

1. 10,522 **2.** 1,979 **3.** 18 **4.** 48 **5.** $506\frac{3}{31}$ **6.** 125 **7.** 47 **8.** 9 **9.** 74 **10.** 19 **11.** 310 **12.** 25.56
13. 71.48 **14.** 14.59 **15.** 4.5 **16.** $\frac{1}{72}$ **17.** 80 **18.** $\frac{7}{18}$ **19.** $\frac{25}{42}$ **20.** $6\frac{3}{4}$ **21.** 25 **22.** 20
23. $P = 42$ in.; $A = 72$ in² **24.** $P = 66$ cm; $A = 150$ cm² **25.** $x = 16$ cm **26.** 20 women **27.** $80\frac{9}{16}$ in. **28.** 97 mi
29. 150 sections **30.** $\frac{11}{20}$ **31.** $\frac{1}{3}$

Chapter 4 Test

1. $\frac{4}{3}$ **2.** $\frac{9}{10}$ **3.** $\frac{3}{2}$ **4.** $\frac{3}{10}$ **5.** $\frac{3}{5}$ **6.** $\frac{40}{7}$ **7.** $\frac{15}{64}$ **8.** 23 mi/gal
9. 16-ounce can: 26¢/ounce; 12-ounce can: 28¢/ounce; 16-ounce can is the better buy **10.** $n = 6$ **11.** $n = 6\frac{1}{4}$ **12.** 36
13. 8 **14.** 24 hits **15.** 135 mi **16.** 27 mg **17.** 33.15 mg **18.** $h = 5$ **19.** 1,047 students

A Glimpse of Algebra

1. $x^2 + 12x + 32$ **3.** $6x^2 + 13x + 6$ **5.** $21x^2 + 34x + 8$ **7.** $x^2 + 7x + 10$ **9.** $2x^2 + 11x + 12$ **11.** $14x^2 + 41x + 15$
13. $9x^2 + 12x + 4$ **15.** $4a^2 + 9a + 5$ **17.** $42y^2 + 111y + 72$ **19.** $8 + 24x + 18x^2$

Chapter 5

Problem Set 5.1

1. $\frac{20}{100}$ **3.** $\frac{60}{100}$ **5.** $\frac{24}{100}$ **7.** $\frac{65}{100}$ **9.** 0.23 **11.** 0.92 **13.** 0.09 **15.** 0.034 **17.** 0.0634 **19.** 0.009 **21.** 23%
23. 92% **25.** 45% **27.** 3% **29.** 60% **31.** 80% **33.** 27% **35.** 123% **37.** $\frac{3}{5}$ **39.** $\frac{3}{4}$ **41.** $\frac{1}{25}$ **43.** $\frac{53}{200}$
45. $\frac{7,187}{10,000}$ **47.** $\frac{3}{400}$ **49.** $\frac{1}{16}$ **51.** $\frac{1}{3}$ **53.** 50% **55.** 75% **57.** $33\frac{1}{3}$% **59.** 80% **61.** 87.5% **63.** 14%
65. 325% **67.** 150% **69.** 48.8% **71. a.** 0.30 **b.** 0.0625 **c.** 0.125 **73. a.** $\frac{41}{100}, \frac{21}{100}, \frac{1}{50}, \frac{3}{50}, \frac{3}{10}$ **b.** 0.41, 0.21, 0.02, 0.06, 0.3
c. About 2 times as likely **75.** 78.4% **77.** 11.8% **79.** 72.2% **81.** 18.5 **83.** 10.875 **85.** 0.5 **87.** 62.5 **89.** 0.5
91. 0.125 **93.** 0.625 **95.** 0.0625 **97.** 0.3125 **99.** 2 **101.** 2 **103.** 2 **105.** 2

Problem Set 5.2

1. 8 **3.** 24 **5.** 20.52 **7.** 7.37 **9.** 50% **11.** 10% **13.** 25% **15.** 75% **17.** 64 **19.** 50 **21.** 925 **23.** 400
25. 5.568 **27.** 120 **29.** 13.72 **31.** 22.5 **33.** 50% **35.** 942.684 **37.** 97.8 **39.** What number is 25% of 350?
41. What percent of 24 is 16? **43.** 46 is 75% of what number? **45.** 4.8% calories from fat; healthy
47. 50% calories from fat; not healthy **49.** 0.80 **51.** 0.76 **53.** 48 **55.** 0.25 **57.** 0.5 **59.** $\frac{5}{8}, \frac{3}{4}, \frac{7}{8}$ **61.** $\frac{1}{4}$
63. 0.25 **65.** 0.75 **67.** 0.125 **69.** 0.375 **71.** $\frac{1}{8}, \frac{1}{4}, \frac{3}{8}, \frac{1}{2}, \frac{5}{8}, \frac{3}{4}, \frac{7}{8}$

Problem Set 5.3

1. 70% **3.** 84% **5.** 45 mL **7.** 18.2 acres for farming; 9.8 acres are not available for farming **9.** $11.20
11. Obama 365; McCain 173 **13.** 21,183,452 people **15.** 3,000 students **17.** 400 students **19.** 1,664 female students
21. 31.25% **23.** 50% **25.** About 19.5 million **27.** 33 **29.** 8,685 **31.** 136 **33.** 0.05 **35.** 15,300 **37.** 0.15
39. $\frac{1}{5}$ **41.** $\frac{5}{12}$ **43.** $\frac{3}{4}$ **45.** $\frac{2}{3}$ **47.** 35.7% **49.** 186 hits **51.** At least 18 hits

Problem Set 5.4

1. $52.50 **3.** $2.70; $47.70 **5.** $150; $156 **7.** 5% **9.** $420.90 **11.** $2,820 **13.** $200 **15.** 14% **17.** 26.9%
19. 47.5 cents or $0.475 **21.** $560 **23.** $11.93 **25.** 4.5% **27.** $3,995 **29.** 1,100 **31.** 75 **33.** 0.16 **35.** 4
37. 396 **39.** 415.8 **41.** $\frac{1}{2}$ **43.** $2\frac{2}{3}$ **45.** $1\frac{1}{2}$ **47.** $4\frac{1}{2}$

Problem Set 5.5

1. $24,610 **3.** $7,020 **5.** $13,200 **7.** 10% **9.** 20% **11.** 5.6% **13.** $45; $255 **15.** 50% **17.** 13.9% **19.** 16%
21. 21.8 in. to 23.2 in. **23.** $46,595.88 **25. a.** 51.9% **b.** 7.8% **27.** 140 **29.** 4 **31.** 152.25 **33.** 3,434.7 **35.** 10,150
37. 10,456.78 **39.** 2,140 **41.** 3,210 **43.** 1 **45.** $\frac{3}{4}$ **47.** $1\frac{5}{12}$ **49.** 6

Problem Set 5.6

1. $2,160 **3.** $665 **5.** $8,560 **7.** $3,300 **9.** $5 **11.** $813.33 **13.** $402.14 **15.** $5,618 **17.** $8,407.56
19. $974.59 **21. a.** $13,468.55 **b.** $13,488.50 **c.** $12,820.37 **d.** $12,833.59 **23.** 30% **25.** $16\frac{2}{3}$% **27.** 108 **29.** 162
31. 8 **33.** 3 **35.** $\frac{1}{2}$ **37.** $\frac{3}{8}$ **39.** $\frac{2}{3}$ **41.** $1\frac{1}{2}$ **43.** $\frac{1}{12}$ **45.** $1\frac{1}{4}$ **47.** $\frac{16}{21}$ **49.** $\frac{64}{525}$ **51.** $30.78
53. Percent increase/decrease in production cost: *Star Wars* 1 to 2 = 38.5%; *Star Wars* 2 to 3 = 80.6%; *Star Wars* 3 to 4 = 253.8%; *Star Wars* 4 to 5 = 0%; *Star Wars* 5 to 6 = 1.7% decrease

Problem Set 5.7

1. a. $\frac{41}{100}$ **b.** $\frac{38}{41}$ **c.** $\frac{31}{50}$ **d.** $\frac{31}{19}$ **3. a.** 11.95% **b.** 18.95% **c.** State of Nevada **d.** 55.91%
5. a. 240 people **b.** 960 people **c.** 750 people **d.** 1,800 people

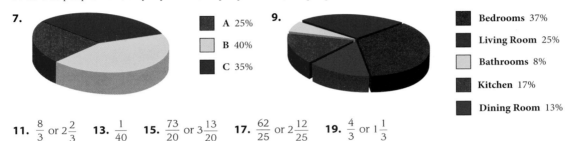

7. A 25%, B 40%, C 35%
9. Bedrooms 37%, Living Room 25%, Bathrooms 8%, Kitchen 17%, Dining Room 13%

11. $\frac{8}{3}$ or $2\frac{2}{3}$ **13.** $\frac{1}{40}$ **15.** $\frac{73}{20}$ or $3\frac{13}{20}$ **17.** $\frac{62}{25}$ or $2\frac{12}{25}$ **19.** $\frac{4}{3}$ or $1\frac{1}{3}$

Chapter 5 Review

1. 0.35 **2.** 0.178 **3.** 0.05 **4.** 0.002 **5.** 95% **6.** 80% **7.** 49.5% **8.** 165% **9.** $\frac{3}{4}$ **10.** $\frac{1}{25}$ **11.** $1\frac{9}{20}$
12. $\frac{1}{40}$ **13.** 30% **14.** 62.5% **15.** $66\frac{2}{3}$% **16.** 475% **17.** 16.8 **18.** 67.1 **19.** 50% **20.** 200% **21.** 80
22. 200 **23.** 75% **24.** $25,200 **25.** $477 **26.** $189.00 **27.** $35; 20% **28.** $4.93 **29.** $3.83 **30.** 55.4% increase
31. $60; 20% off

Chapter 5 Cumulative Review

1. 7,714 **2.** 2,269 **3.** 45,084 **4.** 68.2 **5.** 33 **6.** $\frac{1}{4}$ **7.** $\frac{125}{216}$ **8.** 7.631 **9.** 1.322 **10.** 0.252 **11.** $\frac{1}{6}$
12. $\frac{14}{15}$ **13.** $\frac{23}{24}$ **14.** $2\frac{1}{2}$ **15.** 42 **16.** 5 **17.** 6 **18.** 2(2 + 9) = 22 **19.** $\frac{1}{4}$ **20.** 13,200 feet **21.** 648 in²
22. 12.5% **23.** $\frac{23}{50}$ **24.** 3.2 **25.** 125 **26.** 1.6 **27.** 20% **28.** $58\frac{2}{3}$ **29.** 47¢ **30.** 40¢ **31.** 100°C **32.** 12.5%
33. 345 miles **34.** $1,812.50 **35.** P = 34 in., A = 72.25 in² **36.** 79 **37.** $7 per hour **38.** $\frac{4}{5}$ **39.** $\frac{19.5}{1}$

Chapter 5 Test

1. 0.18 **2.** 0.04 **3.** 0.005 **4.** 45% **5.** 70% **6.** 135% **7.** $\frac{13}{20}$ **8.** $1\frac{23}{50}$ **9.** $\frac{7}{200}$ **10.** 35% **11.** 37.5%
12. 175% **13.** 45 **14.** 45% **15.** 80 **16.** 92% **17.** $960 **18.** $40; 16% off **19.** $1,064 **20.** $220.50 **21.** $100
22. $2,520 **23.** $\frac{7}{10}$

A Glimpse of Algebra

1. 9 **3.** 180 **5.** 8 **7.** $\frac{8}{15}$ **9.** 20 **11.** 36 **13.** $\frac{17}{13} = 1\frac{4}{13}$ **15.** $\frac{22}{23}$ **17.** 140 **19.** 25 **21.** 3 **23.** 1 **25.** 21

Chapter 6

Problem Set 6.1

1. 60 in. **3.** 120 in. **5.** 6 ft **7.** 162 in. **9.** $2\frac{1}{4}$ ft **11.** 13,200 ft **13.** $1\frac{1}{3}$ yd **15.** 1,800 cm **17.** 4,800 m
19. 50 cm **21.** 0.248 km **23.** 670 mm **25.** 34.98 m **27.** 6.34 dm **29.** 69 yd **31.** 20 yd **33.** 80 in.
35. 244 cm **37.** 2,960 chains **39.** 120,000 μm **41.** 7,920 ft **43.** 80.7 ft/sec **45.** 19.5 mi/hr **47.** 1,023 mi/hr
49. 3,965,280 ft **51.** 179,352 in. **53.** 2.7 mi **55.** 18,094,560 ft **57.** 144 **59.** 8 **61.** 1,000 **63.** 3,267,000 **65.** 6
67. 0.4 **69.** 405 **71.** 450 **73.** 45 **75.** 2,200 **77.** 607.5 **79.** $\frac{1}{3}$ **81.** 6 **83.** $\frac{7}{2} = 3\frac{1}{2}$ **85.** 6 **87.** 6 **89.** $\frac{7}{10}$
91. 10,000 steps $\cdot \frac{2.5 \text{ ft}}{1 \text{ step}} \cdot \frac{1 \text{ mi}}{5,280 \text{ ft}} \approx 4.7$ mi; The facts are close.

Problem Set 6.2

1. 432 in²   3. 2 ft²   5. 1,306,800 ft²   7. 1,280 acres   9. 3 mi²   11. 108 ft²   13. 1,700 mm²   15. 28,000 cm²
17. 0.0012 m²   19. 500 m²   21. 700 a   23. 3.42 ha   25. 135 ft³   27. 48 fl oz   29. 8 qt   31. 20 pt   33. 480 fl oz
35. 8 gal   37. 6 qt   39. 9 yd³   41. 5,000 mL   43. 0.127 L   45. 4,000,000 mL   47. 14,920 L   49. 642,560 acres
51. 30 a   53. 5,500 bricks   55. 16 cups   57. 7,200 tiles   59. 48 glasses   61. 20,288,000 acres   63. 3,230.93 mi²
65. 23.35 gal   67. 21,492 ft³   69. 192   71. 6,000   73. 300,000   75. 12.5   77. $\frac{5}{8}$ 79. 9   81. $\frac{1}{3}$ 83. $\frac{17}{144}$

Problem Set 6.3

1. 128 oz   3. 4,000 lb   5. 12 lb   7. 0.9 T   9. 32,000 oz   11. 56 oz   13. 13,000 lb   15. 2,000 g   17. 40 mg
19. 200,000 cg   21. 508 cg   23. 4.5 g   25. 47.895 cg   27. 1.578 g   29. 0.42 kg   31. 48 g   33. 4 g   35. 9.72 g
37. 120 g   39. $9.38   41. 3 L   43. 1.5 L   45. 0.561 grams   47. a. 3 capsules   b. 1 capsule   c. 2 capsules
49. 20.32   51. 6.36   53. 50   55. 50   57. 122   59. 248   61. 40   63. $\frac{9}{50}$ 65. $\frac{9}{100}$ 67. $\frac{4}{5}$ 69. $1\frac{3}{4}$
71. 0.75   73. 0.85   75. 0.6   77. 3.625   79. $\frac{1}{8}$ 81. $6\frac{1}{4}$ 83. 0.125   85. 6.25

Problem Set 6.4

1. 15.24 cm   3. 13.12 ft   5. 6.56 yd   7. 32,200 m   9. 5.98 yd²   11. 24.7 acres   13. 8,195 mL   15. 2.12 qt
17. 75.8 L   19. 339.6 g   21. 33 lb   23. 365°F   25. 30°C   27. 58°C   29. 6,697,600 meters   31. 7.5 mL   33. 3.94 in.
35. 7.62 m   37. 46.23 L   39. 17.67 oz   41. Answers will vary.   43. 91.46 m   45. 20.90 m²   47. 88.55 km/hr
49. 2.03 m   51. 38.3°C   53. 75   55. 82   57. 3.25   59. 22   61. 41   63. 48   65. 195   67. 3.26   69. $90
71. mean = 8, range = 6   73. mean = 6, range = 9   75. median = 21, range = 14   77. median = 40, range = 16
79. mode = 14, range = 7   81. a. 73   b. 71   c. 70   d. 23

Problem Set 6.5

1. a. 270 min   b. 4.5 hr   3. a. 320 min   b. 5.33 hr   5. a. 390 sec   b. 6.5 min   7. a. 320 sec   b. 5.33 min
9. a. 40 oz   b. 2.5 lb   11. a. 76 oz   b. 4.75 lb   13. a. 54 in.   b. 4.5 ft   15. a. 69 in.   b. 5.75 ft   17. a. 9 qt   b. 2.25 gal
19. 11 hr   21. 22 ft 4 in.   23. 11 lb   25. 5 hr 40 min   27. 3 hr 47 min   29. 52 min   31. 10 hr 24 min   33. 33 lb 12 oz
35. 11 ft   37. 0.90 seconds   39. 8:06:58; 8:08:19   41. 00:00:09   43. $104   45. 10 hr   47. $150   49. $6
51.

CAFFEINE CONTENT IN SOFT DRINKS	
Drink	Caffeine (In Milligrams)
Jolt	100
Mountain Dew	55
Coca-Cola	45
Diet Pepsi	36
7 up	0

53. 2.64 sec

Chapter 6 Review

1. 144 in.   2. 6 yd   3. 0.49 m   4. 20,000 dm   5. 435,600 ft²   6. 78 ares   7. 576 in²   8. 14 pts   9. 6 gal
10. 5,000 mL   11. 128 oz   12. 36 oz   13. 5,000 g   14. 5,000 kg   15. 10.16 cm   16. 11.27 km   17. 7.42 qt
18. 18.95 L   19. 141.5 g   20. 19.8 lb   21. 248°F   22. 50°C   23. 8.52 liters   24. 4,551,360 ft   25. 4,383.23 mi²
26. 2,144 oz   27. 402.5 km   28. 3,396 g   29. 600 bricks   30. 26 glasses   31. 80 glasses   32. $2\frac{11}{128}$ ft²
33. 117 mi/hr   34. a. 50 mi/hr   b. In a car   35. 322 km/hr   36. a. 285 minutes   b. 4.75 hr   37. 13 lb   38. $125

Chapter 6 Cumulative Review

1. 16,759   2. 2,001   3. 12   4. 126   5. $490\frac{1}{32}$ 6. 256   7. 21   8. 7   9. 38   10. 126   11. 13   12. 28.35
13. 11.07   14. 42.84   15. 16.2   16. $\frac{1}{256}$ 17. 153   18. $\frac{23}{50}$ 19. $6\frac{2}{5}$ 20. $11\frac{1}{2}$ 21. 0.75   22. 0.6   23. $14\frac{4}{5}$
24. 7.5   25. 11.5   26. 18   27. 72 in., 207 in²   28. $2\frac{11}{12}$ cm   29. 47   30. 56.8 mi/hr   31. 1,788   32. $2 \cdot 3 \cdot 3 \cdot 7$
33. 42   34. 0.65 mi   35. 87%

Chapter 6 Test

1. 21 ft   2. 0.75 km   3. 130,680 ft²   4. 3 ft²   5. 10,000 mL   6. 8.05 km   7. 10.6 qt   8. 26.7° C   9. 2,400 lb
10. 9 lb   11. 3.5 ft²   12. 450 m²   13. 6.5 yd³   14. 104°F   15. 0.26 gal   16. 20,844 in.   17. 0.24 ft³
18. 70.75 liters   19. 74.70 m   20. 7.55 liters   21. 6.30 in.   22. 8.05 km   23. 109.65 cm²   24. 17.29 acres   25. 4.72 L
26. 7.73 kg   27. 90 tiles   28. 64 glasses   29. a. 330 min   b. 5.5 hr   30. 11 lb   31. 9.8 km/L

Chapter 7

Problem Set 7.1

1. 4 is less than 7. **3.** 5 is greater than -2. **5.** -10 is less than -3. **7.** 0 is greater than -4. **9.** $30 > -30$ **11.** $-10 < 0$
13. $-3 > -15$ **15.** $3 < 7$ **17.** $7 > -5$ **19.** $-6 < 0$ **21.** $-12 < -2$ **23.** $-\frac{1}{2} > -\frac{3}{4}$ **25.** $-0.75 < -0.25$
27. $-0.1 < -0.01$ **29.** $-3 < |6|$ **31.** $15 > |-4|$ **33.** $|-2| < |-7|$ **35.** 2 **37.** 100 **39.** 8 **41.** 231 **43.** $\frac{3}{4}$
45. 200 **47.** 8 **49.** 231 **51.** -3 **53.** 2 **55.** -75 **57.** 0 **59.** 0.123 **61.** $-\frac{7}{8}$ **63.** 2 **65.** 8
67. -2 **69.** -8 **71.** 0 **73.** Positive **75.** -100 **77.** -20 **79.** -360 **81.** -450 feet **83.** $-3,060$ feet
85. $-\$5,000$; $-\$2,750$ **87.** $-61°F$, $-51°F$ **89.** $-5°F$, $-15°F$ **91.** New Orleans $-6:00$ GMT **93.** $-7°F$
95. $10°F$ and 25-mph wind **97.** 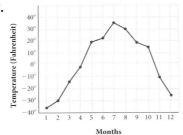   **99.** 25 **101.** 5 **103.** 6 **105.** 19 **107.** 4,313
109. 56 **111.** $5 + 3$ **113.** $(7 + 2) + 6$
115. $x + 4$ **117.** $y + 5$

Problem Set 7.2

1. 5 **3.** 1 **5.** -2 **7.** -6 **9.** 4 **11.** 4 **13.** -9 **15.** 15 **17.** -3 **19.** -11 **21.** -7 **23.** -3 **25.** -16
27. -8 **29.** -127 **31.** 49 **33.** 34

35.

First Number a	Second Number b	Their Sum $a+b$
5	-3	2
5	-4	1
5	-5	0
5	-6	-1
5	-7	-2

37.

First Number x	Second Number y	Their Sum $x+y$
-5	-3	-8
-5	-4	-9
-5	-5	-10
-5	-6	-11
-5	-7	-12

39. 10 **41.** -445 **43.** 107 **45.** -20 **47.** -17 **49.** -50 **51.** -7 **53.** 3 **55.** 50 **57.** -73 **59.** -11 **61.** 17
63. -3.8 **65.** 14.4 **67.** -9.89 **69.** -1 **71.** $-\frac{2}{7}$ **73.** $-\frac{3}{5}$ **75.** -0.86 **77.** -4.2 **79.** $-\frac{1}{4}$ **81.** -21 **83.** -5
85. -4 **87.** 7 **89.** 10 **91.** 380 feet above the trailhead **93.** $\$10$ **95.** -2 **97.** 4 **99.** $-\frac{2}{5}$ **101.** 30
103. -60.3 **105.** 2 **107.** 3 **109.** 604 **111.** 0 **113.** $10 - x$ **115.** $y - 17$

Problem Set 7.3

1. 2 **3.** 2 **5.** -8 **7.** -5 **9.** 7 **11.** 12 **13.** 3 **15.** -7 **17.** -3 **19.** -13 **21.** -50 **23.** -100 **25.** 399
27. -21 **29.** -11.41 **31.** -1.9 **33.** -1 **35.** $-\frac{5}{12}$ **37.** $-\frac{11}{15}$ **39.** -7 **41.** -9 **43.** -14 **45.** -65 **47.** -11
49. 202 **51.** -400 **53.** -17.5 **55.** $-\frac{1}{12}$ **57.** 11 **59.** -4 **61.** 8 **63.** 6 **65.** b **67.** a **69.** -100
71. -55 degrees **73.** 7,603 feet **75.** -100 items **77.** 12.9 watts **79.** -180.2 watts **81.** $-11 - (-22) = 11°$ F
83. $3 - (-24) = 27°$ F **85.** $60 - (-26) = 86°$ F **87.** $-14 - (-26) = 12°$ F **89.** 30 **91.** 36 **93.** 64 **95.** 48
97. 41 **99.** 40 **101.** 17 **103.** 32 **105.** 25 **107.** 72 **109.** $3 \cdot 5$ **111.** $7x$ **113.** $5(3)$ **115.** $(5 \cdot 7) \cdot 8$
117. $2(3) + 2(4) = 6 + 8 = 14$

Problem Set 7.4

1. -56 **3.** -60 **5.** 56 **7.** 81 **9.** -9.03 **11.** $\frac{3}{7}$ **13.** -8 **15.** -24 **17.** 24 **19.** -6 **21. a.** 16 **b.** -16
23. a. -125 **b.** -125 **25. a.** 16 **b.** -16

27.

Number x	Square x^2
-3	9
-2	4
-1	1
0	0
1	1
2	4
3	9

29.

First Number x	Second Number y	Their Product xy
6	2	12
6	1	6
6	0	0
6	-1	-6
6	-2	-12

31. −4 **33.** 50 **35.** 1 **37.** −35 **39.** −22 **41.** −30 **43.** −25 **45.** 9 **47.** −13 **49.** −11 **51.** 19 **53.** 6
55. −6 **57.** −4 **59.** −17 **61.** a **63.** d **65.** a **67.** b **69.** −9°F **71.** −16 degrees **73.** $400 remains
75. 7 **77.** 5 **79.** −5 **81.** 9 **83.** 4 **85.** 17 **87.** 405 **89.** $12 \div 6$ or $\frac{12}{6}$ **91.** $6 \div 3 = 2$ **93.** $5 \cdot 2 = 10$
95. 89 **97.** −54, 162 **99.** 54, −162 **101.** 44 **103.** 19 **105.** −17

Problem Set 7.5

1. −3 **3.** −5 **5.** 3 **7.** 2 **9.** −4 **11.** −2 **13.** 0 **15.** −5

17.

First Number a	Second Number b	The Quotient of a and b $\frac{a}{b}$
100	−5	−20
100	−10	−10
100	−25	−4
100	−50	−2

19.

First Number a	Second Number b	The Quotient of a and b $\frac{a}{b}$
−100	−5	20
−100	5	−20
100	−5	−20
100	5	20

21. −5 **23.** 35 **25.** 6 **27.** 1 **29.** −6 **31.** −2 **33.** −1 **35.** −1 **37.** 2 **39.** −3 **41.** −7 **43.** 30 **45.** 4
47. −5 **49.** −20 **51.** −5 **53.** −1 **55.** c **57.** a **59.** d **61.** −$9,616.67
63. 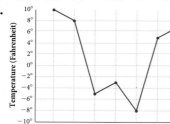   **65.** $x + 3$ **67.** $(5 + 7) + a$ **69.** $(3 \cdot 4)y$ **71.** $5(3) + 5(7)$ **73.** 36
75. 64 **77.** 350 **79.** 7,500 **81.** 4 **83.** −12 **85.** 12
87. −32 **89.** 32 **91.** −2 **93.** −4 **95.** 4 **97.** −1 **99.** −1

Problem Set 7.6

1. $20a$ **3.** $48a$ **5.** $-18x$ **7.** $-27x$ **9.** $-10y$ **11.** $-60y$ **13.** $5 + x$ **15.** $13 + x$ **17.** $10 + y$ **19.** $8 + y$
21. $5x + 6$ **23.** $6y + 7$ **25.** $12a + 21$ **27.** $7x + 28$ **29.** $7x + 35$ **31.** $6a - 42$ **33.** $2x - 2y$ **35.** $20 + 4x$
37. $6x + 15$ **39.** $18a + 6$ **41.** $12x - 6y$ **43.** $35 - 20y$ **45.** $8x$ **47.** $4a$ **49.** $4x$ **51.** $5y$ **53.** $-10a$ **55.** $-5x$
57. $x(\$3,300 - \$300) - \$250 = \$3,000x - \$250$ **59.** $A = 36$ ft²; $P = 24$ ft **61.** $A = 81$ in²; $P = 36$ in. **63.** $A = 200$ in²; $P = 60$ in.
65. $A = 300$ ft²; $P = 74$ ft **67.** 20° C **69.** 5° C **71.** −10° C

Chapter 7 Review

1. −17 **2.** 32 **3.** 4.6 **4.** $-\frac{3}{5}$ **5.** −6 **6.** −8 **7.** 2 **8.** $|-4|$ **9.** 4 **10.** −4 **11.** 6 **12.** 19
13. −1 and −15 **14.** −2 **15.** 5 **16.** −971 **17.** 35 **18.** −12 **19.** −17 **20.** 7 **21.** −12 **22.** −20 **23.** 12
24. −1,736 **25.** −80 **26.** 2 **27.** −5 **28.** 36 **29.** $\frac{9}{16}$ **30.** −8 **31.** 0.0016 **32.** −5 **33.** 2 **34.** −8
35. −42 **36.** −1 **37.** −7 **38.** −4 **39.** −11 **40.** −42 **41.** −129 **42.** $58 + (-\$86) = -\$28$ **43.** −11 **44.** −2
45. 24° **46.** −27 **47.** 39 **48.** 2 **49.** −19 **50.** False **51.** True **52.** True **53.** False **54.** $3x + 12$ **55.** $24x$
56. $-21a$ **57.** $-30y$ **58.** $4x + 12$ **59.** $2x - 10$ **60.** $21y - 56$ **61.** $6a + 15b$ **62.** $3x$ **63.** $2a$ **64.** $4y$ **65.** $16x$

Chapter 7 Cumulative Review

1. $\frac{31}{35}$ **2.** $\frac{1}{3}$ **3.** 316 **4.** $13\frac{5}{12}$ **5.** $\frac{16}{81}$ **6.** 42,771 **7.** $-\frac{1}{16}$ **8.** −6 **9.** 37.65 **10.** $\frac{39}{8}$
11. Thirty-eight thousand, six hundred nine **12.** Distributive property **13.** 126 **14.** $\frac{6}{7}$ **15.** $\frac{23}{24}$
16. 0.098 **17.** −1 **18.** 6 **19.** $\frac{2}{5}$ **20.** $\frac{8}{9}$ **21.** $8\frac{1}{6}$ **22.** 4.875 **23.** 37.5% **24.** $\frac{19}{25}$ **25.** 1 **26.** 40%
27. 0.35 km **28.** 53.06 L **29.** 0.56 **30.** 20 **31.** $\frac{11}{4}$ **32.** $103.20 **33.** 20 women **34.** $80\frac{9}{16}$ in. **35.** 97 mi
36. Area = 33 ft²; Perimeter = $28\frac{1}{6}$ ft **37.** $10.80 **38.** −13 and 5 **39.** −8 **40.** $\frac{2}{3}$

Chapter 7 Test

1. −14 **2.** $\frac{2}{3}$ **3.** $-1 > -4$ **4.** $|-4| > |2|$ **5.** 7 **6.** −2 **7.** −9 **8.** −5.9 **9.** $-\frac{22}{15}$ **10.** −36 **11.** 42 **12.** 6
13. −5 **14.** 5 **15.** −3.6 **16.** $-\frac{7}{12}$ **17.** −14 **18.** −7 **19.** $-\frac{3}{10}$ **20.** $\frac{7}{6}$ **21.** 9 **22.** −8 **23.** −11
24. −15 **25.** 7 **26.** −4 **27.** −74 **28.** −45 **29.** 11 **30.** 4 **31.** $\frac{1}{2}$ **32.** 4 **33.** 19 **34.** $-\frac{5}{2}$ **35.** −61
36. −7 **37.** 24 **38.** −5 **39.** −14 **40.** −5 **41.** −$35 **42.** 25°
43. −3,154; There were 3,154 fewer shows for *A Chorus Line* than for *Phantom of the Opera*.

Chapter 8

Problem Set 8.1
1. $10x$ **3.** y **5.** $3a$ **7.** $8x + 16$ **9.** $6a + 14$ **11.** $x + 2$ **13.** $6x + 11$ **15.** $2x + 2$ **17.** $-a + 12$ **19.** $4y - 4$
21. $-2x + 4$ **23.** $8x - 6$ **25.** $5a + 9$ **27.** $-x + 3$ **29.** $17y + 3$ **31.** $a - 3$ **33.** $6x + 16$ **35.** $10x - 11$
37. $19y + 32$ **39.** $30y - 18$ **41.** $6x + 14$ **43.** $27a + 5$ **45.** 14 **47.** 27 **49.** -19 **51.** 7 **53.** 1 **55.** 18
57. 12 **59.** -10 **61.** 28 **63.** 40 **65.** 26 **67.** 4 **69.** 3 **71.** 0 **73.** 15 **75.** 6 **77.** $6(x + 4) = 6x + 24$
79. $4x + 4$ **81.** $10x - 4$ **83.** $85°$ **85.** $55°$; complementary angles **87.** $20°$; complementary angles
89. a. Yes **b.** No, he should earn $108 for working 9 hours **c.** No, he should earn $84 for working 7 hours **d.** Yes
91. a. $32°F$ **b.** $22°F$ **c.** $-18°F$ **93. a.** $27 **b.** $47 **95.** $3.50 **97.** 0 **99.** -6 **101.** -3 **103.** $\frac{11}{8}$ **105.** 0
107. x **109.** $y - 2$ **111.** -9 **113.** 6 **115.** a **117.** c **119.** c **121.** $-\frac{2}{3}$ **123.** 18 **125.** $-\frac{1}{12}$

Problem Set 8.2
1. Yes **3.** Yes **5.** No **7.** Yes **9.** No **11.** 6 **13.** 11 **15.** -15 **17.** 1 **19.** -3 **21.** -1 **23.** $\frac{7}{5}$ **25.** -10
27. -4 **29.** -3 **31.** 2 **33.** -6 **35.** -6 **37.** -1 **39.** -2 **41.** -16 **43.** -3 **45.** 10 **47.** $x = 4$
49. $x = 12$ **51.** $58°$ celsius **53.** $67°$ **55. a.** 225 **b.** $11,125 **57.** $\frac{1}{4}$ **59.** 2 **61.** $\frac{3}{2}$ **63.** 1 **65.** 1 **67.** 1 **69.** 1
71. x **73.** $7x - 11$ **75.** 2 **77.** $\frac{5}{14}$ **79.** $-\frac{1}{15}$ **81.** $-\frac{7}{6}$ **83.** Equation: $x + 12 = 30; x = 18$
85. Equation: $8 - 5 = x + 7; x = -4$

Problem Set 8.3
1. 8 **3.** -6 **5.** -6 **7.** 6 **9.** 16 **11.** 16 **13.** $-\frac{3}{2}$ **15.** -7 **17.** -8 **19.** 6 **21.** 2 **23.** 3 **25.** 4
27. -24 **29.** 12 **31.** -8 **33.** 15 **35.** 3 **37.** -1 **39.** 1 **41.** 3 **43.** 1 **45.** -1 **47.** $-\frac{1}{3}$ **49.** -14
51. 9 **53.** 8 **55.** $x = 2.8$ miles **57.** 352 ft/sec **59.** 3,600 frames **61.** 111 million viewers **63.** $2x + 5 = 19; x = 7$
65. $5x - 6 = -9; x = -\frac{3}{5}$ **67.** $6a - 16$ **69.** $-15x + 3$ **71.** $3y - 9$ **73.** $16x - 6$

Problem Set 8.4
1. 3 **3.** 2 **5.** -3 **7.** -4 **9.** 1 **11.** 0 **13.** 2 **15.** -2 **17.** 3 **19.** -1 **21.** 7 **23.** -3 **25.** 1 **27.** -2
29. 4 **31.** 10 **33.** -5 **35.** $-\frac{1}{12}$ **37.** -3 **39.** 20 **41.** -1 **43.** 5 **45.** 4 **47.** 2.05 **49.** -0.064 **51.** 12.03
53. 0.44 **55.** -0.175 **57.** 2 **59.** 10 **61.** 5 **63.** 1,250 feet **65.** 13 **67.** 30 hours **69.** $x + 2$ **71.** $2x$
73. $2(x + 6)$ **75.** $x - 4$ **77.** $2x + 5$ **79.** $<$ **81.** $>$ **83.** 0.001, 0.003, 0.01, 0.013, 0.03, 0.031
85. a. $4x + 6(x - 100) = 2,400$ **b.** $x = 300$ **c.** 500 people

Problem Set 8.5
1. $x + 3$ **3.** $2x + 1$ **5.** $5x - 6$ **7.** $3(x + 1)$ **9.** $5(3x + 4)$ **11.** The number is 2. **13.** The number is -2.
15. The number is 3. **17.** The number is 5. **19.** The number is -2. **21.** The length is 10 m and the width is 5 m.
23. The length of one side is 8 cm. **25.** The measures of the angles are $45°$, $45°$, and $90°$. **27.** The angles are $30°$, $60°$, and $90°$.
29. Patrick is 33 years old, and Pat is 53 years old. **31.** Sue is 35 years old, and Dale is 39 years old. **33.** 87 mi **35.** 147 mi
37. 8 nickels, 18 dimes **39.** 16 dimes, 32 quarters **41.** $x = 8, y = 6, z = 9$ **43.** Weekday = 4 hours; Weekend = 6 hours
45. $250 + 150 + 130 + x = 843; x = 313$ acres **47.** 35 **49.** 65 **51.** 2 **53.** 14 **55.** $-\frac{2}{3}$ **57.** 0.75, 75%
59. 1.2, 120% **61.** $\frac{37}{100}$, 0.37 **63.** $\frac{17}{500}$, 0.034 **65.** 20.25 **67.** 75

Problem Set 8.6
1. $y = 7$ **3.** $y = -3$ **5.** $y = -2$ **7.** $x = 3$ **9.** $x = 5$ **11.** $x = 8$ **13.** $y = 1$ **15.** $y = 3$ **17.** $y = \frac{5}{2}$ **19.** $y = 0$
21. $y = \frac{13}{3}$ **23.** $y = 4$ **25.** $x = 0$ **27.** $x = \frac{13}{4}$ **29.** $x = 3$ **31.** 704 ft² **33.** $\frac{9}{8}$ in² **35.** $240 **37.** $285 **39.** 12 ft
41. 8 ft **43.** $140 **45.** 58 in. **47.** $3\frac{1}{4} = \frac{13}{4}$ ft **49.** $C = 100°C$; yes **51.** $C = 20°C$; yes **53.** $0°C$ **55.** $5°F$ **57.** 360 in³
59. 1 yd³ **61.**

Age (Years)	Maximum Heart Rate (Beats per Minute)
18	202
19	201
20	200
21	199
22	198
23	197

63.

Resting Heart Rate	Training Heart Rate
60	144
65	146
70	148
75	150
80	152
85	154

65. a. 4 hrs **b.** 220 miles

67. a. 4 hours **b.** 65 mph   **69.** Complement: 45°; supplement: 135°   **71.** Complement: 59°; supplement: 149°
73. a. 13,330 kilobytes **b.** 2,962 kilobytes   **75. a.** 58° Celsius **b.** 328 K   **77.** $\frac{3}{4}$ **79.** 3   **81.** 2   **83.** 6

Chapter 8 Review

1. $17x$ **2.** $-4x$ **3.** $11a - 3$ **4.** $-3y - 2$ **5.** $5x + 4$ **6.** $-4a + 1$ **7.** $10a - 4$ **8.** $15y - 13$ **9.** 42   **10.** 8   **11.** 1
12. 4   **13.** -9 **14.** 19   **15.** 7   **16.** 10   **17.** Yes   **18.** No   **19.** 9   **20.** -4 **21.** 3   **22.** 2   **23.** 9   **24.** -9
25. 5   **26.** -2 **27.** -1 **28.** 0   **29.** $-\frac{1}{8}$ **30.** 5   **31.** 9   **32.** -2 **33.** 7   **34.** 12   **35.** -16 **36.** 8   **37.** 2
38. $\frac{4}{3}$ **39.** -2 **40.** -4 **41.** The length is 14 m, and the width is 7 m.   **42.** 15° and 75°   **43.** 18°, 54°, 108°   **44.** $y = 6$
45. $y = -6$ **46.** $y = 3$ **47.** $y = \frac{5}{2}$ **48.** $x = 0$ **49.** $x = 4$ **50.** $x = 2$ **51.** $x = 6$ **52.** $x = -3$ **53.** $x = 12$ **54.** $x = -9$

Chapter 8 Cumulative Review

1. 5,996   **2.** $1\frac{24}{55}$ **3.** 4.559   **4.** $2\frac{3}{8}$ **5.** 938,135   **6.** 0.46   **7.** 42   **8.** 1   **9.** 440,000   **10.** $\frac{12}{25}$ **11.** $6\frac{1}{3}$
12. 0.05   **13.** 80%   **14.** 15.4 lb   **15.** $1\frac{6}{25}$ **16.** 35%   **17.** $\frac{4}{81}$ **18.** $-x - 5$ **19.** -7 **20.** 8   **21.** 4   **22.** 9
23. 56   **24.** $-\frac{4}{3}$ **25.** 5   **26.** $\frac{1}{8}$ **27.** 8   **28.** 82 in²   **29.** Ben is 17; Ryan is 9   **30.** 16 mpg   **31.** $80, 20%
32. 13 m   **33.** $14.40   **34.** $21   **35.** 64   **36.** $784   **37.** 240   **38.** 15 million   **39.** $\frac{7}{1}$

Chapter 8 Test

1. $6x - 5$ **2.** $3b - 4$ **3.** $2x - 18$ **4.** $14x - 12$ **5.** -3 **6.** 9   **7.** Yes   **8.** No   **9.** 8   **10.** -6 **11.** 4   **12.** -4.9
13. 27   **14.** 6   **15.** -4 **16.** 1   **17.** $\frac{2}{3}$ **18.** 1   **19.** $\frac{3}{2}$ **20.** 5   **21.** -8 **22.** 22 mi/hr
23. Length = 9 cm; width = 5 cm   **24.** Susan is 11; Karen is 6.   **25.** 20°, 40°, 120°   **26.** 4 quarters, 11 dimes
27. 22.5° and 67.5°   **28.** 4.5 m   **29.** 20 ft, 240 ft

Appendices

Problem Set A

1. $\frac{1}{32}$ **3.** $\frac{1}{100}$ **5.** $\frac{1}{x^3}$ **7.** $\frac{1}{16}$ **9.** $-\frac{1}{125}$ **11.** $\frac{1}{4}$ **13.** 1,000   **15.** $\frac{1}{x^7}$ **17.** 9   **19.** $\frac{1}{8}$ **21.** 1,000
23. $\frac{1}{x^{12}}$ **25.** 2   **27.** 4   **29.** $\frac{1}{4}$ **31.** x **33.** $\frac{1}{x}$ **35.** $\frac{1}{10^7}$ **37.** $\frac{1}{10^7}$ **39.** 1,000   **41.** x^{24} **43.** 4   **45.** 10,000
47. $15x^5$ **49.** $\frac{6}{x^5}$ **51.** $-\frac{12}{x^7}$ **53.** $42x^2$ **55.** x^5 **57.** x^2

Problem Set B

1. 4.25×10^5 **3.** 6.78×10^6 **5.** 1.1×10^4 **7.** 8.9×10^7 **9.** 38,400   **11.** 57,100,000   **13.** 3,300   **15.** 89,130,000
17. 3.5×10^{-4} **19.** 7×10^{-4} **21.** 6.035×10^{-2} **23.** 1.276×10^{-1} **25.** 0.00083   **27.** 0.0625   **29.** 0.3125   **31.** 0.005
33. 4.0×10^4; 51,120,000   **35.** 1.2×10^6; 87,900,000   **37.**   **39.** 1.024×10^3
41. 1.074×10^9

Jupiter's Moon	Period (Seconds)	
Io	153,000	1.53×10^5
Europa	307,000	3.07×10^5
Ganymede	618,000	6.18×10^5
Callisto	1,440,000	1.44×10^6

Problem Set C

1. 6×10^{10} **3.** 1.5×10^{11} **5.** 6.84×10^{-2} **7.** 1.62×10^{10} **9.** 3.78×10^{-1} **11.** 2×10^3 **13.** 4×10^{-9} **15.** 5×10^{14}
17. 6×10^1 **19.** 2×10^{-8} **21.** 4×10^6 **23.** 4×10^{-4} **25.** 2.1×10^{-2} **27.** 3×10^{-5} **29.** 6.72×10^{-1} **31.** 7.81×10^{-3}
33. 1.37×10^{-2} **35.** 2,500,000 stones or 2.5×10^6

Index

A

Absolute value, 467
Acute angle, 533
Addition
　associative property of, 17
　facts, 17
　with carrying, 15
　with common
　　denominator, 146
　commutative property of, 17
　with decimals, 215
　with fractions, 147, 150
　with mixed numbers, 171
　with negative numbers, 475
　with table, 14
　with whole numbers, 14
　with zero, 16
Addition property
　of equality, 541
　of zero, 16
Algebraic expression, 513, 516
　value of, 532
Angle, 528, 532
　acute, 533
　complementary, 533
　measure, 532
　notation, 533
　obtuse, 533
　right, 19, 131, 533
　straight, 533
　sum of (triangle), 570,
　supplementary, 533
　vertex of, 532
Area, 81
　of a circle, 227, 420
　metric units of, 422
　of a parallelogram, 81
　of a rectangle, 81, 420, 516
　of a square, 81, 420, 516
　of a triangle, 129, 420
　U.S. units of, 420
Arithmetic mean, 71
Associative property, 514
　of addition, 17
　of multiplication, 46
Average, 71, 74
Average speed, 581
Axis, 8

B

Bar chart, 8
Base, 67
Blueprint for Problem
　Solving, 566

Borrowing, 35
　with mixed numbers, 174

C

Carrying, 15
Celsius scale, 437
Circle, 227
　area, 227, 420
　circumference, 226
　diameter, 226
　radius, 226
Circumference, 204, 225
Commission, 336, 365
Common denominator, 146
Commutative property
　of addition, 17
　of multiplication, 45
Complementary angles, 533
Complex fraction, 183
Composite numbers, 118
Compound interest, 381
Constant term, 528, 558
Conversion factors
　area, 421, 422
　between metric and U.S.
　　systems, 436
　length, 411, 412
　time, 443
　volume, 423, 424
　weight, 429, 430
Counting numbers, 7
Cube, 84
Cylinder, 227

D

Decimals
　addition with, 214
　converting fractions to, 245
　converting to
　　fractions, 208, 247
　converting percents to, 338
　converting to percents, 339
　division with, 234
　expanded form, 207
　multiplication with, 222
　notation, 205
　numbers, 206
　point, 204, 206
　repeating, 246
　rounding, 208
　subtraction with, 213
　writing in words, 207
Degree measure, 532
Denominator, 104, 108
　least common, 148

Diameter, 204, 226
Difference, 2, 34
Different signs, 478
Digit, 3, 4
Discount, 375
Discount Rate, 373
Distributive property,
　44, 514, 530
Dividend, 56
Division, 55, 506
　with decimals, 234
　with fractions, 138
　with expressions, 296
　long, 57
　mixed numbers, 166
　negative numbers, 506
　with remainders, 59
　with whole numbers, 56
　by zero, 61, 62
Divisor, 56, 118

E

Equation, 541, 527
　in one variable, 18, 46, 295
　involving fractions, 559
　involving percent, 348
　involving proportions, 350
Equivalent fractions, 104, 107
Estimating, 25, 224
Evaluating formulas, 579
Expanded form, 5, 207
Exponents, 67, 69, 118
Extremes, 302

F

Factors, 2, 42
Factoring, 118,
　into prime numbers, 119
Fahrenheit scale, 437, 438
Formula, 579, 528
Fraction bar, 55
Fractions, 106
　addition with, 150
　combinations of operations,
　　181
　complex, 183
　converting decimals to,
　　208, 246
　converting to decimals,
　　208, 245
　converting percents to, 340
　converting to percents, 341
　denominator of, 106
　division with, 109, 138

　and equations, 559
　equivalent, 107
　improper, 107
　least common
　　denominator, 148
　lowest terms, 119
　meaning of, 105
　multiplications with, 125
　and the number one, 109
　numerator of, 108
　and order of operations,
　　140, 142
　proper, 107
　properties of, 108, 109
　as ratios, 281
　reducing to lowest
　　terms, 119
　subtraction with, 150
　terms of, 106
Fundamental property of
　proportions, 302

G

Geometric sequence, 54, 135
Grade point average, 238
Greater than symbol, 466

H

Hemisphere, 204
Horizontal axis, 8
Hypotenuse, 204, 260

I

Improper fraction, 107, 162
　to mixed numbers, 163
Inequality Symbol, 466
Inspection, 18
Integers, 468
Interest, 340, 380
　compound, 381
　principal, 383
　rate, 383
　simple, 380
Irrational number, 204, 259

L

Least common
　denominator, 104, 148
Length, 409
　metric units of, 412
　U.S. units of, 411
Less than symbol, 466
Line graph, 110

I-1

Linear equations in one
    variable, 557
Long division, 57
Lowest terms, 119

M

Mean, 2, 71
Means, 302
Median, 2, 71, 72
Metric system, 412
    converting to U.S.
        system, 421, 423, 429
    units of, 412, 422, 424, 430
Mixed number, 159
    addition with, 171
    borrowing, 174
    combining, 171
    converting fractions to, 161
    converting to fractions, 160
    division with, 162
    multiplication with, 166
    notation, 159
    subtraction with, 176
Mixed units, 444
Mode, 2, 73
Multiplication, 41
    associative property of, 46
    by one, 46
    by zero, 46
    commutative property of, 45
    facts, 42
    of (the word), 129
    with decimals, 223
    with fractions, 125
    with mixed numbers, 166
    with negatives, 495, 497
    with whole numbers, 43, 46
Multiplication property
    of equality, 549
    of one, 45
    of zero, 45

N

Negative numbers, 466
    addition with, 468, 475
    division with, 506
    multiplication with, 495
    subtraction with, 479
Number line, 7, 466
Numbers
    absolute value of, 464, 467
    composite, 118
    counting, 8
    decimal, 206
    integer, 461
    irrational, 204, 259
    mixed, 159
    negative, 466
    opposite of, 467
    positive, 466
    prime, 104, 117
    problems, 566
    rounding, 25, 207
    whole, 7
    writing in words, 6
    writing in digits, 7
Numerator, 104, 106

O

Obtuse angle, 533
Opposite, 467
Order of operations,
    69, 140, 142, 498, 507
Origin, 464, 466

P

Parallelogram
    area of, 81
Percent, 336, 338
    applications, 341, 361
    basic problems, 351
    and commission, 367
    converting decimals to, 339
    converting fractions to, 341
    converting to decimals, 338
    converting to fractions, 340
    and discount, 375, 377
    increase, 375
    decrease, 376
    meaning of, 342
    and proportions, 354
    and sales tax, 367
Perfect square, 258
Perimeter, 19
    of a square, 19, 516
    of a rectangle, 19, 516
Pi, 226
Pie chart, 387, 391, 400
Place value, 4, 205
Polygon, 19
Positive number, 466
Prime number, 104, 117
Product, 2, 42
    of prime factors, 118
    of two fractions, 125
Proper fraction, 107
Proportion, 280, 302
    applications of, 307
    extremes, 302
    fundamental property of, 302
    means, 302
    terms of, 302
Protractor, 390, 534
Pythagorean Theorem, 260

Q

Quotient, 2, 56

R

Radical sign, 204, 257
Radius, 204, 225
Range, 74
Rate equation, 581
Rates, 280, 289
Ratio, 280, 281
Reciprocal, 137
    of a negative number, 551
Rectangle, 19
    area, 81, 420, 516
    perimeter, 19, 516
Rectangular solid, 85
    surface area, 85
    volume, 85
Reducing to lowest terms, 117
Remainder, 59
Right angles, 533
Right circular cylinder, 227
Right triangle, 204, 260
Rounding, 25, 27
    decimals, 208
    whole numbers, 25

S

Sales tax, 363, 367
Scatter diagram, 110
Sets, 7
Similar terms, 515, 531
Similar triangles, 314
Simple interest, 380
Solution, 18, 528, 533, 541
Sphere, 204
    volume of, 249
Square, 19, 249
    area, 81, 420, 516
    perimeter, 19, 420, 516
Square root, 204, 258
    approximating, 259
Straight angles, 533
Subtraction, 33, 486
    with borrowing, 35, 174
    with decimals, 214
    with fractions, 150
    with mixed numbers, 176
    with negative numbers, 485
    with whole numbers, 34
Sum, 62
    of angles in a triangle, 570
    solving equations, 46
Supplementary angles, 533
Surface area, 85

T

Temperature, 437
Terms,
    of a fraction, 106
    of a proportion, 302
    similar, 515, 531
Time, 443
Triangle, 19
    area, 129, 420
    labeling, 314
    perimeter, 19, 516
    right, 204, 260
    similar, 314
    sum of angles, 570

U

Unit analysis, 408
U.S. system of
    measurement, 409

V

Variable, 18, 42, 275, 296
Variable term, 528, 558
Vertical axis, 8
Vertex of an angle, 532
Volume, 84, 85
    metric units of, 424, 436
    of a rectangular
        solid, 85, 424, 436
    of a sphere, 249
    U.S. units of, 423, 436

W

Weight
    metric units of, 430, 436
    U.S. units of, 429, 436
Weighted Average, 238
Whole numbers, 7
    addition with, 14
    division with, 55
    multiplication with, 43
    rounding, 25
    subtraction with, 34
    writing in digits, 7
    writing in words, 6

Z

Zero
    in addition, 16
    in division, 61
    as an exponent, 68
    in multiplication, 46